The Water Framework Directive

Ecological and Chemical Status Monitoring

Water Quality Measurements Series

Series Editor
Philippe Quevauviller
European Commission, Brussels, Belgium

Published Titles in the Water Quality Measurements Series

Hydrological and Limnological Aspects of Lake Monitoring
Edited by Pertti Heinonen, Giuliano Ziglio and Andre Van der Beken

Quality Assurance for Water Analysis
Edited by Philippe Quevauviller

Detection Methods for Algae, Protozoa and Helminths in Fresh and Drinking Water
Edited by Andre Van der Beken, Giuliano Ziglio and Franca Palumbo

Analytical Methods for Drinking Water: Advances in Sampling and Analysis
Edited by Philippe Quevauviller

Biological Monitoring of Rivers: Applications and Perspectives
Edited by Giuliano Ziglio, Maurizio Siligardi and Giovanna Flaim

Wastewater Quality Monitoring and Treatment
Edited by Philippe Quevauviller, Olivier Thomas and Andre Van der Berken

The Water Framework Directive: Ecological and Chemical Status Monitoring
Edited by Philippe Quevauviller, Ulrich Borchers, Clive Thompson and Tristan Simonart

Forthcoming Titles in the Water Quality Measurements Series

Rapid Chemical and Biological Techniques for Water Monitoring
Edited by Richard Greenwood, Catherine Gonzalez and Philippe Quevauviller

Groundwater Monitoring
Philippe Quevauviller, Anne-Marie Fouillac, Johannes Grath and Rob Ward

The Water Framework Directive
Ecological and Chemical Status Monitoring

PHILIPPE QUEVAUVILLER
European Commission, Brussels, Belgium

ULRICH BORCHERS
IWW Water Research Centre, Germany

CLIVE THOMPSON
Alcontrol Laboratories, South Yorkshire, UK

TRISTAN SIMONART
Institut Pasteur de Lille, Lille, France

A John Wiley and Sons, Ltd., Publication

This edition first published 2008
© 2008 John Wiley & Sons, Ltd

Registered office
John Wiley & Sons Ltd, The Atrium, Southern Gate, Chichester, West Sussex, PO19 8SQ, United Kingdom

For details of our global editorial offices, for customer services and for information about how to apply for permission to reuse the copyright material in this book please see our website at www.wiley.com.

Library of Congress Cataloging-in-Publication Data

The water framework directive : ecological and chemical status monitoring /
Philippe Quevauviller ... [et al.].
    p. cm. – (Water quality measurements series)
  Includes bibliographical references and index.
  ISBN 978-0-470-51836-6 (cloth : alk. paper)
    1. Water quality management–Goverment policy–Europe. 2. Water
quality–Europe–Measurement. 3. Water quality monitoring
stations–Europe.
  TD255.W3865 2008
  363.739′463094 – dc22
                                                     2008027398

A catalogue record for this book is available from the British Library.

ISBN: 978-0-470-51836-6 (H/B)

Typeset by Laserwords Private Limited, Chennai, India
Printed and bound in Great Britain by CPI Antony Rowe, Chippenham, Wiltshire

Contents

SECTION 10 CONCLUSIONS   **409**

Series Preface

Water is a fundamental constituent of life and is essential to a wide range of economic activities. It is also a limited resource, as we are frequently reminded by the tragic effects of drought in certain parts of the world. Even in areas with high precipitation, and in major river basins, over-use and mismanagement of water have created severe constraints on availability. Such problems are widespread and will be made more acute by the accelerating demand on freshwater arising from trends in economic development.

Despite of the fact that water-resource management is essentially a local, river-basin based activity, there are a number of areas of action that are relevant to all or significant parts of the European Union and for which it is advisable to pool efforts for the purpose of understanding relevant phenomena (e.g. pollution, geochemical studies), developing technical solutions and/or defining management procedures. One of the keys for successful cooperation aimed at studying hydrology, water monitoring, biological activities, etc., is to achieve and ensure good water quality measurements.

Quality measurements are essential to demonstrate the comparability of data obtained worldwide and they form the basis for correct decisions related to management of water resources, monitoring issues, biological quality, etc. Besides the necessary quality control tools developed for various types of physical, chemical and biological measurements, there is a strong need for education and training related to water quality measurements. This need has been recognized by the European Commission which has funded a series of training courses on this topic, covering aspects such as monitoring and measurements of lake recipients, measurements of heavy metals and organic compounds in drinking and surface water, use of biotic indexes, and methods to analyse algae, protozoa and helminths. In addition, a series of research and development projects have been or are being developed.

This book series will ensure a wide coverage of issues related to water quality measurements, including the topics of the above mentioned courses and the outcome of recent scientific advances. In addition, other aspects related to quality control tools (e.g. certified reference materials for the quality control of water analysis) and monitoring of various types of waters (river, wastewater, groundwater) will also be considered.

The Series Editor – Philippe Quevauviller

Preface

The EU Water Framework Directive (2000/60/EC) is probably the most significant legislative instrument in the water field that was introduced on an international basis for many years. It moves towards integrated environmental management with key objectives to prevent any further deterioration of water bodies, and protect and enhance the status of aquatic ecosystems and associated wetlands. It aims to promote sustainable water consumption and will contribute to mitigating the effects of floods and droughts.

Water management policy, as set out in the WFD, is focussed on water as it flows through river basins to the sea, and its provisions apply to all waters – inland surface waters, ground waters, transitional (estuarine) and coastal waters. An integrated approach is introduced for water quality and water quantity matters, and for surface and groundwater issues, and the Directive introduces a framework for water management based on river basin districts. The overriding objective of the policy is the achievement of "good status" in all waters by the end of 2015.

Linked to the WFD objectives are a series of milestones that have to be complied with (such as an analysis of pressures and impacts, and a characterisation of water bodies in 2005), including monitoring programmes that need to be operational by the end of 2006. This book on Water status monitoring under the WFD will represent a cornerstone for European environmental assessment, which will be closely co-ordinated with the European Environment Agency's State of the Environment (SoE) programme in the context of WISE (Water Information System for Europe). The wide-scale gathering of monitoring data will be of obvious interest to all those involved in environmental sciences, including soil and sediments.

The effectiveness of the monitoring programmes, and hence of the overall WFD implementation, will depend highly on the ability of Member States' laboratories to measure efficiently the status of Community waters (as well as sediments and biota) and changes in this status. Measurement data will, therefore, represent the foundation of the water quality evaluation system, on the basis of which decisions will be taken on the programme of measures required to achieve WFD environmental objectives. This huge challenge will require not only a co-ordination and possible harmonisation at EU level, but also exchanges of expertisc, experiences and best practices among the policy implementers and the practitioners (including the scientific community, industry and environmental NGOs). In this respect, a range of EU-funded research projects, as well as industry-driven initiatives, are contributing to gathering knowledge and developing technical and scientific expertise in direct support to the WFD implementation.

The International Conference "Water Status Monitoring under the WFD", which was held in Lille on 12–14 March 2007, gathered experts from different sectors, disciplines and interests and enabled fruitful exchanges to take place. The success of this event is now reflected in the present book which provides an in-depth analysis of various monitoring features of the WFD. In particular, general monitoring aspects are discussed, as well as case studies concerning different aquatic environments. The book also contains sections on analytical tools in support of WFD monitoring (including modelling), and details aspects of groundwater and sediment monitoring. Finally, risk assessment linked to monitoring as well as data quality and reporting requirements are discussed. The book concludes with discussions about the need for an operational science-policy mechanism and about current activities and perspectives in the context of EU RTD programmes.

The four editors have strived to present state-of-the-art information on WFD monitoring that gives further stimulation to the work of all parties involved in the huge challenges on the way to a good status of all European water bodies.

Ulrich Borchers   Clive Thompson   Tristan Simonart   Philippe Quevauviller

List of Contributors

William Adams
Rio Tinto, 8315 West 3595 South, PO Box 6001, Magna UT 84044, USA

Wolfgang Ahlf
TUHH, Institute of Environmental Technology and Energy Economics, Eissendorferstr. 40, 21073 Hamburg, Germany

Ian J. Allan
Norwegian Institute for Water Research, Gaustadalleen 21, 0349 Oslo, Norway

Damia Barceló
CID-CSIC, Jordi Girona 18, 08034 Barcelona, Spain

Rick Battarbee
Environmental Change Research Centre, University College London, Gower St., London WC1E 6BT, UK

Winfried E.H. Blum
Universität für Bodenkultur, Peter Jordan Str. 82, 1190 Vienna, Austria

Werner Brack
UFZ Centre for Environmental Research Leipzig-Halle, Permoserstr. 15, 04318 Leipzig, Germany

Gyrite Brandt
Copenhagen Energy, Planning Department Water Supply, Ørestads Boulevard 35, 2300 Copenhagen S, Denmark

Jos Brils
TNO Built Environment and Geosciences, PO Box 80015, 3508 TA Utrecht, The Netherlands

Andy Bush
Lead Development International Association, 42 Weymouth Street, London W1G 6NP, UK

Jacob Carstensen
National Environmental Research Institute, Dept. of Marine Ecology, Frederiksborgvej 399, 4000 Roskilde, Denmark

Ana Cristina Cardoso   Joint Research Centre, Institute for Environment and Sustainability, TP 300, 21020 Ispra (VA), Italy

Lynette Chung   Eurometaux, Avenue de Broqueville, 12, 1150 Brussels, Belgium

Agustina De la Cal   CID-CSIC, Jordi Girona 18, 08034 Barcelona, Spain

Katrien Delbeke   European Copper Institute (ECI), Tervurenlaan 168, 1150 Brussels, Belgium

Gert-Jan De Maagd   DGW, Directorate-General for Public Works and Water Management, P.O. Box 20901, 2500 EX The Hague, The Netherlands

Charalambos Demetriou   Water Development Department, Division of Hydrology, 8 Kanary Str., Engomi, Nicosia, 2406, Cyprus

Piet Den Besten   Centre for Water Management, P.O. Box 17, 8200 AA Lelystad, The Netherlands

Valeria Dulio   INERIS, Parc Technologique Alata, B.P. 2, 60550 Verneuil-en-Halatte, France

Ethel Eljarrat   CID-CSIC, Jordi Girona 18, 08034 Barcelona, Spain

Chris Evans   Centre for Ecology and Hydrology, Deiniol Road, Bangor, LL57 2UP, UK

Feiler Ute   Federal Institute of Hydrology, Am Mainzer Tor 1, 56070 Koblenz, Germany

Lisbeth Flindt Jørgensen Geological Survey of Denmark and Greenland, Øster Voldgade 10, 1350 Copenhagen K, Denmark

Stephen Foster   IAH President, c/o P O Box 9, Kenilworth Warwickshire, UK

Ulrich Förstner   Hamburg University of Technology, Dept. of Environmental Science and Technology, Eissendorfer Str. 40, 21071 Hamburg

Gary Free   Joint Research Centre, Institute for Environment and Sustainability, TP 300, 21020 Ispra (VA), Italy

Johannes Grath   Umweltbundesamt GmbH, Spittelauer Lände 5, 1090 Wien, Austria

Richard Greenwood University of Portsmouth, Biological Sciences, King Henry Building, King Henry I Street, Portsmouth PO1 2DY, UK

Joan Grimalt Department of Environmental Chemistry, Institute of Chemical and Environmental Research (CSIC), Jordi Girona, 18, 08034-Barcelona, Spain

Meritxell Gros CID-CSIC, Jordi Girona 18, 08034 Barcelona, Spain

Delphine Haesaerts International Zinc Association- Europe, Tervueren-laan168, Box 4, 1150 Brussels

Georg Hanke Joint Research Centre, Institute Environment and Sustainability TP 290, via Enrico Ferni 1, 21020 Ispra (VA), Italy

Bob Harris 2 Creynolds Close, Cheswick Green, Solihull, West Midlands B90 4 EU, UK

Fred Hatterman Postdam Institute for Climate Impact Research, Telegrafenberg A51, PO Box 60 12 03, 14412 Postdam, Germany

Peter Heininger Federal Institute of Hydrology, Am Mainzer Tor 1, 56070 Koblenz, Germany

Susanne Heise Consulting Centre for Integrated Sediment Management at the TUHH, Eissendorferstr. 40, 21071 Hamburg, Germany

Anna-Stiina Heiskanen Joint Research Centre, Institute for Environment and Sustainability, TP 300, 21020 Ispra (VA), Italy

Daniel Hering Department of Hydrobiology, University of Duisburg-Essen, 45117 Essen, Germany

Niels Jepsen Joint Research Centre, Institute for Environment and Sustainability, TP 300, 21020 Ispra (VA), Italy

Richard Johnson Department of Environmental Assessment, Swedish University of Agricultural Sciences, P.O. Box 7050, SE 750 07, Uppsala, Sweden

Martin Kernan Environmental Change Research Centre, University College London, Gower St., London WC1E 6BT, UK

Roel Knoben   Royal Haskoning, P.O. Box 525, 5201 AM's-Hertogenbosch, The Netherlands

Zbigniew Kundzerwicz   Postdam Institute for Climate Impact Research, Telegrafenberg A51, PO Box 60 12 03, 14412 Postdam, Germany

Anker Lajer Højberg   Geological Survey of Denmark and Greenland, Øster Voldgade 10, 1350 Copenhagen K, Denmark

Peter Lepom   German Federal Environment Agency, Laboratory for Water Analysis, II.2.5, Bismarckplatz 1, 14193 Berlin, Germany

Conor Linstead   Institute for Sustainable Water, Integrated Management & Ecosystem Research, (SWIMMER), University of Liverpool, UK

David M. Livingstone   Water Resources Dept., EAWAG, Swiss Federal Institute of Aquatic Science and Technology, Ueberlandstrasse 133, 8600 Duebendorf, Switzerland

Hannie Maas   Centre for Water Management, P.O. Box 17, 8200 AA Lelystad, The Netherlands

Ed Maltby   Institute for Sustainable Water, Integrated Management & Ecosystem Research, (SWIMMER), University of Liverpool, UK

Hakan Marklund   Naturvårdsverket, Monitoring section (Mm), 106 48 Stockholm, Sweden

Claire Mattelet   European Nickel Industry Association Kunstlaan, 13, 1210 Brussels, Belgium

Graham A. Mills   University of Portsmouth, School of Pharmacy and Biomedical Sciences, St. Michaels's Building, White Swan Road, Portsmouth PO1 2DT, UK

Anne Morin   INERIS, Parc Technologique Alata, B.P. 2, 60550 Verneuil-en-Halatte, France

Brian Moss   School of Biological Sciences, University of Liverpool, Liverpool L69 3GS, UK

Dietmar Müller   Umweltbundesamt GmbH, Spittelauer Lände 5, 1090 Wien, Austria

Philippe Négrel BRGM, Avenue C. Guillemin, 45060 Orléans, France

Uli Nickus Institute of Meteorology and Geophysic, University of Innsbruck, Innrain 52, 6020, Austria

Peeter Noges Joint Research Centre, Institute for Environment and Sustainability, TP 300, 21020 Ispra (VA), Italy

Guiseppe Onorati Environmental Agency of Campania - ARPAC, Via S. Maria del Pianto Torre 1, 80143 Naples, Italy

Didier Pennequin BRGM, 3, avenue C. Guillemin, BP 36009, 45060 Orléans Cedex 2, France

Elena Perez Gallego Plaza San Juan de la Cruz, s/n 28071 Madrid, Spain

Mira Petrovic (1) Environmental Chemistry, IIQAB-CSIC, Jordi Girona, 18, 08034-Barcelona, Spain; (2) Institució Catalana de Recerca i Estudis Avançats (ICREA), Passeig Lluis Companys 23, 80010 Barcelona, Spain

Sandra Poikane Joint Research Centre, Institute for Environment and Sustainability, TP 300, 21020 Ispra (VA), Italy

Philippe Quevauviller (1) European Commission, DG Environment (BU9 3/142), rue de la Loi 200, 1049 Brussels, Belgium; (2) Vrije Universiteit Brussel (VUB), IUW-PARE, Dept; Hydrology and Hydraulic Engineering, Building T, Pleinlaan 2, 1050 Brussels, Belgium

Vala Ragnarsdottir University of Bristol, Department of Earth Sciences, Wills Memorial Building, Queens Road, Bristol BS8 1RJ, UK

Jens Christian Refsgaard Geological Survey of Denmark and Greenland, Øster Voldgade 10, 1350 Copenhagen K, Denmark

Lidia Regoli International Molybdenum Association Kunstlaan, 13, 1210 Brussels, Belgium

Patrick Roose Belgisch Instituut voor Naturwetenschappen, Ostende, Belgium

Wim Salomons Institute for Environmental Studies, De Boelelaan 1087, 1081 HV Amsterdam, The Netherlands

Andreas Scheidleder   Umweltbundesamt GmbH, Spittelauer Lände 5, 1090 Wien, Austria

Cor Schipper   Deltares, P.O. Box 177, 2600 MH Delft, The Netherlands

Richard Skeffington   Aquatic Environments Research Centre, Department of Geography, University of Reading, PO Box 227, Reading RG6 6AB, UK

Jaroslav Slobodnik   Environmental Institute, Okruzna 784/42, 97241 Kos, Slovak Republic

Angelo Solimini   Joint Research Centre, Institute for Environment and Sustainability, TP 300, 21020 Ispra (VA), Italy

Wilhem Struckmeier   BGR, Stilleweg 2, 30655 Hannover, Germany

Andrea Tilche   European Commission, DG Research, Rue de la Loi, 200, 1049 Brussels, Belgium

Thomas Track   DECHEMA Gesellschaft für Chemische Technik und Biotechnologie e.V., Theodor-Heuss-Allee 25, 60486 Frankfurt/Main, Germany

Frank Van Assche   International Zinc Association- Europe, Tervueren-laan168, Box 4, 1150 Brussels

Wouter Van de Bund   Joint Research Centre, Institute for Environment and Sustainability, TP 300, 21020 Ispra (VA), Italy

Martine Van den Heuvel Deltares, P.O. Box 177, 2600 MH, Delft, The Netherlands

Patrick Van Sprang   Euras, Mercatorgebouw, Kortrijksesteenweg 302, 9000 Gent, Belgium

Joop Vegter   VEGTER ADVIES, Amsteldijk Zuid 167, 1189 VM Amstelveen, The Netherlands

Piet Verdonschot   Alterra, Centre for Ecosystem Studies Droevendaalsesteeg 3, Wageningen 6700 AA, The Netherlands

Violaine Veroughstraete Eurometaux, Avenue de Broqueville, 12, BE-1150 Brussels, Belgium

Philippe Vervier	ECOBAG, 15 rue Michel Labrousse, 31023 Toulouse Cedex, France
Rob Ward	Environment Agency, Olton Court, 10 Warwick Road, Olton, Solihull, West Midlands B92 7HX, UK
Hugo Waeterschoot	European Nickel Industry Association Kunstlaan, 13, 1210 Brussels, Belgium
Sue White	Integrated Earth System Sciences Institute, Cranfield University, Building 53, Cranfield, MK43 0AL, UK
Dick Wright	Norwegian Institute for Water Research, Gaustadalleen 21, 0349 Oslo, Norway

Section 1
General WFD Monitoring Features

1.1
Water Status Monitoring under the WFD

Philippe Quevauviller[1]

1.1.1   INTRODUCTION

Monitoring of chemical and/or biological parameters is required in a wide range of environmental policies to evaluate the environmental status of relevant compartments (e.g. water, soil, air), assess environmental risks from different pressures (and follow up the efficiency of e.g. control or remediation measures) and/or carry out trend studies. In this context, monitoring represents a cornerstone of water management systems (Figure 1.1.1).

The soundness of policy decisions is therefore directly related to the reliability of the environmental monitoring programmes. In turn, the design and development of monitoring programmes is directly linked to the availability of recommendations in

---

[1] The views expressed in this chapter are purely those of the author and may not in any circumstances be regarded as stating an official position of the European Commission

The Water Framework Directive - Ecological and Chemical Status Monitoring   Edited by Philippe Quevauviller, Ulrich Borchers, Clive Thompson and Tristan Simonart © 2008 John Wiley & Sons, Ltd

Figure 1.1.1   Monitoring in the context of water management systems

the form of (nonbinding) guides, written standards (e.g. ISO or CEN standards), as well as more generally to scientific and technological progress.

In this context, the Water Framework Directive is certainly the first EU legislative instrument which requires a systematic monitoring of biological, chemical and quantitative parameters in European waters at such a wide geographical scale (covering the territory of the EU and beyond) (European Commission, 2000). The principles are fixed in the legislative text and exchanges of information among experts have enabled the setting out of a common understanding of monitoring requirements in the forms of guidance documents (see paragraph 4). While water monitoring is obviously not a new feature, it should be noted, however, that the WFD monitoring programmes are in their infancy in that they had to be designed and reported by the Member States in March 2007. Monitoring data produced in 2007–2008 under the WFD will form the basis for the design of programmes of measures to be included in the first River Basin Management Plan (due to be published in 2009), and thereafter used for evaluating the efficiency of these measures. Monitoring data will hence obviously be used as a basis for classifying the water status, and they will also be used to identify possible pollution trends. This is an iterative process in that better monitoring will ensure a better design and follow-up of measures, a better status classification and a timely identification of adverse trends (calling for reversal measures), which puts a clear accent on the needs for constant improvements and regular reviews (foreseen under the WFD) and hence on the needs to integrate scientific progress in an efficient way.

Metrological features, including discussions about monitoring data traceability, have been discussed in a previous book of Wiley's Water Quality Measurements Series (Quevauviller, 2007). This chapter is meant as a general introduction of the new volume of the series, which results from the International Conference on Status Monitoring under the Water Framework Directive held in Lille (France) on 12–14 March 2007.

1.1.2 MAIN LEGAL REQUIREMENTS WITH MONITORING IMPLICATIONS

The Water Framework Directive establishes 'good status' objectives to be achieved for all waters by the end of 2015. With regard to surface waters, good status criteria are based on biological parameters (ecological status) and chemical parameters (chemical status). The chemical status is linked to compliance to EU Environmental Quality Standards defined in a 'daughter directive', which negotiation is at its final stage at the time of publication of this volume. For ground waters, good status refers to quantitative levels (balance between recharge and abstraction) and chemistry (linked to compliance to groundwater quality standards established at EU, national, regional or local levels under another 'daughter directive' (European Commission, 2006b)).

Monitoring requirements are detailed in Annex V of the directive. The design of the monitoring programmes had to be developed in 2006 on the basis of an analysis of pressures and impacts and of a characterisation work leading to the delineation of water bodies (reporting units under the WFD) taking into account typologies, systematic classification of types (in the case of surface waters), and to the identification of water bodies 'at risk' of failing the WFD environmental objectives. In this respect, the legal requirements covered by Annex V for surface waters include:

- Quality elements for the classification of ecological status for different types of surface water (rivers, lakes, transitional waters, coastal waters, and artificial and heavily modified surface water bodies).

- Normative definitions of ecological status classifications (high, good and moderate) for the above types of surface water.

- Monitoring provisions for ecological and chemical status for surface waters, covering surveillance monitoring, operational monitoring and investigative monitoring, as well as requirements regarding the frequency of monitoring, protected areas' monitoring and standards for monitoring of quality elements.

- Requirements for the comparability of biological monitoring results, presentation and classification of ecological status and ecological potential, and presentation of monitoring results and classification of chemical status.

In the case of groundwater, requirements include:

- Parameters for the classification of quantitative status, groundwater level monitoring network, including density of monitoring sites, frequency, and interpretation and presentation of results.

- Parameters for the determination of groundwater chemical status, and related monitoring requirements (surveillance and operational), including monitoring for the identification of pollution trends, and interpretation and presentation of results.

This chapter does not aim to provide an extensive overview of monitoring provisions and their interpretation, which are largely described in guidance documents developed by expert groups under the Common Implementation Strategy (European Commission, 2003, 2006a, 2007) and discussed for groundwater in the light of monitoring requirements under a range of parent legislation (Quevauviller, 2005). Furthermore, the present book includes detailed descriptions of chemical monitoring of surface waters (Chapter 1.2) and ecological status monitoring (Chapter 1.3), as well as a series of case studies on the monitoring of different aquatic environments, namely lake monitoring (Chapter 2.1), river monitoring (Chapter 2.2), groundwater monitoring (Chapter 2.3) and coastal and marine monitoring (Chapter 2.4).

EU Member States had to design monitoring programmes before the end of 2006 and report them to the European Commission in March 2007. Basic requirements are that monitoring data have to provide a reliable assessment of status of all water bodies or groups of bodies. This implies that networks have to consider the representativeness of monitoring points as well as frequency. In addition, monitoring has to be designed in such a way that long-term pollution trends may be detected.

The various types of monitoring depend upon the pre-characterisation of pressures and impacts on water bodies (requested under Article 5 of the directive). These are: surveillance, operational and investigative monitoring, which all imply biological, chemical or quantitative measurements, with different frequencies and parameters.

For example, groundwater surveillance monitoring will be used to supplement and validate the impact assessment procedure, and provide information to be used in the assessment of long-term trends both as a result of changes in natural conditions and through anthropogenic activity (European Commission, 2006b). Minimum monitoring parameters include dissolved oxygen content, pH value, electrical conductivity, nitrate and ammonia (for all groundwater bodies). Groundwater bodies which were found to be at risk (following the 2004 impact assessment) will also have to be monitored for those substances which are indicative of the impact of these pressures. In this respect, operational monitoring will have to be undertaken in the periods between surveillance monitoring programmes in order to establish the chemical status of all groundwater bodies determined as being at risk, and the presence of any long-term anthropogenically-induced upward trend in the concentration of any pollutant. The frequency of surveillance monitoring is not strictly defined in the WFD, but operational monitoring will have to be performed at a minimum once per year. Regarding the identification of trends in pollutant concentrations, the monitoring programmes will have to be adapted to local situations and the trends will have to be demonstrated statistically, stating the level of confidence associated with the identification. As discussed in the *Groundwater Monitoring* guidance document (European Commission, 2006a), monitoring obligations also exist in parent legislation, e.g. the nitrates directive (European Commission, 1991a), the pesticide directives (European Commission, 1991b, 1998), etc.

Regarding surface waters, Annex V is more prescriptive concerning monitoring frequencies, in particular for operational monitoring (see table in paragraph 1.3.4 of that annex). Technical specifications are detailed in a guidance document (European Commission, 2007), providing recommendations on key monitoring features.

1.1.3 REPORTING REQUIREMENTS AND THEIR IMPLICATIONS

Monitoring and data reporting for evaluating the environmental status and trends need to be coordinated at EU level in the framework of a common mechanism. This is the goal of the Water Information System for Europe (WISE), which was launched at the end of March 2007 (D'Eugenio *et al.*, 2007) and which will allow making a considerable step forward at the horizon of 2008–2009. Coordinated reporting and data sharing should constitute the core basis for water policy implementation and review within the next decade.

Reporting requirements are closely linked to the need to ensure the quality of measurement data. Recommendations are being developed in this respect in the form of non-legally binding guidance documents and legally binding provisions under a draft Commission Decision on minimum performance criteria for analytical methods. This is also discussed in this book in relation to supporting research and technology development (RTD) projects such as EAQC-WISE (see Chapter 9.1), NORMAN (Chapter 8.2) and networking discussions under SedNET (Chapter 8.3). Reporting requirements for priority substances are also discussed in Chapter 9.1.

1.1.4 SUPPORTING RESEARCH AND DEVELOPMENT

The need to timely and efficiently integrate scientific outputs in policy developments, implementation and review is extensively discussed in the water sector (Quevauviller *et al.*, 2005). This integration is intimately linked to dialogue establishment, transfer mechanisms and intensive multi-stakeholder consultations. The consideration of scientific progress as one of the key aspects for the design of new policies and the review of existing ones is fully embedded into the Sixth Environmental Action programme, which stipulates that 'sound scientific knowledge and economic assessments, reliable and up-to-date environmental data and information, and the use of indicators will underpin the drawing-up, implementation and evaluation of environmental policy' (European Commission, 2002). This requires, therefore, that scientific inputs constantly feed the environmental policy process. This integration also involves various players, namely the scientific and policy-making communities, but also representatives from industry, agriculture, NGOs, etc.

In the context of the above discussions, which are leading to concrete proposals for the development of an operational science-policy mechanism (Chapter 10.1) and enhanced involvement of stakeholders (Chapter 10.2) and researchers (Chapter 10.3), several EU-funded projects are directly or indirectly contributing to the knowledge base for more efficient and scientifically-based monitoring programmes. The 'Status Monitoring under the WFD' conference provided a wide range of examples presented in the form of either posters or keynote lectures. In the present book, examples concern emerging methods for water monitoring issued from the STAMPS and SWIFT-WFD projects (Chapter 3.1), as well as diagnostic water quality instruments (Chapter 3.2). Modelling tools also have a prominent role to play in monitoring programmes, as exemplified by Chapters 4.1 (joint modelling and monitoring of aquatic ecosystems) and 4.2 (harmonised modelling tools), which are derived from the CatchMod cluster.

Other types of research contribution are more specifically linked to groundwater, e.g. regarding hydrogeological science (Chapter 5.1) and georeferencing (Chapter 5.2). A focused EU-funded project has also contributed to build up the foundation for a common methodology for the establishment of groundwater threshold values (environmental quality standards), as described in Chapter 5.3.

Finally, sediment monitoring has been extensively discussed within the SedNet and AQUATERRA projects, as described in Chapters 6.1– 7.2.

Other important features which are closely linked to research and policy are the assessment of metal ecotoxicity (see Chapter 7.3) and climate change impact on aquatic ecosystems and their responses (Chapter 7.4).

1.1.5   CONCLUSIONS, PERSPECTIVES

This chapter provides a general introduction of the overall book, establishing links to specific sections describing in detail various monitoring features, many of them closely linked to RTD developments. Besides the need for an efficient mechanism for transfer of scientific outputs into policy implementation, an EU-wide coordination is needed to ensure that monitoring data produced at the level of more than 180 European river basins will be of comparable quality and fit for the intended purpose. The WFD presents the advantage of offering a very wide testing framework, and the scientific community should take this opportunity to examine how research findings may match the practice and be readily usable in the policy context.

The conference and the resulting book are among the many milestones that will be required to establish strong bridges between the scientific, policy-making and stakeholders' communities, which is one of the challenges to be faced within the forthcoming decades.

REFERENCES

D'Eugenio J., Haastrup P., Jensen S., Wirthmann A. and Quevauviller P. (2006) 'General Introduction into WISE', 7th Int. Conf. Hydroinformatics, Nice, September 2006.

European Commission (1991a) Council Directive 91/676/EEC of 12 December 1991, concerning the protection of waters against pollution caused by nitrates from agricultural sources, *Official Journal of the European Communities*, **L 375**, 31.12.1991, p. 1.

European Commission (1991b) Council Directive of 15 July 1991, concerning the placing of plant protection products on the market, *Official Journal of the European Communities*, **L 230**, 19.8.1991, p. 1.

European Commission (1998) Directive 98/8/EC of the European Parliament and of the Council of 16 February 1998, concerning the placing of biocidal products on the market, *Official Journal of the European Communities*, **L 123**, 24.4.1998, p. 1.

European Commission (2000) Directive 2000/60/EC of the European Parliament and of the Council of 23 October 2000, establishing a framework for Community action in the field of water policy, *Official Journal of the European Communities*, **L 327**, 22.12.2000, p. 1.

European Commission (2002) 6th Environment Action Programme, European Commission, 2002–2012, http://ec.europa.eu/environment/newprg/index.htm.

European Commission (2003) *Monitoring under the Water Framework Directive*, CIS Guidance Document No. 7, European Commission, Brussels.

European Commission (2006a) *Groundwater Monitoring*, CIS Guidance Document No. 15, European Commission, Brussels.

European Commission (2006b) Directive 2006/118/EC of the European Parliament and of the Council of 12 December 2006, on the protection of groundwater against pollution and deterioration, *Official Journal of the European Communities*, **L 372**, 12.12.2006, p. 19.

European Commission (2007) *Surface Water Monitoring*, CIS Guidance Document, European Commission, Brussels, in press.

Quevauviller, P. (2005) Groundwater monitoring in the context of EU legislation: reality and integration needs, *J. Environ. Monit.*, **7**(2), p. 89.

Quevauviller P. (2007) WFD monitoring and metrological implications, in: *Rapid Chemical and Biological Techniques for Water Monitoring*, Water Quality Measurements Series, John Wiley & Sons, Ltd, Chichester.

Quevauviller P., Balabanis P., Fragakis C., Weydert M., Oliver M., Kaschl A. *et al.* (2005) Science-policy integration needs in support of the implementation of the EU Water Framework Directive, *Environ. Sci. Pol.*, **8**, p. 203.

1.2

Chemical Monitoring of Surface Waters

Peter Lepom and Georg Hanke

The Water Framework Directive - Ecological and Chemical Status Monitoring Edited by Philippe
Quevauviller, Ulrich Borchers, Clive Thompson and Tristan Simonart © 2008 John Wiley & Sons, Ltd

1.2.1 INTRODUCTION

A strategy for dealing with pollution of water from chemicals is set out in Article 16 of the Water Framework Directive 2000/60/EC (WFD). As a first step of this strategy, a list of priority substances was adopted (Decision 2455/2001/EC), identifying 33 substances or groups of substances of priority concern at Community level. Recently, the European Commission adopted a proposal for a new Directive to protect surface water from pollution (COM (2006) 397 final). The proposed Directive will set limits on concentrations in surface waters of 41 dangerous chemical substances including 33 priority substances and 8 other pollutants that pose a particular risk to animal and plant life in the aquatic environment and to human health. The proposal is accompanied by a communication (COM (2006) 398 final) which elaborates on this approach and an impact assessment (SEC (2006) 947) which illustrates the choices that the Commission made.

In addition, the WFD requires Member States to identify specific pollutants in their river basins, and to include them in the monitoring programmes. Monitoring of both WFD priority substances and other pollutants for the purpose of determination of the chemical and ecological status shall be performed according to Article 8 and Annex V of the WFD.

Member States have accentuated the need for more guidance on implementation of monitoring requirements for chemical substances. In line with previous documents under the WFD Common Implementation Strategy (WFD CIS), a guidance document has been developed under the mandate of the European Commission within the Chemical Monitoring Activity (CMA) in the period October 2005 to March 2007, the content of which is summarised in this chapter. While not being legally binding, it presents the outcome of the discussion of the CMA working group on how to monitor chemical substances in surface waters. It states best practices, complements existing monitoring guidance and provides links to relevant guidance documents, European and international standards. Certain aspects of chemical monitoring are still under negotiation as the proposal of a Directive on environmental quality standards and the planned Commission Directive adopting technical specifications for chemical analysis and monitoring of water status have not yet been adopted. This might have an impact on the content of the guidance document, which therefore could not be finalised yet. However, an interim version of this document is publicly available via CIRCA, the information platform of the European Commission.[1]

The guidance on chemical monitoring of surface waters covers monitoring design relevant to surveillance, operational and investigative monitoring, techniques for sampling and analysis, as well as aspects of analytical quality assurance and control. It is open to amendments according to the boundary conditions set in the WFD and it is planned to be finalised after adoption of the Directive on environmental quality standards in 2009 at the latest, and to be updated six years thereafter.

---

[1] http://circa.europa.eu/Public/irc/env/wfd/library?l=/framework_directive/chemical_monitoring.

1.2.2 DESIGN OF MONITORING PROGRAMMES

All available information about chemical pressures and impacts should be used for setting up the monitoring strategy. Such information includes substance properties, pressure and impact assessments, and additional information sources, e.g. emission data, data on where and for what a substance is used, and existing monitoring data collected in the past.

Often, a stepwise screening approach enables identification of non-problem areas, problem areas, major sources, etc. This approach may for instance start with providing an overview of expected hot spots and sources to gain a first impression of the scale of the problem. Thereafter, a more focused monitoring can be performed, directed at relevant problem areas and sites. For many substances, screening of levels in water, as well as in biota with limited mobility and in sediment, is the best way to get the optimum information within a given amount of resources.

Monitoring programmes will need to take account of variability in contaminant concentrations in time and space (including depth) within a water body. A sufficient number of samples should be taken and analysed to adequately characterise such variability and to generate meaningful results with proper confidence.

The documentation of progressive reduction in concentrations of priority substances and other pollutants, and the principle of no deterioration, are key elements of the WFD and require appropriate trend monitoring. Member states should consider this when designing their monitoring programmes. Data obtained in surveillance and operational monitoring may be used for this purpose.

1.2.2.1 Sampling Strategy

Important principles of the sampling strategy have been described in the CIS guidance document no. 7. Depending on the objective of the monitoring, the physico-chemical properties of the substance to be monitored and the properties of the water body under study, water, sediment and/or biota samples have to be taken.

The setup of the monitoring strategy includes decisions on sample matrix, sampling locations, frequencies and methods. This selection depends on the purpose of monitoring and usually represents a compromise between a sufficient coverage of samples in time and space and limiting the monitoring costs.

The type of water sample to be taken at each site is part of the strategy for the monitoring programme. For most water bodies, spot samples are likely to be appropriate. In specific situations, where pollutant concentrations are heavily influenced by flow conditions and temporal variation, and if pollution load assessments are to be performed, other more representative types of sample may be beneficial. Flow-proportional or time-proportional samples may be better in such cases. A single depth sample might not be adequate to reflect the situation in stratified water bodies such as lakes, estuaries and coastal areas. Hence, waters samples should be taken at several depths at such locations. For example, multiparameter probes (e.g. CTD probes) can be employed to detect stratifications.

In general, reliable data on emission sources reduce monitoring costs because such data may provide a good basis for choosing proper sampling locations, optimal number of sampling sites and appropriate sampling frequencies.

1.2.2.2   Selection of Sample Matrix

The principle matrix for assessing compliance with respect to Environmental Quality Standards (EQS) for priority substances is whole water; or for metals, the liquid fraction obtained by filtration of the whole water sample. Whole water is a synonym for the original water sample and shall mean that suspended particulate matter (SPM) and the liquid phase have not been separated.

Whole water data may be generated by analysis of the whole water sample, or by separate determinations on liquid and SPM fractions. If it can be justified, for example by consideration of contaminant partitioning, it may be argued that there is no need to analyse a particular fraction. If a sampling strategy is selected involving only liquid or SPM fractions then the Member States shall justify the choice with measurements, calculations, etc. All justifications of practice shall be based on data derived from appropriate quality-control activities.

However, demonstrating compliance with EQS in water may be problematic in some cases. Examples include cases where available analytical methods are not sufficiently sensitive or accurate for quantification of substances at the required concentration level, and water bodies with high and fluctuating SPM content and varying properties (taking a representative water sample may be difficult or even impossible).

Particular EQS for prey tissue have been included in the Commission proposal on environmental quality standards just for mercury, hexachlorobenzene and hex-achlorobutadiene.

For other pollutants, the matrix for analysis should be in line with the matrix for which national EQS have been derived.

1.2.2.3   Sediment and Biota

Besides water status assessment preventing further deterioration of the status of aquatic ecosystems is another important objective of the WFD. Monitoring of contaminants in sediment and biota can be used to assess the long-term impacts of anthropogenic activity and, thus, to assess the achievement of this objective. It includes the determination of the extent and rate of changes in levels of environmental contamination.

Hydrophobic and lipophilic substances that tend to accumulate in sediment and biota should be monitored in these matrices for resource effective trend monitoring in order to:

• To assess compliance with the no deterioration objective (concentrations of sub-stances are below detection limits, declining or stable and there is no obvious risk of increase) of the Water Framework Directive.

- To assess long-term changes in natural conditions and to assess the long-term changes resulting from widespread anthropogenic activity.

- To monitor the progressive reduction in the contamination of water bodies with priority substances and the phasing out of priority hazardous substances.

When using sediment or biota for temporal trend monitoring, it is essential to determine the quantitative objectives of the monitoring before any monitoring programme is started. For instance, the quantified objective could be to detect an annual change of 5 % within a time period of 10 years, with a power of 90 % at a significance level of 5 % with a one-sided test. Guidance on designing monitoring programmes for a specific quantitative objective and statistical treatment of the data is given by the International Council for the Exploration of the Sea (ICES) (http://www.ices.dk).

Sediment samples should be collected at an appropriate frequency, which will have to be defined on a local basis, taking into account the sedimentation rate in the studied water body and hydrological conditions (e.g. flood events). Typical sampling frequency will vary from once every 1–3 years for large rivers or estuaries that are characterised by high sedimentation rates, to once every 6 years for lakes or coastal areas with very low sedimentation rates.

The locations for sediment trend monitoring should be representative of a water body or a cluster of water bodies. Where possible, sampling should be performed in non-erosion areas that are representative of the respective sediment formation. For dynamic systems, it might be useful to collect suspended matter.

In case of using biota in trend monitoring, it is common practice to collect samples at least once a year during the non-spawning season.

Representativeness is a key point, i.e. how well a sample reflects the situation in a given area or what area is represented by the sample given a certain level of statistical significance. For example, it is essential to collect individuals away from the mixing zones when sampling downstream from a discharge.

To improve the significance of monitoring results, samples should be collected from areas characterised by relatively low natural variability.

1.2.2.4 Use of Models

Numeric models are important tools for the planning and designing of monitoring programmes. They can help to understand the spatial and temporal variations in pollutant concentrations. For instance, measurements in sediments and biota combined with models can be used to estimate dissolved water concentrations. Thus, appropriately validated and tested models can provide additional evidence that EQS will not be violated in a specific water body.

However, as modelled contaminant concentrations are associated with considerable uncertainties, they cannot be used for the purpose of compliance checking for water bodies that are at risk of failing WFD provisions. The approach can, however, be used in surveillance monitoring for estimation of concentrations in water bodies that are shown to be not at risk when the uncertainty of the model is considered.

1.2.2.5   Monitoring Frequency

The monitoring frequencies given in WFD, Annex V 1.3.4 of once a month for priority substances or once per three months for other pollutants will result in a certain confidence and precision. More frequent sampling may be necessary, e.g., to detect long-term changes, to estimate pollution load and to achieve acceptable levels of confidence and precision in assessing the status of water bodies. In general, it is advisable to take samples in equidistant time intervals over a year, e.g. every four weeks resulting in 13 samples, to compensate for missing data due to laboratory failure, drought, flood, etc. In case of pesticides and other seasonally variable substances, which show peak concentrations within short time periods, enhanced sampling frequency compared to that specified in the WFD may be necessary in these periods. The results of those measurements should be compared with the environmental quality standards set for maximum allowable concentration for single measurements. For the calculation of the annual average concentrations, results have to be weighted according to the associated time interval (time-weighted average). For example, 12 equidistant values per year with two additional values in November could be accounted for with reduced weights for the three November values.

Collecting composite samples (24 hour to 1 week) might be another option to detect peak concentrations of seasonally variable compounds.

To estimate the pollutant load that is transferred across Member State boundaries and into the marine environment, an enhanced sampling frequency is necessary. In case of spot sampling for substances, which show a wide range of concentrations, biweekly sampling, e.g., 26 samples a year, is advisable. Flow-proportional or time-proportional sampling may be beneficial in such cases.

Reduced monitoring frequencies, and under certain circumstances even no monitoring, may be justified if monitoring has revealed that concentrations of substances are far below the EQS, declining or stable, and if there is no obvious risk of increase.

The monitoring frequencies quoted in the Directive might not be practical for transitional and coastal waters; for Nordic lakes, which can be iced for months; or for Mediterranean rivers, which can run dry several months a year.

1.2.3   TYPES OF WFD MONITORING

Implementation of the WFD requires three different types of monitoring: surveillance, operational and investigative monitoring, each of which serves its own purpose.

1.2.3.1   Surveillance Monitoring

Surveillance monitoring has to be performed in a sufficient number of surface water bodies to provide an assessment of the overall surface water status within each catchment or sub-catchment within the river basin district. The location of monitoring stations within a water body should provide information that is representative of the general conditions of the water body and which specifically addresses the objectives of the surveillance monitoring programme. Therefore, surveillance monitoring must enable the assessment of long-term changes resulting from changes in natural conditions or anthropogenic activity and provide sufficient information to both supplement the Annex II risk assessments and assist with design of future monitoring programmes. For this purpose, a wide range of quality elements is being assessed at a low frequency in the most important waters.

It is recommended to consider available monitoring data which has to be reported according to other European Directives, international river and seas conventions (e.g., 76/464/EWG, Nitrates Directive 91/676/EEC, OSPAR JAMP) for the purpose of surveillance monitoring where appropriate.

The criteria for selecting sampling points for the surveillance monitoring are given in WFD Annex V 1.3.1. Water bodies probably at risk, probably not at risk and not at risk of failing the environmental objectives should be covered adequately.

Sampling points should include major rivers as well as points at the downstream end of relevant sub-catchments. Sampling points for general physico-chemical parameters supporting the biological quality elements shall be identical with those for the biological elements. For priority substances and other pollutants, other sampling points may be selected.

It is recommended to establish surveillance monitoring sites with fixed monitoring stations and automatic samplers allowing the collection of mixed samples. If not available, spot samples should be collected. Water level and flow should be recorded, as well as pH, conductivity and temperature.

In case of transboundary waters, consultations about the proposed water body and surveillance monitoring sites should be held between the Member States involved.

Monitoring sites to be used for pollution load estimation (country boundaries and transition from inland waters to marine environment) should include representative water quantity as well as quality monitoring.

Representative approaches related to diffuse and widespread sources are often relevant in surveillance monitoring. In such cases, a sufficient number of monitoring points must be sampled within a selection of water bodies in order to assess the magnitude and impact of the pressures.

1.2.3.2 Operational Monitoring

Unlike surveillance monitoring, operational monitoring is characterised by spatial and temporal flexible monitoring networks and by problem-oriented parameter selection and sampling strategy. This type of monitoring programme must be established for those water bodies that have been identified as being at risk of failing to meet environmental objectives. Operational monitoring focuses on quality elements, which characterise the most important pressures that are present in a water body, and on locations where measures are being taken. This type of monitoring allows assessment of significant short-term changes in concentrations of hazardous substances and other pollutants within a management period, and of any changes in the status of such bodies resulting from the programme of measures.

The operational monitoring programme may be modified during the planning period (six years) if indicated by monitoring results. The monitoring frequency can be reduced, for example, when a certain effect is no longer deemed to be significant or the pressure in question has been eliminated. This applies when at least the good chemical and ecological status have been achieved. As soon as the good status has actually been achieved, the operational monitoring can be stopped and surveillance monitoring will suffice. If operational monitoring aims at the assessment of changes in the status of water bodies resulting from a programme of measures, it might be justifiable to reduce monitoring frequencies or suspend monitoring for a certain time period, as long as no change in the status can be expected.

The criteria for selecting operational monitoring sites are given in Annex V 1.3.2 of Directive 2000/60/EC. In case of significant chemical pressures from point sources, sufficient locations must be selected to assess the magnitude and impact of these point sources according to Annex V of the WFD.

In case of significant chemical pressures from diffuse sources, the water body selected for operational monitoring must be representative of the occurrence of the diffuse pressures and of the relative risk of failure to achieve good surface water status. However, it should be taken into account that water bodies can only be grouped where the type and magnitude of pressure are similar.

Small water bodies, $<0.5\,km^2$ (lakes) or $<10\,km^2$ river basin (rivers), do not need to be included in the operational monitoring unless they are of considerable importance for the total river basin so that chemical pressures would affect the major part of the river basin.

Aggregation of water bodies is possible if the water bodies resemble each other in respect of geography, hydrology, geomorphology, trophic level and extent of human pressures. In such cases, Member States shall provide evidence that the water body where monitoring is carried out is indeed representative of the group of water bodies.

In order to assess the magnitude of the chemical pressure to which bodies of surface water are subjected, Member States shall monitor for all priority substances and other pollutants discharged in significant amounts. In addition, physico-chemical parameters relevant for reliable interpretation of the results of chemical measurements (e.g., DOC, Ca, SPM content) should be measured.

1.2.3.3 Investigative Monitoring

Investigative monitoring may be required under certain circumstances (Directive 2000/60/EC, Annex V.1.3.3), e.g.,

- Where the reasons for any exceedance of environmental objectives are unknown.

- Where surveillance monitoring indicates that the objectives set under Article 4 for a water body are not likely to be achieved and operational monitoring has not yet been established, in order to ascertain the causes of a water body or water bodies failing to achieve the environmental objectives.

• To ascertain the magnitude and impacts of accidental pollution.

Investigative monitoring is therefore even more specific and intensive than operational monitoring. As it aims at finding the reason why a water body exceeds the objectives of the WFD, requirements as regards frequency of sampling, selection of sampling location and analysis may differ from those for surveillance and operational monitoring.

Investigative monitoring might also include alarm or early warning monitoring, for example, for the protection of water bodies subject to drinking water abstraction against accidental pollution.

Investigative monitoring can also be triggered when a water body has been identified as being at risk of failing the objectives due to chemical pressures on the basis of the assessment of biological elements.

The starting point of investigative monitoring will often be that results of surveillance or operational monitoring have revealed that the EQS values are exceeded, but the causes are unknown or poorly understood. It is, however, very difficult to give general guidance on how to proceed in investigative monitoring since a case-by-case approach seems to be the only way forward. Local conditions, the types of pressure and the specific aim of the investigation have to be taken into account. This will in general require expert knowledge and judgment. The necessary monitoring points, the matrix and parameters to be monitored, as well as the frequency of sampling and the duration of the monitoring, have to be adjusted to the specific case or problem under investigation. Investigative monitoring is characterised by spatial and temporal flexible sampling and can be stopped as soon as the cause of non-compliance has been identified. When a programme of measures is in operation and its effect can be expected to be measurable, a suitable operational monitoring has to be established. In case of accidental pollution, investigative monitoring can be ceased as soon as the magnitude of accidental pollution has been ascertained.

Before starting investigative monitoring, in-depth pressure analysis may be required. In particular, it has to be clarified whether point or diffuse sources have to be taken into account as potential cause of non-compliance.

In order to identify the causes of exceedance of EQS in a water body or several water bodies, Member States shall monitor the priority substance(s) or other pollutant(s) of which the water concentrations exceed EQS.

1.2.4 TECHNIQUES FOR SAMPLING

The quality of assessments based on results of chemical analyses is dependent on the quality of sampling and on understanding the inherent variability in the media from which samples are taken. The variability of contaminant concentrations in water bodies is often difficult to quantify and can be higher than uncertainties associated with the analyses. This should be considered in the data evaluation and needs to be addressed in the design of a representative monitoring programme. The design of a monitoring programme includes the selection of sampling points and matrices, as well as sampling frequencies, as described earlier in this chapter. For example, in case of water sampling, the definite selection of sampling points including sampling depths depends on local conditions, e.g. vertical and lateral mixing, water homogeneity, and on the possibility of using appropriate sampling equipment.

A key factor in reducing the uncertainties related to the sampling technique itself as much as possible is that the personnel in charge of the sampling is sufficiently educated and trained in the sampling procedures and in the risks and consequences of taking inappropriate samples. This includes knowledge of the objectives of the monitoring programme, the further treatment of the samples taken and a certain understanding of the hydro-geochemical processes in the water body. A sufficiently detailed sampling report needs to be prepared for each sampling. It should include any observations relevant to the assessment of the monitoring results.

QA/QC procedures should be established to ensure the quality of the sampling activities of a monitoring programme, including taking care to preserve sample integrity (see ISO 5667-14 and other guidelines). Quality assurance of sampling including pretreatment, sub-sampling, preservation, storage and transport is essential for the quality of final results of the chemical analyses. Quality control of sampling should comprise measures that enable estimation of sampling precision. Other measures could be participation in sampling intercomparison trials and regular education of personnel.

Guidance on sampling can be found in the ISO Standard on Water Quality – Sampling 5667 (http://www.iso.org), and for the marine environment in the guidelines for the Joint Assessment and Monitoring Programme (JAMP) of OSPAR (http://www.ospar.org) or in the HELCOM COMBINE Manual (http://www.helcom.fi).

1.2.5 TECHNIQUES FOR ANALYSIS

Article 8, paragraph 3 of the WFD requires that technical specifications and standardised methods for analysis and monitoring of water status be laid down in accordance with the procedure given in Article 21. Moreover, Annex V.1.3.6 of the WFD states that the standards for monitoring of quality elements for physico-chemical parameters shall be any relevant CEN/ISO standards or such other national or international standards, which will ensure the provision of data of an equivalent scientific quality and comparability.

The strengths of such methods are that they are well established and have often been subjected to collaborative trials to demonstrate their interlaboratory comparability and applicability. They may not represent the current state of the art in all cases and usually represent a compromise in performance that is tailored to a number of different users' goals and operational needs.

In general, performance-based methods shall be used in surveillance and operational monitoring. They shall be described clearly, be properly validated and give laboratories the flexibility to select from several options when possible and meaningful. Irrespective of what method is applied in chemical monitoring, certain minimum performance criteria have to be met, which are laid down in the draft proposal for a Commission Directive adopting technical specifications for chemical analysis and monitoring of water status in accordance with Directive 2000/60/EC.

According to this draft Commission Directive, the laboratories may choose any analytical method for the purpose of monitoring under Article 8 and Annex V of Directive 2000/60/EC provided they meet the minimum performance criteria set out in this document or by the national competent authorities.

Suitable methods for monitoring of priority substances and some other pollutants are summarised in Table 1.2.1.

In order to assist Member States in selecting properly validated methods for surveillance and operational monitoring of 41 dangerous chemical substances, substance guidance sheets have been provided for each of those chemicals in Annex II of the guidance document on chemical monitoring of surface waters. These substance guidance sheets summarise basic information on physico-chemical properties of a substance and preliminary environmental quality standards expressed as annual average or maximum allowable concentration for inland and other surface waters. Available EN or ISO standard methods for the analysis in water, and where appropriate in sediment or biota, are specified, including information on sampling, storage and pre-treatment, performance characteristics and a short description of the principle. Where required, other analytical methods are mentioned, and references are given.

1.2.5.1   Method Performance Criteria

Minimum performance criteria have been defined as the limit of quantification, LoQ, and measurement uncertainty, U (expanded uncertainty of measurement). They are linked to the EQS where possible. Guidance on how to determine/estimate these parameters in a pragmatic way is provided in international standards and guidelines.

In case no proper analytical method, i.e. meeting the minimum performance criteria laid down in the Draft Commission Directive 'adopting technical specifications for chemical analysis and monitoring of water status in accordance with Directive 2000/60/EC', is currently available for a particular priority substance, e.g. tributyltin compounds or short-chain. Member States shall ensure that monitoring is carried out using best available techniques not entailing excessive costs. The use of more resource-intensive methodologies, if these can provide the needed performance, at reduced frequencies, is encouraged in these cases.

1.2.5.2   Group Parameters and Definition of Indicator Substances

Some substances of interest are described in generic terms only. These generic substances may be composed of a finite number of isomeric forms, where the potential number of different individual isomers can range from two (e.g. endosulfan) to more than 200 (e.g. polybrominated diphenyl ethers), of which only a few are of environmental relevance. Moreover, it is often difficult or impossible to analyse all these isomers.

Table 1.2.1 Standardised methods applicable to the analysis of priority substances and other pollutants in surface water

Priority Substance[1,2]	Applicable Standard	Principle
Alachlor[3]	EN ISO 6468:1996	GC-ECD
Anthracene	ISO 17993: 2002	HPLC-Fluorescence Detection
Atrazine	EN ISO 11369:1997	HPLC-UV
	EN ISO 10695: 2000	GC-NPD or GC-MS
Benzene	EN ISO 15680:2003	Purge and Trap + Thermal Desorption
	ISO 11423-1:1997	Headspace-GC-FID
Cadmium and its compounds	ISO 17294-2:2003	ICP-MS
Chlorfenvinphos	EN 12918:1999	GC-FPD, GC-NPD, GC-MS, GC-AED or GC-ECD
Chlorpyrifos (-ethyl, -methyl)	EN 12918:1999	GC-FPD, GC-NPD, GC-MS, GC-AED or GC-ECD
1,2-Dichloroethane	EN ISO 10301:1997	GC or Headspace-GC or other
	EN ISO 15680:2003	Purge and Trap + Thermal Desorption
Dichloromethane	EN ISO 10301:1997	GC or Headspace-GC/ECD or other
	EN ISO 15680:2003	Purge and Trap + Thermal Desorption
Di(2-ethylhexyl)phthalate (DEHP)[4]	ISO 18856:2004	GC-MS
Diuron	EN ISO 11369:1997	HPLC-UV
DDT (4 Isomers)[5]	EN ISO 6468:1996	GC-ECD
Fluoranthene	ISO 17993: 2002	HPLC-Fluorescence Detection
Hexachlorobenzene[5]	EN ISO 6468:1996	GC-ECD
Hexachlorobutadiene[3]	EN ISO 10301:1997	GC-ECD or Headspace-GC-ECD or other
	EN ISO 6468:1996	Purge and Trap + Thermal Desorption
	EN ISO 15680:2003	GC-ECD
Hexachlorocyclohexane[5]	EN ISO 6468:1996	GC-ECD
Isoproturon	EN ISO 11369:1997	HPLC-UV
Lead and its compounds	ISO 17294-2:2003	ICP-MS
	ISO 15586:2003	ET-AAS
Mercury and its compounds	EN 12338:1998	CV-AAS with Amalgamation
	EN ISO 17852:2006	Atomic Fluorescence Spectrometry
Naphthalene	ISO 17993: 2002	HPLC-Fluorescence Detection
	EN ISO 15680:2003	Purge and Trap + Thermal Desorption
Nickel and its compounds	ISO 17294-2:2003	ICP-MS
	EN ISO 11885:2007	ICP-AES*
	ISO 15586:2003	ET-AAS
Nonylphenols [6]	ISO 18857-1:2005	GC-MS
Octylphenol (4-(1,1,3,3)-Tetramethylbutylphenol	ISO 18857-1:2005	GC-MS
Pentachlorophenol	EN 12673:1998	GC-ECD or GC-MS after Derivatisation
Benzo(a)pyrene	ISO 17993: 2002	HPLC-Fluorescence Detection
Benzo(b)fluoranthene[7]	ISO 17993: 2002	HPLC-Fluorescence Detection
Benzo(k)fluoranthene[7]	ISO 17993: 2002	HPLC-Fluorescence Detection

Table 1.2.1 (*continued*)

Priority Substance[1,2]	Applicable Standard	Principle
Simazine	EN ISO 11369:1997	HPLC-UV
	EN ISO 10695: 2000	GC-NPD or GC-MS
Tetrachloroethene	EN ISO 10301:1997	GC-ECD or Headspace-GC-ECD or other
	EN ISO 15680:2003	Purge and Trap + Thermal Desorption
Tetrachloromethane	EN ISO 10301:1997	GC-ECD or Headspace-GC-ECD or other
	EN ISO 15680:2003	Purge and Trap + Thermal Desorption
Trichlorobenzenes	EN ISO 6468:1996	GC-ECD
	EN ISO 15680:2003	Purge and Trap + Thermal Desorption
Trichloroethene	EN ISO 10301:1997	GC-ECD or Headspace-GC-ECD or other
	EN ISO 15680:2003	Purge/Trap + Thermal Desorption
Trichloromethane	EN ISO 10301:1997	GC-ECD or Headspace-GC-ECD or other
	EN ISO 15680:2003	Purge and Trap + Thermal Desorption
Trifluralin	EN ISO 10695: 2000	GC-NPD or GC-MS

*axial viewing

[1] For the analysis of pentabromodiphenyl ether and C_{10}-C_{13} chloroalkanes in water there is no standard method available.

[2] The existing standards for the analysis of certain organochlorine pesticides (aldrin, endrin, isodrin, dieldrin, endosulfan, pentachlorobenzene), PAHs (benzo(ghi)perylene, indeno(1,2,3-cd)pyrene) and tributyltin compounds are not sensitive enough to conduct compliance monitoring.

[3] Alachlor and hexachlorobutadiene are not within the scope of the standard but national monitoring laboratories reported that EN ISO 6468 may be used for the determination of these compounds.

[4] Although the method is applicable to the analysis of DEHP in surface water and allows achievement of a sufficiently low LoQ to conduct compliance checking in principle, many laboratories have serious blank problems and are hence, not able to meet the LoQ performance criterion (LoQ $\leq$ 30 % EQS).

[5] According to the results of a survey conducted within the Chemical Monitoring Activity, LoQ low enough to allow compliance checking is difficult to achieve or even impossible for DDT (due to the fact that 4 isomers have to be determined), hexachlorocyclohexane and hexachlorobenzene.

[6] Although the method is applicable to the analysis of NP in surface water and allows achievement of a sufficiently low LoQ to conduct compliance checking in principle, many laboratories have serious blank problems and are hence, not able to meet the LoQ performance criterion (LoQ $\leq$ 30 % EQS).

[7] Although benzo(k)fluoranthene and benzo(b)fluoranthene are mentioned in the scope of the standard, LoQ low enough to allow compliance checking is difficult to achieve or even impossible.

Hence, analysis of indicator substances representative of the entire group is common practice. Indicator substances which have to be analysed have been specified in the proposal for a Directive of the European Parliament and of the Council on environmental quality standards in the field of water policy, amending Directive 2000/60/EC, and will be added to the final version of the guidance document on chemical monitoring as soon as the negotiations on this Directive have been completed.

1.2.6 WATER ANALYSIS

According to the proposal for a Directive on environmental quality standards, EQS are expressed as total concentrations in the whole water sample, except for cadmium, lead, mercury and nickel. In the case of metals, the EQS refers to the dissolved concentration measured in the liquid fraction of a water sample obtained by filtration through a 0.45 µm filter.

This suggests reporting monitoring results except for metals as whole water concentrations. Whole water data may be generated by analysis of the whole water sample, or by separate analyses of the liquid and SPM fractions.

Unfortunately, most available analytical methods have not been validated for water samples containing substantial amounts of SPM. Incomplete extraction of hydrophobic organic contaminants adsorbed to SPM can, therefore, result in an underestimation of the whole water concentration.

The SPM content of the water sample is not critical for the analyses of water-soluble compounds such as some pesticides (e.g., alachlor, atrazine, simazine, diuron, isoproturon) and volatile compounds (benzene, dichloromethane, 1,2-dichloromethane, trichloroethane, tetrachloroethene, trichloroethene, tetrachloromethane, trichlorobenzene, naphthalene). These compounds can be analysed in the whole water or in the filtered sample.

In case of hydrophobic compounds, which strongly adsorb to particles, e.g. pentabromodiphenylether or 5 and 6 ring polycyclic aromatic hydrocarbons, special care is required to ensure complete extraction of the particle bound fraction. Separate analysis of SPM and of the liquid would be a good option. If it can be justified, for example, by consideration of contaminant partitioning, analysis of the SPM fraction as surrogate for whole water might be appropriate. Nevertheless, in water bodies with extremely low SPM content (<3 mg/L) the dissolved fraction of those contaminants contributes significantly to the total concentration, and hence, has to be taken into account.

Depending on the SPM content of the sample and its organic carbon content, medium polar compounds can adsorb to a variable degree to SPM. In such cases, both fractions (dissolved and adsorbed concentrations) have to be considered.

For the determination of dissolved metal concentrations, water samples have to be passed through a membrane filter of 0.45 µm pore size. The filtration should be done in the field in order to prevent artefacts due to adsorption to the container walls. If expected metal concentrations are far below the suggested EQS it would be possible to analyse whole water samples and compare the results for the total content with the EQS for the dissolved fraction. In such cases, samples shall be acidified in the field and sample containers shall be pre-treated accordingly.

Bioavailable concentrations of metals depend on various parameters, including pH, Ca and Mg concentrations, as well as dissolved organic carbon content. Hence, measuring these parameters in parallel with the metals can assist with interpretation of results where appropriate. In case of cadmium, the measurement of hardness is mandatory because EQS values have been derived for five classes of hardness.

Bioavailability and natural background concentrations of metals can be taken into account when assessing the monitoring results against EQS.

1.2.7 SEDIMENT/SPM ANALYSIS

Methods for the analysis of hydrophobic priority substances in sediments are well established in many European monitoring laboratories even though there are no European or international standard methods specifically developed for the analysis of sediments/SPM except for PBDE.

Comprehensive guidance on the analysis of marine sediments, including sample pre-treatment, storage and normalisation of results, is given in OSPAR (2002).

In general, organic contaminants should be analysed in the <2 mm fraction of the sediment, and metals in the <63 μm fraction. If the specific purpose of the monitoring requires analysis of the fine sediment fraction, the sample should be split using appropriate sieving techniques.

The accumulation of a contaminant depends on the sediment and SPM characteristics (grain size, composition and surface properties). Hence, it is essential only to compare analytical results from sediments and SPM with similar properties, or to compare normalised results to assess the contamination. Particle size analyses, measurements of organic carbon content, or measurement of other frequently applied normalisation parameters, such as Li and Al, are advised. Detailed guidance on the use of normalising parameters for sediments is given in Annex 5 of OSPAR (2002).

1.2.8 BIOTA ANALYSIS

At present, formally approved standard methods for the analysis of priority pollutants and other contaminants in biota are scarce and only available for metals, PAH, PCB and some other organic contaminants.

Comprehensive guidance on the analysis of marine biota (seabird eggs, fish, shell-fish), including selection of species and suitable tissue, sampling, sample pre-treatment and storage, is given in OSPAR (1999) and the *Manual for Marine Monitoring in the COMBINE Programme of HELCOM*.

Lipophilic organic contaminants accumulate in the lipid tissue of the species studied. Therefore, concentrations should be provided on wet weight as well as lipid weight basis or the lipid content of the sample should be provided together with the analytical results. It is important to state whether total lipids or extractable lipids have been determined, and to specify the method for lipid determination. Whether or not a normalisation should be performed has to be adjusted to the objective of the monitoring.

1.2.9 COMPLEMENTARY METHODS

While checking compliance with the WFD provisions is currently based on laboratory chemical analysis of spot samples taken in a defined frequency, it is desirable to introduce other techniques for improving the quality of the assessment and to benefit from resource saving developments as they become available. Currently, advanced methods for environmental assessment, referred to as complementary methods in this chapter, are under development and evaluation. Examples of these techniques are:

- Biological assessment techniques (e.g. biomarker analysis, bioassays, biosensors, biological early warning systems).

- Sampling and chemical analytical methods (e.g. sensors, passive sampling devices, test kits (see e.g. ISO 17381:2003 'Water quality – Selection and application of ready-to-use test kit methods in water analysis'), GC-MS or LC-MS screening methodologies).

Details on complementary chemical and biological methods, as well as on *in situ* sampling techniques, are given in Chapter 3.1.

1.2.10   ACKNOWLEDGEMENTS

The excellent cooperation of a sizable number of colleagues from all over Europe, who have significantly contributed to the guidance document on chemical monitoring of surface water, is gratefully acknowledged, namely Ian Allan (University of Portsmouth, United Kingdom), John Batty (Environment Agency, United Kingdom), Anders Bignert (The Swedish Museum of Natural History, Sweden), Katrine Borgå (Norwegian Institute for Water Research, Norway), Susanne Boutrup (National Environmental Research Institute, Denmark), Bruce Brown (Environment Agency, United Kingdom), Mario Carere (Istituto Superiore di Sanità, Italy), Gert-Jan de Maagd (RWS-RIZA, The Netherlands), Elena Dominguez (European Commission, DG Research), Anja Duffek (Federal Environment Agency, Germany), Ola Glesne (Norwegian Pollution Control Authority, Norway), Norman Green (Norwegian Institute for Water Research, Norway), Richard Greenwood (University of Portsmouth, United Kingdom), Robert Loos (EC Joint Research Centre, Ispra) Amparo Martin (Ministry of Environment, Spain), Jens Møller Andersen (National Environmental Research Institute, Denmark), Elisabeth Nyberg (The Swedish Museum of Natural History, Sweden), Ciaran O'Donnel (Environmental Protection Agency, Ireland), Stefano Polesello (CNR-IRSA, Italy), Alejandra Puig (Ministry of Environment, Spain), Philippe Quevauviller (European Commission, DG Environment), Alfred Rauchbüchl (Federal Agency for Water Management, Austria), Joan Staeb (RWS-RIZA, The Netherlands), Celine Tixier (IFREMER, France), Gert Verreet (European Commission, DG Environment) and Jan Wollgast (EC Joint Research Centre, Ispra).

REFERENCES

Allan, I.J., Vrana, B., Greenwood, R., Mills, G.A., Roig, B. and Gonzalez, C. (2006) A 'toolbox' for biological and chemical monitoring requirements for the European Union's Water Framework Directive, *Talanta* **69**, pp. 302–322.
COM (2006) 397 final, Proposal for a Directive of the European Parliament and of the Council on environmental quality standards in the field of water policy and amending Directive 2000/60/EC.
COM (2006) 398 final, Communication from the Commission to the Council and the European Parliament Integrated – Prevention and control of chemical pollution of surface waters in the European Union.

Common Implementation Strategy for the Water Framework Directive (2000/60/EC).

European Commission (2003a) *Monitoring under the Water Framework Directive*, CIS Guidance Document No. 7, European Commission, Luxembourg.

European Commission (2003b) *Analysis of Pressures and Impacts*, CIS Guidance Document No. 3, European Commision, Luxembourg,

Directive 2000/60/EC of the European Parliament and of the Council of 23 October 2000, establishing a framework for Community action in the field of water policy.

Decision 2455/2001/EC of the European Parliament and of the Council of 20 November 2001 establishing the list of priority substances in the field of water policy and amending Directive 2000/60/EC.

Ellison, S.L.R., Rosslein, M. and Williams, A. (eds) (2000) *EURACHEM/CITAC Guide – Quantifying Uncertainty in Analytical Measurement*, 2nd edition.

EN Standard (1998) Water quality – determination of mercury – methods after enrichment by amalgamation, EN 12338:1998.

EN Standard (1999a) Water quality – gas chromatographic determination of some selected chlorophenols in water, EN 12673:1999.

EN Standard (1999b) Water quality – determination of parathion, parathion-methyl and some other organophosphorus compounds in water by dichloromethane extraction and gas chromatographic analysis, EN 12918:1999.

EU Report Contributions of the expert group on analysis and monitoring of priority substances, AMPS, to the Water Framework Directive Expert Advisory Forum on Priority Substances and Pollution Control, EUR 21587 EN.

International Council for the Exploration of the Sea (1995) Report of the ICES/HELCOM Workshop on temporal trend assessment of data on contaminants in biota from the Baltic Sea.

International Standards Organisation (1987) Water quality – sampling, part 4: guidance on sampling from lakes, natural and man-made, ISO 5667-4:1987.

International Standards Organisation (1992) Water quality – sampling, part 9: guidance on sampling from marine waters, ISO 5667-9:1992.

International Standards Organisation (1995) Water quality – sampling, part 12: guidance on sampling of bottom sediments, ISO 5667-12:1995.

International Standards Organisation (1996) Water quality – determination of certain organochlorine insecticides, polychlorinated biphenyls and chlorobenzenes – gas chromatographic method after liquid-liquid extraction, ISO 6468:1996.

International Standards Organisation (1997a) Water quality – determination of highly volatile halogenated hydrocarbons – Gas-chromatographic methods, ISO 10301:1997.

International Standards Organisation (1997b) Water quality – determination of benzene and some derivatives – part 1: head-space gas chromatographic method, ISO 11423-1:1997.

International Standards Organisation (1997c) Water quality – determination of selected plant treatment agents – method using high performance liquid chromatography with UV detection after solid-liquid extraction, ISO 11369:1997.

International Standards Organisation (1998) Water quality – sampling, part 14: guidance on quality assurance of environmental water sampling and handling, ISO 5667-14:1998.

International Standards Organisation (1999) Water quality – sampling, part 15: guidance on preservation and handling of sludge and sediment samples, ISO 5667-15:1999.

International Standards Organisation (2000a) Water quality – determination of selected organic nitrogen and phosphorus compounds – gas chromatographic methods, ISO 10695:2000.

International Standards Organisation (2000b) Water quality – sampling, part 17: guidance on sampling of suspended sediments, ISO 5667-17:2000.

International Standards Organisation (2002) Water quality – determination of 15 polycyclic aromatic hydrocarbons (PAH) in water by HPLC with fluorescence detection after liquid-liquid extraction, ISO 17993:2002.

International Standards Organisation (2003a) Water quality – application of inductively coupled plasma mass spectrometry (ICP-MS) – part 2: determination of 62 elements, ISO 17294-2:2003.

International Standards Organisation (2003b) Water quality – determination of trace elements using atomic absorption spectrometry with graphite furnace, ISO 15586:2003.

International Standards Organisation (2003c) Water quality – gas-chromatographic determination of a number of monocyclic aromatic hydrocarbons, naphthalene and several chlorinated compounds using purge-and-trap and thermal desorption, ISO 15680:2003.

International Standards Organisation (2003d) Water quality – sampling, part 3: guidance on the preservation and handling of water samples, ISO 5667-3:2003.

International Standards Organisation (2004) Water quality – determination of selected phthalates using gas chromatography/mass spectrometry, ISO 18856:2004.

International Standards Organisation (2004b) Water quality – sampling, part 19: guidance on sampling of marine sediments, ISO 5667-19:2004.

International Standards Organisation (2005a) Water quality – determination of selected alkylphenols – part 1: method for non-filtered samples using liquid-liquid extraction and gas chromatography with mass selective detection, ISO 18857-1:2005.

International Standards Organisation (2005b) Water quality – sampling, part 6: guidance on sampling of rivers and streams, ISO 5667-6:2005.

International Standards Organisation (2006) Water quality – sampling, part 1: guidance on the design of sampling programmes and sampling techniques, ISO 5667-1:2006.

International Standards Organisation (2007a) Water quality – determination of mercury – method using atomic fluorescence spectrometry, EN ISO 17852:2006.

International Standards Organisation (2007b) Water quality – determination of selected elements by inductively coupled plasma optical emission spectrometry (ICP-OES), ISO 11885:2007.

International Standards Organisation (1997d) Water quality – guidance on analytical quality control for chemical and physico-chemical water analysis, ISO/TR 13530:1997.

International Standards Organisation/International Electric Commission (1998) Guide to the expression of uncertainty in measurement (GUM), ISO/IEC Guide 98:1995.

Manual for Marine Monitoring in the COMBINE Programme of HELCOM, http://www.helcom.fi/groups/monas/en_GB/main/.

Nicholson, M.D. and Fryer R. (1995) *Techniques in the Marine Environmental Sciences – A Robust Method for Analysing Contaminant Trend Monitoring Data*, International Council for the Exploration of the Sea.

Nordtest Report TR537 (2004) *Handbook for Calculation of Measurement Uncertainty in Environmental Laboratories*, 2nd edition.

OSPAR (1999) JAMP guidelines for monitoring contaminants in biota, http://www.ospar.org/asp/ospar/dra.asp?id=11.

OSPAR (2002) JAMP guidelines for monitoring contaminants in sediments, http://www.ospar.org/asp/ospar/dra.asp?id=11.

SEC (2006) 947, Impact assessment – Proposal for a Directive of the European Parliament and of the Council on environmental quality standards in the field of water policy and amending Directive 2000/60/EC.

1.3

The Monitoring of Ecological Status of European Freshwaters

Angelo G. Solimini,[1] Ana Cristina Cardoso, Jacob Carstensen, Gary Free, Anna-Stiina Heiskanen, Niels Jepsen, Peeter Nõges, Sandra Poikane and Wouter van de Bund

---

[1] The views expressed in this chapter are purely those of the authors and may not in any circumstances be regarded as stating an official position of the European Commission.

The Water Framework Directive - Ecological and Chemical Status Monitoring Edited by Philippe Quevauviller, Ulrich Borchers, Clive Thompson and Tristan Simonart © 2008 John Wiley & Sons, Ltd

1.3.1 ECOLOGICAL STATUS AND ITS ASSESSMENT IN THE WATER FRAMEWORK DIRECTIVE CONTEXT

The fact that anthropogenic activities alter the physical and chemical environment, impair the conditions of the biota, and thus degrade the functioning of a given aquatic ecosystem provides an impetus for the assessment of ecological status. By measuring the deviation of the biological parameters from predefined reference values, the assessment of ecological status provides an integrated, holistic and powerful tool for the monitoring of surface waters. Following this concept, the Water Framework Directive (WFD or 'the Directive' from now on) prioritised the measurement of ecological status, requiring that all European water bodies meet the environmental objective of 'good surface water status' by 2015. In the Directive, the term 'ecological status' is defined as an expression of the quality of the structure and functioning of aquatic ecosystems, and the different quality classes are described with narrative criteria. As many of the European river basins are international, crossing administrative and territorial borders, a common understanding and approach for the assessment of ecological status is crucial for the successful implementation of the Directive.

The Directive explicitly requires that ecological status is assessed through the analysis of various characteristics of aquatic flora and fauna (quality elements; see Cardoso *et al.*, 2006). Through the intercalibration process, different countries sharing common types of water body have set comparable numerical class boundaries for classification of each quality element, in accordance with the narrative criteria given in the Directive. To ensure comparability between different assessment methods, thus providing a common scale of ecological quality, the results of the classification schemes are to be expressed using a numerical scale between 0 and 1, the so-called 'Ecological Quality Ratio' (EQR). The EQR value 1 represents a quality near reference conditions (e.g. the conditions observed in absence of anthropogenic alterations), and values close to 0 represents bad conditions. The results of the classification schemes of the various biological elements, along with the supporting physical and chemical elements, will be combined in determining the surface water status of a given water body.

Although simple enough in theory, the EQR concept is rather difficult to put into practice in the pragmatic implementation of the WFD (Van de Bund and Solimini, 2007). It requires that several key issues are addressed, including the choice of appropriate indicators, typology of water bodies, reference conditions, and agreement on common principles for setting quality class boundaries (see Heiskanen *et al.*, 2004). All these issues need to be taken into consideration and incorporated into the biological assessment methods adopted into the monitoring programmes. For this reason, in the last few years several new biological assessment methods or degrees of modification of already existing methods have been adopted in European countries. In several cases, the results of ECRP were used in the development of new indicators and/or classification schemes (e.g. Charm, Aqem, Star, Rebecca, Fame; see Heiskanen *et al.*, 2005; Hering *et al.*, 2006; Pont *et al.*, 2006; Solimini *et al.*, 2006a).

In the following pages we provide a general overview of the current developments concerning the approaches to the monitoring of ecological status in European freshwaters. We have organised the information by water body type (lakes and rivers) and by the biological quality elements. In each combination of these, we summarise the current information available at the moment of writing on the assessment methods in use and on the general characteristics of the biological monitoring programmes, including spatial and temporal issues. We also shortly address the issues of the uncertainty associated with classification and of the ongoing work on the standardisation of assessment methods. As the biological assessment methods are at different stages of progress, our analysis gives a somewhat incomplete picture. Nevertheless, it is useful to summarise the achievements to date (i.e. before the first round of WFD-compliant monitoring programmes) and to point out which areas are lagging behind and need major effort in the near future.

1.3.2 THE ASSESSMENT OF ECOLOGICAL STATUS IN LAKES

Traditionally, lake monitoring has focused on physico-chemical parameters (nutrients, oxygen profiles, etc.) and on phytoplankton biomass as indicated by chlorophyll *a*, on which several classification schemes exist (e.g. OECD, 1982; Carlson, 1977). Only recently, following the new requirement introduced by the WFD to assess lake ecological status (see Cardoso *et al.*, 2006 for a review) have most European countries included several other biological quality elements in their routine monitoring programmes, such as phytoplankton, macrophytes and phytobenthos, benthic invertebrates and fish.

In general, all Member States have indicated their intention to fulfil the Directive's requirements, including all biological elements in their surveillance monitoring programmes. Some countries will also monitor zooplankton, which is an important component of the lake food web but which the WFD does not require to be monitored. The majority of Member States have decided that the minimum frequencies given in the Directive may not be adequate to achieve an acceptable level of confidence in the resulting classification. Therefore, a higher frequency of sampling has been adopted, especially for nutrients and phytoplankton (typically 4–6 times per year). According to the WFD, all Member States (MS) should have developed assessment systems before the end of 2006. Nevertheless, the development of indicators and setting of class boundaries has turned out to be one of the most critical and difficult tasks of the WFD implementation and is not yet completed.

1.3.2.1 Lake Phytoplankton

Phytoplankton is widely used as an important water quality indicator because of its high species differentiation and sensitivity to environmental factors. Murphy *et al.* (2002) list the following main advantages of using phytoplankton in lake monitoring:

(1) As primary producers, algae are directly affected by physical and chemical factors, and changes in phytoplankton community status have direct implications for the biointegrity of the lake ecosystem as a whole.

(2) Algae generally have rapid reproduction rates and very short life cycles, making them valuable indicators of short-term (scales of days–weeks) impacts.

(3) Phytoplankton provides a good indication of lake trophic state, measurable for example as chlorophyll *a* concentration, and responds quickly and predictably to changes in nutrient status. Relatively standard methods exist for evaluation of functional and non-taxonomic structural (biomass, chlorophyll measurements) characteristics of algal communities.

(4) Sampling is easy, inexpensive, and creates minimal impact to resident biota. Laboratory chlorophyll *a* analysis is quick and cheap.

(5) Algal assemblages are sensitive to some pollutants which may not visibly affect other aquatic assemblages, or may only affect other organisms at higher concentrations (e.g. herbicides).

(6) Changes in community composition can provide finer-scale assessment of changes due to ecological impacts.

Given these features, phytoplankton was included in the WFD monitoring requirements as a relevant quality element for all surface water categories. As parameters to be studied, the WFD prescribes composition and abundance, biomass, and the frequency of blooms. All these parameters are considered to undergo changes along the pressure gradient, and the extent of this degradation can be 'translated' into WFD normative definitions. Phytoplankton is especially suitable for detecting eutrophication, the impact of excessive nutrient loading. Phytoplankton has been widely used in MS national monitoring schemes and was included as an obligatory element for all lake types in Europe used for WFD intercalibration (Cardoso *et al.*, 2005). At the moment, the standard for the routine analysis of phytoplankton abundance and composition using inverted microscopy (Utermöhl technique; European Committee for Standardisation, CEN, 2004) is the only standard available for phytoplankton analysis at a pan-European scale.

In selecting sampling sites, MS shall ensure that monitoring is carried out at points sufficient to indicate overall surface water status within catchments and sub-catchments. Sites should also be monitored where the volume of water present is significant within the river basin district, including large lakes and reservoirs, but also in significant transboundary bodies of water. For operational monitoring, sampling points shall be selected in such a way as to enable the assessment of the magnitude and impact of point sources and diffuse sources of pollution. The spatial distribution of phytoplankton, especially in large lakes, is often patchy and the concentration can vary considerably in different parts of a lake. Vertical and horizontal sample profiles are required due to spatial heterogeneity. Remote sensing (Kallio *et al.*, 2003) and the use of vertical probes (Gregor *et al.*, 2005; Mehner *et al.*, 2005) are powerful tools to register blooms and to overcome the problems of horizontal and vertical heterogeneity in phytoplankton distribution.

The WFD stipulates for surveillance monitoring that phytoplankton should be monitored at intervals not exceeding six months. For operational monitoring, sampling frequency shall be determined by MS so as to provide sufficient data for a reliable assessment of the status of the relevant quality element. High temporal variability in community structure and biomass of phytoplankton requires frequent sampling. The minimum frequencies suggested by the WFD for surveillance monitoring are generally lower than those currently applied in most MS. According to a survey in 2004, lake sampling frequencies in MS varied between 2 and 12 times per year (Cardoso *et al.*, 2005) with spring and summer being the most common sampling seasons. Also, the monitoring guidance (Anonymous, 2003a) agrees that more frequent sampling will be necessary to obtain sufficient precision in supplementing and validating risk assessments in many cases.

Among other points, the WFD calls for an assessment of the duration and intensity of phytoplankton blooms to be used to indicate the ecological status of lakes. To handle temporal changes of phytoplankton, the basic principle is that the sampling frequency should be at least double the highest significant frequency (Dubelaar *et al.*, 2004). As the turnover of both algae and cyanobacteria are rather short, 0.5–7 days (Reynolds, 1984), a rough picture of the dynamic can only be gained with at least weekly sampling (Honti *et al.*, 2007), while daily sampling is needed to follow the phytoplankton community in detail (Kallio *et al.*, 2003). Such frequencies are unrealistic, although scientifically supported. One possibility to diminish the unexplained variance between seasonal 'snapshot' samples is to apply a multivariate approach in which abiotic variables like water temperature and nutrient concentrations determined at the sampling dates and locations are used as covariates to phytoplankton (Mehner *et al.*, 2005).

The metrics used by MS in their national monitoring schemes (Table 1.3.1) can be roughly divided into taxonomy-based and non-taxonomy-based metrics. The use of phytoplankton taxa for water quality assessment dates back to the 1940s and since then numerous indices have been developed, some of which have been included in the WFD monitoring schemes. For example, Austria uses the Brettum (1989) index modified by Dokulil *et al.* (2005), which is based on trophy scores of a large number of planktonic algae. Three new phytoplankton trophic indices (PTI) where elaborated for deep subalpine lakes (Salmaso *et al.*, 2006). A slightly different approach is the one based on functional groups (Reynolds *et al.*, 2002). Species frequently found to co-exist and to increase or decrease in number simultaneously were delimited and given association identities. Algae forming a single functional group also have similar morphologies, and dimensions of the algal cells or colonies such as surface area, volume and maximum linear dimension are powerful predictors of optimum dynamic performance (Reynolds and Irish, 1997). Padisák *et al.* (2003, 2006) developed further the functional group approach, which is applied now in the WFD monitoring scheme in Hungary.

The general disadvantages of the taxonomy-based metrics are that they are time consuming and that the analysis of the community composition requires identification expertise, which may not be available. According to the WFD, MS shall identify the appropriate taxonomic level for biological quality elements required to achieve adequate confidence and precision in the classification. The EU FP-5 project ECOFRAME tested a simplified taxonomic metric for phytoplankton based on the domination

Table 1.3.1 Phytoplankton sampling and assessment methods used by selected EU Member States and Norway in WFD water quality monitoring. Table compiled from Cardoso *et al.* (2005), GIG Milestone Reports from June 2007 (http://circa.europa.eu/Public/irc/jrc/jrc_eewai/library) and published sources. Months marked with roman numerals

Country	Sampling method	Assessment system or metric
Austria	Integrated sample over the euphotic zone or epilimnion or fixed depth range at the lake's deepest point at least 4 times a year.	Biovolume and modified Brettum index (Dokulil *et al.*, 2005). Planktonic blooms are not regarded as they occur too rarely and irregularly, if at all.
Belgium	Monthly samples from V to X.	% of cyanobacteria in total phytoplankton biovolume applied. An indicator-based approach (Van Wichelen *et al.*, 2005) has been proposed for Flanders.
Cyprus	2–4 depth integrated samples from IV to IX over the euphotic layer (2.5* Secchi depth).	Total biovolume, % of bloom-forming cyanobacteria, Algae Group Index (InGa; Agència Catalana de l'Aigua, 2006), Med PTI (Marchetto *et al.*, 2007), maximum of Chl *a* and algae blooms used for intercalibration.
Germany	Integrated sample over the euphotic zone or epilimnion or fixed depth range at the lake's deepest point at least 6 times during vegetation period.	Multimetric assessment system (Nixdorf *et al.*, 2006) consists of three metrics: biomass, algal class, PTSI (phytoplankton lake index). The class boundaries are derived by using a pre-assignment of ecological quality of the lakes based on the German LAWA-index (LAWA, 1999). Method still under change.
Denmark	Under development.	Proposed method (Søndergaard *et al.*, 2005): Median biomass values of phytoplankton phyla within predefined TP classes separately for deep and shallow lakes (Jeppesen *et al.*, 2000). Under development.
Estonia	4 times a year (V, VII, VIII, IX) 2–3 vertical samples from the deepest point using a sampler (for quantitative samples) and a net (for adjustment of species list).	For large lakes Nõges and Nõges (2006) proposed: Chl *a*, biomass of cyanobacteria (VII-VIII) and diatoms (IX). For small lakes: multimetric method based on four parameters: Chl *a*, modified Pielou's evenness index (Pielou, 1966), Phytoplankton Compound Quotient (modified Nygaard's index; Ott and Laugaste, 1996), and domination structure.
Spain	Methods differ by autonomous communities.	Total biovolume, % of bloom-forming cyanobacteria, Algae Group Index (InGa; Agència Catalana de l'Aigua, 2006), Med PTI (Marchetto *et al.*, 2007), maximum of Chl *a* and algae blooms used for intercalibration.

Table 1.3.1 (*continued*)

Country	Sampling method	Assessment system or metric
Finland	1–2 depth integrated samples from 0 to 2 m from V to IX (X).	Biomass classification method based on statistical analysis of reference sites almost finalised for some types. Method for plankton composition under development (proposed method: Lepistö *et al.*, 2006). Method for blooms under development, difficulties with defining reference conditions. The % of cyanobacteria intercalibrated within the Northern GIG.
France	3 times a year depending on thermocline development. One horizontal and one vertical sample with a 10 μm net.	The French phytoplankton index (Barbe *et al.*, 1990) is still sometimes used, but it is not an agreed method in France and was not included in the IC exercise. France is currently working on developing a WFD compliant national method.
United Kingdom	Sampled 3 times a year (VI/VII, VIII, IX).	% of cyanobacteria to total phytoplankton biovolume. Chroococcales with the exception of *Microcystis* are excluded from the calculation. Total biovolume of 0.5 mg/l is considered as a threshold below which a lake cannot be at worse than 'good' ecological status regardless of the proportion of cyanobacteria in the sample.
Greece	2–4 depth integrated samples from IV to IX over the euphpotic layer (2.5* Secchi depth).	Total biovolume, % of bloom-forming cyanobacteria, Algae Group Index (InGa; Agència Catalana de l'Aigua, 2006), Med PTI (Marchetto *et al.*, 2007), maximum of Chl *a* and algae blooms used for intercalibration.
Hungary	Monthly sampling V–IX. Integrated sample from euphotic zone of deep lake, for the whole depth in shallow lakes.	The phytoplankton assemblage index developed for WFD ecological status assessment (Padisák *et al.*, 2006) includes the relative biomass share of 33 functional groups and factor numbers established for each functional group for each lake type. Class boundaries established by expert judgement.
Ireland	Sampled 3 times a year (VI/VII, VIII, IX).	Total biovolume and % of cyanobacteria to total phytoplankton biovolume. Identical to the British approach.

(*continued overleaf*)

Table 1.3.1   (*continued*)

Country	Sampling method	Assessment system or metric
Italy	2–4 depth integrated samples from IV to IX over the euphotic layer (2.5* Secchi depth).	Three plankton trophic indices using the algal orders (PTI_{orders}), species ($PTI_{species}$) and groups of algal orders with opposite trophic characteristics (PTI_{oe}) applied for deep lakes in the Alpine ecoregion (Salmaso *et al.*, 2006). Another method (Buzzi *et al.*, 2007) developed for small- and medium-sized lakes. None of those is fully WFD compliant. Total biovolume, % of bloom-forming cyanobacteria, Algae Group Index (InGa; Agència Catalana de l'Aigua, 2006), Med PTI (Marchetto *et al.*, 2007), maximum of Chl *a* and algae blooms used for intercalibration exercise.
Lithuania	Sampling according to LST EN 25667-2 2001.	Total biomass, abundance and indicator species. Under development.
Latvia	Under development.	WFD compliant classification system under development.
Netherlands	4–6 times from IV to IX.	Cell counts, blooming. Bloom types are defined based on blooming species and bloom density. To each bloom type a specific EQR is assigned, ranging from 0.1 to 0.7, depending on its relation to eutrophication.
Norway	4–6 depth integrated samples from 0 to 6 m from V to IX (X).	Only % of cyanobacteria was intercalibrated within the Northern GIG.
Poland	Under development.	% of cyanobacteria in total phytoplankton biovolume, under development.
Portugal	2–4 depth integrated samples from IV to IX over the euphotic layer (2.5* Secchi depth).	Total biovolume, % of bloom-forming cyanobacteria, Algae Group Index (InGa; Agència Catalana de l'Aigua, 2006), Med PTI (Marchetto *et al.*, 2007), maximum of Chl *a* and algae blooms.
Romania	2–4 depth integrated samples from IV to IX over the euphotic layer (2.5* Secchi depth).	Total biovolume, % of bloom-forming cyanobacteria, Algae Group Index (InGa; Agència Catalana de l'Aigua, 2006), Med PTI (Marchetto *et al.*, 2007), maximum of Chl a and algae blooms.

Table 1.3.1 (*continued*)

Country	Sampling method	Assessment system or metric
Sweden	1–2 depth integrated samples during the period from V to IX (X).	Metrics: 1) total biomass, 2) % of cyanobacteria, 3) trophic plankton index (TPI), 4) species richness. Indicators 1–3 refer to a trophic gradient and no. 4 to a gradient of acidity (Willén, 2007). Only % of cyanobacteria was intercalibrated within the Northern GIG.
Slovenia	Integrated sample over the euphotic zone or epilimnion or fixed depth range at the lake's deepest point at least 4 times a year.	Slovenia decided not to develop a national method, as only two large lakes are situated in the country. The national method from Austria will be adopted for the Slovenian lakes.

pattern (Moss *et al.*, 2003). The observer was supposed to distinguish only between the communities with essentially only one dominant species, accounting for more than 95 % of the total numbers, and those having more than one proportionately abundant species. Scanning the samples requires neither much expertise in taxonomy nor much time. The disadvantage is that it does not result in a WFD-compatible 5-step scale.

Methods based on instrumental detection of pigment composition can also be considered semi-taxonomic but require calibration with microscopically counted samples. Modern scanning flow cytometry (SFC), in which phytoplankton is detected by its native chlorophyll autofluorescence, offers information close to the taxonomic level (Dubelaar *et al.*, 2004). SFC is used most efficiently complementary to microscopical analyses for mutual validation. It presents a realistic solution to generate the essential high-frequency observations required to assess ecosystem variability and for the early detection of cyanobacterial blooms.

Phytoplankton abundance and occurrence of blooms are the parameters for which a not necessarily taxonomic determination is required. The abundance can be measured as the total count of cells and/or colonies in a unit volume of water or recalculated further into biovolume or biomass. The WFD allows use of chlorophyll *a* as a surrogate for phytoplankton biomass, thus it is considered a biological parameter. In fact, chlorophyll *a* is the most frequently measured phytoplankton metric in lakes. Not all countries have included the bloom occurrence in routine monitoring as in some areas (e.g. countries belonging to the Alpine GIG) they occur too rarely and irregularly (if at all). Other non-taxonomy-based metrics, like size composition and primary productivity, are successively less considered in lake monitoring schemes.

Besides eutrophication as the main pressure monitored by means of phytoplankton, other pressures like acidification, loads of toxic substances, hydromorphological modifications and climate change will also have an impact on phytoplankton parameters. Acidification (Geelen and Leuven, 1986) and several toxic substances may have specific effects on phytoplankton composition, distinguishable from the effect of changed nutrient loadings. Climate change affects phytoplankton, often through mechanisms related to nutrient availability. Altered processes in catchment hydrology and soils and in lake sediments may increase external and internal loadings of nutrients, a longer stagnation period alters nutrient exchange between the hypo- and epilimnion, while higher temperatures increase nutrient turnover and favour the development of bloom-forming cyanobacteria. For this reason, it is often difficult to distinguish between the impacts of climate change and those of direct human activities within the river basin. Within the timeframe relevant to achieving the water quality targets set by the WFD, a considerable change in climate can be expected.

1.3.2.2 Other Lake Quality Elements

Macrophytes and Phytobenthos

Macrophytes and phytobenthos are important components of lake food webs, although their contribution to the overall primary production decreases with increasing lake depth. The algal assemblage is influenced by both nutrients and habitat quality but it has received relatively little attention as an indicator of ecological status compared to other primary producers (e.g. phytoplankton and macrophytes). Like phytoplankton, phytobenthos abundance is generally measured as chlorophyll *a*, but it is more difficult to refer the sample to a standard sampling area. The combined response of phytoplankton and phytobenthos communities is probably the best indicator of the lake response to eutrophication (Carvalho *et al.*, 2006). Some Member States are undertaking preliminary study (see Diatom Assessment of Lake and Loch Ecological Status (DALES) in England) to set up WFD-compliant classification schemes of lake ecological status based on phytobenthos (King *et al.*, 2005).

Macrophytes are a key component of lake ecological assessment, being important in their own right but also through their influence on other species; they provide habitat for macroinvertebrates (Jónasson, 1978; Weatherhead and James, 2001), influence plankton dynamics (Scheffer, 1999) and are the most useful group in explaining multi-group taxa richness in lakes (Declerck *et al.*, 2005). Annex V of the WFD requires that the composition and abundance of macrophytes are used in the ecological assessment of lakes. Good status, according to the WFD, can be characterised by the absence of changes 'resulting in undesirable disturbance to the balance of organisms present in the water body'. Restoration from moderate to at least good status is required when there is a moderate change in composition and abundance from the type-specific reference community. These status definitions are to be better characterised following an intercalibration exercise of macrophyte data from EU states.

Classification schemes for macrophytes are currently being developed by many Member States. One approach being followed in Germany (Schaumburg *et al.*, 2004) and England (Willby *et al.*, 2006) is to designate macrophytes as reference, impacted or indifferent for specific lake types. The classification of a lake is then based on the proportions of macrophytes that are indicative of reference and impacted conditions. The method used in northern Belgium also incorporates aspects of this approach together with metrics describing the diversity of growth forms and changes in abundance (Leyssen *et al.*, 2005). The Dutch method incorporates information on the percentage cover of submerged macrophytes (for a depth range of 0–3 m), shoreline emergent vegetation cover and species composition (divided into three indicative groups weighted by their abundance) (Van der Molen, 2004; Coops *et al.*, 2007). In

Sweden, assessment is based on taxa richness and the assignment of a trophic ranking score for different lake types (Swedish EPA, 2007). In Ireland, a multimetric approach is followed, incorporating several of the aforementioned parameters as well as the depth of colonisation of macrophytes (Free *et al.*, 2007).

Although there appears to be some concordance of assessment approaches across Europe, the methods used to collect such data are diverse. One of the best methods of conducting a macrophyte survey is by SCUBA diving (Melzer, 1999), which is preferable to a grapnel-based method used in many assessments, which tends to under-represent strongly rooted taxa with a growth form close to the sediment (e.g. *Isoetes* spp.). The question of where to sample is also approached differently. Typically there is a sampling bias towards the shallower areas where macrophytes tend to occur more frequently. Sampling is done along transects perpendicular to the shore (although some countries also sample parallel to the shore), with sampling points occurring at either set depth or set distance intervals. The location of sampling sites is of importance in describing the response of lake macrophytes to, for example, changes in transparency owing to eutrophication. In eutrophic lakes, macrophytes can still be abundant in the 0–1.5 m depth zone compared to deeper depths (Coops *et al.*, 2007), so that sampling strategies strongly biased towards shallow areas may not fully reflect the influence of pressure. Future work needed includes the gathering of an extensive standardised dataset matched with important environmental parameters, including sediment characteristics, and obtaining further understanding of the role of macrophytes in lake ecosystem functioning.

Benthic Invertebrates

Annex V of the WFD lists the three main parameters of the macroinvertebrate community to be focused on for ecological assessment: diversity, taxonomic composition and abundance, and the ratio of disturbance-sensitive taxa to insensitive taxa. Restoration from moderate to at least good status is required when these parameters differ from type-specific reference conditions to an extent coincident with the absence of major taxonomic groups. These parameters are not exclusive and others may be included to achieve the overall aim of an ecological status designation of the WFD as an 'expression of the quality of the structure and functioning of aquatic ecosystems' (WFD, Article 2).

Benthic macroinvertebrates play an important role in helping provide a comprehensive ecological assessment alongside other WFD biological elements. They form an important intermediary link between primary producers/microbial decomposers and fish, which is key to a functioning ecosystem, thereby providing a reflection of ecosystem health useful in assessment. At a recent conference timed to coincide with the commencement of WFD monitoring programmes in 2007, most EU countries tabled proposals to monitor macroinvertebrates in lakes as required by the WFD. The UK, for example, identified their potential in indicating a response to chemical, hydromorphological, acidification and organic loading pressures, whereas eutrophication was considered to be better reflected by phytoplankton and phytobenthos communities. The response of lake macroinvertebrates to eutrophication may be complex, especially in shallow littoral zones, where Brauns *et al.* (2007) found that while the response to eutrophication was detectible it needed to be interpreted in the context of habitat type alongside biotic interactions.

At present, there is little harmonisation in the sampling of benthic macroinvertebrates. Cardoso *et al.* (2005) found that the sampling designs of EU Member States differed substantially in the frequency of sampling and in the equipment used to collect samples. Samples were taken using a variety of devices, such as an Ekman grab, sediment corer and triangle bottom dredge. Where hand nets were used there was also variation in the mesh size (100–670 μm) sampling effort (kick-sampling duration varied from 1 to 3 minutes) and whether sampling was conducted in one or several littoral habitats. The zone focused on for sampling also varied among countries, with some sampling either the littoral or the profundal while other countries sample both. Probably these different strategies are designed to help detect the impact of different pressures, either acidification or eutrophication, but they also reflect the lack of common/harmonised sampling procedures as encouraged by the WFD. To help harmonise methods, the European Committee for Standardisation (CEN) is preparing new standards to cover sampling methods, surveying, devices used and the processing of samples. A more specific method relating to the sampling and processing of pupal exuviae of Chironomidae is also in preparation.

While practical (and WFD-compliant) assessment tools using macroinvertebrate parameters are already in use to assess the ecological quality of rivers, for lakes such assessment systems are either nonexistent or in a developmental stage in most EU countries. In Sweden, ecological assessment in exposed littoral zones is based on four indices (Swedish EPA, 2007): Shannon's diversity index (Shannon, 1948), the average score per taxa index (Armitage *et al.*, 1983), the Danish fauna index (Skriver *et al.*, 2000) and an acidity index (Henrikson and Medin, 1986). Assessment in the profundal zone is based on chironomid and oligochaete communities following Wiederholm's (1980) BQI and O/C indices. Solimini *et al.* (2006) summarised current trends in classification as being largely based on indicator taxa, scores associating taxa to trophic state, and multimetric and multivariate models. Approaches that hold promise for future classification tools include those based on functional groups or guilds, as well as comparative analysis of body size distributions across a pressure gradient. Species traits were also considered potentially useful as they may provide insight into the role of macroinvertebrates in ecosystem functioning, but currently information on traits is lacking for many species present in lakes.

A key requirement is to determine whether an assessment system can effectively differentiate response to anthropogenic pressure from natural variation associated with year, season, lake-zone, habitat and site. Johnson (1998) found that an indice's ability to detect acidification impact (based on standardised effect size and statistical power) varied strongly with choice of indicator metric and habitat. The sub-littoral zone was sampled at 4–6 m, typically above the thermocline, and several metrics performed best in that zone, having higher standardised effect sizes, largely owing to the low among-year variation in the metrics tested. An effective selection of habitats, metrics and temporal windows is one method to control variation and better detect change owing to anthropogenic pressure. While such an approach to monitoring is attractive, there is also the danger that a narrow focus, on a specific metric or temporal window for example, may lead to deterioration going undetected. Furthermore, if certain habitats are unmonitored then assessment may fail to detect change there or represent only a partial assessment of a lake, as well as miss out on zones that are of importance for their biodiversity, such as the littoral (Jónasson, 1978). While intensive monitoring programmes may seem costly, they can be designed to meet the objectives of several policies. In this context, a potential for synergies exists between monitoring for the WFD, the Habitats Directive and the recent EU commitment to halt biodiversity loss by 2010 (European Community, 1992, 2000, 2006).

Fish Fauna

In comparison with the other ecological quality elements, the monitoring of fish constitutes an extra challenge. Owing to their high mobility, their distribution is far from uniform both on the spatial and on the temporal scale, and often the probability of capture varies highly between species and age classes. Therefore the collection from a lake (or river) of a representative sample of the actual fish community is not easy, if at all possible. Thus, there are many empirical and theoretical problems in using fish communities as an indicator for environmental quality, but in many EU Member States there are also institutional problems, because fish traditionally is 'placed' in sectors performing research on food, agriculture or fisheries and not on the environment.

The fish community of lakes responds to a number of human pressures and thus can be used as an indicator for the presence and magnitude of such pressures (Jeppesen *et al.*, 2000; Mehner *et al.*, 2005). Despite the fact that the status of lake fish populations has been monitored in several Member States for long periods, the monitoring and use of fish as a biological quality element, in the context of the WFD, is still far from being operational. The specific descriptions of the characteristics of the biological quality elements given in the WFD identify many of the key elements to be used in ecological assessment. For fish, these are listed as the composition and abundance of the fish community and the age structure of the fish populations (identifying sustainable populations).

There are major problems in establishing sensitive methods for evaluation such as multimetric indices due to a general lack of knowledge of basic pressure-response relationships. In addition, the development of effective comparable monitoring programmes poses enormous challenges and to date no single sampling method has been agreed upon. Probably the most widely used method is sampling with multimesh gill nets, fishing passively overnight in the lakes (Table 1.3.2). For this method a CEN-standard has been developed (European Committee for Standardisation; EN 14757). Gill netting is a passive sampling method and thus very dependent on the activity of the fish, making the results subject to high variation and statistical uncertainty. The method is also problematic due to the fact that it is lethal for the fish and this is not acceptable in many cases. It is widely agreed that a good picture can only be obtained by using a combination of methods, i.e. netting and hydroacoustic surveys, providing results for age structure, abundance and composition of the fish populations. Such a combined approach is, however, rather resource demanding and could be a significant economic problem for MS with many lakes.

Most countries are already carrying out some fish monitoring or plan to commence such programmes soon. The planned sampling frequencies range from twice a year to

Table 1.3.2   Examples of Lake fish sampling methods used for different parameters in 6 member states.

	Multi-mesh gill nets	Electro-fishing	Hydro-acoustics (large lakes)	Fyke netting	Trawling	Seine netting
Species composition	ES, FR, SE	FR, NL, SE, UK, SE (littoral zones)		ES, UK, NL	NL	NL, UK
Abundance	SE, FR, ES, AT	NL	AT, UK FR* SE**		NL	NL, UK
Population – age structure	FR, SE, ES	NL, UK	UK			NL, UK

*certain deep lakes in France
**commercially fished lakes in Sweden

once every six years. The results will mainly be used to develop local metrics or an index and will be evaluated by experts. For shallow lakes multimesh gill netting does not give an accurate estimate of biomass; trawling and seine netting are more accurate. Electro-fishing gives a good picture of the situation in the littoral zone, but does not work well in the open water. Hydroacoustics is a promising method for estimating abundances, especially in large and deep lakes. The recent development of DIDSON technology (Dual Frequency Identification Sonar) provides promise of a future ability to use sound to directly obtain high-quality images of fish in the water.

1.3.3   THE ASSESSMENT OF ECOLOGICAL STATUS IN RIVERS

The biological monitoring of water quality in rivers has a long tradition in Europe (see Ziglio *et al.*, 2006 for a recent review). However, the fulfilment of the WFD's requirements imposed a revision of many old assessment methods, which were either adapted to meet WFD specifications or resulted in the setup of new classification systems.

It is surprisingly difficult to obtain a clear overview of methods that will be used by the EU Member States for monitoring ecological status, in spite of the fact that there is an obligation for monitoring systems to be established by December 2006. One of the best sources of information is the reports of the WFD intercalibration exercise where the high/good and good/moderate boundaries of the classification systems were set and harmonised. These reports include descriptions of the methods that were intercalibrated. At the time this chapter was written they were only available in draft form. The methods that were intercalibrated will be used when initiating river (and lake) monitoring for the WFD, but it is expected that in the next few years, when actual monitoring data will be available, many countries will further refine their monitoring methods.

The river national monitoring networks of European countries are quite complex in many cases; often they have exceptions to the basic sampling regime for particular water bodies, so that it is not possible to summarise them here in reasonable detail. In general terms, the frequency of monitoring should take into account the variability of biological parameters resulting from natural as well as anthropogenic conditions. Most Member States decided to undertake a more intensive sampling than the minimum frequency required by the Directive, at least for some biological elements and for a subset of sites.

1.3.3.1 Benthic Invertebrate Fauna

Macroinvertebrates have been used as indicators for the quality of rivers for many decades in Europe and elsewhere (Mancini, 2006). Historically, the emphasis of macroinvetebrate assessment methods has been on the effects of organic pollution, but it is clear that this quality element can also be used effectively for detecting the effects of other pressures, including hydromorphological alterations, acidification and perhaps also toxic substances. The history, principal approaches and available methods for river monitoring and assessment based on macroinvertebrates in the context of the WFD were recently reviewed extensively by De Pauw *et al.* (2006), and will not be further discussed here.

The WFD explicitly provides definitions for high, good and moderate status in rivers (WFD Annex V, 1.2.1). For benthic macroinvertebrates, the factors that need to be taken into account when establishing monitoring and assessment method are taxonomic composition, abundance, the ratio of disturbance-sensitive taxa to insensitive taxa, and diversity. Only three methods are shared between two or three countries, while Belgium and Spain apply different methods for different river types. Virtually all methods that are reported are multimetric indices, together covering all (or at least several) aspects that are required by the WFD.

The frequency of inclusion in assessment systems of four commonly used groups of metrics is shown in Figure 1.3.1. The metrics based on the occurrence of sensitive and tolerant taxa (e.g. saprobic indices, tolerance score, ASPT, BMWP, are included in all assessment methods. The metrics based on richness and diversity (e.g. number of taxa, number of EPT taxa, Shannon-Wiener diversity, Margalef index) are included in 65 % of the assessment methods. The metrics based on taxonomic composition and metrics based on the composition in terms of functional groups (e.g. feeding type index) were used less frequently (in 30 % and 20 % of the assessment methods, respectively).

The intercalibration exercise has so far only focused on macroinvertebrate methods that are aimed at detecting the effects of 'general pressure' caused by intensified land use, combining increased nutrient and organic substances loads and hydromorpholog-ical alterations. In the north of Europe, benthic invertebrates are also used to assess the effects of acidification, but these methods have yet to be intercalibrated and are not included in this overview.

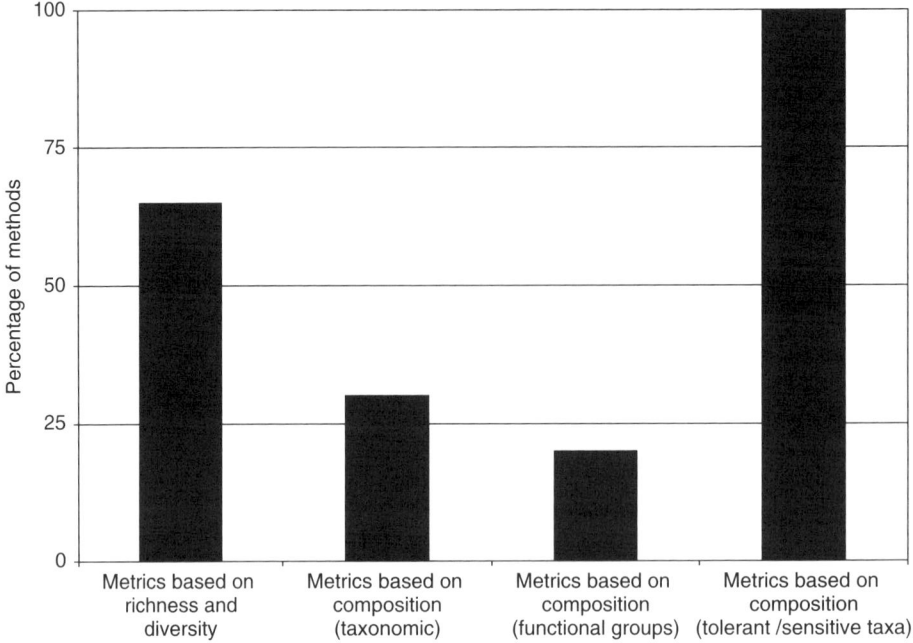

Figure 1.3.1 Percentage of Member States river invertebrate assessment methods that include the four main categories of metrics (see text)

Most of the EU Member States have river invertebrate assessment methods in place that are to a large extent compliant with the requirements of the WFD, and the comparability of their high-good and good-moderate class boundaries has been shown in the intercalibration exercise. The general focus is on multimetric indices covering most of the requirements of the WFD. From this overview one might conclude that there is no need for further development of river macroinvertebrate methods, but in fact there are several open issues and uncertainties that need to be addressed. Issues that will need further clarification include the development of more harmonised procedures and criteria for setting reference conditions, specific methods for large rivers, and methods aimed at specific pressures (including acidification). Also, many of the methods have been developed only recently, often based on limited data. It can be expected that in the next few years, with the start of the monitoring programmes, data of sufficient quality will become available, covering all European rivers, allowing us to fill these gaps.

1.3.3.2   Other River Quality Elements

Phytoplankton, Macrophytes and Phytobenthos

Phytoplankton is an important component of riverine food webs in large rivers (Prygiel and Haury, 2006). The short generation time makes this group of organisms highly reactive to changes in flow conditions, light and nutrients. However, phytoplanktonic biomass is strongly linked to the water residence time and significant amount of biomass may be reached only in low-gradient tracts of large rivers and canals when the residence time is long enough for algal development (e.g. more than six days). Routine monitoring using phytoplankton is foreseen by those countries where such river types are present in a relevant number but may often be limited to the measure of chlorophyll *a* as an indicator of phytoplanktonic biomass.

Annex V of the WFD refers to 'macrophytes and phytobenthos' as a single biological element and identifies four characteristics (taxonomic composition, abundance, likelihood of undesirable disturbances and presence of bacterial tufts) that need to be considered for the purpose of ecological status assessment. Most countries decided to develop separate methods for macrophytes and phytobenthos. Notably, some MS included larger algae such as *Cladophora* in their macrophyte methods, while others included the latter as part of the phytobenthos. Additionally, most of the countries decided to use diatoms as a representative group for the whole phytobenthos.

The use of macrophytes for the biological monitoring of European rivers has a long tradition (Prygiel and Haury, 2006). The composition of macrophyte assemblages and their coverage has been related to water quality and the structure of the (physical) habitat. Channel morphology, texture and stability of sediments, flow regime, depth and light are all factors influencing the species composition of macrophyte assemblages (Janauer and Dokulil, 2006). For this reason, biological metrics based on macrophytes are suitable for determining the effect of hydromorphological changes, such as water level fluctuations downstream of an impoundment (Mancini, 2006). Several WFD-compliant classification systems based on macrophytes are in use in Europe, especially in relation to general degradation (Table 1.3.3).

The term 'phytobenthos' refers to a highly diverse group of organisms (diatoms, filamentous algae, blue-green, etc.) with heterogeneous growth forms on many different river substrates. For the assessment of ecological status, diatoms are the most frequently used indicators included in monitoring programmes (Table 1.3.4). Almost all metrics

Table 1.3.3 Overview of national assessment methods based on river macrophytes in selected European countries

Country	Method	Reference
Austria	Austrian Index for Macrophytes in Rivers (AIM Rivers)	BMLFUW (2006); Pall & Moser (2006)
Belgium (Flanders)	MAFWAT (Macrofyten Waterlopen)	Leyssen *et al.* (2005)
Belgium (Wallonia)	Indice Biologique Macrophytique en Rivière (IBMR)	NF T90-395:2003; Galoux (2007)
France	Indice Biologique Macrophytique en Rivière (IBMR)	NF T90-395:2003
Germany	Reference Index (RI)	Meilinger *et al.* (2005); Schaumburg *et al.* (2006)
Great Britain	LEAFPACS	Willby *et al.* (2006)
Netherlands	Maatlatten	Molen & Pot (2007)
Poland	Macrophyte Index for Rivers (MIR)	Szoszkiewicz *et al.* (2006)

Table 1.3.4 National metric/assessment methods for phytobenthos used in the intercalibration exercise

Country	National metric
Austria	Multimetric method consisting of three modules/metrics: a) trophic status module (based on Tropic Index; Rott *et al.*, 1999), b) saprobic status module (based on Saprobic Index; Rott *et al.*, 1997) and c) reference species module (portion of defined reference and bioregion-specific species in total abundance and species number)
Belgium – Flanders	Proportions of negative (impact-associated) and positive (impact-sensitive) indicator taxa (Hendrickx & Denys, 2005)
Belgium – Wallonia	Specific Polluosensibility Index (IPS; Lenoir & Coste, 1996)
Germany	Diatom module: WFD Diatom Index (following Schaumburg *et al.*, 2005) and Trophic Index (Rott *et al.*, 1999) or Saprobic Index (Rott *et al.*, 1997). Non-diatom module: WFD reference species index depends on type-specific taxa and abundances (following Schaumburg *et al.*, 2005)
Estonia	Specific Polluosensibility Index (IPS; Lenoir & Coste, 1996)
Spain	MDIAT: multimetric index composed by simple average of six indices and two sensitive taxa metrics
France	IBD (Biological Diatom Index; Lenoir & Coste, 1996), Standardised Diatom Index (IBD; AFNOR NF T90-354, 2000), Specific Polluosensibility Index (IPS; Lenoir & Coste, 1996)
Ireland	Revised form of Trophic Diatom Index (TDI; Kelly *et al.*, 2006)
Luxembourg	Specific Polluosensibility Index (IPS; Lenoir & Coste, 1996)
Netherlands	WFD Compliant Index (KRW-maatlat; Van der Molen, 2004)
Poland	Average of Trophic Index (Rott *et al.*, 1999) and Saprobic Index (Rott *et al.*, 1997)
Portugal	CEE index (Descy and Coste, 1991)
Sweden	Specific Polluosensibility Index (IPS; Lenoir & Coste, 1996)
United Kingdom	Revised form of Trophic Diatom Index (TDI; Kelly *et al.*, 2006)

rely on the taxonomic composition of the assemblages, often relating the metric value to the pressure gradient with weighted averaging, like in the Trophic Diatom Index (TDI; Kelly, 2001) or the Indice de Polluosensibilité (IPS; Coste in Cemagref, 1981). For example, the TDI relies on the fact that in theory at least the diatom assemblages characteristic of low, moderate and high phosphorus concentrations can be defined (Kelly, 2001). In practice, there are many other factors that can also influence the composition of the diatom assemblage, making assessment difficult. The phytobentos abundance is also highly temporally and spatially variable and its assessment, also in relative terms, is problematic. Few MS have developed new methods based on the relative abundance of positive and negative indicator species (Table 1.3.4).

Fish Fauna

The fish communities of rivers are highly sensitive to several anthropogenic pressures and as such make good indicators, especially in regards to physical modifications and disruptions of connectivity. The fish fauna present can reflect a rather long-term history of pressures experienced in their habitat as well as, for migratory species, the environment encountered over an extended geographic range and life history.

This, in combination with the fact that fish receive considerable focus due to their recreational value and use as a food source, results in the use of fish as indicators having both scientific and sociopolitical support. However, the very same issues, spatial and temporal scale, fishing activities and stocking, make it difficult to establish solid pressure-response relationships.

In contrast to the situation regarding monitoring of fish in lakes, fish populations in rivers have been routinely sampled for decades in many European countries, where the results have mainly been used for fisheries management, but also as indicators of ecological quality (e.g. Schmutz *et al.*, 2000; Oberdorff *et al.*, 2002). Thus, there is a large body of results available, making it possible to test some pressure-response relationships and subsequently develop indices for the measurement of ecological quality. Since the FAME project developed a European Fish Index in 2004 (Pont *et al.*, 2006), the same approach has been used in several countries, and at present there are 13 'national methods' in use, of which eight can be considered official, but more are under development (Table 1.3.5). Most methods (all but one) are multimetric indexes, using functional groups of fish (guilds), measuring the response of such guilds to various pressures.

Sampling fish in European rivers is almost exclusively done by electro-fishing. The principle is that a gasoline- or battery-driven generator provides the current, which is led through the water via a fixed cathode and a mobile anode. The fish that are hit by the electrical field become stunned and/or attracted to the anode and can then be sampled with a hand-operated net. The procedure is relatively simple and safe and most fish can be released unharmed after recording. There are a vast number of variations in the fishing methods, but the general rules outlined in the CEN standard (EN 14011) are widely used in the monitoring programmes, ensuring a level of standardisation.

Monitoring is relatively easy in small wadeable streams, where two persons can sample several stations in one day. Electro-fishing in small rivers and streams provide a very robust and efficient sampling method because most fish species can be caught with high and stable efficiency. However, when the method is applied in larger and deeper (non-wadeable) rivers using a boat, the results become subject to much more variation and bias.

Most European countries are monitoring their streams by sampling a number of stations/sites within each river system. The appropriate length (or area) of one sample site has been discussed and it appears that while 50 or 100 m sites can provide a good picture of the fish community in some rivers, such short stretches do not produce sufficient catches in fast-flowing 'Scandinavian' type rivers, where the fish populations are less dense. In some cases, stretches of up to 1 km are electro-fished to constitute one sample. Furthermore, there seems to be a relatively high number of 'naturally fish-free' rivers in the Nordic countries, as well as in high-altitude streams, and these obviously cannot be evaluated using fish as a quality element.

Monitoring programmes usually cover a network of sampling sites, each to be sampled from annually to once every six years. From each site, the species present, number and size of fish are recorded, providing information on population composition, abundance and age structure. Also, total weight of the catch and even individual weight of fish are sometimes recorded to give estimates of fish biomass.

Table 1.3.5 Overview of the national methods used in the fish-river intercalibration pilot exercise, source: Jepsen and Pont (2007)

National method	Type	Status	Comments
Austria FIA	Type-specific multimetric index (9 metrics)	Official	A method with considerable detailed adjustment to local areas
Germany FIBS	Multimetrics (6 or 9)	Official	Very detailed and elaborate method, clear stepwise construction of a reference community
Spain IBICATS	Multimetrics (1 to 3)	Official in Catalan watersheds	Under revision
Sweden VIX	Multimetrics (6)	Under approval	Adjustment to population type of trout (migrating or resident). Similar to EFI
Netherlands FI	Multimetrics (8)	Official	
France FBI	Multimetrics (7)	Official	A statistical based method, close to the American IBI
Lithuania LZI	Multimetrics (12)		Based on fish ecological guilds and 'sentinel' species
Belgium and Luxembourg IBIP	Multimetrics (6)		IBI-based method
Finland FIFI	Multimetric (5)	Non-official	IBI-based method
Czech Republic FI	Multimetric (3)		New indices should be tested in consecutive years
Belgium Flanders IB	Multimetrics (8 or 9)	Official	A statistical based method explained in two papers; one an adoption of Karr's approach and one a modification of the trisection method
Italy FIDESS	Decision support system		The Decision Support System is based on a neural network that mimics the expert judgment on the basis of geomorphological and faunistic data
Portugal POFI	Multimetrics (5 or 6)	Official	Index includes metrics which were tested for responsiveness to degradation in each river type

Though the monitoring of fish populations in smaller streams is relatively straight-forward using electro-fishing, there are major problems in developing a suitable method to do the same in the larger rivers (>20 m wide and 2 m deep). Here conditions are rather similar to a lake and while electro-fishing may provide good insight into the fishes found in the littoral zone, it cannot give a picture of the whole community. Thus, at present there is no common method for sampling fish in large rivers, but in MS with such rivers, like Germany, the Netherlands, Austria and the UK, efforts are being made to test the efficiency of various methods. The methods tested are similar to the ones used in lake sampling and probably the best solution will be a combination of advanced hydroacoustic and netting/electro-fishing. In some river stretches, trawling has been shown to give good results.

1.3.4 REDUCING UNCERTAINTY, INCREASING CONFIDENCE AND PRECISION

Ecological status should be assessed by comparing values for indicators to those values defined for reference conditions and status class boundaries, and confidence and precision in the classification should also be reported (Anonymous, 2005). Indicators are calculated from data obtained through WFD-compliant monitoring programmes, and as such give only a partial representation of the actual status, since it is practically infeasible to sample a water body at all locations and at all times throughout an assessment period. Consequently, all indicators used for ecological status classification will contain elements of uncertainty, and thus there will always be a risk of misclassification. It is therefore of paramount importance that the statistical aspects involved in indicator estimation are thoroughly considered.

First, indicators should be calculated such that they represent an unbiased estimate of the true status, i.e. the estimate should not be biased by uneven sampling in time and space and specific monitoring conditions underlying individual observations. Second, indicator estimation should aim at reducing uncertainty to its minimum by separating variations in the monitoring data into structural (systematic) and random variations. Reduced uncertainty has direct impact on the indicator precision, which is defined as the half-width of the indicator confidence interval (Anonymous, 2003a) and is determined by 1) the chosen confidence level that should be determined prior to classification, 2) the number of observations, and 3) the magnitude of random variation (uncertainty). Thus, large uncertainty in monitoring data can be compensated for by an increased number of observations, but it must be recognised that the minimum sampling requirements of the directive itself and the WFD monitoring guidelines (Anonymous, 2003a) are most likely insufficient to obtain reasonable indicator precision (Carstensen, 2007).

1.3.4.1 Indicator Estimation and Precision

Monitoring data characterising the biological and physico-chemical elements underlie large sources of uncertainty, which are normally partitioned into 1) spatial variations, 2) temporal variations, 3) sampling and analytical error, in addition to the uncertainty introduced by the classification system. The confidence of classification should include sampling and analytical errors only (Anonymous, 2003b), but it is difficult to partition variations in monitoring data into the three sources and quantify the magnitude of sampling and analytical variances.

If y_i are the observations (n samples) from a monitoring programme used for estimating a relevant indicator, these are considered the outcome of a stochastic process in time and space:

$$Y_{s,t} = \mu_{s,t} + e_{s,t}$$

where $\mu_{s,t}$ is the true value at location s and at time t, and $e_{s,t}$ is the random variation associated with observing the true process, i.e. the sampling and analytical error. In order to quantify the precision of an indicator, the variance of $e_{s,t}(\sigma^2)$ must be determined.

In theory we are interested in the value of an indicator calculated from the true process $I = f(\mu_{s,t})$, but since we do not observe the true process we have to estimate the indicator from a finite number of monitoring data, i.e. $\hat{I} = g(y_1, y_2, \ldots, y_n)$, and potentially also including covariates, i.e. $\hat{I} = g(y_1, y_2, \ldots, y_n; \underline{x}_1, \underline{x}_2, \ldots, \underline{x}_n)$. For example, if the indicator is the mean chlorophyll level of an entire water body over the entire assessment period then the true value of the indicator is $I = \int_S \int_T \mu_{s,t} ds dt$, i.e. the process integrated over the entire water body and assessment period. If the expectation value of the indicator estimator, $E[\hat{I}]$, equals the true value of the indicator, I, the indicator estimate is said to be unbiased. Moreover, the indicator estimator is efficient if there are no other estimators that can provide a more precise and unbiased estimate of the true indicator value.

The simplest and probably most commonly used estimator for an indicator is the average of the observations:

$$\hat{I} = \frac{1}{n} \sum_{i=1}^{n} y_i$$

which has a $1 - 2\alpha$ confidence interval:

$$[\hat{I} + t(n-1)_{\alpha} \cdot \hat{\sigma}/\sqrt{n}; \hat{I} + t(n-1)_{1-\alpha} \cdot \hat{\sigma}/\sqrt{n}]$$

where $t(n-1)_x$ is x-percentile of the t-distribution having $n-1$ degrees of freedom, and $\hat{\sigma}$ is the estimated standard error of the random variation of the true process. In mathematical terms, the indicator precision equals $t(n-1)_{1-\alpha} \cdot \hat{\sigma}/\sqrt{n}$.

This indicator estimator is unbiased and the most efficient (precise) estimator of the true value only, provided that the true process is constant over time and space, i.e. $\mu_{s,t} = \mu$. However, there are no processes for the biological elements in the WFD that can justifiably be assumed constant. Consequently, indicator estimation by averaging monitoring data is far from optimal in terms of precision and will in most cases also result in biased indicator values, i.e. not representing the true indicator value of the water body. The lack of indicator efficiency is characterised by an estimated random variation, which in many situations will be much larger than the true random variation, i.e. $\hat{\sigma}^2 \gg \sigma^2$. The bias of the indicator depends on how unevenly samples are taken over time and how representative the sampling stations are for the entire water body.

1.3.4.2 Recommendations

Bias can be reduced by improving the model describing the true process in time and space. For example, annual indicators of chlorophyll means were 10–15 % lower when including a seasonal model, because there most data were sampled in periods with generally higher chlorophyll levels (Carstensen, 2007). Thus, it is important to consider whether all observations should be given equal weight or whether some observations should have larger weight because they represent longer time periods. Similarly, stations should be given different weight if they represent different portions of the water body. This could be volume weights for pelagic indicators and area weights for benthic indicators. An alternative to spatial weights is to use salinity measurements and standardise indicators to a mean salinity level for the water body, as shown in Carstensen (2007). In that study, yearly means of total nitrogen were adjusted up to 25 % when variations in salinity were taken into account. Another approach for future development is to create a mechanistic model to describe the true process, adjust this model to the observations, and integrate over time and space by means of the model. The development of such an approach is still premature for practical implementation in relation to the WFD, with a great risk of obtaining misleading indicator values (strong bias).

Indicator precision, as given mathematically in the equation for the indicator confidence interval above, depends on 1) confidence level $(1 - 2\alpha)$, 2) random variation $(\hat{\sigma})$, and 3) the number of monitoring observations (n). The confidence level used for classification should, as stated above, be predetermined to the assessment and have a general value across all of Europe. In statistical testing it has become standard to use 95 % confidence, although this value should not be interpreted rigorously. Anonymous (2003a) mentioned a lower confidence level of 90 % but the confidence level should not be set lower than this value as a statement with less than 90 % confidence has little informational value. Random variation should be reduced by improving the models to describe the variations of the true process $\mu_{s,t}$. The more variation that can be explained by including explanatory variables, the less variation will be termed random. The estimate of the random variation should not be based on the assessment data only, if more data are available. In fact, it is beneficial to estimate σ from as many data as possible, including experimental design aiming at quantifying the sampling and analytical uncertainty. Actually, it is recommended that part of the WFD monitoring programme should be devoted to quantifying the magnitude of sampling and analytical uncertainty. If indicator precision is still insufficient after reducing the estimated random variation, the only option left is to increase the number of observations, acknowledging an increased cost for monitoring. This should advocate improving the statistical methods for indicator estimation before expanding the monitoring programme.

Although the underlying theoretical concepts, as described above, are well known to the statistical community, these ideas have not yet materialised into simple useful tools widely used by environmental managers (Fox, 2001). With the increased use of information technology and improved access to data, standardised assessment methods, including proper statistical processing of data, could become a reality in the near future.

1.3.5 STANDARDISATION NEEDS

The effectiveness of the monitoring programmes, and hence of the overall WFD implementation, will greatly depend on the ability of Member States' laboratories to measure and detect change in the chemical and ecological status of the Community's aquatic ecosystems. The measured data will represent the foundation of the water quality evaluation system, on the basis of which decisions will be taken on the programme of measures required to achieve WFD environmental objectives. To ensure the scientific quality and comparability of the data collected by the Member States, the Directive requires (Annex V 1.3.6) the use of standard methods for the monitoring of water quality elements.

At the time of publication of the Directive there were only a few pertinent biological standard methods published and a large amount of work had still to be done on the development of methods for all biological elements in the different water categories. Comprehensibly, Article 20 of the WFD explicitly indicates that technical adaptations to Section 1.3.6 of Annex V can be made in accordance with the procedure laid down in Article 21, i.e. the WFD Regulatory Committee.

Standardisation needs have been highlighted on various occasions over the last three years within WFD-Common Implementation Strategy (CIS) working groups. An activity was initiated in 2004 by a working group on ecological status (ECOSTAT) to support the standardisation of methods for the monitoring and assessment of ecological status. Information gathered at the time on the Member States' biological monitoring systems (see Cardoso *et al.*, 2005) revealed that for all biological elements, with the partial exception of phytoplankton (in lakes and transitional and coastal waters) and macroinvertebrates (in rivers), only limited biological data had been collected by national monitoring programmes. Moreover, even when available, the comparison of those monitoring data was hampered by differences in sampling frequencies and sampling methods, and in how the results were expressed (e.g. biological metrics). Further evidence of the need for additional efforts towards standardisation of biological methods came from the WFD- CIS Intercalibration Working Group, which raised the issue of the lack of comparability of national biological assessment systems as an obstacle to a complete intercalibration exercise.

In 2005, in the framework of bilateral discussions among the European Commission and the European Committee for Standardisation (CEN), an agreement was reached on a modus operandi to foster the development standard methods to support the assessment of the water status following the WFD. This includes links with ad hoc CIS expert groups (on ecological status) regarding technical exchanges, and consultation with the WFD Regulatory Committee (under Article 21) to confirm that identified needs are indeed recognised by Member States. The main objective is to assure the identification and prioritise the development of standard methods to be added to the list on Section 1.3.6 of Annex V of the WFD, following Article 20 of the WFD.

According to the procedure proposed, Member States were extensively consulted during 2006 for revision of CEN biologic standard methods (Table 1.3.6) for their relevance in the context of the WFD and to identify priority areas for future standardisation activities. It became apparent from this process that sampling and quality assurance methods are mostly needed at community level, and also that published standard methods often do not take into account water category type-specific features, as is required in the Directive. This consultation process enabled the identification of a number of candidate sampling and analytical methods fulfilling short-term standardisation needs for lakes, rivers and coastal waters.

Table 1.3.6 List of standard methods consulted among Member States for their applicability regarding the WFD biological monitoring requirements

Standard No.	Title of Standard	Date Published
EN 14757	Water quality – Sampling of fish with multi-mesh gill nets	2005
EN 14011	Water quality – Sampling of fish with electricity	2003
EN 13946	Water quality – Guidance standard for the routine sampling and pre-treatment of benthic diatoms from rivers	2003
EN 14407	Water quality – Guidance standard for the identification, enumeration and interpretation of benthic diatom samples from running waters	2003
EN 14184	Water quality – Guidance standard for the surveying of aquatic macrophytes in running waters	2003
EN 14614	Water quality – Guidance standard for assessing the hydromorphological features of rivers	2004
EN ISO 16665	Water quality – Guidelines for quantitative investigations of marine soft-bottom benthic fauna in the marine environment	2005 in ISO
ISO/CD 19493 prEN	Water Quality – Guidance on marine biological surveys on hard substrate communities	CEN lead
prEN 14996	Water Quality – Guidance on assuring the quality of biological and ecological assessments in the aquatic environment	Awaiting publication
prEN 15110	Water Quality – Guidance standard for the routine sampling of zooplankton from standing waters	Approved
prEN 15204	Water Quality – Guidance standard for routine analysis of phytoplankton abundance and composition using inverted microscopy (Utermöhl technique)	Formal vote procedure
prEN 15196	Water Quality – Guidance on the sampling and processing of the pupal exuviae of Chironomidae (Order Diptera) for ecological assessment	Formal vote procedure
EN 14962	Water Quality – Guidance on the scope and selection of fish-sampling methods	Published 2006
prEN 15460	Water Quality – Guidance standard for the surveying of macrophytes in lakes	Draft prEN
EN ISO5667-1 (REV)	Water quality – Sampling – Part 1: Guidance on the design of sampling programmes and sampling techniques	Publication 2006
EN ISO 27828	Water quality – Methods of biological sampling – Guidance on hand net sampling of aquatic benthic macro-invertebrates (ISO 7828:1985)	1994 Under revision
EN ISO 28265	Water quality – Methods of biological sampling – Guidance on the design and use of quantitative samplers for benthic macro-invertebrates on stony substrata in shallow freshwaters (ISO 8265:1988)	1994 Under revision

(*continued overleaf*)

Table 1.3.6   (*continued*)

Standard No.	Title of Standard	Date Published
EN ISO 9391	Water Quality – Sampling in deep waters for macro-invertebrates – Guidance on the use of colonization, qualitative and quantitative samples (ISO 9391:1993)	1995 Under revision
EN ISO 8689-1	Water quality – Biological classification of rivers – Part 1: Guidance on the interpretation of biological quality data from surveys of benthic macro-invertebrates in running waters (ISO 8689-1:2000)	2000 Under review
EN ISO 8689-2	Water quality – Biological classification of rivers – Part 2: Guidance on the presentation of biological quality data from surveys of benthic macro-invertebrates (ISO 8689-1:2000)	2000 Under review

At the moment of writing, a mandate to CEN to develop European standards in support of the ecological and chemical status monitoring requirements of the WFD is in preparation. In addition, a guidance document on WFD standardisation is expected by the end of 2008. Among other important aspects, this guidance document should provide an opportunity to consolidate a more comprehensive analysis of the long-term standardisation needs for WFD implementation, building on the experience from the intercalibration exercise. This outcome may eventually serve as a basis for future mandates for CEN.

1.3.6 SUMMARY AND CONCLUSIONS

The WFD implementation process is a unique example of collaboration among different countries, which have their own sovereignty but share at the same time a common vision on the sustainable use and protection of water resources that goes beyond national borders. In the Directive, 'good' ecological status defines a balanced biological structure of communities as part of a well-functioning aquatic ecosystem. Legally, it represents the ambitious objective to be reached for all European water bodies by 2015. Although biological classification has a long tradition in Europe (e.g. the saprobic system of Kolkwitz and Marson (1908) or the trophic paradigm of Naumann (1919) cited by Nõges *et al.*, (2006)), few comprehensive datasets have been collected so far within national monitoring programmes (Nõges *et al.*, 2005). Another recent review on biological assessment methods used in European countries revealed a high heterogeneity of methods and approaches for many quality elements and water bodies (Cardoso *et al.*, 2005).

Mainly driven by the intercalibration exercise, a great amount of work has been carried out in Europe in the past few years to accommodate WFD requirements in biological methods suitable for the monitoring and assessment of ecological status. It should be underlined that many of the most recent methods were developed based on limited data and it is very difficult at this point to evaluate their potential in fulfilling the

WFD's monitoring requirements. In general, the degree of maturity of the classification schemes varies according to the country, the quality element and the specific water type. Thus, there is a strong necessity to harmonise the assessment results at European level and to reduce the uncertainty of the classification. When the results of the first WFD-compliant monitoring programmes are available, it is likely that a revision of those methods using data of sufficient quality from all water types will be undertaken.

REFERENCES

AFNOR (2000) Détermination de l'Indice Biologique Diatomées (IBD), Norme NF T 90–354.

Agència Catalana de l'Aigua (2006) Protocol d'avaluació de l'estat ecològic dels estanys, Barcellona, Spain.

Anonymous (2000) Directive 200/60/EC of the European Parliament and of the Council of 23 October 2000 establishing a framework for Community action in the field of water policy, *Official Journal of the European Communities*, **L 327/1**.

Anonymous (2003a) Monitoring under the Water Framework Directive, Common Implementation Strategy for the Water Framework Directive (2000/60/EC), Guidance Document No. 7, available at: http://forum.europa.eu.int/Public/irc/env/wfd/library.

Anonymous (2003b) Typology, reference conditions, and classification systems for transitional and coastal waters, Common Implementation Strategy for the Water Framework Directive (2000/60/EC), Guidance Document No. 5, available at: http://forum.europa.eu.int/Public/irc/env/wfd/library.

Anonymous (2005) Overall approach to the classification of ecological status and ecological potential, Common Implementation Strategy for the Water Framework Directive (2000/60/EC), Guidance Document No. 13, available at: http://forum.europa.eu.int/Public/irc/env/wfd/library.

Armitage, P.D., Moss, D., Wright, J.F. and Furse, M.T. (1983) The performance of a new biological water quality score system based on macroinvertebrates over a wide range of unpolluted running waters, *Water Research*, **17**, pp. 333–347.

Barbe, J., Lavergne, E., Rofes, G., Lascombe, M., Rivas, B.C. and De Benedittis, J. (1990) Diagnose rapide des plans d'eau, *Informations Techniques du Cemagref*, **79**, 1–8.

BMLFUW (2006) Leitfaden für die Erhebung der Biologischen Qualitätselemente, Arbeitsanweisung Fließgewässer, A4-01a Qualitätselement Makrophyten: Felderhebung, Probenahme, Probenaufbereitung und Ergebnismitteilung, Dezember, Bundesministerium für Land- und Forstwirtschaft, Umwelt und Wasserwirtschaft, Wien.

Brauns, M., Garcia, X.-F., Pusch, M.T. and Walz, N. (2007) Eulittoral macroinvertebrate communities of lowland lakes: discrimination among trophic states, *Freshwater Biology*, **52**, pp. 1022–1032.

Brettum, P. (1989) *Algen als Indikatoren für die Gewässerqualität in Norwegischen Binnenseen*, NIVA, Trondheim, Norway.

Buzzi, F., Dalmiglio, A., Garibaldi, L., Legnani, E., Marchetto, A., Morabito, G. *et al.* (2007) Indici fitoplanctonici per la valutazione dello stato qualità ecologica dei laghi della regione alpine, Ministero dell'Ambiente e della Tutela del Territorio e del Mare, Rome, Italy.

Cardoso, A.C., Solimini, A.G., Premazzi, G., Birk, S., Hale, P., Rafael, T. and Serrano, M.L. (2005) Report on harmonisation of freshwater biological methods, European Commission, Joint Research Centre, Ispra, EUR 21769 EN.

Cardoso, A.C., Solimini, A.G., Premazzi (2006) Biological monitoring of rivers and the European Water Legislation, In: Ziglio, G., Flaim, G., Siligardi, M. (eds), *Biological Monitoring of Rivers*, John Wiley & Sons, Ltd, pp. 229–240.

Carlson, R.E. (1977) A trophic state index for lakes, *Limnology and Oceanography*, **22**, pp. 361–369.

Carstensen, J. (2007) Statistical principles for ecological status classification of Water Framework Directive monitoring data, *Marine Pollution Bulletin*, **55**, pp. 3–15.

Carvalho, L., Lepisto, L., Rissanen, J., Pietiläinen, O.-P., Rekolainen, S., Torok, L. *et al.* (2006) Nutrients and eutrophication in lakes, In: Solimini A.G., Cardoso A.C, Heiskanen A.-S. (eds) *Indicators and Methods for the Ecological Status Assessment under the Water Framework Directive*, European Commission, Joint Research Centre, Ispra, EUR report 22314EN.

CEN (2004) Water quality guidance standard for the routine analysis of phytoplankton abundance and composition using inverted microscopy (Utermohl technique), CEN/TC230/WG2/TG3/N83, CEN, Brussels, Luxembourg.

Coops, H., Kerkum, F.C.M., van den Berg, M.S. and van Splunder, I. (2007) Submerged macrophyte vegetation and the European Water Framework Directive: assessment of status and trends in shallow, alkaline lakes in the Netherlands, *Hydrobiologia*, **584**, pp. 395–402.

Coste in CEMAGREF (1982) Etude des méthodes biologiques d'appréciation quantitative de la qualité des eaux, In: Rapport, Q.E., Lyon, A.F. (eds) *Bassin Rhône-Méditérannée-Corse*, p. 218.

Declerck, S., Vandekerkhove, J., Johansson, L., Muylaert, K., Conde-Porcuna, J.M., Van der Gucht, K. *et al.* (2005) Multi-group biodiversity in shallow lakes along gradients of phosphorus and water plant cover, *Ecology*, **86**, pp. 1905–1915.

De Pauw, N., Gabriels, W. and Goethals, P.L.M. (2006) River monitoring and assessment methods based on macroinvertebrates, In: Ziglio, G., Siligardi, M. and Flaim, G. (eds) *Biological Monitoring of Rivres: Applications and Perspectives*, John Wiley & Sons, Ltd, pp. 113–134.

Descy, J.P. and Coste, M. (1991) A test of methods for assessing water quality based on diatoms, *Verhandlungen der Internationale Vereinigung für Theoretische und Angewandte Limnologie*, **24**, pp. 2112–2116.

Dokulil, M.T., Teubner, K. and Greisberger, J. (2005) Typenspezifische Referenzbedingungen für die integrierende Bewertung des ökologischen Zustandes stehender Gewässer Österreichs gemäß der EU-Wasserrahmenrichtlinie, Modul 1: Die Bewertung der Phytoplankton struktur nach dem Brettum-Index. Projektstudie Phase 3, Abschlussbericht, Im Auftrag des Bundesministeriums für Land- und Forstwirtschaft, Umwelt und Wasserwirtschaft, Wien.

Dubelaar, G.B.J., Geerders, P.J.F. and Jonker, R.R. (2004) High frequency monitoring reveals phytoplankton dynamics, *Journal of Environmental Monitoring*, **6**, pp. 946–952.

European Community (1992) Directive 92/43/EEC of the Council of 21 May 1992 on the conservation of natural habitats and of wild fauna and flora, *Official Journal of the European Communities*, **L206**, pp. 7–50.

European Community (2000) Directive 2000/60/EC of the European Parliament and of the Council of 23 October 2000 establishing a framework for Community action in the field of water policy, *Official Journal of the European Communities*, **L327**, pp. 1–72.

European Community (2006) Halting the loss of biodiversity by 2010 – and beyond: sustaining ecosystem services for human well-being, COM (2006) 216.

Free, G., Little, R., Tierney, D., Donnelly, K. and Caroni, R. (2007) A reference based typology and ecological assessment system for Irish lakes, preliminary investigations, EPA, Wexford, http://www.epa.ie.

Fox, D.R. (2001) Environmental power analysis – a new perspective, *Environmetrics*, **12**, pp. 437–449.

Galoux, D. (2007) Reference conditions, high status definition and intercalibration exercise in Wallonia (Belgium) for R-C3 type rivers – macrophytes, Centre de Recherche de la Nature, des Forêts et du Bois, Gembloux.

Geelen, J.F.M. and Leuven, R.S.E.W. (1986) Impact of acidification on phytoplankton and zooplankton communities, *Cellular and Molecular Life Sciences*, **42**, pp. 486–494.

Gregor, J., Geriš, R., Maršálek, B., Heteša, J. and Marvan, P. (2005) In situ quantification of phytoplankton in reservoirs using a submersible spectrofluorometer, *Hydrobiologia*, **548**, pp. 141–151.

Heiskanen, A.-S., Carstensen, J., Gasiūnaitė, Z., Henriksen, P., Jaanus, A., Kauppila, P. *et al.* (2005) Monitoring strategies for phytoplankton in the Baltic Sea coastal waters, European Commission, Joint Research Centre, Ispra, EUR 21583 EN.

Heiskanen, A.S., van de Bund, W., Cardoso, A.C. and Noges, P. (2004) Towards good ecological status of surface waters in Europe – interpretation and harmonisation of the concept, *Water Science and Technology*, **49**, pp. 169–177.

Hendrickx, A. and Denys, L. (2005) Toepassing van verschillende biologische beoordelingssystemen op vlaamse potentële interkalibratielocaties overeenkomstig de europese kaderrichtlijn water: partim 'fytobenthos', Rapport van het Instituut voor Natuurbehoud, IN.R.2005.06, Instituut voor Natuurbehoud, Brussels, Belgium, p. 107.

Henrikson, L. and Medin, M. (1986) Biologisk bedömning av försurningspåverkan påLelångens tillflöden och grundomrÅden, Aquaekologerna: report to Älvsborgs County Administrative Board.

Hering, D., Johnson, R.K., Kramm, S., Schmutz, S., Szoszkiewicz, K. and Verdonschot, P.F.M. (2006) Assessment of European streams with diatoms, macrophytes, macroinvertebrates and fish: a comparativemetric-based analysis of organism response to stress, *Freshwater Biology*, **51**, pp. 1757–1785.

Honti, M., Istvánovics, V. and Osztoics, A. (2007) Stability and change of phytoplankton communities in a highly dynamic environment – the case of large, shallow Lake Balaton (Hungary), *Hydrobiologia*, **581**, pp. 225–240.

Janauer, G. and Dokulil M. (2006) Macrophytes and algae in running waters, In: Ziglio, G., Flaim, G. and Siligardi, M. (eds), *Biological Monitoring of Rivers*, John Wiley & Sons, Ltd, pp. 89–110.

Jeppesen, E., Jensen, J.P., Søndergaard, M., Lauridsen, T. and Landkildehus, F. (2000) Trophic structure, species richness and biodiversity in Danish lakes: changes along a nutrient gradient, *Freshwater Biology*, **45**, pp. 201–218.

Jepsen, N. and Pont, D. (2007) Intercalibration of fish-based methods to evaluate river ecological quality, European Commission report – Scientific and Technical Research Series, EUR 22878 EN.

Johnson, R.K. (1998) Spatiotemporal variability of temperate lake macroinvertebrate communities: detection of impact, *Ecological Applications*, **8**, pp. 61–70.

Jónasson, P.M. (1978) Zoobenthos of lakes, *Verhandlungen der Internationale Vereinigung für Theoretische und Angewandte Limnologie*, **20**, pp. 13–37.

Kallio, K., Koponen, S. and Pulliainen, J. (2003) Feasibility of airborne imaging spectrometry for lake monitoring – a case study of spatial chlorophyll a distribution in two meso-eutrophic lakes, *International Journal of Remote Sensing*, **24**, pp. 3771–3790.

Kelly, M.G. (2001) Role of benthic diatoms in the implementation of the Urban Wastewater Treatment Directive in the River Wear, NE England, *Journal of Applied Phycology*, **14**, 9–18.

Kelly, M.G. (2006) A comparison of diatoms with other phytobenthos as indicators of ecological status in streams in northern England, In: Witkowski, A. (ed.), *Proceedings of the 18th International Diatom Symposium*, Biopress, Bristol, pp. 139–151.

King, L., Clarke, G., Bennion, H., Kelly, M. and Yallop, M. (2005) Sampling littoral diatoms in lakes for the ecological status assessments: a literature review, Environmental Agency, Science Report SC030103/SR1, Bristol, UK.

LAWA (1999) Gewässerbewertung Stehende Gewässer, Vorläufige Richtlinie für eine Erstbewertung von natürlichen entstandenen Seen nach trophischen Kriterien, Kultur-Buch Verlag, Berlin.

Lenoir, A. and Coste, M. (1996) Development of a practical diatom index of overall water quality applicable to the French National Water Board network, In: Whitton, B.A. and Rott, E. (eds) *Use of Algae for Monitoring Rivers*, Institut für Botanik, Universität Innsbruck, pp. 29–43.

Leyssen, A., Adriaens, P., Denys, L., Packet, J., Schneiders, A., Van Looy, K. and Vanhecke, L. (2005) Toepassing van verschillende biologische beoordelingssystemen op Vlaamse potentiële interkalibatielocaties overeenkomstig de Europese Kaderrichtlijn Water – Partim 'Macrofyten', Rapport van het Instituut voor Natuurbehoud IN.R.2005.05 in opdracht van VMM, Brussels.

LST EN 25667-2 (2001) Vandens kokybė, Mėginių ėmimas, 2 dalis, Nurodymai, kaipimti mėginius, Lietuvos Respublikos Sveikatosapsaugos Ministro, Vilnius.

Lepistö, L., Holopainen, A.-L., Vuoristo, H. and Rekolainen, S. (2006) Phytoplankton assemblages as a criterion in the ecological classification of lakes in Finland, *Boreal Environment Research*, **11**, pp. 35–44.

Mancini, L. (2006) Organisation of biological monitoring in the European Union, In: Ziglio, G., Flaim, G. and Siligardi, M. (eds), *Biological Monitoring of Rivers*, John Wiley & Sons, Ltd, pp. 171–202.

Marchetto, A., Lugliè, A., Padedda, B.M., Mariani, M.A. and Sechi, N. (2007) Indice per la valutazione della qualità ecologica dei bacini artificiali mediterranei (MedPTI) a partire dalla composizione del fitoplancton, Documento presentato al Ministero dell'Ambiente, available at: http://www.ise.cnr.it/ftp/medpti.pdf.

Mehner, T., Hölker, F. and Kasprzak, P. (2005) Spatial and temporal heterogeneity of trophic variables in a deep lake as reflected by repeated singular samplings, *Oikos*, **108**, pp. 401–409.

Meilinger, P., Schneider S. and Melzer, A. (2005) The Reference Index Method for the macrophyte-based assessment of rivers – a contribution to the implementation of the European Water Framework Directive in Germany, *International Revue of Hydrobiology*, **90**, pp. 322–342.

Melzer, A. (1999) Aquatic macrophytes as tools for lake management, *Hydrobiologia*, **395/396**, 181–190.

Molen, D.T. and Pot, R. (2007) Referenties en maatlatten voor rivieren ten behoeve van de Kaderrichtlijn Water, update February 2007, Stowa-rapport 2004-43b, Stowa, Utrecht/RIZA Lelystad.

Moss, B., Stephen, D., Alvarez, C., Becares, E., Van de Bund, W., Collings, S.E. *et al.* (2003) The determination of ecological quality in shallow lakes – a tested system (ECOFRAME) for implementation of the European Water Framework Directive, *Aquatic Conservation: Marine and Freshwater Ecosystems*, **13**, pp. 507–549.

Murphy, K.J., Kennedy, M.P., McCarthy, V., Ó'Hare, M.T., Irvine, K. and Adams, C. (2002) A review of ecology based classification systems for standing freshwaters, SNIFFER Project Number: W(99)65, Environment Agency R&D Technical Report: E1-091/TR.

NF T90-395:2003 (2003) Water quality – determination of the macrophytes biological index for rivers (IBMR), Association Française de Normalisation (AFNOR), Saint Denis La Plaine.

Nixdorf, B., Mischke, U., Hoehn, E., Riedmüller, U., Rücker, J. and Schönfelder, I. (2006) Leitbildorientierte Bewertung von Seen anhand der Teilkomponente Phytoplankton im Rahmen der Umsetzung der EU-Wasserrahmenrichtlinie, Endbericht zum LAWA – Projekt: Bad Saarow, Berlin, Freiburg.

Nõges, P. and Nõges, T. (2006b) Indicators and criteria to assess ecological status of the large shallow temperate polymictic lakes Peipsi (Estonia/Russia) and Võrtsjärv (Estonia), *Boreal Environment Research*, **11**, pp. 67–80.

Nõges, P., Van de Bund, W., Cardoso, A.C. and Heiskanen, A.-S. (2005). Setting ecological quality class boundaries for the Water Framework Directive: the lake intercalibration network. Verhandlungen der Internationale Vereinigung für Theoretische und Angewandte Limnologie 29(1): pp. 265–267.

Nõges P., S. Poikane, A. C. Cardoso, W. van de Bund, 2006a. Water Framework Directive – The way to water ecosystems sustainability in Europe. Lakeline 37: pp. 36–43.

Oberdorff, T., Pont, D., Hugueny, B. and Porchers, J.-P. (2002) Development and validation of a fish-based index for the assessment of 'river health' in France, *Freshwater Biology*, **47**, pp. 1720–1734.

OECD (1982) Eutrophication of waters: monitoring, assessment and control, OECD, Paris, pp. 154.

Ott, I. and Laugaste, R. (1996) Fütoplanktoni koondindeks (FKI), üldistus Eesti järvede kohta, *Eesti Keskkonnaministeeriumi Infoleht*, **3**, pp. 7–8.

Padisák, J., Borics, G., Fehér, G., Grigorszky, I., Oldal, I., Schmidt, A. and Zámbóné-Doma, Z. (2003) Dominant species and frequency of equilibrium phases in late summer phytoplankton assemblages in Hungarian small shallow lakes, *Hydrobiologia*, **502**, pp. 157–168.

Padisák, J., Borics, G., Grigorszky, I. and Soróczki-Pintér, É. (2006) Use of phytoplankton assemblages for monitoring ecological status of lakes within the Water Framework Directive: the assemblage index, *Hydrobiologia*, **553**, pp. 1–14.

Pall, K. and Moser, V. (2006) Leitbildbezogenes Bewertungsverfahren für Österreichische Fließgewässer anhand der Makrophyten gemäß EG-Wasserrahmenrichtlinie, Studie im Auftrag des Bundesministeriums für Land- und Forstwirtschaft, Umwelt und Wasserwirtschaft, Systema, Wien.

Pielou, E.C. (1966) The measurement of diversity in different types of biological collections, *Journal of Theoretical Biology*, **13**, pp. 131–144.

Pont, D., Hugueny, B., Beier, U., Goffaux, D., Melcher, A., Noble, R. *et al.* (2006) Assessing the biotic integrity of rivers at the continental scale: a European approach using fish assemblages, *Journal of Applied Ecology*, **43**, pp. 70–80.

Prygiel, J. and Haury, J. (2006) Monitoring methods based on algae and macrophytes, In: Ziglio, G., Flaim, G. and Siligardi, M. (eds), *Biological Monitoring of Rivers*, John Wiley & Sons, Ltd, pp. 155–170.

Reynolds, C.S. (1984) Phytoplankton periodicity: the interactions of form, function and environmental variability, *Freshwater Biology*, **14**, pp. 111–142.

Reynolds, C.S. and Irish, A.E. (1997) Modelling phytoplankton dynamics in lakes and reservoirs: the problem of in-situ growth rates, *Hydrobiologia*, **349**, pp. 5–17.

Reynolds, C.S., Huszar, V., Kruk, C., Naselli-Flores, L. and Melo, S. (2002) Towards a functional classification of the freshwater phytoplankton, *Journal of Plankton Research*, **24**, pp. 17–428.

Rott, E., Hofmann, G., Pall, K., Pfister, P. and Pipp, E. (1997) Indikationslisten für Aufwuchsalgen, Teil 1: Saprobielle Indikation, *Wasserwirtschaftskataster, BMfLF*, pp. 1–73.

Rott, E., Van Dam, H., Pfister, P., Pipp, E., Pall, K., Binder, N. and Ortler, K. (1999) Indikationslisten für Aufwuchsalgen, Teil 2: Trophieindikation, geochemische Reaktion, toxikologische und taxonomische Anmerkungen, *Wasserwirtschaftskataster, BMfLF*, pp. 1–248.

Salmaso, N., Morabito, G., Buzzi, F., Garibaldi, L., Simona, M. and Mosello, R. (2006) Phytoplankton as an indicator of the water quality of the deep lakes south of the Alps, *Hydrobiologia*, **563**, pp. 167–187.

Schaumburg, J., Schranz, C., Hofmann, G., Stelzer, D., Schneider, S. and Schmedtj, U. (2004) Macrophytes and phytobenthos as indicators of ecological status in German lakes – a contribution to the implementation of the Water Framework Directive, *Limnologica*, **34**, pp. 302–314.

Schaumburg, J., Schmedtje, U., Köpf, B., Schranz, C., Schneider, S., Meilinger, P. *et al.* (2005) Makrophyten und Phytobenthos in Flüssen und Seen, Leitbildbezogenes Bewertungsverfahren zur Umsetzung der EG-Wasserrahmenrichtlinie, *Informationsbericht Heft*, January, Bayerisches Landesamt für Wasserwirtschaft, München.

Scheffer, M. (1999) The effect of aquatic vegetation on turbidity: how important are the filter feeders?, *Hydrobiologia*, **409**, pp. 307–316.

Schmutz, S., Kaufmann, M., Vogel, B., Jungwirth, M. and Muhar, S. (2000) A multi-level concept for fish-based, river-type-specific assessment of ecological integrity, *Hydrobiologia*, **422/423**, pp. 279–289.

Shannon, C.E. (1948) A mathematical theory of communication, *Bell System Technical Journal*, **27**, pp. 379–423.

Skriver, J., Friberg, N. and Kirkegaard, J. (2000) Biological assessment of running waters in Denmark: introduction of the Danish Stream Fauna Index (DSFI), *Verhandlungen der Internationale Vereinigung für Theoretische und Angewandte Limnologie*, **27**, pp. 1822–1830.

Solimini A.G., Cardoso A.C and Heiskanen A.-S. (eds) (2006a) Indicators and methods for the ecological status assessment under the Water Framework Directive, European Commission, Joint Research Centre, Ispra, EUR report 22314.

Solimini, A.G., Free, G., Donohue, I., Irvine, K., Pusch, M., Rossaro, B. *et al.* (2006b) Using benthic macroinvertebrates to assess ecological status of lakes: current knowledge and way forward to support WFD implementation, European Commission, Joint Research Centre, Ispra, EUR report 22347 EN.

Søndergaard, M., Jeppesen, E. and Jensen, J.P. (2005) Water Framework Directive: ecological classification of Danish lakes, *Journal of Applied Ecology*, **42**, pp. 616–629.

Szoszkiewicz, K., Zbierska, J., Jusik, Sz. and Zgoła, T. (2006) Opracowanie podstaw metodycznych dla monitoringu biologicznego wód w zakresie makrofitów i pilotowe ich zastosowanie dla części wód reprezentujących wybrane kategorie i typy, Etap II, tom II – rzeki, Instytut Ochrony Środowiska – Akademia Rolnicza im. A. Cieszkowskiego w Poznaniu – Uniwersytet Warmińsko-Mazurski w Olsztynie, Warszawa-Poznań-Olsztyn.

Swedish EPA (2007) Lakes and watercourses, aquatic plants in lakes, available at: http://www.internat.naturvardsverket.se.

Van de Bund, W. and Solimini, A.G. (2007) Ecological quality ratios for ecological quality assessment in inland and marine waters, European Commission, Joint Research Centre, Ispra, EUR report 22722.

Van der Molen, D.T. (ed.) (2004) Referenties en concept-maatlatten voor rivieren voor de Kaderrichtlijn Water, STOWA-rapport 2004/43, STOWA, Utrecht, p. 365.

Van Wichelen J., Denys, L., Lionard, M., Dasseville, R. and Vyverman, W. (2005) Ontwikkelen van scores of indices voor het biologisch kwaliteitselement fytoplankton voor de Vlaamse rivieren, meren en overgangswateren overeenkomstig de Europese Kaderrichtlijn Water, Universiteit Gent in opdracht van VMM, Gent.

Weatherhead, M.A. and James, M.R. (2001) Distribution of macroinvertebrates in relation to physical and biological variables in the littoral zone of nine New Zealand lakes, *Hydrobiologia*, **462**, pp. 115–129.

Wiederholm, T. (1980) Use of benthos in lake monitoring, *Journal of the Water Pollution Control Federation*, **52**, pp. 537–547.

Willby, N., Hilton, J., Pitt, J.-A. and Philipps, G. (2006) Summary of approach used in LEAFPACS for defining ecological quality of rivers and lakes using macrophyte composition, Interim Report June 2006, University of Stirling, Stirling.

Willén, E. (2007) Växtplankton i sjöar, Bedömningsgrunder, Institutionen för Miljöanalys Rapport 5, ISSN 1403-977X.

Wolfram, G., Dokulil, M.T., Donabaum, K., Reichmann, M. and Schulz, L. (2006) Handbuch zur Bewertung des ökologischen Zustandes stehender Gewässer in Österreich gemäß EU-Wasserrahmenrichtlinie: Phytoplankton, Bundesministerium für Land- und Forstwirtschaft, Umwelt und Wasserwirtschaft, Wien.

Ziglio, G., Flaim, G. and Siligardi, M. (2006) *Biological Monitoring of Rivers: Applications and Perspectives*, John Wiley & Sons, Ltd, Chicester, UK.

Section 2
Case Studies on Monitoring Different Aquatic Environments under the WFD

2.1
Lake Monitoring in Sweden

Håkan Marklund

2.1.1   INTRODUCTION

This chapter is an overview of the Swedish system of monitoring the aquatic environment, with a focus on lake monitoring. It summarises the goals of monitoring, the legal authorities concerned with monitoring, and the actual lake monitoring programme. The first part gives an introduction to the premises that have to be considered to understand the programme design. The second part presents the lake monitoring programme in Sweden after adjustment to the Water Framework Directive.

The Water Framework Directive - Ecological and Chemical Status Monitoring   Edited by Philippe Quevauviller, Ulrich Borchers, Clive Thompson and Tristan Simonart © 2008 John Wiley & Sons, Ltd

Monitoring in the aquatic systems of Sweden has been going on for about eight decades, if one considers the first investigative monitoring efforts concerning water chemistry, macrophytes and phytoplankton. The aims of these early studies were to characterise different types of lake and to investigate relationships between water chemistry and species composition of aquatic ecosystems. Even older data are available for fish captures and distribution of species, collected in the interest of food production and for commercial purposes.

The early aims of monitoring usually emanated from questions put forward by scientific demands or for other specific reasons. Many of the investigations were not designed to answer questions in a more general way. Starting from around 1950, monitoring changed focus to pollution control of downstream factories and other potential point sources. Thanks to this monitoring, several legal decisions were taken on measures to prevent pollution of lakes and watercourses in Sweden. The two most visible effects of these efforts are the expansions of waste water treatment plants in the whole country and reduced pollution loads from pulp mills. From around 1970 onward, most waste water treatment plants have had secondary (biological) and tertiary (chemical) treatment. Several also have nitrogen reduction functions, especially those adjacent to seas, including the Baltic Sea.

In the next phase in the history of monitoring, more attention was given to coordination and integration of monitoring. This led to a better understanding of how elements flow through the ecosystems. Soon, energy and nutrient budgets could be calculated, which made it possible to model how pollutions spread through the system, and to answer the question of where to perform measurements and place monitoring stations. During this time, Sweden focused on acidification and long-range transports of pollutants, thereby stressing the importance of monitoring reference or baseline conditions.

These findings led to two conclusions: 1) some but not all monitoring is cost-effectively done by integrated monitoring and not all parameters have to be monitored at the same site; 2) it is important to establish environmental quality criteria for biological as well as for chemical elements. This again led to a spatial broadening of the monitoring net, with the aim of revising and developing quality criteria, a development that started in the aquatic areas and mainly in the surface fresh waters. National surveys of lakes, and later on water courses, were performed every fifth year.

This was the situation in the year 2000 when the Water Framework Directive was put in force and surveillance, operational and investigative monitoring were introduced in Swedish legislation.

2.1.2 DEMANDS ON MONITORING

The design of a monitoring programme is usually a compromise between different demands and a limited budget in Sweden, as in any other country. The presentation below compiles the demands on the Swedish monitoring system, with focus on lakes.

2.1.2.1 Environmental Objectives

The Swedish Parliament has decided upon sixteen national environmental quality objectives which are to be reached within one generation, or at the latest by the year 2020. Of these sixteen goals, five are applicable for lake environments: Zero Eutrophication, Natural Acidification Only, Flourishing Lakes and Watercourses, a Rich Diversity in Plant and Animal Life, and a Non Toxic Environment. Since these goals are set on a national level, monitoring data need to be collected to represent the whole population of lakes in Sweden. Data collected should be indicators for acidification such as pH and alkalinity/acidity, and for eutrophication such as phosphorus and nitrogen. For all the goals there is a demand for suitable biological variables.

More information on the Swedish environmental objectives can be found on the Internet at http://www.miljomal.nu.

2.1.2.2 Official Statistics

In Sweden, different authorities are responsible for generating data sets for official purposes, for example to make national state of the environment reports, and for annually updating figures and facts published at various web sites. The official statistics data on lakes are reported at the Swedish Environmental Protection Agency's web site: http://www.naturvardsverket.se.

2.1.2.3 Swedish Management of Waters

In addition to the demands mentioned above, there is a continuous demand for reference data, mainly for three different purposes:

(1) Data are used to estimate and follow trends in the environment due, for instance, to climate change or changes in long-range transports of pollutions.

(2) In developing and revision of national ecological quality criteria there is a continuous demand for data on inter- and intra-annual variations.

(3) Annually collected reference data are used to determine normative values and whether deviations can be explained by natural causes or are of local or regional anthropogenic origin.

2.1.3 INTERNATIONAL OBLIGATIONS

2.1.3.1 Not Legally Binding Reporting

Sweden is a partner in several international conventions (e.g. HELCOM, PARCOM). By joining these conventions, a country is morally obliged to report the state of its environment on different chemical and biological parameters. Although this reporting is not always mandatory it often has the same status as legally binding reporting.

Reporting to EEA to make pan-European state of the environment reports also sets up a list of obligations on parameters to be gathered and reported.

2.1.3.2   Legally Binding Reporting

Having been a member of the European Union since 1995, Sweden is legally obliged to monitor its environment and report monitoring data to the commission. These obligations are regulated in different directives, such as the Fish Water Directive (78/659/EEC) and the Bathing Water Directive (2006/7/EC), but above all the Water Framework Directive (2000/60/EC). Some of the directives are already operational, but the Water Framework Directive has just entered its legally binding schedule on monitoring. A monitoring network had to be established by 22 December 2006 and reported to the Commission by 22 March 2007.

This need for reporting according to the Water Framework Directive puts additional demands on monitoring beyond those previously stated. For instance, the monitoring must be performed in different lake types; also, the monitoring must preferably be done by intercalibrated quality elements. In Sweden we have adjusted our aquatic monitoring system to be in accordance with the obligations of the Water Framework Directive. The outcome of this revision is presented in Section 2.1.6.

2.1.4   LEGAL SYSTEM OF MONITORING

The environmental monitoring of Sweden is performed by different authorities with slightly different aims. The general goals are set by the Swedish Government, and the decision about the amount of money for national and regional monitoring (see below) is taken by the Ministry of Environment. The money is transferred to the Swedish Environment Protection Agency but the final decisions on how the money is to be distributed are taken by the authority called the Environmental Objectives Council (Figure 2.1.1).

After these decisions are made, the Swedish Environmental Protection Agency is responsible for the national environmental monitoring programmes, and also for the distribution of money to the County Administrative Boards, which in turn are responsible for the regional monitoring. These monitoring programmes correspond mostly to surveillance monitoring according to the Water Framework Directive. This means that most of the monitoring is performed in water bodies that are not locally impacted by point sources. Some municipalities have their own surveillance monitoring, but this is not regulated by the government and can be designed in many different ways depending on their specific purposes.

The above-mentioned monitoring is not the only monitoring that is performed, but it stands for the main contribution of data on reference and not locally impacted sites. When there is impact or risk of impact on a water body, monitoring is carried out according to Swedish environmental laws. This monitoring is called *recipient control* or *combined recipient control*. The latter is a compilation of several recipient

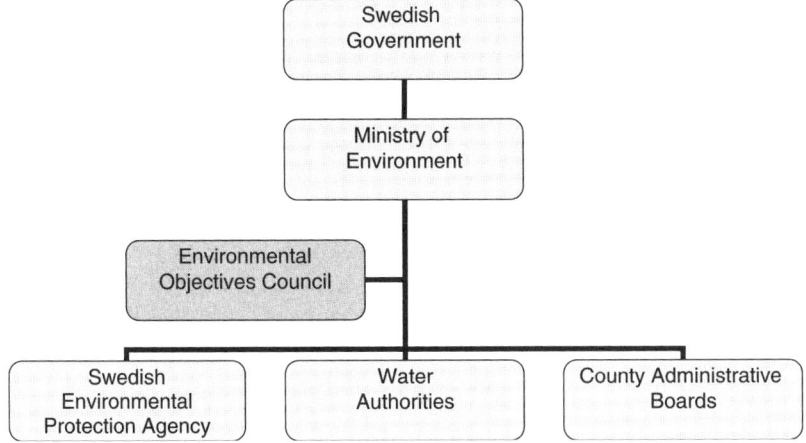

Figure 2.1.1 The decision process of monitoring recourses

control programmes connected to the same water body. A combined recipient control programme is one way to ensure monitoring of a water body is cost-effective compared to monitoring of the whole run-off area. This part of our monitoring system corresponds to operational monitoring that will be performed in water bodies at risk of or already classed as not meeting the demands of good ecological status. The operational monitoring is mainly decided upon by county administrative boards but can also be carried out by municipalities. Some of the data collected in operational monitoring can be of use for surveillance monitoring.

Another rather unique operational monitoring system in Sweden is the monitoring connected with those acidified objects that are limed to mitigate the biological effects of acidity. This monitoring gathers a large number of data. Most of the data, however, are chemical (pH, alkalinity, etc.), but in a few cases there are also data on benthic invertebrates and fish.

Concerning investigative monitoring, there is no specific requirement on the respective responsibilities of the three main authorities. This is decided on a case-by-case design, and depends on, among other things, where the impacted water body is situated, what kind of water body it is and the size of the impact.

Scientific research and other inventory activities are not included in freshwater monitoring if not carried out in a repetitive and quality-assured way.

2.1.5 SPECIFIC PREMISES

Sweden is a large country. In central and western Europe it is second only to Spain and France. Sweden also contains a large number of lakes. The lakes are not evenly distributed and are of different sizes and characters, from the biggest one, Lake Vänern ($5660\,km^2$), which is the third largest lake in Europe, down to very small lakes with areas of $0.01\,km^2$. Altogether, the 97 500 Swedish lakes have a combined surface coverage of 10 %.

Fjållområdet
- *Många små och medelstora sjöar.*
- *Stor brutenhet.*
- *Kalfjåll och morön.*

Förfjållens slåttområden
- *Relativt få och stora sjöar.*
- *Liren brutenhet.*
- *Myrar och morån.*

*Norra Norrlands kust- och
slåttområden*
- *Relativt få sjöar.*
- *Liten brutenhet.*
- *Sedimentära slättområden.*

*Mellersta och norra
Skogsområdena.*
- *Många små sjöar.*
- *Blandad brutenhet.*
- *Myrar och morån.*

De stora slätterna
- *Få och mycket stora sjöar.*
- *Liten brutenhet.*
- *Lera.*

Vasijusen
- *Få och smd sjöar.*
- *Blandad brutenhet.*
- *Strandslåtter och
  berg i dagen.*

Smålåndska höglander
- *Relativt många och
  stora sjöar.*
- *Slåttlandskap*
- *Myrar och morån*

Öland och Gorland
- *Mycket få sjöar.*
- *Liten brutenhet.*
- *Kalksten i dagen.*

Höglandets sluttningsumrådea
- *Många och små sjöar.*
- *Stor brutenhet.*
- *Morån och kalt berg.*

Skåne
- *Fü vch smů sjöar.*
- *Back och Slättandskap.*
- *Moränlera på kalkberggrund.*

Kalmarslättnen
- *Mycket fů sjöar.*
- *Slämlandskap.*
- *Morån och sand.*

Figure 2.1.2   Lake distributions in Sweden

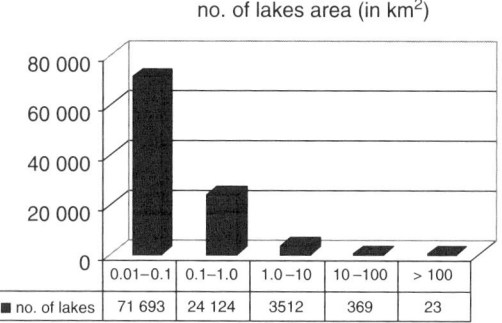

no. of lakes area (in km^2)

	0.01–0.1	0.1–1.0	1.0–10	10–100	> 100
■ no. of lakes	71 693	24 124	3512	369	23

Figure 2.1.3   Size distribution of lakes in Sweden

Figure 2.1.2 gives an indication of how the lakes are distributed. The largest lakes, Vänern, Vättern and Mälaren, are situated in a region which is characterised by not having many small lakes. This is an effect of the ice age of Scandinavia. Once, the main outlet of the Ancylus Lake (a lake much the same size as the Baltic Sea is today) to the Atlantic ran through this part of Sweden. Later on, this part of Sweden became a sound, when the Ancylus Lake was transformed to the Littorina Sea. Another effect of the glaciation on our lake composition was that the region above the highest coast line is now rich in small lakes (Figure 2.1.3). This is evident in the central part of the south of Sweden and also in the north-western part of the country. The ice age also explains the nutrient-poor state of the lakes above the highest coast line. Only the southernmost part of Sweden and the two largest islands of the Baltic, Öland and Gotland, have calcareous bedrock and can be compared to central Europe.

Due to limitations in nutrients and small lake size, the average species number is low. The total number of fish species found in Swedish lakes is around 50, but the number of species in a normal lake varies from about 14 in southern Sweden to 1, arctic char or brown trout, in the most northern parts. Some lakes in the alpine region are naturally fishless, since fish have not been reintroduced since the ice age.

In short, the most common lake type is a fairly small dimictic lake, with ice coverage from the beginning of November to the end of April. It is situated in forested surroundings, has a high content of humic matter compared to most European lakes, low alkalinity, four species of fish, and the biodiversity of other animals and plants is also low.

2.1.6   DESIGN OF THE LAKE MONITORING SYSTEM

The revision of the monitoring system in Sweden started from our previous monitoring programme and all the obligations of monitoring described above, and also from the monitoring requirements of the Water Framework Directive. When we revised our programme, the intention was to get a cost-efficient system with as many multipurpose functions as possible.

Table 2.1.1   The sieve for reference conditions

Kind of impact	Limits for reference objects
Acidification	• Annual mean pH > 6
	• If pH < 6, then the F-factor was used to correct for natural acidity according to present EQC
Eutrophication	• <8 µg total P/l in the northern part arctic/alpine region)
	• <10 µg total P/l in the other parts
	• If total P > 8 or 10 µg/l, respectively, then a correction was made for humic substances using regional relationships between total P and water colour (absorbance of filtered water)
Land use	• <10 % arable land
	• <10 % clear cuts (lasts for 5 years in southern Sweden, 10 years in the southern part)
	• <0.1 % urban areas (estimate of point sources)
Hydro morphology	To be developed

The revision started with an overview of all national and regional monitoring sites in lakes. All data from these water bodies were tested to determine pristine conditions. To do this, we created a filter or sieve for some chemical and other data, to sort out anthropogenic impact. If a lake did not exceed the limits of the sieve, it might pass as a reference lake. The criteria for this sieve are listed in Table 2.1.1.

To the data from the preliminary reference sites coming out as a result of this impact sieve, we added information from monitoring data on biological quality elements. If the data only showed slight deviation from what is described as high status, the lake was classified as a true reference lake according to the Water Framework Directive.

This analysis showed that around 50 % of all selected lakes were true reference lakes. These lakes were decided as reference lakes according to all standards. In some parts of Sweden it has not been possible to find truly pristine water bodies, but it is still important to have monitoring stations in these areas since it is necessary to have lake objects in as many as possible of the nominated lake types. Figure 2.1.4 and Table 2.1.2 show the selected lake types for Sweden. It is also important to have regional references to estimate spatial changes due to changes in climate and other large-scale effects. Monitoring stations selected for this reason are referred to as *trend stations*. To be selected as a trend station, the lake normally has to reach at least good status for all biological quality elements monitored. At some sites it is not possible to evaluate all quality elements. For example, in the alpine region, where fish are scarce, it might not be possible to use fish for status evaluation.

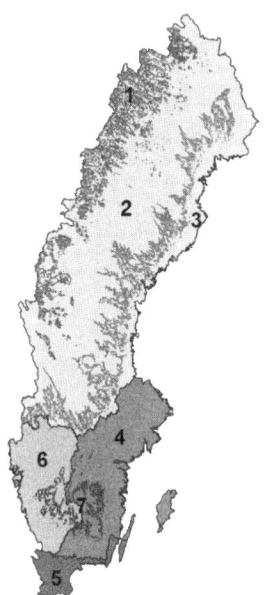

Figure 2.1.4 The seven eco regions in Sweden: 1) the alpine region, above tree line; 2) northern boreal region, above highest coast line; 3) northern region below highest coast line; 4) Baltic region, below highest coast line; 5) south region; 6) south-western region, drainage to the North Sea; 7) southern boreal region, 200 m above sea line

Table 2.1.2 Selected lake types in Sweden. To determine the type for a given lake, one combines the eco region given in Figure 2.1.4 with the factors listed below. This gives 112 theoretical different lake types in Sweden. However, not all of these are found in Sweden. For instance, most of the alpine lakes have low content of humic substances

Depth (mean)	Area	Humic substances	Alkalinity
>4 m	>10 km^2	>50 mg Pt/l	>1.0 mekv alk
≤4 m	≤10 km^2	≤50 mg Pt/l	≤1.0 mekv alk

When selecting monitoring sites according to the lake types, it is important to note that one cannot always differentiate between all lake types when using different biological quality elements. Two possible explanations for this are:

(1) The indicator (fish, macrophytes, etc.) might not respond to the difference between one lake type and another, i.e. lake depth does not have to change the distribution of benthic invertebrates or perifyton.

(2) The data collected so far might not be enough to detect significant differences and changes in index values between lakes.

Figure 2.1.5   Map of fish regions in Sweden (see Plate 1)

Table 2.1.3   Description of biological quality element regions

Quality element	Number of eco regions	Comments
Fish	5	See Figure 2.1.5
Benthic invertebrates	3	
Phytoplankton	3	
Macrophytes	3	

For the biological quality criteria, the indices do not change according to the selected types as shown in Figure 2.1.4, but rather according to larger eco regions as described in Table 2.1.3 and shown in Figure 2.1.5.

2.1.7 DEVELOPMENT OF THE NEW NETWORK

2.1.7.1 Surveillance Network

At this point in our work, we had a set of obligations for a monitoring system that was the sum of all obligations: first, the eco regions and the different lake types within the regions (Figure 2.1.4 and Table 2.1.2); second, the geographical borders between our river basin districts; third, the different regions emanating from specific regions for different quality elements. On this overlay of geographical information we added the information that had been gathered from our ongoing monitoring lakes that had passed through the sieve.

We ended up with a station net containing 110 national monitoring stations in lakes, where some were reference stations and the rest were nominated as trend stations. The geographical distribution of the lakes is shown in Figure 2.1.6.

This net of 110 lakes corresponds to national monitoring stations. Some more lake stations will be added by the county administrative boards when they make their decisions on regional monitoring. This regional monitoring effort is not of the same extent as the national net, but about 50 % more lakes are likely to be added. This regional revision process will not be completed until the end of 2008.

Not all quality elements are monitored in all the selected lakes. Table 2.1.4 gives an overview of the distribution of the monitoring efforts in the national monitoring programme.

This system of annually-monitored stations will not be sufficient to be in compliance with the Water Framework Directive due to lack of data from randomly selected lakes and from lakes with lower status than high and good. To compensate for this bias, Sweden has selected a system of randomised surveillance monitoring. This implies selecting a number of lake stations every year to be monitored. This selection is not completely randomised due to the distribution of lake size and clustering tendencies in some regions. Also, by chance whole regions or main river basins might become excluded if the selection was totally randomised. To compensate for this, a stratification is made that ensures a fairly homogenous distribution over the country, as well as between river basin districts. Table 2.1.5 shows how the proportion of different lake sizes is selected.

Since Sweden is more densely populated in the south, more possible pollutants are handled in that part. The deposition of long-range transported pollutants is also higher in the south. This is the reason why we have chosen to have a stratification with 60 % of the lakes in the south of Sweden (the black area in Figure 2.1.7).

Since the annual survey sampling is done by helicopter, there is a focused chemical monitoring. When the selection of stations takes place, a request is put forward to the county administration board to see if they are willing to take part in the survey, in order to get even more water bodies sampled and monitored.

Trendstationer:
sjöar

Figure 2.1.6   Reference and trend lakes in Sweden

Specimen Banking

To understand and evaluate the distribution and availability of organic substances and metals, fish are collected from 32 lake sites and taken to a specimen bank. Some of these fish sampling stations have been run continuously since 1964, but others are more recent. The species normally used are perch and arctic char, and in a few cases pike and roach. The sites for specimen banking are, with two exceptions, situated at stations for reference or trend monitoring to get information from supporting parameters. A

Table 2.1.4 Numbers and frequencies of the parameters monitored in national lakes

Type of variable monitored	No. of sites	Frequency per year
Chemistry	110	4–8
Phytoplankton	108	1–4
Benthic invertebrates	108	1
Fish	45	1–1/6
Macrophytes	10–30 (to be developed)	1–1/6
Specimen banking (see text)	32	1

Table 2.1.5 Selection of lake sizes in the annual survey

Size class area in km^2	10–100	1–10	0.1–1	0.01–0.1
Proportion selected	1	5	10	15
Number of lakes	2	130	254	388

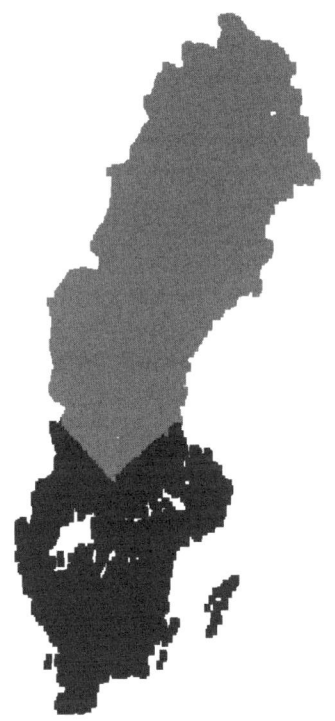

Figure 2.1.7 Northern and southern parts of Sweden

part of the collected material is analysed every year for organic substances and metals. Some years this is performed as a screening activity and other years the focus might be on retrospective studies of certain substances. The specimen bank is hosted by the Swedish Museum of Natural History, http://www.nrm.se.

In short, the national surveillance monitoring net consists of 110 annually monitored reference or trend stations. In addition, 800 stations per year are monitored once every sixth year, which adds up to a total of 4910 monitored lakes in the six year monitoring cycle of the Water Framework Directive.

2.1.7.2 Operational Monitoring

The monitoring net described so far is surveillance monitoring. In addition to this, operational monitoring is performed. As mentioned above, the operational monitoring is called 'recipient control' and is regulated by Swedish law. The operational monitoring being carried out in a certain area depends on what activities are taking place. Most of the recipient control in the southern and central parts of Sweden is combined recipient control, which has the whole river basin or sometimes a part thereof in focus.

The methods used in operational monitoring are often but not always the same as the methods used in surveillance monitoring. The Swedish Environmental Protection Agency has published a handbook on monitoring, where methods for each biological and chemical quality element are described and recommended, and EN or ISO standards are listed. The handbook also gives information on how to design a monitoring programme, how to select a monitoring site, etc. This handbook is only in Swedish and is available at our homepage, http://www.naturvardsverket.se.

All data collected in national monitoring systems are available on the Internet at different data hosts. Web links to data hosts may be found at our homepage, under 'Monitoring'.

2.1.8 FUTURE DEVELOPMENT

During the period of implementation of the Water Framework Directive we have also performed some screening activities for the priority hazardous substances and priority substances. The next step in developing our monitoring system will focus on biodiversity, one of the Swedish environmental objectives mentioned in Section 2.1.2. This monitoring should be designed to show the presence, nature and persistence of possible threats to biodiversity.

We are investigating two different ways to monitor biodiversity. One concerns the status for red-listed species and early warnings if anything is happening. The other focuses on invasive alien species and discovering introductions at an early stage.

To see if the selected biodiversity monitoring stations were correctly chosen, the station net decided upon will also be revised in a few years, when it is possible to evaluate the data collected. It is being discussed whether it is possible to make use of community-based monitoring for some species that are more easy to detect and identify.

2.2
River Monitoring

Elena Pérez Gallego

2.2.1   INTRODUCTION

Article 8 of the Water Framework Directive sets out that Member States shall ensure the establishment of programmes for the monitoring of water status in order to establish a coherent and comprehensive overview of water status within each river basin district. According to the Water Framework Directive, these programmes had to be operational at the latest by the end of 2006 (six years after the date of entry in force of this directive).

This chapter includes the main issues of a guidance document developed by the Ministry of Environment of Spain for the design of monitoring networks in compliance with Article 8 of the Water Framework Directive.

The Water Framework Directive - Ecological and Chemical Status Monitoring   Edited by Philippe
Quevauviller, Ulrich Borchers, Clive Thompson and Tristan Simonart © 2008 John Wiley & Sons, Ltd

Spain is divided into 15 river basin districts. Each river basin district has a river basin management plan which will have to be adapted to the Water Framework Directive. There are 3116 water bodies in rivers in Spain. The Ministry of Environment has taken into account three criteria for the identification of the water bodies in rivers, two hydrological criteria (average flow is greater than 100 L/s, and more than 75 % of the months have a flow not equal to zero) and a geographical criterion (the catchment area is greater than $10 \, \text{km}^2$).

2.2.2 PROJECTS CARRIED OUT BY SPAIN IN COMPLIANCE WITH ARTICLE 8 OF THE WATER FRAMEWORK DIRECTIVE

The ministry of environment of Spain has carried out several projects in compliance with Article 8 of the Water Framework Directive. Some of the more important projects are described below:

- Creation of a drafting group related to the surface water bodies status. This drafting group is an interesting group for the information exchange related to the design of surface waters status monitoring networks and for the exchange of the results of these networks.

- Elaboration of an inventory of the current water quality monitoring networks in Spain.

- Creation of a guidance document for the design of monitoring networks in compliance with the Water Framework Directive.

- Assessment of the current water quality monitoring networks by comparing them with the criteria of the guidance document mentioned before. This comparison has allowed the Ministry of Environment of Spain to identify the deficiencies, excesses and current needs of Spain concerning the design of water quality monitoring networks.

- Redefinition of the surface water monitoring networks in compliance with the Water Framework Directive, taking into account the results of the comparison mentioned in the previous point.

2.2.2.1 Guidance Document for the Design of Monitoring Networks in Compliance with the WFD

The guidance document for the design of monitoring networks developed by the Ministry of Environment of Spain mainly includes general criteria for the design of the water quality monitoring networks and the programmes that must be established for the monitoring of water status.

In relation to the general criteria for the design of the water quality monitoring networks, the guidance refers to:

● Three types of statistical design of the monitoring programmes, these being:

  – *Deterministic* – design based on the existing knowledge or on experts' judgement. It is not possible to extrapolate the results.

  – *Probabilistic* – design for the random selection of the sampling points. It is possible to extrapolate the obtained information.

  – *Census* – the results are directly obtained from sampling results, without need to extrapolate.

● Statistical criteria of design, which include the target population, the number of water bodies to control and the sampling frequency.

  – Three types of statistical...

  – Statistical criteria of design, which include...

  – Management criteria, which include sampling and analysis...

Management criteria, which include sampling and analysis protocols developed by different river basin districts in Spain, the parameters to control and the most suitable areas for the location of sampling points inside the water bodies.

The Water Framework Directive sets out three types of monitoring programme – surveillance, operational and investigative monitoring programmes – as well as an additional monitoring for protected areas. The guidance document for the design of monitoring networks developed by the Ministry of Environment of Spain proposes several monitoring sub-programmes for each monitoring programme (see Table 2.2.1).

Table 2.2.1 Monitoring sub-programmes

MONITORING PROGRAMMES	MONITORING SUB-PROGRAMMES
SURVEILLANCE MONITORING	Assessment of the overall surface water status and assessment of trends due to human activity
	Assessment of trends in natural conditions
	Information exchange decision
	Estimation of trans-boundary pollutant loads and emissions to the sea
OPERATIONAL MONITORING	Operational monitoring
INVESTIGATIVE MONITORING	Assessment of the need to establish operational monitoring
	Monitoring of accidental pollution
PROTECTED AREAS MONITORING	Bodies of surface water used for the abstraction of drinking water which provide more than $100\,m^3$ a day

2.2.3   SURVEILLANCE MONITORING

The main objectives of the surveillance monitoring are the assessment of the overall surface water status and the assessment of trends due to human activity and in natural conditions.

2.2.3.1   Monitoring Stations

In Spain, there are several water quality monitoring networks that permit assessment of the overall surface water status and the trends due to human activity and in natural conditions. The monitoring stations belonging to these water quality monitoring networks, which are proposed by the guidance document for the design of monitoring networks, are shown in Table 2.2.2.

Next, a brief description of some of the water quality monitoring networks in Spain is given:

- *General water quality control* – the objective of this network is the control of the general quality and surveillance of the quality of the potentially contaminated sections. It measures 40 physical-chemical parameters, divided into four groups. Some of these parameters belong to the list of priority substances of Article 16 of the Water Framework Directive. Sampling frequency depends on the kind of station and the parameter to analyse. A total of almost 700 stations compose the general water quality control in Spain.

- *Dangerous substances network* – the objective of this network is the surveillance of the contamination caused by the discharge of dangerous substances. The reference law comes from Directive 2006/11/EC, following up of the recently abolished Directive 76/464/EEC. The dangerous substances network is composed of

Table 2.2.2   Monitoring stations in the surveillance monitoring

MONITORING SUB-PROGRAMMES	MONITORING STATIONS
Assessment of the overall surface water status and assessment of trends due to human activity	General water quality control (representative monitoring points) Dangerous substances network (representative monitoring points) Biological monitoring network (representative monitoring networks) Reservoirs and lakes (representative monitoring points)
Assessment of trends in natural conditions	Reference network
Information exchange decision	Information exchange network
Estimation of trans-boundary pollutant loads and emissions to the sea	Albufeira Convention Ospar Convention Barcelona Convention

two sub-networks, the preferential substances sub-network and the pesticides sub-network. The preferential substances sub-network is composed of 115 sampling points and measures the substances of Royal Decree 995/2000, which sets out environmental quality standards for certain pollutant substances (a total of 28), selected taking into account their toxicity, persistence and bioaccumulation in the Spanish aquatic environment. The pesticides sub-network is composed of 45 sampling points located in agricultural zones.

- *Biological monitoring network* – the objective of this network is the surveillance and monitoring of the biological quality elements, such as invertebrates, diatoms, fish and macrophytes. The first biological monitoring networks started working in Spain in 1993. Today there are more than 2000 biological stations, of which 441 are reference stations.

- *Reference network* – the three networks mentioned above enable assessment of the overall surface water status and trends due to human activity, while the reference network permits assessment of trends in natural conditions. As already stated, there are 441 reference stations identified in Spain and 33 types of water body in rivers. Of the 33 types of water body, 25 have reference sites and 16 have a sufficient number of reference sites.

- *Information exchange network* – the objective of this network is the establishment of a common procedure for the exchange of information about the quality of surface fresh water in the European Community. In Spain there are 15 stations identified under the Identification Exchange Decision 77/795/EEC.

- *Albufeira Convention* – the signatory countries of the Albufeira Convention are Spain and Portugal, and its objective is the estimation of trans-boundary pollutant loads.

- *Ospar Convention* – the objective of this convention is the protection of the marine environment of the north-east Atlantic. Spain is one of the 16 contracting parties to the Ospar Convention.

2.2.3.2 Types of Statistical Design and Statistical Criteria of Design

As mentioned above, the guidance document for the design of monitoring networks developed by the Ministry of Environment of Spain includes different types of statistical design and different statistical criteria of design. The statistical criteria of design refer to the target population, the number of water bodies to control and the sampling frequency.

Table 2.2.3 shows these criteria for each monitoring sub-programme.

Related to the sampling frequency, the Water Framework Directive requires that surveillance monitoring shall be carried out at each monitoring site during one year in the period covered by the river basin management plan.

However, the Ministry of Environment of Spain proposes a higher frequency than required by the Water Framework Directive for the first river basin management plan in order to satisfy information needs. In this way, surveillance monitoring shall be

Table 2.2.3 Types of statistical design and statistical criteria of design

MONITORING SUB-PROGRAMMES	TYPES OF STATISTICAL DESIGN	STATISTICAL CRITERIA OF DESIGN		
		Target population	Number of WB to control	Sampling frequency
Overall assessment	Probabilistic	All WB in the river basin district	Different approaches	
Assessment of trends in natural conditions	Probabilistic or census	All reference sites	Now all reference sites. In the near future only a probabilistic selection	
Information exchange Decision	Census	15 stations designated under Decision 77/795/EEC	15 stations designated	Every year
Estimation of trans-boundary pollutant loads and emissions to the sea	Deterministic	Main trans-boundary rivers	It depends on the hydrological and geographical characteristics of the river basin district	

carried out every year in the period covered by the first river basin management plan, with the following frequency:

- *For the biological quality elements* – sampling frequency will be twice a year for macrophytes, benthic invertebrates, phytobentos and fish. Phytoplankton is not a relevant biological quality element for rives.

- *For the hydromorphological quality elements* – sampling frequency will be twice a year, as it will for the continuity and the morphology. The hydrology will be measured continuously.

- *For the physical-chemical quality elements* – sampling frequency will be monthly in water.

The Ministry of Environment of Spain also proposes to measure semi-volatile and hydrophobic compounds in sediment every year.

2.2.3.3 Management Criteria

Management criteria include sampling and analysis protocols, the parameters to control and the most suitable areas for the location of sampling points inside the water bodies.

Some river basin districts in Spain have developed guidance documents on biological sampling and analysis protocols in accordance with the Water Framework Directive.

This is the case for the Ebro River Basin District, which has prepared a guidance document[1] including the following:

- sampling procedures

- handling and preparation of physical samples

- analytical methods to obtain results

- metrics.

Related to the physical-chemical quality elements, the Ministry of Environment of Spain has recently published a technical instruction[2] with the objective of standardising the parameters of the water quality legislation and the analytical trials carried out by the water quality laboratories.

The guidance document for the design of monitoring networks also indicates where the sampling points must be located inside the water bodies in the surveillance monitoring. The objective of the surveillance monitoring is the assessment of the surface water status in general, so stations must be located in zones which are representative of the status of the water body.

2.2.4 OPERATIONAL MONITORING

The objective of the operational monitoring is to establish the status of those water bodies identified as being at risk of failing to meet their environmental objectives, and to assess any changes in the status of such bodies resulting from the programmes of measures. Therefore, operational monitoring shall be carried out for all those bodies which, on the basis of either the impact assessment or surveillance monitoring, are identified as being at risk. According to the results of the impact assessment of Article 5 of the Water Framework Directive, 13.1 % of water bodies are at risk in Spain.

2.2.4.1 Monitoring Stations

The monitoring stations belonging to the water quality monitoring networks, which enable establishment of the status of the water bodies identified as being at risk of failing to meet their environmental objectives and assessment of any changes in the status of such bodies resulting from the programmes of measures, are shown in Table 2.2.4.

In this case, and unlike surveillance monitoring, stations must be located in impact zones instead of representative zones, that is, downstream of the industrial areas and behind the mixing zone.

---

[1] http://oph.chebro.es/DOCUMENTACION/Calidad/dma/indicadoresbiologicos/protocolos.htm.

[2] http://www.mma.es/portal/secciones/acm/aguas_continent_zonas_asoc/vertidos_aguas/entidades.htm.

Table 2.2.4 Monitoring stations in the operational monitoring

MONITORING SUB-PROGRAMMES	MONITORING STATIONS
Operational monitoring	General water quality control (Impact monitoring points) Dangerous substances network (priority substances) (Impact monitoring points) Point and diffuse sources Biological monitoring network (Impact monitoring networks) Point and diffuse sources Reservoirs and lakes (Impact monitoring points) Point and diffuse sources Protected areas designated for the protection of habitats or species (Representative monitoring points)

2.2.4.2   Types of Statistical Design and Statistical Criteria of Design

The type of statistical design in operational monitoring will be deterministic, that is, a design based on the existing knowledge or on experts' judgement. In this case, it is not possible to extrapolate the results.

The target population in operational monitoring will be:

- All water bodies which, on the basis of either the impact assessment or surveillance monitoring, are identified as being at risk.

- Water bodies in which programmes of measures have been applied to assess changes in the status of such bodies.

- Water bodies into which priority substances are discharged.

The number of water bodies to control will be all water bodies of the target population with possibilities of grouping. There is no need to control at-risk water bodies for classification of status because their status is worse than good. Such water bodies will be assessed to establish programmes of measures.

In relation to the sampling frequency, quality elements will be measured every year until a water body is classified with a frequency similar to the frequency of surveillance monitoring. Therefore, for the biological quality elements, sampling frequency will be twice a year; for the hydromorphological quality elements, the sampling frequency will be twice a year as well; and for the physical-chemical quality elements, sampling frequency will be monthly.

In operational monitoring and apart from sediment, the Ministry of Environment of Spain proposes to measure the bioaccumulative compounds in biota every year.

2.2.5 INVESTIGATIVE MONITORING

According to the Water Framework Directive, investigative monitoring shall be carried out:

- Where the reason for any exceedance is unknown.

- Where surveillance monitoring indicates that the objectives set out in Article 4 of the Water Framework Directive for a body of water are not likely to be achieved and operational monitoring has not already been established.

- To ascertain the magnitude and impacts of accidental pollution.

The guidance document for the design of monitoring networks developed by the Ministry of Environment of Spain suggests two monitoring sub-programmes for the investigative monitoring, that is, an investigative monitoring to assess the need to establish operational monitoring, and an investigative monitoring of accidental pollution.

Some of the stations that enable assessment of the need to establish the operational monitoring are those belonging to the pesticides sub-network (see Section 2.2.3.1). Also, Spain has an alert automatic stations network for the monitoring of accidental pollution.

The pesticides sub-network is composed of 45 stations located in agricultural zones. There are two kinds of campaign, a routine campaign and an intensive campaign. Sampling frequency in the routine campaign is eight times a year, and it measures 17 pesticides which have an environmental quality standard. The objective of the intensive routine is, however, to find new pesticides which are not usually measured. In this case, sampling frequency is twice a year.

The alert automatic stations network provides a global and instantaneous view of the surface water quality status in Spain in 200 points. The objectives of this network are:

- The control and surveillance of spillages.

- The monitoring of accidental pollution.

- The protection of areas of interest to the environment and the protection of drinking water catchment areas.

Table 2.2.5 Cost estimation of the WFD monitoring networks (1000 €)

	2005	2006	2007	2008
Design of monitoring programmes	0.500	2.750	2.750	0
WFD monitoring networks	11.335	31.100	29.672	25.507
TOTAL	**11.835**	**33.850**	**32.422**	**25.507**

2.2.6 COST ESTIMATION OF THE WATER FRAMEWORK DIRECTIVE MONITORING NETWORKS IN SPAIN

The Ministry of Environment of Spain has invested a lot of money in the design of monitoring programmes and in the Water Framework Directive monitoring networks.

Table 2.2.5 shows the expenses incurred by the Ministry of Environment of Spain from 2005 until 2008 in compliance with Article 8 of the Water Framework Directive.

2.3

Groundwater Monitoring: Implementation in Two Member States

Rob Ward, Johannes Grath and Andreas Scheidleder

2.3.1   INTRODUCTION

The Water Framework Directive (WFD) (European Union, 2000) establishes a legal requirement for the Member States of the European Union to implement groundwater chemical and level monitoring programmes. These programmes must be designed and operated so that they can meet the needs of the WFD and its associated Groundwater Daughter Directive (GWD) (European Union, 2006), and provide reliable and comparable information on groundwater across the whole of the European Union.

The Water Framework Directive - Ecological and Chemical Status Monitoring   Edited by Philippe Quevauviller, Ulrich Borchers, Clive Thompson and Tristan Simonart © 2008 John Wiley & Sons, Ltd

This chapter focuses on the development of groundwater quality (chemical) monitoring programmes and their role in supporting the delivery of the WFD's environmental objectives. The WFD, in conjunction with the GWD, defines the requirements and objectives of the individual groundwater monitoring programmes but does not provide any detail on how the programmes should be designed and operated. Therefore, to assist Member States, a working group of experts from across Europe (the Common Implementation Strategy (CIS) Groundwater Working Group) has developed guidance on establishing and operating the monitoring programmes to meet directly the needs of the WFD and GWD (Working Group C, 2007). In developing the guidance it was recognised that there is considerable variation in hydrogeological, climatic and socioeconomic conditions across Europe. The outcome was guidance that recommends a risk-based approach, that takes into account the variability across Europe and that will enable comparable cost-effective monitoring programmes to be implemented. Some of the key elements of the recommended approach, endorsed by the Water Directors of the European Union, are described in this chapter and illustrated by examples from two EU Member States.

2.3.2   OBJECTIVES OF MONITORING

The monitoring programmes are needed to provide information to support the achievement of a number of specific WFD objectives. These include:

- Supporting the characterisation of (ground) water bodies and the assessment of risks.

- Enabling the chemical status of groundwater bodies to be determined.

- Demonstrating compliance with protected area objectives, especially drinking water protected areas (DWPA).

- Identifying trends in groundwater quality and in particular upward trends in pollutant concentrations.

- Assisting in the targeting and design of programmes of measures and assessment of their effectiveness.

To meet the monitoring objectives, a two-tier monitoring approach is required. The tiers are referred to as *surveillance monitoring* and *operational monitoring*. The purpose of the surveillance monitoring is principally to support characterisation of groundwater bodies, carry out risk assessment and inform the design of the operational monitoring. The operational monitoring is more targeted, with the aim of providing the information needed to quantify the impacts of pressures on the groundwater body, enable status and trend assessment, and demonstrate the effectiveness of measures put in place to meet WFD environmental objectives. Additional monitoring is also required to support achievement of DWPA objectives, but in practice this is likely to form part of surveillance and/or operational monitoring. The relationship of the monitoring programmes to WFD and GWD objectives is illustrated in Figure 2.3.1.

Monitoring objective(s)	WFD Specified Monitoring Programmes		
	Surveillance Monitoring	Operational Monitoring	Drinking Water Protected Area (DWPA) Monitoring
Supplement and validate the risk assessment (initial and further characterisation)	✓	(✓ [1])	
Identify saline or other intrusions resulting from alterations of flow within the groundwater body	✓	✓	
Assess chemical trends in natural conditions	✓		
Assess chemical trends caused by anthropogenic activity	✓	✓	✓
Trans-boundary groundwater bodies	✓		
Status assessment – determining status of bodies that are at risk		✓	✓ [2]
Status assessment – confirming that bodies not at risk are at good status	✓		✓ [2]
Assess the effectiveness of Programmes of measures		✓	✓

[1]) Results will support characterisation in future River Basin Plan cycles

[2]) New Groundwater Directive requires DWPA objectives to be met for groundwater body to achieve good status

Figure 2.3.1 The relationship between the different WFD groundwater quality monitoring programmes and the directive's objectives

2.3.2.1 General Principles

In order to design an effective monitoring programme, careful consideration must be given to the groundwater system to be monitored. Groundwater bodies are three-dimensional hydraulically-active systems. In many cases they are heterogeneous, with changes in property occurring over small distances. Superimposed on this are physical alterations and pressures resulting from human activity. It is important therefore that all these factors are taken into account when designing monitoring programmes and that a clear understanding of the prevailing environmental conditions is developed. This understanding is referred to as the *conceptual model*.

Conceptual models are simplified representations, or working descriptions, of the hydrogeological system (groundwater body) to be investigated and monitored. They can be used to not only aid monitoring network design and operation but also establish the relationship between pressures and receptors, e.g. ecosystems, surface water bodies and drinking water abstractions. As the amount of, and confidence in, environmental information on the groundwater body increases, including groundwater monitoring data, the accuracy of the conceptual model will improve.

The improved knowledge should be used to refine the design and operation of the monitoring programmes on a regular basis, at least every river basin cycle (six years). This will ensure that the programmes continue to meet their objectives.

Although the monitoring relates to the groundwater body, it must be recognised that the water environment is a continuum. This is reflected in the WFD by the linkage between good groundwater chemical status and the conditions for surface water status and protection of groundwater-dependent terrestrial ecosystems (groundwater-fed wetlands). The monitoring programmes for the different water body types need to be designed and operated in an integrated way where interactions exist, and use the best available scientific and technical knowledge to meet the new water policy needs that the WFD and GWD bring (Grath *et al.* 2007).

2.3.2.2   Surveillance Monitoring

A surveillance monitoring programme is required for all groundwater bodies whether at risk of failing WFD objectives or not. This monitoring must be undertaken in each river basin plan period to the extent necessary to support the risk assessment/characterisation process, confirm that groundwater bodies are at good status and assess trends in groundwater quality. An additional benefit of this programme will be to provide information to define natural background concentrations in groundwater and their variability across groundwater bodies. This is needed to provide reference data to support derivation of groundwater quality standards (or threshold values) for groundwater bodies. For practical and cost-efficiency reasons, the WFD allows bodies to be grouped for monitoring purposes where their characteristics are sufficiently similar and they are subject to similar pressures and impacts. In these cases the monitoring data in one body can be used to indicate groundwater quality in another body and confirm status or monitoring results across the two (or more) bodies combined.

The selection of sites for monitoring must take into account the three-dimensional nature of the groundwater body, flow characteristics, variability of land use, groundwater vulnerability and the potential receptors. All these should have been identified in the conceptual model. An effective network of monitoring sites will be one that is able to detect the impacts from pressures and the evolution in groundwater quality along flow paths within the groundwater body.

At every surveillance monitoring site a core set of parameters must be measured. These are: dissolved oxygen, pH, electrical conductivity, nitrate, ammonium and temperature. The EU guidance (Working Group C, 2007) also recommends that this core set is supplemented by a set of major and trace ions to adequately characterise the groundwater composition and contribute to quality assurance. Additional indicators of anthropogenic contaminants typical of the land use activities and range of pressures associated with the groundwater body will also be required. These data will support the risk assessment and identify any impacts from new emerging pressures.

The frequency of monitoring must be sufficient to meet information needs with adequate reliability and confidence. Unlike surface water, there is no universal fixed frequency for surveillance monitoring. Instead, frequency should be selected by taking into account the conceptual model and existing knowledge of groundwater quality and its variability over time. In dynamic groundwater systems that are at risk, more monitoring will be required than in those that are less vulnerable. The results of monitoring should be reviewed regularly and frequency of monitoring modified as needed to ensure that a reliable and cost-effective programme can be maintained.

2.3.2.3 Operational Monitoring

The principles underlying the design of the operational monitoring programme are identical to those for surveillance monitoring. The main difference is the objective of the monitoring. Operational monitoring is only required in bodies identified as being at risk of failing to achieve their environmental objectives. In other words, those bodies that may potentially be at poor status, have upward trends in pollutant concentrations or be subject to measures aimed at reversing trends or avoiding deterioration in status.

The selection of sites to monitor should be carefully considered by taking into account the characteristics of the pressures actually, or potentially, impacting on the groundwater body and the receptors at risk. Where specific receptors such as ecosystems are at risk, additional monitoring may be needed. In the case of drinking water abstractions in drinking water protected areas, monitoring will also need to include representative potable abstraction points to demonstrate compliance with the protected area objectives.

As with surveillance monitoring, groundwater bodies can be grouped to optimise the monitoring and ensure cost-effectiveness, but in this case there should ideally be at least one monitoring point in each of the component bodies, or else monitoring should be focused in the component body that is most sensitive to impacts from the identified pressures.

Parameter selection is focused more on those pollutants or parameters that are putting the groundwater body at risk. Adequate monitoring is needed to ensure that the information needs can be met, i.e. the impacts of all risks can be characterised and groundwater body classification and trends assessed with the required level of confidence.

The frequency of monitoring also needs to be adequate, especially to support trend assessment. The identification of trends will require the application of statistical methods and their data requirements will inform the frequency of monitoring. However, there are additional factors that will need to be considered, particularly those relating to the behaviour of groundwater systems and pollutants. The minimum frequency dictated by the WFD is once per year but it is widely acknowledged that this will be too low in many hydrogeological situations. The conceptual model, once again, has a very important role to play.

2.3.2.4 Quality Requirements

Because hydrogeological systems, and hence groundwater bodies, are complex and often hard to monitor, it is difficult to establish formal quality control criteria for the whole programme. It is therefore recommended to establish criteria for each stage of the process as part of a quality plan. Certain elements are well controlled, e.g. sample handling, laboratory analysis and data management, but others, such as conceptual model development and sampling, are less so. Recent international and national standards have improved sampling methods and procedures, e.g. ISO 5667-18, but the use of conceptual models is relatively new and there is limited experience in their use. For conceptual models the key thing is to clearly establish the questions that the model is being designed to answer and what tolerance(s) will be acceptable for the outcomes. The development of the model should be iterative and data and information should be collected to refine the model until it is able to answer the questions with the required level of confidence. This process should be recorded and reviewed regularly as knowledge improves to ensure that the best outcomes possible are maintained.

2.3.2.5 Programme Review

Once established, the monitoring programme should also be regularly reviewed, not just because objectives change but because our knowledge improves. Continuous review will ensure that monitoring programmes and the information they produce remain cost-effective, risk-based and targeted. Because the WFD is the first directive to formally establish groundwater quality monitoring requirements, there has been a lot of rapid development of the programmes. The requirements for the use of the data are also new and so it will take time to establish fully-effective monitoring across Europe. However, with the approach outlined and the feedback mechanisms in place, this can be achieved in a relatively short period of time.

2.3.3 CASE STUDIES

To illustrate some of the elements that form part of the monitoring programme design, two case studies follow. One is from the United Kingdom and the other is from Austria. Each describes how parts of the guidance are being applied within the Member State.

2.3.3.1 Case Study 1: United Kingdom (England and Wales)

Background

The current groundwater quality monitoring programme for England and Wales was developed over the period 2000–2006. The programme is operated by the Environment Agency for England and Wales and comprises approximately 3400 monitoring sites.

In order to deliver the required programme, a national strategy was developed (Environment Agency, 2004). This strategy clearly set out the monitoring objectives and the steps required to deliver a programme that could meet statutory and environmental data and information needs in a cost-effective way. In particular, it has been designed to meet the requirements of the WFD and the EU Nitrates Directive (91/676/EEC). The design and operation of the network has built on existing good practice by introducing the use of conceptual models and risk assessment to optimise the programme.

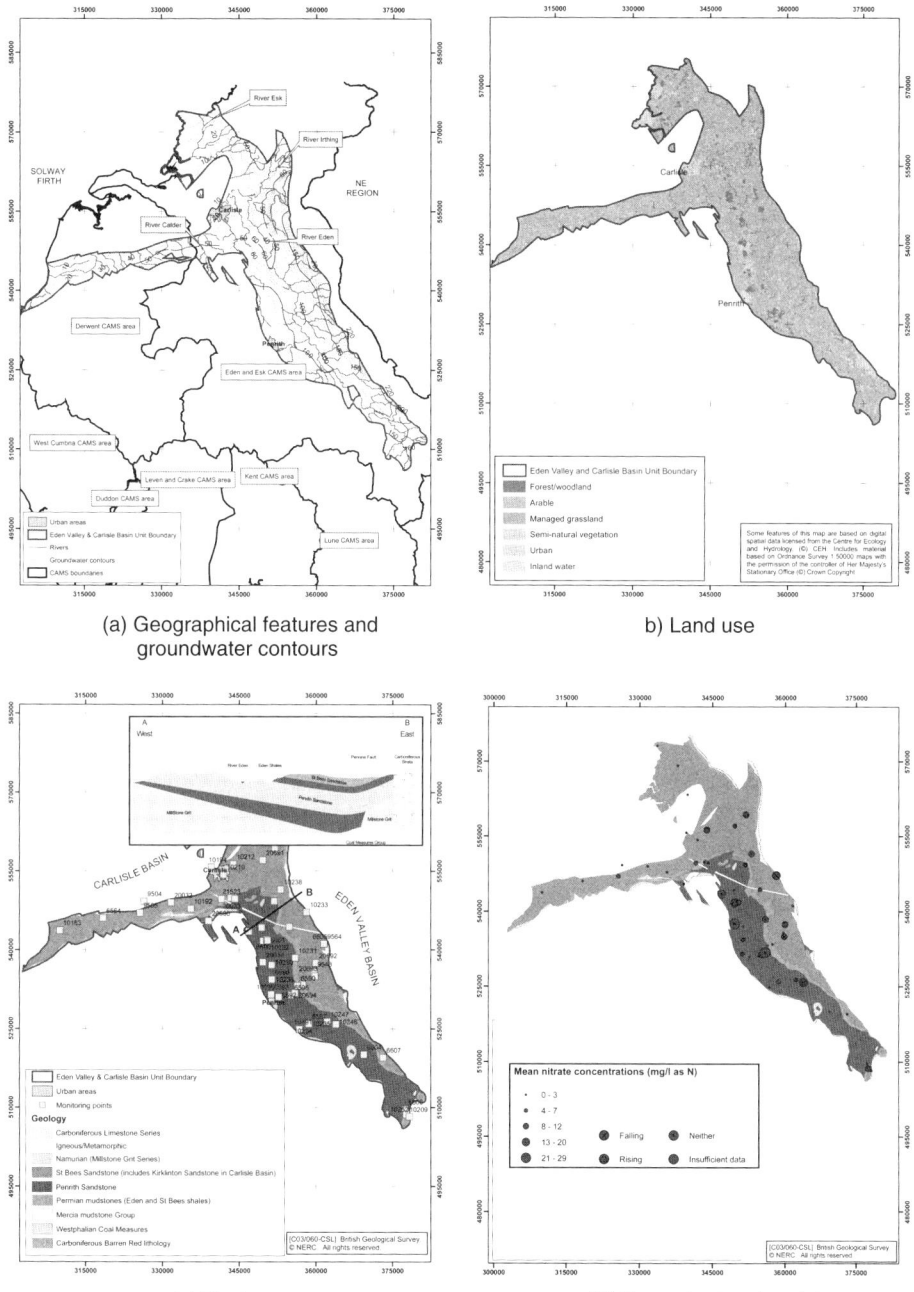

(a) Geographical features and groundwater contours

b) Land use

(c) Geology

(D) Groundwater chemistry

Figure 2.3.2 A selection of the components making up the conceptual model for a groundwater body in north-western England (see Plate 2)

The Role of Conceptual Models in Network Site Selection

The United Kingdom has extremely complex geology, with the age of rocks making up its aquifers ranging from Quaternary (glacial) deposits through to pre-Cambrian. The wide range of rock types results in a diverse range of aquifer types with very different hydrogeological characteristics. This diversity makes design of a monitoring network difficult and so a process has been adopted that requires the development of a conceptual model for each aquifer or sub-division (e.g. groundwater body). This brings together a wide range of information, such as:

- Physical characteristics of the aquifer and any overlying/underlying strata.

- Hydraulic properties such as flow direction and rates.

- Interactions with surface water bodies.

- Existing groundwater quality information.

- Land use activities that may affect groundwater quality or flow.

- Results of any groundwater risk assessments carried out, such as those for the WFD or designation of groundwater protection zones.

An example of some of the components used in developing the conceptual model for one aquifer is shown in Figure 2.3.2. The information has been compiled as a series of GIS layers for easy viewing and analysis.

Once the information has been compiled, the network of monitoring sites is selected. The approach chosen was to establish a distributed network that had monitoring along selected flow paths and in the different land use/pressure areas (Figure 2.3.3). The number of sites in a body/aquifer is determined by the complexity of the hydrogeological system and the degree to which the groundwater is at risk. The objective is to have sufficient monitoring to characterise groundwater quality with adequate confidence.

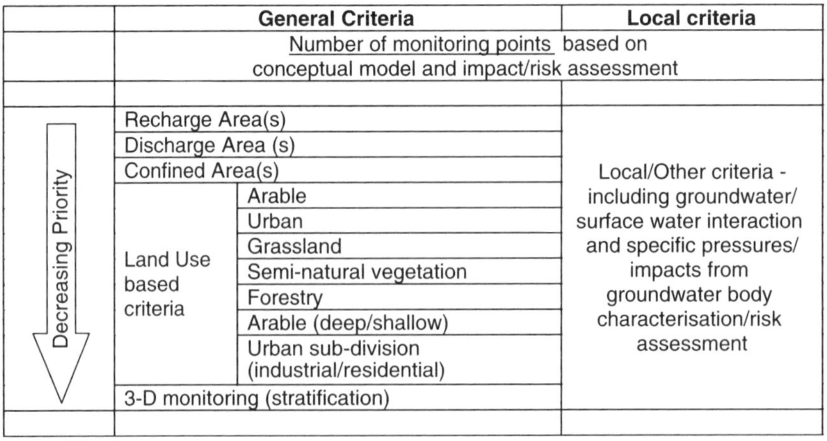

	General Criteria		Local criteria
	Number of monitoring points based on conceptual model and impact/risk assessment		
	Recharge Area(s)		Local/Other criteria - including groundwater/ surface water interaction and specific pressures/ impacts from groundwater body characterisation/risk assessment
Decreasing Priority	Discharge Area (s)		
	Confined Area(s)		
	Land Use based criteria	Arable	
		Urban	
		Grassland	
		Semi-natural vegetation	
		Forestry	
		Arable (deep/shallow)	
		Urban sub-division (industrial/residential)	
	3-D monitoring (stratification)		

Figure 2.3.3   Monitoring network site selection matrix

It was not economically viable to install purpose-designed monitoring boreholes, so to minimise cost and maximise coverage, existing groundwater abstractions (pumped boreholes) and springs were assessed for suitability as monitoring points. Due to the large numbers of public and private abstractions in England and Wales, it has been relatively easy to find enough suitable sites.

Before being selected, an assessment is made of each site by examining all the available information, e.g. constructions details, geological logs, hydrogeological conditions and local operational factors. This information is recorded in a national database to assist management, audit and reporting of the network.

Parameter Selection and Monitoring Frequency

A risk-based approach has been used for selecting which parameters to monitor and how frequently. Before this was carried out, however, careful consideration was given to the logistics and practicalities of monitoring. Discussions between the network managers, sampling teams and the laboratory resulted in a national set of parameter suites being defined. These took into account the types of parameter indicative of the main (pollutant) pressures on groundwater, numbers of sample bottles required, and laboratory capability. These suites have been adopted nationally and are regularly reviewed to ensure they remain adequate and operationally efficient.

The selection of which parameter suite or suites to be monitored at each site is determined by examining the conceptual model, results of previous monitoring, legal requirements and the outputs from the WFD risk assessment work. This has resulted in a mandatory suite of field and inorganic parameters being measured at all sites, supplemented by additional inorganic and organic suites after considering the risks to groundwater and other information needs (Table 2.3.1).

To assist in better targeting of parameter selection, low-cost semi-quantitative GCMS (gas chromatography–mass spectrometry) screening of samples is also carried out. This has proved extremely successful and means that monitoring is not only risk-based but also cost-effective. The GCMS screening is also identifying compounds in groundwater that were previously not considered as a threat, and these results are now feeding back into the risk assessments and conceptual models to improve confidence.

Table 2.3.1 Land use/risk-based parameter suite selection for groundwater monitoring

Parameter Suite	Land use					
	Arable	Managed grassland	Managed woodland	Urban/ industrial	Sheep	Amenity
Inorganic and field parameters	✓	✓	✓	✓	✓	✓
ONP Pesticides	✓	✓	✓	✓	✓	✓
OCP Pesticides	✓					✓
Acid Herbicides	✓	✓		✓		✓
Uron Pesticides	✓			✓		✓
Phenols				✓		
VOCs	✓	✓	✓	✓	✓	✓
PAHs				✓		
Special Organics	✓				✓	

The frequency of monitoring ranges between one and four times per year, depending on the aquifer type and behaviour, and its susceptibility to pollution pressures. Because sampling frequency is low, sampling takes place at the same time each year as far as possible to allow comparability of results between years.

Surveillance, Operational and DWPA Monitoring

The monitoring programme is designed and operated to meet the requirements of the WFD, with a single network of sites meeting surveillance and operational monitoring requirements. The differentiation between the two is made by the way in which the data are used to meet each programme's objectives. This makes management of the programme easier. For drinking water protected area (DWPA) monitoring, the same network again provides the bulk of the information, as all groundwater bodies in England and Wales have been designated as groundwater DWPAs and many of the monitoring points on the network are also (public or private) drinking water abstractions. Additional data from other drinking water supplies are also used to supplement the data as needed.

Data Analysis and Results

The results of the monitoring programme are being used in a variety of ways. Some of these are well established, but some introduced by the WFD and GWD are new. For the WFD, the monitoring results have informed initial and further characterisation and are now supporting the determination of the chemical status of each groundwater body. The UK approach for determining chemical status comprises a number of tests, each relating to one or more of the criteria which define good chemical status. Monitoring data from individual monitoring points, and aggregated across a groundwater body (or group of bodies), will be used to assess compliance with each relevant test for the body.

Another important use of the monitoring results will be to assess trends in groundwater quality, particularly those caused by pollutants. The data will be used in the first instance to identify adverse trends so that measures can be put in place to reverse them, and subsequently to monitor the improvements in quality as a result of the measures. It is recognised that much of the monitoring network is quite new and data will be limited when trend assessment is carried out for the first time. Over time the data will improve and trend assessment will become more reliable. However, this requires long-term commitment to maintaining the monitoring programme and to ensuring that its operation is continuous. In the meantime, to promote the programme, the opportunity is being taken to use the data to raise the profile of groundwater and highlight both successes and concerns about protecting groundwater.

2.3.3.2   Case Study 2: Austria

In Austria, standardised groundwater quality monitoring, based on legal provisions, was established in 1991. Its aim was to ensure the collection of consistent and reliable data in order to assess the current status of Austrian groundwaters and detect increasing concentrations at an early stage. This information was also to be used as the basis for designing and implementing measures for the protection of groundwater.

The resulting monitoring programme covers groundwater in porous media and in karst and fractured (fissured) rock systems. In total about 2000 groundwater sites are investigated and monitored. Groundwater areas were delineated as monitoring units and the monitoring was carried out on a quarterly basis (four times per year) for the whole of Austria.

To take account of the new requirements of the WFD, the Austrian Federal Water Act was amended and provided the basis for a new Ordinance for Water Quality Monitoring (Austrian Federal Law Gazette, 2006). Consequently the groundwater quality monitoring network in Austria was assessed for compliance with the new requirements and, where necessary, the network was amended accordingly. The most important impact resulted from the introduction of WFD groundwater bodies as groundwater management units.

Groundwater is the major source of drinking water in Austria (99 %). It is therefore a general aim of the Austrian Federal Water Act to keep all groundwater at a quality that makes it suitable for drinking water purposes. As a result, the whole territory of Austria is assigned to groundwater bodies. All porous medium groundwater bodies larger than $50 \, km^2$ of economic importance or with considerable risk potential were treated as single groundwater bodies. All other groundwater bodies were grouped together by taking into account the hydrogeological conditions and the borders of the sub-river basin districts. In total, 127 shallow groundwater bodies covering the whole territory of Austria, as well as nine deep groundwater bodies, have been defined.

Monitoring Network: Conceptual Model

The basic requirements for surveillance and operational monitoring network design are specified in the Ordinance on Water Quality Monitoring. They cover all main aspects of the monitoring cycle, i.e. the establishment of sites, the consideration of the conceptual model, parameter selection, duration and frequency of monitoring, methods for sampling and analyses, quality assurance, data management and publication of results.

The conceptual model is of major importance for the selection and drilling of surveillance and operational monitoring sites. The ordinance requires the following factors to be considered:

- Hydrological, hydrogeological and hydrochemical characterisation of the groundwater body (or group of groundwater bodies).

- Groundwater–surface water interaction.

- Groundwater retention time; respectively the age of groundwater, permeability of overlying strata, pressures and groundwater time lag.

- Sustainability concerning the identification of long-term natural trends.

- Information concerning land use (e.g. urban area, industrial sites, forest, agricultural use).

- Results of the pressure-impacts assessment, including effects of significant anthropogenic pressures.

- Sustainability concerning the identification of long-term anthropogenically-induced trends in pollutant concentrations.

- Requirements under the legal provisions concerning the compliance regime for chemical status assessment.

- Special consideration of existing groundwater uses at trans-boundary groundwater bodies.

- Efficient integration of (chemical and ecological) surface water monitoring sites, as well as groundwater quantity monitoring sites.

- Available information and results from previous monitoring activities.

- Local conditions regarding e.g. accessibility and safety during sampling and the suitability for taking representative samples.

The pre-WFD groundwater quality monitoring network was assessed in the light of the delineation of groundwater bodies and the WFD characterisation exercise that assessed pressures and impacts (Article 5 of the WFD). The network was redesigned taking into account the principles outlined above. The total number of monitoring sites remains at about 2000, comprising investigation (observation) boreholes, private wells, industrial wells, public water supply wells and springs.

Monitoring Frequency and Investigated Parameters

To comply with the WFD and the Austrian Ordinance on Water Quality, groundwater monitoring is carried out according to a six-year cycle (Figure 2.3.4). The cycle starts with an 'initial investigation' under the surveillance monitoring programme. This includes monitoring for an extensive number of parameters.

Depending on the results of this 'initial investigation' and the validated assessment of the risks of not meeting the WFD environmental objectives at the end of the River Basin Management Plan (RBMP) period, either the surveillance monitoring is continued as 'repeated investigation' or operational monitoring is carried out. The 'repeated investigation' is continued for the whole period of an RBMP for groundwater bodies at good status. It consists of monitoring for a reduced list of parameters and/or at a reduced frequency of sampling compared to operational monitoring.

The parameters monitored in groundwater, which total about 120, are grouped into two blocks:

- *Block 1* – important inorganic parameters with relevance to the environment, e.g. nitrate, nitrite, ammonium, phosphate, boron, alkali metal and alkaline earth metal (e.g. potassium, calcium, magnesium).

- *Block 2* – the heavy metal group (e.g. arsenic, mercury, cadmium) and lightly volatile halogenated hydrocarbons (e.g. tetrachloroethylene), the broad group of pesticide substances (e.g. triazine, phenoxy alkane carbon acids) and polycyclic aromatic hydrocarbons (PAHs).

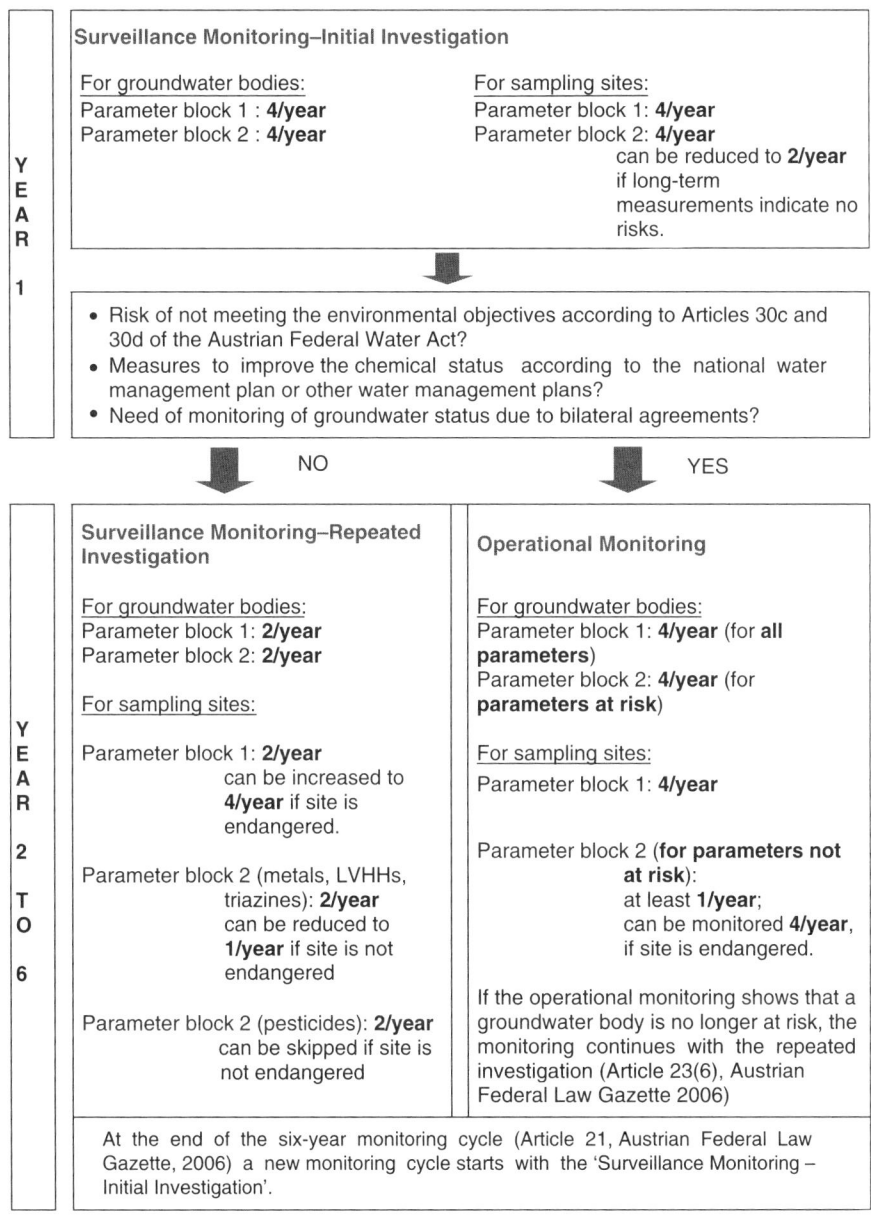

Figure 2.3.4 Monitoring cycle of the Austrian Groundwater Quality Monitoring (Schramm, 2006). The assessment of whether a monitoring site is endangered or not is based on the Ordinance on Groundwater Threshold Values (Austrian Federal Law Gazette, 1991). The monitoring frequency at monitoring sites refers to each respective parameter responsible for endangering a site, i.e. showing an exceedance of quality targets.

For 'Surveillance Monitoring – Initial Investigation' (first year in the cycle), parameters in Blocks 1 and 2 are included. For the 'Surveillance Monitoring – Repeated Investigation', parameters in Block 1 plus parameters which appeared to be relevant based on results from the initial investigation are included.

Operational monitoring is relevant to groundwater bodies at risk, and the monitoring of parameters in Block 1 plus additional parameters indicative of the risks to the groundwater body are obligatory.

In addition, the option for 'extra investigations' exists. This is intended to allow for consideration of chemical parameters not mentioned in the Ordinance on Water Quality Monitoring.

Analytical Quality Assurance

To ensure confidence in the analytical results, various elements of quality assurance were introduced in the monitoring programme (Umweltbundesamt, 2006). They include:

- Compulsory participation in training courses in groundwater sampling.

- Provision of key figures of the analytical performance within the bidding files.

- Inspection of laboratories prior to contracting their services.

- Inspection of laboratories during the contract periods.

- Compulsory participation in international quality control ('round robin') testing.

- Compulsory participation in the permanent spiked samples system performed by the Institute for Agrobiotechnology (IFA-Tulln).

Implementation

The implementation of the Austrian Water Quality Monitoring System is a shared responsibility between the federal and provincial authorities (Bundesländer). At the federal level, the Federal Ministry for Agriculture, Forestry, Environment and Water Management, Department Water Management Register, is responsible for:

- The integrated assessment of data.

- The bi-annual publication of results.

- Ensuring uniform procedures across Austria.

- Main funding of the monitoring programme.

Based on an agreement, the Federal Environment Agency (Umweltbundesamt) is responsible for:

- IT development and data management.

- Providing technical support on analytical requirements and data assessment.

- Reporting.

The Provincial Governor (Landeshauptmann) is responsible for:

- Operational management (calls for tender, tendering, inspection of contractors during sampling and analyses, quality assessment of received data and data delivery to the federal level).

- Contributing to programme costs.

- Cooperation in development and amendment of guidance documents.

- Remediation measures.

Publication/Data Access

Results of the monitoring programme are published as bi-annual reports. Access to the data as well as to the reports is available via the web page of the Federal Environment Agency: http://www.umweltbundesamt.at/en/umweltschutz/wasser/.

REFERENCES

Austrian Federal Law Gazette (1991) Grundwasserschwellenwertverordnung – GSwV (Ordinance on Groundwater Threshold Values), *Austrian Federal Law Gazette*, **502**.

Austrian Federal Law Gazette (2006) Grundwasserzustandsüberwachungsverordnung – GZÜV (Ordinance for Water Quality Monitoring), *Austrian Federal Law Gazette*, **479**.

Environment Agency (2004) Groundwater quality: a framework for improved monitoring, Environment Agency – England and Wales, available at: http://www.environment-agency/gv.uk/publications.

European Union (2000) Establishing a framework for community action in the field of water policy, European Parliament and Council Directive 2000/60/EC, *OJ*, **L237**, 22 December.

European Union (2006) Protection of groundwater against pollution and deterioration, European Parliament and Council Directive 2006/118/EC, *OJ*, **L372**, 12 December.

Grath, J., Ward, R.S. and Scheidleder, A. (2007) Groundwater monitoring, In: Quevauviller, P. (ed.), *Groundwater Science and Policy: An International Overview*, p. 648.

Schramm, C. (2006) Monitoring cycle of the Austrian Groundwater Quality Monitoring, unpublished report of the Federal Environmental Agency (Umweltbundesamt).

Umweltbundesamt (2006) *Wassergüte in Österreich – Jahresbericht 2006*, available at: http://www.umweltbundesamt.at/jb2006.

Working Group C (2007) Guidance on groundwater monitoring, Common Implementation Strategy for the Water Framework Directive 2000/60/EC, Guidance Document No. 15.

2.4
Coastal and Marine Monitoring

Patrick Roose

2.4.1   INTRODUCTION

The EC Water Framework Directive (WFD) essentially combines the efforts in protecting groundwater and all surface waters on land and in the territorial waters of the EU Member States, and therefore also transitional and coastal marine waters. As a result,

The Water Framework Directive - Ecological and Chemical Status Monitoring   Edited by Philippe Quevauviller, Ulrich Borchers, Clive Thompson and Tristan Simonart © 2008 John Wiley & Sons, Ltd

this holistic approach to monitoring overlaps not only regionally but also thematically with programmes carried out by the existing Marine Environmental Conventions. This overlap holds both risks and opportunities. If insufficiently coordinated, the current situation could easily result in an unnecessary duplication of monitoring efforts or even a reduction of the marine monitoring obligations for the conventions in favour of the much more stringent legal obligations towards the WFD. It would also be regrettable to neglect the valuable experience that has been gained during these programmes, particularly for the marine component of the WFD. The long-term commitment of many European countries to the marine monitoring programmes has resulted in e.g. important datasets, innovative approaches towards monitoring and assessment of results. It's precisely in this extensive experience that the opportunities for the WFD lie.

Monitoring of the marine environment has now been ongoing for several decades at both national and international levels. The latter can already be illustrated by having a closer look at the principal contaminants in Table 2.4.1 that are being measured in some well-known long-term marine monitoring programmes. The example of contaminants is well merited because it's precisely the realisation of the potential danger of certain, or rather, many, chemicals that resulted in a call for measures to regulate their input into the seas and also for long-term monitoring to evaluate their effectiveness. Not surprisingly, heavy metals (HM), polycyclic aromatic hydrocarbons (PAHs), organotins (OTINs) and a number of organochlorine pesticides (OCPs) are hazardous substances under the WFD (European Commission, 2000a).

Although contamination and also eutrophication of the marine environment were the initial driving forces behind these programmes, there has, in recent years, been an evolution towards a more holistic, ecosystem-orientated approach. For instance, OSPAR and HELCOM (see below) jointly adopted in 2003 a statement on the ecosystem approach to the management of human activities. The ecosystem approach can be defined as 'the comprehensive integrated management of human activities based on the best available scientific knowledge about the ecosystem and its dynamics, in order to identify and take action on influences which are critical to the health of marine ecosystems, thereby achieving sustainable use of ecosystem goods and services and maintenance of ecosystem integrity.' The application of the precautionary principle

Table 2.4.1 Overview of major long-term monitoring programmes, and the contaminants and matrices measured (after Roose and Brinkman, 2005)

Organisation or programme[1]	Start of the programme	Parameters[2,3]	Sample types
AMAP	1978	HM, PCBs, PAHs, OCPs	biota, sediment, water
HELCOM	1979	HM, PCBs, PAHs, OCPs, OTINs	biota, sediment
NS&T	1986	HM, PCBs, PAHs, OCPs	biota, sediment
IMW	1965	HM, PCBs, PAHs, OCPs	biota (bivalves)
OSPAR	1978	HM, PCBs, PAHs, OCPs, OTINs	biota, sediment

[1] AMAP; HELCOM; NS&T;The IMW, International Mussel Watch (actually started in 1991–1992, but data were already available from earlier programmes with a different name as early as 1965).
[2] Not all parameters measured during entire period.
[3] HM: heavy metals; PCBs: polychlorinated biphenyls; PAHs: polycyclic aromatic hydrocarbons; OCPs: organochlorine pesticides; OTINs: organotins.

is equally a central part of the ecosystem approach (OSPAR, 2006a). The ecosystem approach is also reflected in the WFD.

2.4.2 MARINE MONITORING PROGRAMMES IN EUROPE

2.4.2.1 Introduction

For Europe, the situation is such that five international marine conventions or programmes predate the WFD, namely AMAP, BSC, HELCOM, MEDPOL and OSPAR. This is illustrated for the major regional seas in Figure 2.4.1. Each of these important conventions will be highlighted in the sections below. Also, the European Marine Strategy Directive (MSD) is expected shortly and some attention will be given to that as well.

2.4.2.2 AMAP

The Arctic Monitoring and Assessment Programme (AMAP) was established in 1991 to implement certain parts of the Arctic Environmental Protection Strategy (AEPS), primarily 'providing reliable and sufficient information on the status of, and threats to, the Arctic environment, and providing scientific advice on actions to be taken in order to support Arctic governments in their efforts to take remedial and preventive

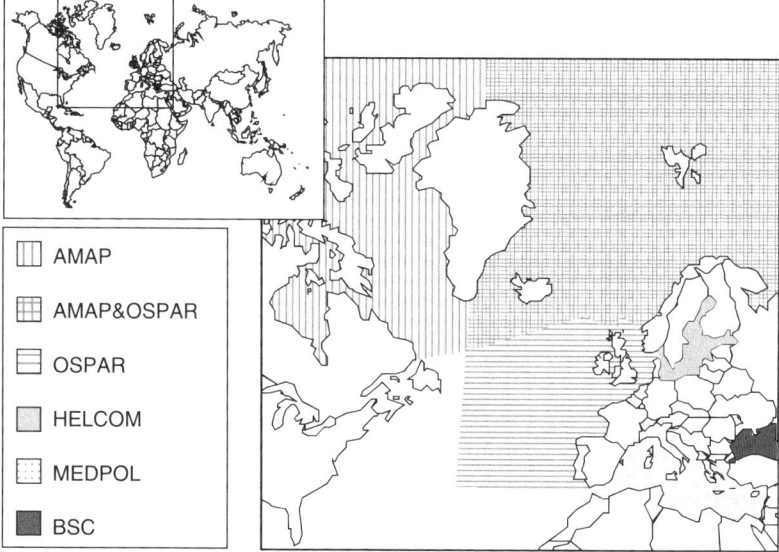

Figure 2.4.1 Regional Marine Conventions/Commissions and Programmes that are of relevance for Europe. The programmes, given by their acronyms, are described in the text

actions relating to contaminants' (AMAP, 2007). The Arctic Council, established in 1996 by the eight Arctic countries (CA, DK, FI, IS, NO, RU, SE and the US), coordinates AMAP activities (ISO 3166 codes for countries). AMAP was conceived as a programme which integrates both monitoring and assessment activities in relation to pollution issues and provides information and reports on the state of the Arctic environment. The AMAP Trends and Effects Monitoring Programme is designed to monitor the levels of pollutants and their effects in all compartments of the Arctic environment. There are five sub-programmes, which deal with atmospheric, terrestrial, freshwater and marine environments, and with human populations with respect to human health. The sub-programmes are defined in terms of essential and recommended parameters and media (matrices) to be monitored on a circumpolar or sub-regional level. The programme includes both monitoring and research components, and special studies that yield information which is vital for the valid interpretation of monitoring data.

2.4.2.3   BSC

The Black Sea Commission (BSC), or the Convention on the Protection of the Black Sea Against Pollution, was signed in Bucharest in April 1992, and ratified by the legislative assemblies of all six Black Sea countries (BG, GE, RO, RU, TR and UA) in early 1994 (BSC, 2007). The convention aims at 1) control of land-based sources of pollution, 2) control of dumping of waste, and 3) establishing a framework for joint actions in case of accidents such as oil spills. Specifically for the assessment and monitoring of pollutants, a 'State of Pollution of the Black Sea' report will be prepared and published every five years, beginning in 2006. It will be based on the data collected through the coordinated pollution monitoring and assessment programmes.

2.4.2.4   HELCOM

The Baltic Marine Environment Protection Commission, or the Helsinki Commission (HELCOM), is the governing body of the Convention on the Protection of the Marine Environment of the Baltic Sea Area, signed in 1992 (HELCOM, 2007). HELCOM's main goals are to protect the marine environment of the Baltic Sea from all sources of pollution, and to restore and safeguard its ecological balance. The present contracting parties to HELCOM are DE, DK, EE, EC, FI, LV, LT, PL, RU and SE. The setup is very similar to that of OSPAR (see below), and many of the principles – such as the 'best environmental practices', 'best available technologies' and 'the polluter pays' – are adopted and applied by HELCOM. Monitoring and assessment are an integral part of the convention, and according to the convention 'Emissions from both point sources and diffuse sources into water and the air should be measured and calculated in a scientifically appropriate manner by the Contracting Parties.' Every five years, the Commission publishes a 'Periodic Assessment of the State of the Environment of the Baltic Marine Area' based on monitoring activities going on in the area.

2.4.2.5 MEDPOL

MEDPOL, or the Programme for the Assessment and Control of Pollution in the Mediterranean region, was initiated in 1975 in Barcelona as the environmental assessment component of the Mediterranean Action Plan (MAP) and is now in Phase III (MEDPOL, 2007). Its task is to assist Mediterranean countries in the implementation of pollution assessment programmes (marine pollution trend monitoring, compliance monitoring and biological effects monitoring). In parallel, MEDPOL provides assistance in the formulation and implementation of pollution control and regional and national action plans addressing pollution from land-based sources and activities. It also formulates and carries out capacity-building programmes related to the analysis of contaminants and treatment of data and to technical and management training. MEDPOL-collected data and information directly contribute to the implementation of the LBS (land-based sources) and Dumping Protocols. The countries which signed the Barcelona Convention are AL, DZ, BA, HR, CY, EG, ES, FR, GR, IL, IT, LB, LY, MT, MC, MA, SI, SY, TN and TR, as well as the EU.

2.4.2.6 OSPAR

Founded in 1992 as a merger of two earlier conventions, both dating back to the seventies, i.e. the 1972 Oslo Convention – also called the Convention for the Prevention of Marine Pollution by Dumping from Ships and Airplanes – and the Paris Convention or Convention for the Prevention of Marine Pollution from Land-based Sources, established in 1974, the Convention for the Protection of the Marine Environment of the North-east Atlantic (OSPAR) (OSPAR, 2007a) entered into force on 25 March 1998. The convention has been signed and ratified by all contracting parties (CPs) to the Oslo and the Paris Conventions (BE, CH, DE, DK, ES, FI, FR, IE, IS, LU, NL, NO, PT, SE, UK and the EC). The central theme behind the convention is the agreement 'to take all possible steps to prevent and eliminate pollution and to take the necessary measures to protect the maritime area against adverse effects of human activities so as to safeguard human health and to conserve marine ecosystems and, when practicable, restore marine areas which have been adversely affected.' The guiding principles behind the convention are the precautionary principle, that preventive action should be taken whenever possible, that environmental damage should, as a priority, be rectified at source, that the polluter should pay, and that best available techniques and best environmental practice, including, where appropriate, clean technology, should be applied whenever possible. In order to achieve these goals, the major issues were identified as being: the conservation of the ecosystems and biological diversity, eutrophication, hazardous substances, the effects of offshore activities, and radioactive substances, each resulting in a thematic strategy. The OSPAR Convention requires the Contracting Parties, amongst other things, to 'cooperate in carrying out monitoring programmes' and to develop quality assurance methods and assessment tools.

2.4.2.7   ICES

The tradition in collaboration in matters relating to the marine environment even pre-dates these conventions. ICES (the International Council for the Exploration of the Seas) was established in 1902 and claims to be the oldest intergovernmental organisa-tion in the world concerned with marine and fisheries science for the North Atlantic (ICES, 2007a). ICES is a leading scientific forum for the exchange of information on the sea and its living resources, and for the promotion and coordination of marine research by scientists in its member countries. Since the 1970s, a major area of ICES work has been providing information and giving advice to member country govern-ments (BE, CA, DE, DK, EE, ES, FI, FR, IE, IS, LV, NL, NO, PL, PT, RU, SE, UK and the US) and international regulatory commissions on the protection of the marine environment and on fisheries conservation. In support of these activities, the ICES Secretariat in Denmark maintains three databanks – the oceanographic data-bank, the fisheries databank and the environmental (marine contaminants) databank. Both HELCOM and OSPAR have extensively used ICES's advice and its database for their monitoring programmes and assessment work. This link has ensured extensive similarities in their approach to monitoring and a sound scientific base. ICES was also heavily involved in the consultations process for the European Marine Strategy Directive.

2.4.2.8   The European Marine Strategy Directive

On 2 October 2002, the EC published a Communication to the Council of the EU and the European Parliament entitled 'Towards a strategy to protect and conserve the marine environment' (COM (2002) 539), which sets out objectives and related actions (European Commission, 2004a). The Commission Communication represents the first step in the incremental development of the European Marine Strategy for the protection and conservation of the marine environment. The Environment Council Conclusions of 4 March 2003 welcomed the Commission Communication, endorsed the approach and the outline of its objectives and requested an ambitious Strategy by 2005. Currently a proposal for a directive establishing a Framework for Community Action in the field of Marine Environmental Policy (Marine Strategy Directive or MSD) is on the table awaiting final approval by the parliament and the Commission (European Commission, 2005).

   With this directive, the Commission wants to install a strong, integrated, EU policy on marine protection. While the commission recognises that progress has been made in certain areas, e.g. in reducing nutrient inputs or pollution from hazardous substances, in particular heavy metals, it also recognises that the state of the marine environment has been deteriorating significantly over recent decades. However, the current policy framework is not delivering the high level of protection of the marine environment that is needed, hence this initiative.

   The Strategy has been prepared with the help of an extensive consultation process from 2002 to 2004, which included all EU Member States, candidate countries, the European Parliament, European Economic Area (EEA) States (Norway and Iceland), the various, mainly regional, international organisations engaged in different sectoral aspects of the marine environment (such as OSPAR, ICES and IMO (International Maritime Organisation)), and with environmental non-governmental organisations and various sectoral industry associations.

Coordination with existing programmes is thus an inherent part of the MSD. From the onset, it has been recognised that the Regional Marine Conventions/Commissions and Programmes, discussed above, play an important role at the interface between marine research and policy, both in the context of regional marine assessments and in the development of measures for marine management. The draft directive states that 'In order to achieve the coordination referred to in Article 4 (2), Member States shall, where practical and appropriate, use existing regional institutional cooperation structures, including those under Regional Seas Conventions, covering that Marine Region or Sub-Region' (European Commission, 2005).

According to the draft, Marine Strategies shall apply an ecosystem-based approach to the management of human activities while enabling the sustainable use of marine goods and services. This is certainly in line with policies of the existing regional commissions. Similarly to the WFD, the ultimate aim of the MSD is achieving or maintaining good environmental status in the marine environment by the year 2021 at the latest. 'Good environmental status' is the status of marine waters where they provide ecologically diverse and dynamic oceans and seas which are clean, healthy and productive within their intrinsic conditions, and the use of the marine environment is at a level that is sustainable, thus safeguarding the potential for uses and activities by current and future generations.

It has also been recognised that monitoring and assessment have a vital role when the ecosystem approach is applied to the management of human activities affecting the marine environment. In fact, the establishment and implementation of a monitoring programme for ongoing assessment and regular updating of targets of a monitoring programme is foreseen six years after date of entry into force, except where otherwise specified in the relevant Community legislation. This policy must initially be based on an assessment or evaluation of the state of the marine environment, and the implementation of the latter must be followed by observation and assessment of what has, and has not, been achieved. Two working groups, the Working Group on Strategic Goals and Objectives (SGO) and the Working Group on European Marine Monitoring and Assessment (EMMA), have been created to work out the monitoring component of the EMS. The SGO was, as its name suggests, identifying strategic objectives, while EMMA had and still has the task of working out practical solutions to the latter. Ongoing discussions in EMMA make clear that these regional assessments will play an important part in the context of pan-European assessments to be made under the framework of the EMS. Where there exist regional-seas monitoring and assessment programmes, these should be used as far as possible for new developments on EU and pan-European levels. Likewise, in developing existing EU measures – especially the EC Water Framework Directive – attention should be given to the links to both the pan-European and the regional-seas levels (European Commission, 2004b, 2004c).

It is by no means the objective of this chapter to give an extensive overview of all relevant aspects of the marine monitoring programmes for European waters, but rather, by taking the OSPAR Joint Monitoring and Assessment Programme (JAMP) as an example, to illustrate how marine environmental monitoring is approached. OSPAR has played a leading role in the development and execution of monitoring programmes, as well as in the assessment of the resulting data. OSPAR has during recent years tried to align its programme with those of the EC and HELCOM. Nevertheless, certain specific aspects merit a closer look and could certainly initiate or feed future developments. It is precisely on these aspects that this chapter will focus in more detail.

2.4.3 THE OSPAR JAMP, A CASE STUDY

2.4.3.1 Introduction

OSPAR has, under the name of the Joint Assessment and Monitoring Programme (JAMP) instigated numerous programmes, activities and tools to monitor and assess the quality of the marine environment and the progress on the implementation of its five thematic Strategies (OSPAR, 2003a). For a number of monitoring and assessment activities, detailed guidance has been developed to harmonise national practices. In particular, under the Hazardous Substances Strategy and the Eutrophication Strategy an elaborate and harmonised monitoring and assessment regime exists. The main objective of the JAMP is to provide arrangements for preparing periodic assessments of the environmental quality status of the OSPAR Convention area and for progress assessments on the implementation of the five thematic OSPAR Strategies. In this, the objectives of the JAMP are very similar to those of other monitoring programmes and the MSD, and some essential components can be identified. First, a monitoring strategy as such needs to be implemented. In the case of an international programme, this strategy will require collective elements such as common guidelines for sampling and analysis, assessment tools and, very importantly, quality assurance protocols. Second, it will require the preparation of environmental data and information products needed to implement the initial strategies (e.g. data reporting protocols, data assessments, collection of scientific information).

As stated before, the highest level of coordination has been reached for that part of the JAMP that deals with the Hazardous Substances Strategy and the Eutrophication Strategy. The necessary monitoring for these strategies is effectively carried out under three sub-programmes that cover the major routes of input of hazardous substances and nutrients to the marine environment, as well as the environment as such. These programmes are:

- The Coordinated Environmental Monitoring Programme (CEMP), which can be described as that part of monitoring under the JAMP where the national contributions overlap and are coordinated. It covers temporal trend and spatial monitoring for concentrations of selected chemicals and nutrients and for biological effects.

- The Comprehensive Atmospheric Monitoring Programme (CAMP), which covers monitoring at coastal stations of the concentrations of selected contaminants (including nitrogen) in precipitation and air and their depositions.

- The Comprehensive Study on Riverine Inputs and Direct Discharges (RID), which assesses, on an annual basis, all riverborne and direct inputs of selected contaminants (including nutrients) to the OSPAR Convention area and determines the long-term trends of such inputs.

It is also on the topics of hazardous substances and particularly on eutrophication that the greatest overlap exists with the WFD. With regard to the monitoring that is envisaged under the WFD, the CEMP bears the most similarities, and it will therefore be discussed in somewhat more detail below.

2.4.3.2   The CEMP

As stated above, the CEMP has a temporal trend part, i.e. to detect long-term trends in concentrations or effects of substances in the maritime area – this involves continuous monitoring and data assessment (approximately every five years); and a spatial part, i.e. spatial distribution monitoring for concentrations of selected chemicals, nutrients and biological effects (OSPAR, 2006b). Both aspects have also been taken up in the WFD. So far, the list includes the following parameters:

- Mercury, cadmium and lead in biota and sediments.
- PCBs in biota and sediments.
- PAHs (parent and alkylated) in biota and sediments.
- Nutrients in seawater.
- Direct and indirect eutrophication effects.
- PAH- and metal-specific biological effects.
- Organotins in sediments and TBT-specific effects.
- Polybrominated diphenylethers (planned for 2008).

In order to implement a common programme as the CEMP requires – as it does for any coordinated programme – a number of steps are necessary, starting with the development of a strategy/working scheme on how its elements should be tackled practically. This is schematically represented in Figure 2.4.2. For instance, for most hazardous substances selected through the OSPAR DYNAMEC process (OSPAR, 2002a, 2006c), background documents have been produced, which include advice on monitoring. Specifically for the CEMP, but again these could well be seen as universal requirements, three further steps or elements are essential for its realisation. These are:

- Presence or development of common guidelines.
- Presence or development of common quality assurance tools.
- Presence or development of common assessment tools.

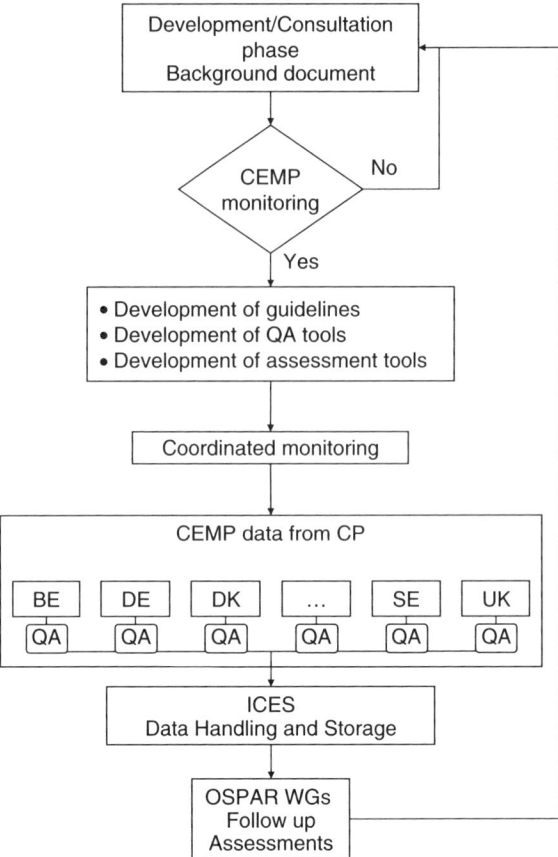

Figure 2.4.2 Schematic overview of the processes and products involved in monitoring under the CEMP (after Roose and Brinkman, 2005)

In others words, measurement of any given parameter is only demanded of contracting parties if these elements are in place. This is quite essential and involves a considerable amount of work. For the specialised technical guidelines and assessment criteria mentioned above, OSPAR has, like HELCOM, relied heavily on the scientific expertise of ICES. The latter has a broad range of expert working groups, including topics such as statistics, chemistry, oceanography, etc. If necessary, ad hoc working groups of experts could be installed to tackle a specific request. An overview of the JAMP guidelines for the major CEMP parameters is given in Table 2.4.2. As can be observed from the table, these include the practical aspect of the analysis as well as statistical aspects and normalisation issues. The latter are quite crucial in sediment analysis because of the vast differences in sediment types and composition, and have involved a considerable amount of discussion. However, the recent CEMP assessment has clearly revealed their importance (Roose and Brinkman, 2005). The table also shows that the status of a given guideline reflects the inevitable link with quality assurance (QA). The latter and the assessment tools will be discussed in more detail below.

Table 2.4.2 JAMP guidelines for monitoring of hazardous substances under the CEMP

Title	Year of adoption	Year of revision	Status[1]
JAMP guidelines for monitoring contaminants in biota	1997		Category I
Technical Annex 1 – determination of organic contaminants	1997		Category I
Technical Annex 2 – determination of metals	1997		Category I
Technical Annex 3 – determination of PAHs	1999		Category I
JAMP guidelines for monitoring contaminants in sediments	1997		Category I
Technical Annex 1 – statistical aspects	1997		
Technical Annex 2 – determination of CBs	1997		Category I
Technical Annex 3 – determination of PAHs	1998		Category I
Technical Annex 4 – determination of TBT	1999		Category I
Technical Annex 5 – normalisation of contaminant concentrations	2002		Category I
Technical Annex 6 – determination of metals – analytical methods	2002		Category I
JAMP guidelines for general biological effects monitoring	1997		Category II
Technical Annex 1 – whole sediment bioassays	1997		Category II
Technical Annex 2 – sediment pore-water bioassays	1997		Category II
Technical Annex 3 – sediment sea water elutriates	1997		Category II
Technical Annex 4 – water bioassays	1997		Category II
Technical Annex 5 – CYP1a	1997		Category II
Technical Annex 6 – lysosomal stability	1997		Category II
Technical Annex 7 – liver neoplasia/ hyperplasia	1997		Category I
Technical Annex 8 – liver nodules	1997		Category I
Technical Annex 9 – externally visible fish diseases	1997		Category I
Technical Annex 10 – reproductive success in fish	1997		Category II
JAMP guidelines for contaminant-specific biological effects monitoring	1997		Category II
Technical Annex 1 – metal-specific biological effects monitoring	1997		Category II
Technical Annex 2 – PAH-specific biological effects monitoring	1997		Category II
Technical Annex 3 – TBT-specific biological effects monitoring	1997	1998, 2002, 2003	Category I

[1]Category I guidelines are those for which quality assurance procedures are in place. Category I guidelines may be used for monitoring and the data obtained are appropriate for Convention-wide assessments. Category II guidelines are those for which quality assurance procedures are not yet in place. Category II guidelines may be used for monitoring, although caution should be exercised when making comparisons of the data obtained between different Contracting Parties.

Once these three elements have been assured, the planning of activities in space and time can be started, followed by a period of actual monitoring which results in data. The CEMP contains provisions for the submission and management of data in a common database, which is crucial for any monitoring programme. Data reporting and data storage are handled for OSPAR by ICES. Furthermore, a system has been set up that allows the identification of gaps in coverage that need to be filled. Specialised OSPAR working groups continuously review the CEMP and make adjustments to any aspect of it as required. Finally, joint assessments are made by specialised OSPAR working groups and it is these that eventually become the backbone of the quality status reports that are produced by OSPAR. The validity of the basic setup has been demonstrated in recent years. For instance, the recent CEMP assessments clearly show how concentrations of contaminants have evolved in the last decade. They also illustrate how assessment tools, such as the background assessment concentrations, can effectively be used to obtain an idea about the status of the marine environment. Most importantly, it shows the value of large and qualitative data collected through a well-organised marine monitoring programme. This will be discussed in the following sections.

2.4.3.3 Quality Assurance

The quality assurance of analytical measurements is, today, receiving increasing attention, and it will continue to be a most important aspect in the future. At present, OSPAR does not require accreditation, but it does encourage it. However, OSPAR does demand participation in proficiency testing schemes (PTS) and expects the results to be submitted to the ICES database with the monitoring results. Also, additional QA information, such as analysis of certified reference materials and laboratory reference materials, is also asked for. In short, relevant QA/QC data should be reported together with the data. OSPAR has also made arrangements with the QUASIMEME PTS concerning reporting of results from OSPAR labs. This information is and has been used in the OSPAR assessments in essentially two ways.

A few years ago, during the 1998 OSPAR assessment of trends in the concentrations of some metals, PAHs and other organic micro-contaminants in the tissue of various fish species and mussels, some 30 % of the data had to be rejected because of a lack of, or insufficiency of, quality assurance information (OSPAR, 1998). During this assessment, data screening for analytical quality resulted in two classes of data: 'acceptable' and 'unacceptable'. Only data with 'acceptable' QA were used, and these were assessed for trends using smoothers, with each observation given equal statistical weight. However, many data were rejected as 'unacceptable' and this led to the shortening or loss of many time series. It has since been argued (Nicholson *et al.*, 2001) that the QA acceptance criteria were too stringent and that some data, previously rejected as unacceptable, could be used in future assessments if they were appropriately down-weighted in the statistical analysis. Nicholson and Fryer (2002) suggested an alternative where the available QA information categorises the analytical quality of data as good, poor, unknown and unacceptable and allocates statistical weights $1 > W_{poor} > W_{unknown} > 0$ accordingly. Although simple and intuitively appealing, the choice of statistical weights is arbitrary and takes no account of the relative importance of the analytical variance to the total environmental and analytical variance. OSPAR finally opted for a third approach, which provides a compromise between the two methods described above. It is assumed that available QA information can be used to construct an analytical weight for each datum, ranging from 0 (totally unacceptable) through to 1 (totally acceptable). An iterative procedure is then used to convert these analytical weights into statistical weights that account for the relative magnitudes of the environmental and analytical variances (OSPAR, 2003b). A graphical representation can be found in Figure 2.4.3. As an additional benefit, this can easily be applied routinely to data in the ICES databank.

The approach has been successfully used in the latest trend assessment of the CEMP data. Valuable older data have been succesfully incorporated in the assessment, where this was previously impossible. These older data are not necessarily bad as such, it's just that at the time there were no demands for QA information or it could not be traced back. The effect of the weighting is illustrated in Figure 2.4.4 for data from a recent CEMP trend assessment (OSPAR, 2005a). The size of the bullets is proportional to their weight in the statistical analysis. This simple graphical approach allows an immediate appreciation of the statistical strength of the dataset for even an untrained eye and has great merits. It also shows the importance of long datasets with older data irrespective of the availability of QA data.

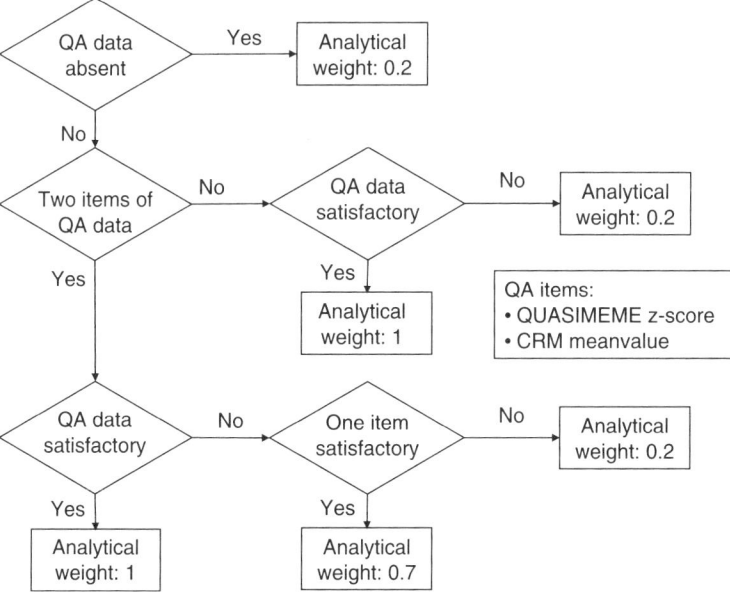

Figure 2.4.3 Weighting of the data of CPs against the provided QA information during OSPAR assessments

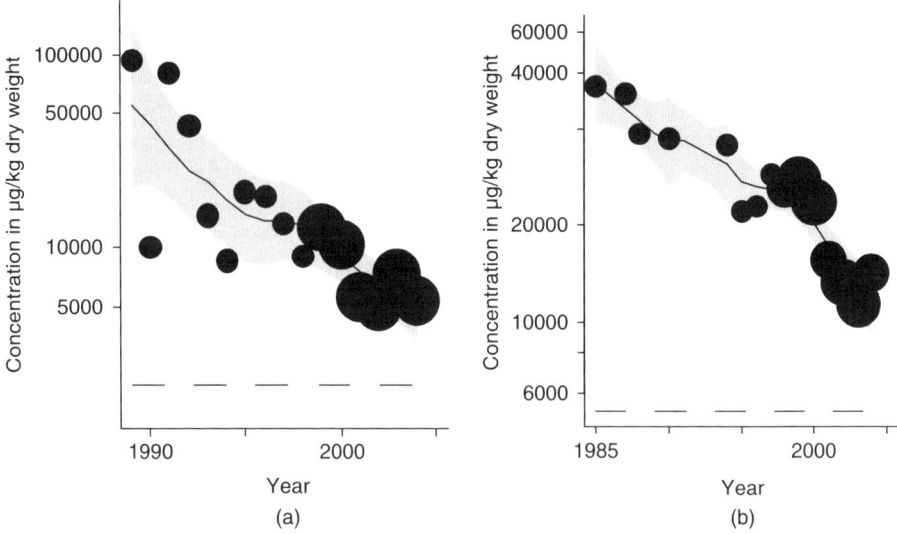

Figure 2.4.4 CEMP trend assessment of Cd in blue mussel from a location in Norway (A) and in Pacific oyster from a location in France (B)

2.4.3.4   Assessment Tools

Monitoring, or, in other words, the collection of measurements, always takes place with a certain end goal in mind, and the produced numbers will ultimately need to be compared with this. The challenge is therefore not only to provide data of sufficient quality, but also to have clearly defined end goals and the tools to compare the data with them. This principle holds whether one is dealing with chemical measurements, biological effect monitoring or biological monitoring. The WFD, for instance, has quite clearly set an end goal for chemical measurement in the proposed EQS values, where the chemical status of a water body will become bad the moment the concentration of one of the chemicals from the list of priority substances exceeds the EQS according to the one all-out principle (see Chapter 1.1). However, no advice or instruction is given on how to compare the measured concentrations with the EQS. Also, the WFD states that it should be demonstrated that the situation in a water body doesn't deteriorate, but again no advice is given on how this should be done (see Chapter 1.1).

Marine monitoring programmes have faced the same challenges and in the following paragraphs it will be demonstrated how OSPAR has approached these challenges in close collaboration with ICES. This section will also focus on the measurement of hazardous substances, not because it is deemed that this is the most important aspect of monitoring, but mainly because of the relative ease with which good quality measurements can be obtained. Chemical measurements are nowadays backed by an impressive and comprehensive system of QA, particularly in laboratories that have an ISO 17025 accreditation. This means that statisticians actually have a good handle on the uncertainty of a chemical measurement, which makes the development of statistical comparisons more straightforward.

To stick with the example of hazardous substances (OSPAR, 2003c), the OSPAR strategy can easily be translated into the following questions:

• What are the concentrations in the marine environment, and the effects, of the substances on the OSPAR list of chemicals for priority action?

• Are they at, or approaching, background levels for naturally occurring substances and close to zero for manmade substances?

These questions contain a spatial (What are ...) and a temporal part (are they approaching ...), and an end goal ('... background levels for naturally occurring substances and close to zero for manmade substances ...') for hazardous substances and their effects. In other words, the concentration of chemicals and their evolution has to be compared against two criteria, both reflecting the natural condition, i.e. background levels and no effects.

The trend assessment part has been well studied and has reached a state of maturity in recent years. This is not surprising, since trend assessments have been an essential part of all marine monitoring programmes. Moreover, the non-deterioration principle of the WFD can effectively be translated into a trend analysis. Trend detection programmes in OSPAR – and therefore also the statistics – are based on an annual sampling programme for a given region (Nicholson *et al.*, 1997, 1998). The statistics have mainly been developed by scientists active in the ICES working group on statistical aspects of environmental monitoring, and will not be further discussed here given their advanced state. The discussion here will mainly focus on the criteria that have been developed for the end goal described above, because this is very similar to the challenges that face the practical execution of the WFD.

To assess progress towards the objectives of the OSPAR Hazardous Substances Strategy, two assessment tools have been developed: Background Concentrations (BCs) and associated Background Assessment Criteria (BACs), and Environmental Assessment Criteria (EACs) (OSPAR, 2004).

Background Concentrations, formerly Background Reference Concentrations (BRCs), are intended to represent the concentrations of certain hazardous substances that would be expected in the north-east Atlantic if certain industrial developments had not happened. They represent the concentrations of those substances at 'remote' sites or in 'pristine' conditions, based on contemporary or historical data respectively, in the absence of significant mineralisation and/or oceanographic influences. In this way they relate to the background levels referred to in the OSPAR Hazardous Substances Strategy and are used to assess whether the concentrations in the marine environment are at, or approaching, background levels for naturally occurring substances and close to zero for manmade substances. Two questions that follow are:

• How do we quantify near background and close to zero?

• How do we test whether the objective has been met?

The OSPAR solution was the introduction of Background Assessment Criteria. BACs are statistical tools that enable precautionary testing of whether mean observed concentrations can be considered to be near background concentrations. A more detailed

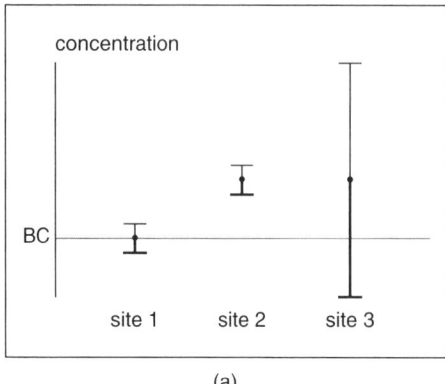 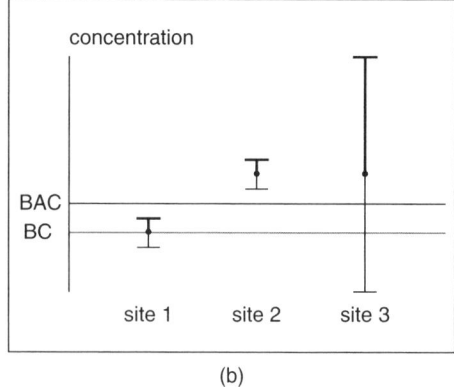

(a)   (b)

Figure 2.4.5 Graphical illustration of the brown test and the modified green test for environmental concentrations from three hypothetical sites

introduction to BACs can be found in Fryer (2004) but essentially they can be explained as follows. Suppose three environmental measurements at different sites, each with a known measurement uncertainty (Figure 2.4.5). In order to test whether the concentrations at these sites are at or near the background, one approach, the so-called brown test, is to use the lower confidence limit (Figure 2.4.5 A). One can conclude that concentrations are at background if the lower confidence limit is below the BC (e.g. Sites 1 and 3 in Figure 2.4.5). However, Site 2 has the same concentration as Site 3 but is 'punished' for its superior (reduced/improved?) uncertainty limits. This is contrary to the precautionary principle that has been endorsed by OSPAR. As an alternative – the so-called green test – the upper confidence limit can be used. However, the green test also concludes that concentrations are above background at Site 1 (Figure 2.4.5), so the test is too precautionary. The solution brought forward was to introduce a modified green test in which some environmental protection is sacrificed for a more effective test. The modified green test assumes one can establish a BAC below which concentrations can be considered near background. The assumption is that the mean concentration is above background unless there is statistical evidence to show that it is near background. The test again uses the upper confidence limit in the sense that concentrations near background if the upper confidence limit is below the BAC. To illustrate, we now conclude that concentrations at Site 1 are near background, but that concentrations at Sites 2 and 3 are above background (Figure 2.4.5).

The BAC should be both relevant – low enough to reflect near-background concentrations – and effective – high enough that we are likely to conclude that concentrations are near background when they are. Essentially, this approach can be used to compare virtually any environmental measurement against a reference or limit value. To give a concrete example, it suffices to replace the BC with an EQS and calculate an assessment concentration on the EQS that takes into account the variability of the environmental data for that particular parameter.

OSPAR's BACs have been calculated based on the residual variability of the data in the CEMP dataset and on the basis of QA information provided by CPs. The outcome of this method is that, on the basis of what is known about variability in observations, there is a 90 % probability that the observed mean concentration will be below the BAC when the true mean concentration is at the BC (OSPAR, 2005a).

The BCs themselves have a somewhat different history. For xenobiotic compounds, the BC is assumed to be zero as the substance is normally not present. For substances with a natural background, the discussion on the approach in still ongoing for biota but a decision has been reached for sediments. For these, deep core sediment data representing deposition layers of pre-industrial times are used as the basis for determination of BCs. Essentially, the median of median concentrations of cores from the entire OSPAR area was used to determine the BC (WGMS, 2004). An overview of the currently applicable BCs/BACs (OSPAR agreement 2005-6), the EACs agreed by OSPAR (agreement 1997-14) and those proposed in their review and trialled in the assessments, are given in Tables 2.4.3 and 2.4.4 for metals and organic pollutants, respectively.

During the CEMP assessments the fitted concentration in the last of the trend analysis, as given above, is compared with the BC. In order to facilitate the interpretation of the data, a graphical presentation of the assessment is made for each parameter, as shown in Figure 2.4.6 for Cd in sediment. This allows a rapid assessment for the situation in each OSPAR region. The approach has shown its merits in the recent assessments (CEMP overview report). During these, it became obvious that in the large majority of cases, concentrations of heavy metals are above background levels. In biota, for example, over 85 % of concentrations in blue mussels in the last year of each time series were above background levels for lead and cadmium.

For the assessment of the potential effects of hazardous substances, Environmental Assessment Criteria (EACs), formerly Ecotoxicological Assessment Criteria, are used. Their main purpose is to identify potential areas of concern and to indicate which substances could be considered as a priority. Regardless of similarities with EQS, they should not be used as firm standards or as triggers for remedial action in place of the EQS. EACs link chemical monitoring data and/or joint chemical/biological effects monitoring data and are based on toxicity tests for individual substances. They particularly relate to the questions if there are any unintended/unacceptable biological responses, or unintended/unacceptable levels of such responses, being caused by exposure to hazardous substances. The EACs adopted in 1997 are provisional and need further refinement and updating. Their review is still ongoing, taking into account the approach taken for the development of the EQS. The set of revised provisional EACs is given in Table 2.4.3 for metals and in Table 2.4.4 for organics. These have been developed and trialled in assessments but have not yet been agreed by OSPAR (OSPAR, 2005a).

However, EACs are not the only OSPAR assessment tool for effects of contaminants. The OSPAR CEMP also contains actual biological effect monitoring and, for the effects of tributyltin (TBT), a set of guidelines and assessment tools has been developed. This is particularly interesting because the effect of TBT on certain species of marine snails occurs at levels that cannot or can hardly be measured (ICES, 2005). More generally, organisms are exposed to a range of substances which have the potential to cause metabolic disorders, an increase in disease prevalence and, potentially, population effects such as changes in growth, reproduction and survival. Monitoring

Table 2.4.3 BC, range of background reference concentrations (BRCs), provisional BACs, and provisional environmental assessment criteria (EACs) for metals in sediment, blue mussel and fish. Bold text indicates metals that are OSPAR chemicals for priority action

Parameter	SEDIMENT (mg/kg dry weight; normalised to 5 % Al for BC/BAC)				BIOTA – blue mussel (mg/kg wet weight)			BIOTA – fish (mg/kg wet weight)		
	Range of BRC	BC	BAC	EAC	Range of BCs	BAC	EAC	BRC- fillet*	BAC- fillet	EAC-whole
Arsenic		15	25	1–10 (0.71)						
Cadmium		**0,2**	**0,31**	**0.1–1 (0.06)**	**0.07–0.11**		**(55.9)**			**7.35***
Chromium		60	81	10–100 (21)						
Cobalt	7–23									
Copper		20	27	5-5–(0.22)	0.76–1.1					
Iron	0.6–6.3									
Lead		**25**	**38**	**5-5–(2.22)**	**0.01–0.19**		**(1690)[3]**			**(300)[3]**
Lithium	22–44									
Mercury		**0.05**	**0.07**	**0.05–0.5 (0.22)**	**0.005–0.01**		**(1.7)[3]**	**10–50[1]** **30–70[2]**		**(3.5)[3]**
Nickel	0.2–0.35	30	36	5–50 (2.8)						
Titanium	0.2–0.35									
Vanadium	60–110									
Zinc		90	122	50–500 (1.48)	11.6–30					

Note: BRC values are those adopted in 1997 with a conversion to the appropriate units (OSPAR agreement 2005–6). EACs are those agreed by OSPAR in agreement 1997–15 and, in brackets, those proposed in the review and trialled in assessments. [1]BRC for round fish; [2]BRC for flat fish; [3]Provisional EACs for secondary poisoning, whole fish.

Table 2.4.4 Background concentration (BC) and provisional background assessment criteria (BAC) (OSPAR agreement 2005-6) and provisional environmental assessment criteria (EAC) for organochlorines and PAHs in sediment, blue mussel and fish

Parameter	SEDIMENT (μg/kg dry weight normalised to 2.5 % carbon)			BIOTA – blue mussel (μg/kg dry weight)			BIOTA – fish (μg/kg wet weight)		
	BC	BAC	EAC	BC	BAC	EAC	BC – liver	BAC – liver	EAC – whole
DDE			0.5–5 (4)	0		10			5–50
Dieldrin			0.5–5 (19.75)	0		10			5–50
Lindane			(2.75)			(0.29)			0.5–5 (1.1)
TBT			(0.025)			(2.4)			
CB 153	0	0.2		0	0.4	2.5	0	0.2	2.5
ΣCB₇[1]	0	1.5		0	0.7	10	0	1.2	1–10
Naphthalene	5	8	50–500 (95)	0.2	1.1	(91)			
Phenanthrene	17	32		0.9	4.9				
Anthracene	3	5		0.2	0.4				
3 rings (PA+ANT)			(77.5)			(1290)			
Fluoranthene	20	39		1.4	2.5				
Pyrene	13	24		1.1	1.8				
Benz[a]anthracene	9	16		0.3	1.1				
Chrysene	11	20		1.3	3.4				
4 rings (FLU+PYR+BAA+CHR)			(352.5)			(6900)			
Benzo[a]pyrene	15	30		0.2	0.7				
5 rings (BAP+BKF)			(52.5)			(1069)			
Benzo[ghi]perylene	45	80		0.5	2.7				
Indeno[123–cd]pyrene	50	103		0.4	1.6				
6 rings (BGHIP+ICDP)			(9.25)			(73)			

Note: EACs are those agreed by OSPAR in agreement 1997–15 and – in brackets – those proposed in the review and trialled in assessments. [1]Sum of chlorinated biphenyl congeners CB 28, CB 52, CB 101, CB 118, CB 138, CB 153, CB 180

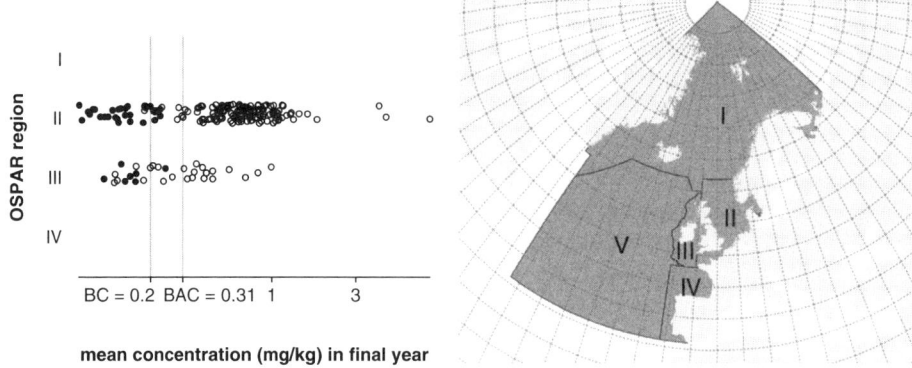

Figure 2.4.6 Comparison of the mean concentration in the last year of the different time series for Cd in sediment against the BC and BAC. Filled-in dots are significantly below the BC (source: OSPAR, 2005a)

to assess the 'impact' of hazardous substances has been and is primarily based on chemical measurements. However, the combined effect of the chemicals is unknown and chemical monitoring will only allow you to see what you're looking for. Biological effects techniques (BETs) have the potential to fill this gap and have become increasingly important in recent years. The revised OSPAR JAMP now specifies requirements for information on the effects of the substances on the OSPAR List of Chemicals for Priority Action and on any emerging problems related to the presence of hazardous substances in the marine environment.

In Table 2.4.2, the BETs can be found which have been included in the CEMP. The major holdup for BETs has been the availability of appropriate assessment criteria, which is an essential prerequisite for coordinated monitoring (ICES, 2007b). For certain BETs an approach similar to the BAC has actually been suggested. Furthermore, there is general agreement that, with respect to hazardous substances, the best way to assess their impact is to use a suite of chemical and biological measurements in an integrated fashion. OSPAR has been actively working towards that goal in recent years through a series of workshops in collaboration with ICES (ICES/OSPAR Workshops on Integrated Monitoring of Contaminants and their Effects in Coastal and Open-sea Areas 2005–2007). Work has progressed towards guidelines for integrated chemical and biological effect monitoring and proposals for assessment criteria for a number of BETs (ICES, 2007b). This approach links chemistry with the health of the ecosystem, which is also one of the objectives of the WFD. WFD-related organisations or institutions may well benefit, in a later stage, from these developments.

2.4.4   FINDING SYNERGIES AND HARMONISING DIFFERENCES

The introduction of the WFD has caused quite a stir in the marine monitoring community in Europe. In the near future the MSD could potentially further complicate matters.

The challenge will be to find modi operandi that are both practical and achievable for all parties involved. This will require a high level of harmonisation on matters of quality assurance, sampling, matrix selection and possibly even assessment procedures. Fortunately, the tendency to move towards mutually supporting programmes exists.

To take the example of the OSPAR JAMP and the WFD, both aim at monitoring the levels of contaminants in coastal and transitional waters, describing their spatial or temporal trends, and using this information to develop, or improve, programmes of action or to advise decision makers. But there are differences. The WFD, for instance, does not require monitoring beyond the twelve-mile zone. On the other hand, monitoring of temporal trends and spatial distribution of hazardous substances under the CEMP is comparable to the surveillance monitoring under the WFD. OSPAR currently does not require monitoring equivalent to operational monitoring under the WFD. Likewise, the monitoring requirements of the WFD are restricted to the monitoring of the status of surface (and ground) waters, while monitoring of direct discharges (into the sea), riverine inputs and atmospheric deposition is required by OSPAR (RID and CAMP programmes).

Similar differences and communalities are also found at the lower level, i.e. the practical execution of the monitoring. The highest level of agreement between the JAMP and WFD can be found in eutrophication-related monitoring. Both OSPAR and the WFD have combating eutrophication as an objective, whether to 'achieve and maintain a healthy marine environment' or a 'good ecological status'. Both have very similar definitions of eutrophication (OSPAR, 2005b). The OSPAR requirements for eutrophication monitoring have been historically laid out in the Eutrophication Monitoring Programme (OSPAR Agreement 2005-04, OSPAR, 2005c) and have been developed through the OSPAR Common Procedure (OSPAR, 2005d). The monitoring requirements are linked to the eutrophication status of an area. Areas that are eutrophic or have the potential to become eutrophic will require more intensive monitoring (OSPAR, 2005b). The parameters and the monitoring frequency are laid down in the Common Procedure. The WFD requires Member States to monitor phytoplankton, macroalgae, angiosperms, benthic invertebrate fauna and supporting physico-chemical quality elements like nutrients, turbidity, salinity, etc. as part of the determination of ecological status. Monitoring frequencies are related to the degree of risk that a water body will fail to meet good ecological status. Surveillance monitoring is the minimal requirement, with operational and investigative monitoring being required where a body is at risk of failing to meet good status. Not surprisingly there is broad agreement on the main parameters for monitoring. A summary of the synergies in monitoring are outlined in Table 2.4.5.

The similarities between the parameters are obvious. The only thing that remains to be assured is that the methods used for the measurement of these parameters will give the same results. Only in this way can a serious duplication of efforts be avoided. True to its principles, OSPAR has a number of clear monitoring guidelines to assist in eutrophication monitoring and assessment. These are:

- JAMP Eutrophication Monitoring Guidelines: Nutrients 97/02.

- JAMP Eutrophication Monitoring Guidelines: Oxygen 97/03.

Table 2.4.5 Synergies in monitoring for eutrophication-related parameters in the WFD and the OSPAR-JAMP

Monitoring Requirement	OSPAR	WFD
Category I Degree of Nutrient Enrichment		
Riverine inputs	✓	✓ (as a pressure)
Winter nutrient concentrations	✓	Physico-chemical quality element
Increased N/P ratio	✓	Physico-chemical quality element
Category II Direct Effects of Nutrient Enrichment (during growing season)		
Chlorophyll a concentration	✓	✓
(Phytoplankton biomass)	NS	NS
Phytoplankton indicator species	✓	✓
Macrophytes including macroalgae	✓	✓
Category III Indirect Effects of Nutrient Enrichment (during growing season)		
Degree of oxygen deficiency	✓	✓
Changes/kills in zoobenthos	✓	✓
Fish kills	✓	Transitional waters only
Organic carbon/organic matter	✓	NS but implied as suitable supporting det.
Category IV Other Possible Effects of Nutrient Enrichment (during growing season)		
Algal toxins (DSP/PSP)	✓	NS

NS: not specified

- JAMP Eutrophication Monitoring Guidelines: Chlorophyll a in Water 97/04.

- JAMP Eutrophication Monitoring Guidelines: Phytoplankton Species Composition 97/05.

- JAMP Eutrophication Monitoring Guidelines: Benthos 97/06.

- JAMP Guidelines for General Biological Effects Monitoring 97/07.

- JAMP Guidelines on Quality Assurance for Biological Monitoring in the OSPAR Area.

In contrast, there are not yet any monitoring guidelines of that level under the WFD. The WFD states that 'Methods used for the monitoring of type parameters shall conform to the international standards listed below or such other national or international standards which will ensure the provision of data of an equivalent scientific quality and comparability' (Annex V 1.3.6 – Water Framework Directive), and this is where programs can converge again. There are currently ISO/CEN marine methods and both of these are identical to their OSPAR equivalents. Discussions are currently underway between DG–ENV and ISO/CEN to agree a way forward for the development of marine methods (SIAM). Also, the Common Implementation Strategy COAST Group (2.4), who developed guidance on 'Typology, Reference Conditions and Classification Schemes in Transitional and Coastal Waters', made a recommendation that new monitoring guidelines should not be developed under the WFD as this may invalidate long-term biological data sets available for national assessments or for the conventions.

For hazardous substances, the approaches are less synchronised, with synergies and differences both in the selection of priority chemicals and in the approach to monitoring. A more thorough comparison of the prioritisation of chemicals for marine monitoring can be found elsewhere (Roose and Brinkman, 2005) and is not the objective of this chapter. It suffices to say that there is much overlap, but that there are also several striking differences. Most surprisingly, not a single compound is common for all existing lists. In Table 2.4.6 the WFD list of priority substances (PS) is compared with the OSPAR selection. For the latter, a difference can be made between substances for which monitoring is already under way (CEMP) and those that are considered to be a priority but for which no coordinated monitoring has been organised (OSPAR PS). OSPAR has an additional 30 PS, which can be found in Roose and Brinkman (2005) or OSPAR (2007b).

The most striking difference remains PCBs, which are a mandatory CEMP parameter (see above). Although the initial COMMPS selection procedure included them (even as top-ranking substances), they were not considered as priority substances because of the fact that there is no current production or usage, or use is strictly regulated or forbidden (Klein *et al.*, 1999; European Commission, 2000b). Exclusion is therefore not based on their toxicological properties and/or presence in the environment. In contrast, the latter is precisely the reason why organisations such as OSPAR and HELCOM consider them as priority substances, which seems a sounder approach.

The main discussion between the WFD and marine monitoring programmes such as the OSPAR-CEMP for monitoring of hazardous substances has thus far focused on the matrix selection. The principle matrix for assessing compliance with respect to Environmental Quality Standards (EQS) for priority substances is whole water; or for metals, the liquid fraction obtained by filtration of the whole water sample. This is mainly because environmental quality criteria (based on toxicological tests) are mostly available for the water phase and cannot be converted directly to sediment or biota. However, this is not a fortunate choice in view of the many apolar compounds listed in Table 2.4.6. Marine scientists have amply demonstrated that sediments and biota are much more suitable sample types for such analytes, while water is a proper matrix for more polar compounds, such as e.g. the 'new' pesticides, which are indeed frequently detected in coastal and estuarine waters. Recently, the Commission proposal on environmental quality standards stated that biota may be used as matrix for compliance monitoring in case of hexachlorobenzene, hexachlorobutadiene and mercury (Ref Commission proposal EQS). Also, one purpose of the WFD is to prevent further deterioration of the status of aquatic ecosystems. Monitoring of contaminants in sediment and biota can therefore be used to assess the long-term impacts of anthropogenic activity. This is very compatible with the trend assessment part of the OSPAR-JAMP. Furthermore, for a number of hydrophobic and lipophilic PS, analytical difficulties demonstrating compliance with EQS in water may be problematic because the available analytical methods are not sufficiently sensitive or accurate. Sediment and/or biota are in that case a viable alternative, provided that EQS values become available. The latest version of the Commission proposal on environmental quality standards quite rightly foresees the use of these matrices in specific cases. The marine environment

Table 2.4.6   Synergies between the list of WFD PS and the OSPAR list

No	Name of substance	CAS number	OSPAR PS	CEMP
PART A Proposal 2006/0129 (COD)				
1	Alachlor	15972-60-8		
2	Anthracene	120-12-7	X	X
3	Atrazine	1912-24-9		
4	Benzene	71-43-2		
5	Penta BDE	32534-81-9	X	X
6	Cadmium and its compounds	7440-43-9	X	X
7	Chloroalkanes, C10-13	85535-84-8	X	
8	Chlorfenvinphos	470-90-6		
9	Chlorpyrifos	2921-88-2		
10	1,2-Dichloroethane	107-06-2		
11	Dichloromethane	75-09-2		
12	Di(2-ethylhexyl)phthalate (DEHP)	117-81-7	X	
13	Diuron	330-54-1		
14	Endosulfan	115-29-7	X	
15	Fluoranthene	206-44-0	X	
16	Hexachlorobenzene	118-74-1		
17	Hexachlorobutadiene	87-68-3		
18	Hexachlorocyclohexane (lindane)	608-73-1	X	
19	Isoproturon	34123-59-6		
20	Lead and its compounds	7439-92-1	X	X
21	Mercury and its compounds	7439-97-6	X	X
22	Naphthalene	91-20-3	X	X
23	Nickel and its compounds	7440-02-0		
24	Nonylphenols	25154-52-3	X	
25	Octylphenols	1806-26-4	X	
26	Pentachlorobenzene	608-93-5		
27	Pentachlorophenol	87-86-5	X	
28	Polyaromatic hydrocarbons			
	(Benzo(a)pyrene)	50-32-8	X	X
	(Benzo(b)fluoranthene)	205-99-2	X	X
	(Benzo(g,h,I)perylene)	207-08-9	X	X
	(Benzo(k)flouranthene)	191-24-2	X	X
	(Indeno(1,2,4-cd)pyrene)	193-39-5	X	X
29	Simazine	122-34-9		
30	Tributyltin compounds	688-73-3	X	X
31	Trichlorobenzenes	12002-48-1	X	
32	Trichloromethane (chloroform)	67-66-3		
33	Trifluralin	1582-09-8	X	
PART B Proposal 2006/0129 (COD)				
1	DDT total			
2	para-para-DDT	50-29-3		
3	Aldrin	309-00-2		
4	Dieldrin	60-57-1		
5	Endrin	72-20-8		
6	Isodrin	465-73-6		
7	Carbontetrachloride	56-23-5		
8	Tetrachloroethylene	127-18-4		
9	Trichloroethylene	79-01-6		

or coastal water, for the WFD, is certainly such a case, and this would result in a substantial additional level of harmonisation.

The importance of the selection of the monitoring matrix cannot be stressed enough given the implications it has on the monitoring frequencies, costs and scientific relevance. The monthly sampling required by the WFD makes sense for river systems with relatively well-known hydrodynamic properties, but it is a different thing altogether for coastal waters with a much more dynamic nature. Representative sampling under these conditions is much more difficult and certainly much more expensive. Sediment and/or biota sampling requires less frequent sampling to get a representative picture. The trend detection programmes for hazardous substances in biota and sediment in OSPAR – and therefore also the statistics – are based on an annual sampling programme for a given region (Nichols *et al.*, 1997, 1998). Generally, it is expected that the programme should be able to detect a significant trend, i.e. log-linear trend, after 10 years of monitoring with a power of 90 % using a test at the 5 % significance level. Next to the frequency, advice is given on the general aspects of selecting sampling stations, e.g. in relation to sedimentation (OSPAR, 2002b). Unfortunately, advice on the required frequency and number of locations for spatial distribution monitoring is still lacking. Clearly the CMA working groups have, at least partly, been inspired by this, as similar advice can be found in the latest version of their guidelines (European Commission, 2007).

As far as the actual sampling and analysis is concerned, a high level of harmonisation can be expected simply because of the scientific consensus that exists on these matters. As was mentioned earlier, OSPAR has quite detailed guidelines on these topics, and these, as well as the HELCOM COMBINE guidelines, have been taken up in the proposed technical guidelines for WFD monitoring (European Commission, 2007).

2.4.5 CONCLUSIONS

The WFD overlaps regionally and also thematically with programmes carried out by the existing Marine Environmental Conventions. Matters are being further complicated by the introduction of the Marine Strategy Directive (MSD). The principal challenge will be to obtain a sufficient level of harmonisation, as the current situation could easily result in an unnecessary duplication of monitoring efforts or even a reduction of efforts at the expense of the programmes that have a less stringent legal obligation.

The existing marine monitoring programmes, such as the OSPAR JAMP introduced in this chapter, often have a long history and therefore a unique experience that cannot be taken lightly. Generally, a much different approach towards monitoring will be required for marine and coastal waters than for fresh water systems, not only because of the entirely different dynamics in coastal systems but because of the practical considerations of sampling at sea. Furthermore, all of Europe's seas have a totally different character, and the regional approach suggested by the MSD is the way forward.

The harmonisation mentioned above will therefore have to take place at all levels, being: the parameters that need to be measured, sampling (frequency, spatial distribution, operational procedures), selection of matrices, interpretation and assessment of results, and even reporting. Although the latter seems obvious, it's absolutely imperative that different programmes do not come to different conclusions concerning the state of the same (marine) environment. Fortunately, initiatives have already been taken to assure the required level of harmonisation. A particularly good example is the harmonisation between the OSPAR eutrophication monitoring and the WFD. However, this level of harmonisation has not been reached for other topics such as hazardous substances, where the discussion on the appropriate matrices for monitoring of hydrophobic compounds is a case in point.

Clearly, the ultimate goal is the same for everybody, whether it is 'clean and healthy seas' or a 'good environmental status'. Future developments should therefore not only focus on the harmonisation of the existing programmes but also on their shortcomings.

REFERENCES

AMAP (2007) http://www.amap.no.

BSC (2007) http://www.blacksea-commission.org.

European Commission (2000a) Directive 2000/60/EC of the European Parliament and of the Council of 23 October 2000 establishing a framework for Community action in the field of water policy, *Official Journal of the European Communities*, **L327**, 22.12.2000, p. 1.

European Commission (2000b) Modified proposal for a procedure for the identification of priority hazardous substances in accordance to Article 16(3) of the Water Framework Directive, final draft (Working Document ENV/191000/01 of 19 October 2000), European Commission, DG Environment, Unit E.1, Brussels.

European Commission (2004a) SGO(3) 04/3/1, The European Marine Strategy.

European Commission (2004b) DG ENVIRONMENT, Directorate B, EMMA(2) 04/1/1 Towards a Common Approach on European Marine Monitoring and Assessment.

European Commission (2004c) DG ENVIRONMENT, Directorate B, EMMA(1) 04/2/1 The Progress Report.

European Commission (2005) Proposal for a directive of the European Parliament of the Council, establishing a framework for Community action in the field of marine environmental policy (Marine Strategy Directive), COM (2005) 505 final.

European Commission (2007) Guidance document on surface water monitoring, in press.

Fryer, R. (2004) Appendix 6.2: the use of background concentrations in the assessment of CEMP data, OSPAR Commission 2004: OSPAR/ICES workshop on evaluation and update of BRCs and EACs, OSPAR Commission, London.

HELCOM (2007) http://www.helcom.fi.

ICES (2005) Report of the ICES/OSPAR workshop on integrated monitoring of contaminants and their effects in coastal and open-sea areas (WKIMON), 10–13 January 2005, International Council for the Exploration of the Seas, Copenhagen, Denmark, ICES CM 2005/ACME:01, p. 231.

ICES (2007a) http://www.ices.dk.

ICES (2007b) Report of the ICES/OSPAR workshop on integrated monitoring of contaminants and their effects in coastal and open-sea areas (WKIMON III), 16–18 January 2007, International Council for the Exploration of the Seas, Copenhagen, Denmark, ICES CM 2007/ACME:01, p. 209.

Klein, W., Denzer, S., Herrchen, M., Lepper, P., Müller, M., Sehrt, R. *et al.* (1999) Declaration ref.: 98/788/3040/DEB/E1, Fraunhofer-Institut, Schmallenberg, Germany.

MEDPOL (2007) http://www.unep.ch/seas/main/med.

Nicholson, M.D., Fryer, R.J. and Ross, C. (1997) *Mar. Poll. Bull.*, **34**, p. 821.

Nicholson, M.D., Fryer, R.J. and Larsen, J.R. (1998) ICES techniques in marine environmental sciences, No. 20, ICES, Copenhagen, Denmark.

Nicholson, M.D., Fryer, R.J., Law, R. and Davies, I. (2001) ACME report Annex 8, ICES, Copenhagen, Denmark.

Nicholson, M.D. and Fryer, R.J. (2002) Weighted smoothers for assessing trend data of variable analytical quality, ICES Working Group on Statistical Aspects of Environmental Monitoring.

OSPAR (1998) Report of an assessment of trends in the concentrations of certain metals, PAHs and other organic compounds in the tissues of various fish species and blue mussels: OSPAR ad hoc Working Group on Monitoring, Oslo and Paris Commissions, London, UK.

OSPAR (2002a) Dynamic selection and prioritisation mechanism for hazardous substances (DYNAMEC), hazardous substances series, OSPAR Commission, London.

OSPAR (2002b) JAMP guidelines for monitoring contaminants in sediments, OSPAR Commission, London, 2002-16.

OSPAR (2003a) Draft strategy for a Joint Assessment and Monitoring Programme (JAMP), OSPAR Commission, London, MMC 2003/4/2-E.

OSPAR (2003b) Implementing weighted smoothers in trend assessments, OSPAR Commission, London, MON 03/2/3-E.

OSPAR (2003c) 2003 strategies of the OSPAR Commission for the Protection of the Marine Environment of the North-East Atlantic, OSPAR Commission, London, 2003-21.

OSPAR (2004) OSPAR/ICES workshop on the evaluation and update of background reference concentrations (B/RCs) and ecotoxicological assessment criteria (EACs) and how these assessment tools should be used in assessing contaminants in water, sediment and biota, OSPAR Commission, London.

OSPAR (2005a) Assessment and monitoring series no. 235, OSPAR Commission, London, p. 115.

OSPAR (2005b) Assessment and monitoring series no. 230, OSPAR Commission, London, p. 67.

OSPAR (2005c) Agreement on the eutrophication monitoring programme, OSPAR Commission, London, 2005-4.

OSPAR (2005d) Common procedure for the identification of the eutrophication status of the OSPAR maritime area, OSPAR Commission, London, 2005-3.

OSPAR (2006a) Assessment and monitoring series no. 287, OSPAR Commission, London, p. 89.

OSPAR (2006b) Revised OSPAR Coordinated Environmental Monitoring Programme (CEMP), OSPAR Commission, London, 2006-1.

OSPAR (2006c) Dynamic selection and prioritisation mechanism for hazardous substances (new DYNAMEC manual), hazardous substances series, OSPAR Commission, London, 256/2006.

OSPAR (2007a) http://www.ospar.org.

OSPAR (2007b) OSPAR list of chemicals for priority action (update 2007), OSPAR Commission, London, 2004-12.

Roose, P. and Brinkman, U.A.T. (2005) *Trends in Anal. Chem.*, **24**, p. 897.

Section 3
Analytical Tools in Support of WFD Monitoring

3.1

Emerging Methods for Water Monitoring in the Context of the WFD

Richard Greenwood and Graham A. Mills

3.1.1   INTRODUCTION

The aims of the European Union's Water Framework Directive (2000/60/CE) (WFD) are similar to those of the equivalent legislation in Australia, Canada and the United States of America and have been discussed thoroughly in earlier chapters of this book. However, it is worth stressing that the main aims are to improve and protect the quality of all water bodies across Europe, and to achieve sustainable use of water resources. The quality of surface waters is to be managed at river basin level, and there is a tight deadline (2015) for the achievement of the main aims. In this context, this chapter will deal with the methods that are available, or in some cases under development, for use in monitoring the chemical and biological quality of water, and

The Water Framework Directive - Ecological and Chemical Status Monitoring   Edited by Philippe Quevauviller, Ulrich Borchers, Clive Thompson and Tristan Simonart © 2008 John Wiley & Sons, Ltd

which could complement or partially replace the methodologies that are currently employed. Some of these 'emerging methods' could be used in one, two, or all three modes of monitoring (surveillance, operational and investigative) that are specified in the WFD. Currently a limited range of tools is available to those responsible for the successful implementation of the WFD, and it is important that this is expanded to allow the use of the most appropriate tool for any job. A tool box that contained only a hammer would severely limit the range and quality of work that an artisan could achieve.

Current practice depends heavily on the use of spot (bottle or grab) sampling in combination with classical laboratory-based analytical techniques (mostly based on chromatography linked to an appropriate detector). Compliance with the legislation is generally monitored by comparing levels of priority substances measured in infrequent (monthly) spot samples with legislatively defined levels – Environmental Quality Standards (EQS) – of these substances. Two types of EQS (annual average concentration and maximum allowable concentration) are defined on the basis of the protection of ecosystems from the effects of short-term (acute) and long-term (chronic) exposure to pollutants (Proposal for a Directive of the European Parliament and of the Council on Environmental Quality Standards in the Field of Water Policy and Amending Directive 2000/60/EC (COM (2006) 397 final)). Some biologically-based tools have been used in some special contexts for some years and the information that they provide has become accepted as a valuable aid in determining water quality or in protecting sensitive water resources. Biomonitoring involves measuring the accumulation of substances of interest or of concern in caged organisms placed at strategic sites. The most commonly used organisms are bivalve molluscs since these are sessile, robust and easy to maintain in a confined space (Salazar and Salazar, 2006). Biological early warning systems (BEWS) use changes in the behaviour or biochemistry of confined organisms to provide an instant indication of any deterioration in water quality (van der Schalie, 2001), and generally have been deployed at facilities such as drinking water capitation sites. They are normally linked to an alarm system that allows the operator, or in some cases an automatic system, to cut off the water intake until the problem has been investigated and it is safe to continue taking water from the affected source.

The methods currently in use have a number of strengths. The laboratory-based methods used for analysing spot samples are thoroughly validated, and standards are available for many methods. The techniques and instrumentation available to the analytical chemist have improved in terms of robustness, sensitivity, specificity and reliability over the last twenty years. Further, these methods are underpinned by sound quality assurance and control methodology, supported by the availability of certified reference materials and data from inter-laboratory trials. Most laboratories involved in monitoring water quality in a legislative context are accredited to ISO 17025, and the data obtained can be used to support compliance monitoring. However, these methods also have a number of weaknesses. It is often not possible to detect trace levels of organic contaminants using standard low volume ($<5\,L$) sampling techniques, and there are physical difficulties in handling and processing large volumes of water. This has implications for quality control of the sampling and sample preparation processes. These methods do not measure the truly dissolved (biologically

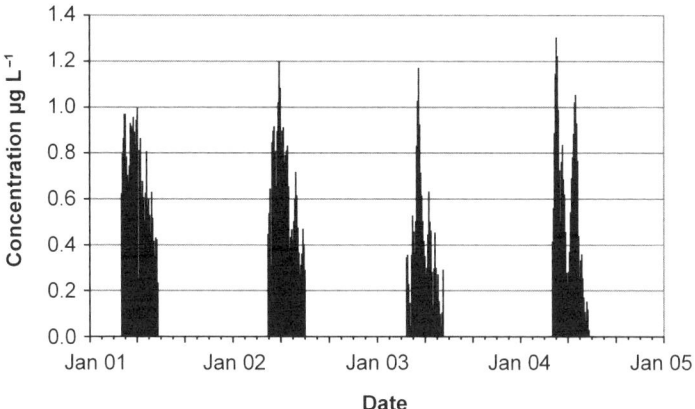

Figure 3.1.1 A schematic illustration of temporal variation in the concentration of a pesticide in a river over a number of years. The pollution events correspond to spraying and run-off caused by rainfall

relevant) fractions of contaminants. Systematic error (bias) can be introduced by changes during transport, storage and preparation (e.g. filtration) of bottle samples. Further, where there is marked temporal variation (Figure 3.1.1) due to factors such as seasonal use of pesticides, intermittent industrial discharges, illegal disposal, accidents or weather-related sources such as road wash-off, storm drain discharges or leaching from agricultural sources following rain, then intermittent spot sampling may not give an accurate measure of average concentrations or of peak concentrations (Allan *et al.*, 2006a, 2006b). Despite the high quality of the analytical data, such methods may give a very precise wrong answer. A further problem with the use of spot sampling is that it is expensive, because of the personnel and transport costs involved, and so is not an appropriate tool for mapping spatial variation in concentrations of pollutants such as occurs when there are many discharges into, for instance, a river, or where there is incomplete mixing of discharged material with the bulk flow of the river (Figure 3.1.2). In some systems where there are local differences in flow there is a tendency for discharges to remain close to the bank of a river for several kilometres before homogeneous mixing is achieved (Environment Canada, 1995; Louch *et al.*, 2003).

BEWS provide continuous monitoring and give an immediate warning of any adverse change in water quality. They react to peak levels, not average concentrations. Biomonitoring provides a measure of average levels of contaminants and biologically-relevant information on water quality, since it is only the biologically-available fraction of a substance that will be accumulated. Biomonitoring therefore provides useful information, but biological factors such as the excretion either of the parent compound or of detoxification products resulting from metabolism can result in a failure to accumulate a pollutant that is assimilated by the organism. The levels accumulated may therefore not provide a reliable reflection of the environmental concentration of the bioavailable fractions of all pollutants of concern. A further limitation is that living organisms can be deployed only at sites where environmental conditions are suitable

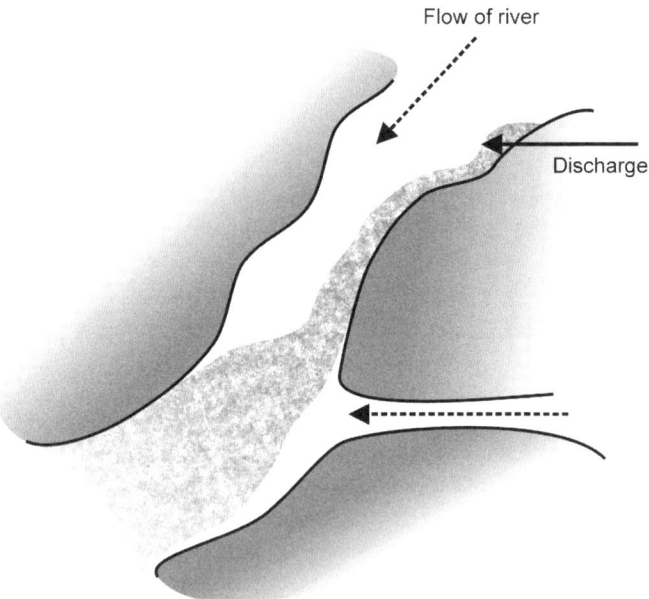

Figure 3.1.2 Spatial variation in the concentration of pollutants caused by restricted mixing of a discharge with the bulk flow of a river

(e.g. an appropriate salinity, pH, temperature) and where pollutants are not present in toxic concentrations.

Since the tools that are currently used have limitations as well as advantages, alternative or complementary methods have been sought to provide data that can reduce the uncertainty of information used in risk assessments or in following the effects of remedial actions. This is important, since the costs of incorrect decisions that lead either to inappropriate actions being taken or to no action being taken when it is needed could be very high. Some of the so-called *emerging methods* have been developed for many years, but not implemented in the context of regulatory monitoring; others are still under development. However, most share common problems, a lack of proper validation studies and of support for quality assurance and control. These aspects need to be addressed as a matter of urgency if a full range of appropriate tools is to be made available to those responsible for achieving good status in all European water bodies within the timeframe of the WFD. This chapter will examine some of these emerging techniques, and describe their properties and potential utility within the context of the WFD.

3.1.2 EMERGING METHODS

It has been difficult to find a suitable term that encompasses all of the methods (sampling or analytical) that have been proposed recently for use in monitoring the quality of water. The term 'emerging methods' can be misleading, since it includes methods that have been available for many years but are only now being considered for use

Table 3.1.1 Summary of the emerging methods and their modes of application

Type of Method	Analytical	Sampling	In laboratory	On site	*In situ*	Online	Use spot sampling
Biomonitoring		✓			✓		
Passive sampling		✓			✓		
Immunoassays	✓		✓	✓			✓
Test kits	✓		✓	✓			✓
Sensors	✓		✓	✓	✓	✓	✓
Toxicity assays	✓		✓	✓			✓
Biomarkers	✓				✓		
Biological early warning systems	✓				✓	✓	

in a regulatory context, and methods that are still under development. However, it provides a useful shorthand that can be used when describing the wide range of methods involved (Table 3.1.1). This category includes sampling methods, such as passive sampling and biomonitoring, that provide alternatives to spot sampling procedures. It also includes methods that can be used instead of classical analytical techniques. Some of these emerging methods give quantitative information and others only qualitative information. Some depend on spot sampling and are laboratory-based, but with advances in technology that have enabled miniaturisation and the production of robust instrumentation, some can now be applied in the field (e.g. on a river bank or in a boat), or have been adapted to operate online. Some methods (e.g. handheld sensors) can be used *in situ*, allowing measurements that avoid the need for the sampling, transport, storage and sample preparation stages (Honeychurch and Hart, 2003; Killard *et al.* 2001). Biological methods such as toxicity assays, biomarkers and BEWS can provide qualitative rather than quantitative information on water quality, and interpretation and evaluation of the data provided needs a different approach from that taken with the quantitative methods.

A range of methods is available to provide an alternative to infrequent spot sampling and more representative information, particularly where there is marked temporal variation in water quality. One approach is simply to increase the frequency of spot sampling, but this is expensive because of the transport and labour involved in taking the samples in the field and then carrying them to the laboratory for analysis. Automated methods have been developed for taking sequential samples over a period of time (e.g. 24 hours) to provide composite samples. Some of the equipment can be triggered by storm events and can take flow-weighted samples. The drawbacks with these techniques are that they provide a measure of quality over only a limited timescale, and that they are expensive and so need to be deployed ideally in a secure site. A range of sampling technologies can be installed online to provide frequent or continuous measurements. Here water is pumped continuously from a sampling site and flows through a system where some pretreatment (e.g. settlement of suspended matter and/or filtration) can take place before the analytical step. Some classical chromatographic methods have been automated and installed online and some, such as the SAMOS systems which use a liquid chromatograph linked to a UV diode array detector, can be programmed to send an alarm message if concentrations of some key pollutants rise above a preset threshold level. These systems perform an analysis every few hours, and between sampling points carry out cleaning and calibration checks. In some systems, sensors have been installed to provide continuous measurements of individual analytes. BEWS can operate online or *in situ*, depending on the circumstances, and give all-or-nothing signals concerning water quality. The disadvantage of the online analytical instruments and BEWS is that they need to be deployed in secure sites and so are not suitable for widespread deployment. The other approaches available for obtaining representative information on water quality are biomonitoring and passive sampling. Originally, passive samplers such as solvent-filled bags and the semi-permeable membrane device (SPMD) containing a lipid (triolein) were designed to mimic living organisms such as those used in biomonitoring. Both of these sampling approaches have been widely used.

3.1.2.1   Sampling Methods

Both biomonitoring and passive sampling can provide measures of time-weighted average (TWA) concentrations of contaminants (both metals and organics) to which they have been exposed over a deployment period (Booij *et al.*, 2006; Gunther *et al.*, 1999; Ouyang and Pawliszyn, 2007; Seethapathy *et al.*, 2008). These methods are purely sampling methods and are linked, as with spot samples, with classical laboratory analysis. Biomonitoring involves placing caged animals, typically bivalve molluscs, at sites of interest, and measuring the accumulation of substances of concern. In order to achieve this, it is necessary to collect the organisms from a clean site and to allow them to depurate in order to remove as far as possible contaminants from the animals. It is necessary to take a large, representative sample before deployment to obtain a reliable measure of background levels of contaminants at the beginning of the exposure period. After some time, the animals are removed from the environment, killed and extracted to measure the concentrations of contaminants present, and these are usually expressed relative to the lipid content of the organisms. It is difficult to determine in some cases whether the concentrations represent equilibrium concentrations or not. After correction for the background levels, these figures give a measure of the rate of uptake, and hence the environmental concentrations of pollutants. However, there is some uncertainty concerning some substances since they may be eliminated by the test organisms. If the rate of elimination of a contaminant matched the uptake rate then there would be no evidence of the presence of that substance in the water. It is therefore necessary to have reliable 'calibration' data for the pollutants of interest or concern. This is one of the weaknesses of this approach. Another is the limited range of environments in which organisms can be placed. It would not be possible to place them in areas receiving discharges with, for instance, high concentrations of toxic substances, extreme pH values, low oxygen tensions or high carbon dioxide concentrations. The method does have one major advantage; the information that it provides on concentrations of contaminants is directly biologically relevant.

Passive samplers were originally designed to mimic living organisms, and early versions comprised dialysis membranes containing organic solvents. One of the longest-established samplers, SPMD, comprises a bag made of lay-flat polyethylene tubing that contains triolein (Huckins *et al.*, 2006). The lipid was intended to accumulate nonpolar organic pollutants in a way similar to that observed in aquatic animals. In recent years a range of passive sampling devices has been developed to monitor metals, non-polar organics, polar organics and organometallics (Table 3.1.2). All of these devices have the same basic structure; a receiving phase with a high affinity (and ideally a high capacity) for the analytes of interest, separated from the bulk aquatic environment by a diffusion-limiting layer that includes the water boundary layer. The natures of the diffusion-limiting layer and receiving phase are different for the different sets of analytes. Receiving phases for samplers for nonpolar organics include triolein, low-density polyethylene, silicon rubber, C_{18} chromatographic phase and solvents such as

Table 3.1.2 Examples of passive sampling devices available for monitoring the chemical quality of environmental waters (Greenwood *et al.*, 2007; Ouyang and Pawliszyn, 2007; Seethapathy *et al.*, 2008). LDPE: low-density polyethylene; PDMS: polydimethylsiloxane; SDP-RPS: styrenedivinyl-benzene – reverse phase sulphonate

Class of substances monitored	Sampler	Receiving phase	Diffusion limiting layer
Nonpolar organics	SPMD	Triolein	LDPE
	LDPE strips	LDPE	Water boundary layer
	Silicone strips	Silicone rubber, PDMS	Water boundary layer
	Ecoscope	Hexane	Microporous cellulose
	MESCO	PDMS	LDPE, microporous regenerated cellulose
	Chemcatcher®	C_{18} Empore™ disk	Nonporous LDPE
	Ceramic dosimeter	Amberlite IRA-743, Dowex Optipore L-493	Ceramic membrane
	Fibre in needle SPME	PDMS rod or membrane	Still water gap
Organotin compounds	Chemcatcher®	C_{18} Empore™ disk	Microporous cellulose acetate
Polar organics	POCIS	Mixtures or single sorbents including Oasis HLB, ENV⁺, Ambersorb 1500 dispersed on S-X3 BioBeads	Microporous polyether sulphone
	Chemcatcher®	C_{18} Empore™ disk SDP-RPS Empore™ disk	Microporous polyether sulphone
Volatile organic compounds	Diffusion samplers	Hydrophobic carbonaceous resins, polymeric resins	LDPE, Gore-Tex®, regenerated cellulose, polysulphone, silicone polycarbonate, porous polyethylene
Inorganics	DGT	Chelating agent (e.g. Chelex 100)	Polyacrylamide hydrogel with protective filter membrane
	Chemcatcher®	Chelating Empore™ disk	Microporous cellulose acetate

n-hexane. For polar organics, polymers such as styrenedivinylbenzene (SDB), various derivatives of this polymer with polar substituents, and carbon have been used. For metals, chelating agents are widely used. The diffusion-limiting membranes include the water boundary layer, low-density polyethylene, polyethersulphone, cellulose acetate and polyacrylamide hydrogels.

All of the variants of passive samplers function in the same way. The receiving phase maintains a concentration of the analytes of interest close to zero, and so the analyte diffuses from the bulk environmental phase at a rate determined by the concentration in the bulk phase and the dimensions and properties of the sampler. The curves describing the accumulation of analyte with time therefore have the form of an exponential approach to a maximum. In the region of the curve between zero time and the time to reach half of the equilibrium level the samplers operate in kinetic mode, and a linear relationship is assumed. In the region between the half time and equilibrium it is necessary to fit a nonlinear function for which more calibration parameters are needed. When equilibrium is reached, any information about the environmental concentrations of pollutants at early times during the deployment are lost. Equilibrium samplers are mainly used where environmental concentrations change only slowly, and have been successfully deployed to monitor a range of organic pollutants in ground water. Where concentrations of pollutants fluctuate markedly, samplers are generally used in the kinetic mode to measure TWA concentrations (Booij *et al.*, 2007).

In order to be able to use passive samplers quantitatively in the field it is necessary to obtain calibration data. This is usually provided from laboratory experiments, and several approaches have been used. For the diffusive gradients in thin films (DGT) samplers that are used to monitor inorganic pollutants, the calibration data supplied are in the form of diffusion coefficients of the analytes of interest in the hydrogel diffusion-limiting layer over a range of temperatures (Zhang and Davison, 1999). For the Chemcatcher® sampler that is used to monitor inorganics, polar organics and non-polar organics, the calibration data are in the form of a series of sampling rates (the volume of water cleared of a particular analyte per hour) for the analytes of interest, measured in a range of temperatures and turbulence conditions in flow-through calibration tanks (Vrana *et al.*, 2006). Similar approaches to calibration are used with the SPMD and the membrane-enclosed sorptive coating (MESCO) devices used for monitoring nonpolar organic pollutants. In recent years the use of performance reference compounds (PRCs) with Chemcatcher®, MESCO and SPMD samplers for nonpolar organic compounds has increased confidence in the field monitoring data obtained using them. PRCs are used to compensate for variations in temperature and turbulence during deployment. This approach involves measuring the offloading rates of labelled standard compounds loaded on to the receiving phase before deployment. Since the offloading and uptake processes are isotropic, the offloading rates can be used as an *in situ* calibration that can predict uptake rates without measuring temperature and turbulence (Booij *et al.*, 2007; Vrana *et al.*, 2007). The PRC approach may also be able to compensate for the impact of biofouling on uptake rates.

Both passive sampling and biomonitoring have strengths and weaknesses. Their main advantages are that they can provide information on average concentrations of pollutants over periods of time from weeks to months, and because they accumulate analytes over a period of time, they accumulate larger amounts than would be present in standard low-volume (< 5 L) spot samples. This has the advantage of bringing some contaminants that are present at only trace levels above the levels of detection and quantification to provide quantitative data, where in spot samples the only information that would be available would be 'below the level of detection'. Biomonitoring and passive sampling provide representative data on the freely dissolved (biologically relevant) fraction of pollutants. Where filter feeders are used in biomonitoring, the uptake may reflect material taken in with the food as well as that diffusing through the body surface (mainly the respiratory surface). This information is provided without the repeated field visits that would be necessary to obtain the equivalent information by frequent spot sampling.

There are some disadvantages associated with biomonitoring. One of the main problems is that the sample preparation for analysis of analytes of interest is difficult and expensive because of the high lipid and protein content of tissues. Since representative samples need to be analysed before and after deployment, this involves large amounts of tissue, and matching volumes of organic solvents. Calibration of uptake by living organisms is difficult because of large variation between individuals, seasonal variation in physiological status (e.g. throughout the reproductive cycle), and changes in tolerance and rates of excretion because of previous exposure to the pollutants of interest. Biofouling of the diffusion-limiting membrane surface is a problem in passive sampling. The growth of mainly microorganisms to produce a biofilm over the sampling surface can reduce the rate of uptake of pollutants by increasing the length of the diffusion path, and possibly by increasing the resistance to diffusion. A further complication is that the biofilm may accumulate or metabolise the pollutants, thus further reducing the uptake by the sampler. Since biofilms can vary enormously between sites and within sites with time, depending on, amongst other factors, water quality and temperature, it is difficult to account for this in any laboratory-based calibration procedures.

In passive sampling, as with biomonitoring, there is a need to establish levels of contaminants present in the samplers before deployment, and it is essential to use adequate numbers of laboratory and field blanks. However, in contrast with biomonitoring, for many of the passive sampler designs the preparation for analysis is simple, and requires only small volumes of solvents. There is a wide range within the various designs. On one extreme, the MESCO uses no solvents, and the receiving phase (a small silica fibre coated with stationary phase) is introduced directly into the injection port of a gas chromatograph for thermal desorption of the sample. On the other extreme is the SPMD, which requires large volumes of solvents and a number of preparatory steps to extract analytes from the triolein receiving phase, and subsequently remove all traces of the lipid before analysis. However, even the latter is much simpler and less costly than the equivalent process for biomonitoring organisms. There is a range of methods for calibrating the various passive samplers. The availability and validation of calibration data is a limiting area for passive sampling, and although much progress has been made, there is an urgent need for further work in this area, particularly for samplers for polar organic analytes where PRCs are not available. The development of a standard (BSI PAS-61) (British Standards Institute, 2006) for field deployment of passive samplers has been helpful in spreading the reliable application of this technology.

3.1.2.2   Emerging Methods for Replacing Classical Analysis

A wide range of instruments is commercially available for measuring physical and physicochemical variables such as dissolved organic carbon (DOC), pH, temperature and turbidity in water. Many of these have been available in the laboratory for many years, and have become more robust and reliable with technological developments in electronics, fibre optics and electrode design, which have enabled miniaturisation and the achievement of a stable performance. These instruments have been modified for use in the field. Special protective housings with built-in robust power supplies, computer controllers and communication systems have been designed to permit long *in situ* deployments of the instruments with minimal maintenance. Since these are well established and widely commercialised they are beyond the scope of this chapter.

Other methods of measuring water quality are far less mature and well established, but have the potential to provide data less expensively and/or more quickly than conventional laboratory analytical methods, and some can provide information in the field or *in situ*. However, like all methods they have strengths and weaknesses, and for most of them there is a lack of reliable validation data, and robust correlations with standard methods are rare. These methods can be separated into two main groups: those involving the use of living organisms or isolated biological systems, and those based on non-living systems. Biological assessment techniques, such as toxicological assays, the use of biomarkers to assess exposure to pollutants, and BEWS that detect deleterious changes in water quality to provide an alert, provide information on water quality but not on concentrations of individual pollutants. Other methods, such as immunochemical methods, test kits and sensors (including biosensors), can provide quantitative information on the levels of specific contaminants present in water. Some depend on spot sampling and are laboratory-based or field-based; others can be used *in situ*.

Immunochemical methods offer a means of analysing small volumes of sample in a high-throughput screening format that enables simultaneous analysis of multiple samples. These methods can be automated. They therefore have the potential for some analytes to replace conventional analytical methods in the laboratory. Since robust, portable instruments are now available for use with multiple well plate or tube formats in which immunoassays can be effected, it is possible to use these methods in the field. Although these methods depend on spot sampling, they require the transport of much smaller samples than are needed for classical analytical methods, and when used in the field avoid the potential changes that can take place during transport and storage of bottle samples. Specific antibodies are becoming available for a wide range of priority substances, including pesticides and some metals (Blake *et al.*, 1998; Farré *et al.*, 2005).

Test kits are normally based on methods that use reagents to produce a coloured product that can be measured with a spectrophotometer or colorimeter. These have been widely applied for components such as nitrate and ammonia. These methods often use reagents and equipment provided by a supplier who is responsible for ensuring the necessary quality and consistency of the components of the kit. In some cases, advantage has been taken of the miniaturisation of the analytical instruments and improvements in their robustness to develop systems that can be used in the field.

Some direct spectroscopic methods are available for direct measurement of contaminants in water. Some, such as the laser-induced fluorescence spectroscopy, are available in miniaturised format because of developments in laser fibre optics, and have been applied to a range of aromatic hydrocarbons, including some of the priority pollutants. UV spectroscopy has also been applied for analysing contaminants (including nitrate and the detergent dodecyl benzene sulphonate) in water samples, and again robust, miniaturised instruments are available for field applications (Roig *et al.*, 2007).

In recent times, a range of electrochemical sensors based on voltamperometric probes and selective or ion-selective electrodes has become available for the routine measurement of metals and some pesticides. These methods require little or no sample treatment, and the development of miniaturised screen-printed electrodes (with working electrodes of either graphite or gold, and silver reference electrodes) has permitted the production of robust handheld equipment for deployment on-site (Hanrahan *et al.*, 2004; Honeychurch and Hart, 2003). Usually the electrodes are used with spot samples, but some can be used *in situ*. In recent times there has been a large research effort put into developing biosensors for measuring a wide range of environmental contaminants. The early work in this area focused on biomedical applications. These sensors combine a biological component (e.g. enzymes, DNA, antibodies, receptors) that provides the specific recognition and an electronic transducer that converts changes in the biological component into a detectable signal (electrical or optical). Some sensors have sensitivities in the nanomolar to micromolar range, and excellent specificity (Kim *et al.*, 2006; Tschmelak *et al.*, 2005). Examples include an enzyme-based biosensor that uses inhibition of an acetylcholine esterase linked to an amperometric system to measure the organophosphate (OP) insecticide dichlorvos (Gosh *et al.*, 2006). This sensor has been demonstrated to measure this OP at nanomolar concentrations in river water. Many biosensors use an immunological element as the receptor linked to fluorescence detectors, and have been demonstrated to measure concentrations of a range of pesticides, including atrazine and isoproturon (Krämer *et al.*, 2005, 2007), and a range of phenols. This technology shows great promise, and if developed to produce robust sensors that can be deployed remotely over extended periods with telemetric reporting, as are available for measuring physical and physicochemical variables, would provide an ideal monitoring system. However, currently most of this technology has been shown to work only in the laboratory, usually in clean samples, but only a few have been transferred from the laboratory to deal with field samples. The potential impact of the wide range of matrix effects found in field samples has not been fully investigated.

All of these emerging analytical methods can be used in the laboratory with relatively small sample volumes and with little sample pretreatment, but the main potential advantages are associated with the availability of rapid, low-cost measurements in the field. These technologies can deliver data rapidly. In contrast with classical methods where transport to the laboratory is involved, these methods can provide rapid mapping of the distribution of pollutants in both space and time, and can follow rapid changes after, for instance, a pollution event. However, whilst these may be used in investigative monitoring, currently they all have limitations that prevent them from being incorporated into routine monitoring in support of legislation. One drawback is that most still depend on spot samples. Some handheld devices can be used directly in a water body, but for these it is even more difficult to standardise operating procedures (e.g. speed of stirring), and hence to achieve a full and thorough validation.

The most important problems are that the uncertainties associated with the measurements are not well defined, and it is difficult to transfer laboratory calibrations to the field because of matrix effects and the difficulty in controlling operating conditions in the field. This is an area where research and development activity is urgently needed to develop a framework for the thorough field validation of these new technologies. Without this information, this potentially valuable set of tools will not be available to the regulators.

3.1.2.3   Biological Methods

Some tools provide a different type of information and not concentrations of individual pollutants. These methods are based on biological responses to pollutants in the aquatic environment, and include the use of biomarkers, measures of ecological status and bioassays. The latter tend to be laboratory-based and to depend on spot sampling.

Biomarkers are changes in an organism that can be attributed to exposure to or the toxic effects of environmental contaminants. Three categories of biomarkers are recognised: of exposure, of effect and of susceptibility (Livingstone *et al.*, 2000). The first category includes detection and measurement of an exogenous substance, its metabolites, or the result of its impact on target molecules or cells in an organism. Molecular biomarkers of exposure are mainly proteins involved in the defence of the living system from the potentially harmful effects of toxicants and include: transporters that remove toxic molecules from cells or body compartments surrounded by epithelial layers, detoxification enzymes and chaperon-proteins involved in binding toxic molecules. Examples include heat shock proteins, which increase in response to a wide range of stressors; cytochrome P-450-linked oxygenases, which are induced by a range of important organic pollutants (e.g. polycyclic aromatic hydrocarbons, polychlorinated biphenyls, dioxins and some classes of pesticides); and metallothioneins, which are induced in response to exposure to heavy metals, and bind heavy metals to regulate their concentrations in body compartments (Vasseur and Cossu-Leguille, 2003). The second category includes biochemical, physiological and other alterations within tissues or body fluids of an organism that are associated with an impairment of health or with a disease state resulting from interactions with exogenous chemicals (Allan *et al.*, 2006b; Galloway *et al.*, 2004). Molecular biomarkers of effect include changes in the integrity of DNA molecules or in the integrity of cell membranes (e.g. through peroxidation of lipids) that can cause, for instance, deleterious changes in intracellular redox state. Changes in the size and numbers of lysosomes, and decreases in the activity of antioxidant enzymes (e.g. superoxide dismutase, glutathione peroxidase or glutathione reductase), are other biomarkers in this category. Inhibition of acetylcholine esterase is a specific marker of exposure to carbamate and OP compounds that are used as insecticides, and the induction of vitellogenin production in male fish is a specific marker of endocrine disruption produced by compounds with feminising properties (e.g. oestrogenic hormones and some pesticides). The third category, biomarkers of susceptibility, comprises the inherent or acquired ability of an organism to respond to exposure to a specific exogenous substance, and includes genetic factors and changes in receptors that alter the susceptibility of an organism to that substance (Dixon *et al.*, 2002). This is a new and rapidly developing area that is benefiting from developments in molecular biological technologies. It is based on knowledge of signal transduction or mechanisms involved in protecting cells that have been exposed to xenobiotics or a mixture of such substances. It involves detecting the induction of genes and variations in gene expression, or the modulation of associated enzymatic activity. An example is provided by paraoxonase (PON1). This is a liver and plasma enzyme involved in the oxidation of lipids, and has been shown to be a sensitive marker of exposure to OP insecticides. Since these changes are found at concentrations of toxicants below those causing cytotoxicity, they can provide early warnings of potential harm long before they would be detected at the whole organism level, and certainly well in advance of damage at the population level. Biomarkers have the potential to be incorporated into high-throughput screen formats using robotics. However, a large amount of work will be required to fully validate these methods and to provide regulators with a sound understanding of the significance of the results before they can be used routinely in support of legislation.

Whole-organism bioassays have been used for many years in the pharmaceutical and pesticide industries, and in ecotoxicological studies. They have been based on a wide range of organisms, including animals, plants and microorganisms, and can provide information on the effects of short-term (acute) and long-term (chronic) exposure to toxicants. The endpoints used include mortality and sub-lethal effects such as changes in locomotory or feeding activity. Since organisms will respond only to that fraction of a toxicant that is bioavailable, these assays provide direct measures of the biological impact of contaminants present in a water sample, and it is not necessary to have knowledge of the substances present or of the concentrations of the individual contaminants. Further interactions between components of complex mixtures will be detected. There is, however, great variability due to intra-specific (inter-individual) variation in tolerance to toxicants, and this can be affected by previous exposure of individuals to specific substances. In laboratory-based tests this is usually minimised by using carefully selected individuals from standard laboratory-bred strains of test organisms. The danger with this approach is that the organisms may no longer be representative of organisms in the environment. Toxicity can also be affected markedly by factors such as pH, water hardness and dissolved oxygen. In laboratory measurements of the toxicity of individual substances these matrix effects can be tightly controlled; however, when field samples are used this is not possible. In order to minimise expense, and to assay a large number of samples, these methods tend to be short-term (24–48 hours) and can miss deleterious effects of chronic exposure.

There is a very large variation in tolerance between species, and where direct toxicological methods are used to protect bodies of water it is important to choose a test species that shows the lowest possible tolerance (maximum sensitivity) to as wide as possible a range of pollutants. In practice, several test species from different trophic levels are used in order to fingerprint the water samples in terms of the differential tolerances of the various species, and it is necessary to use different sets of organisms for the main divisions (freshwater, mixohaline waters and marine) of the aquatic environment. Several microbiological assay systems are available. These have the advantage of using large numbers of organisms for each dilution of the sample. These assays include *Vibrio fisheri* (Microtox® from SDI Europe Ltd) and *Pseudomonas putida*, or microorganisms present in activated sludge. In these cases, the toxicological end-points are depression of bioluminescence, changes in metabolic status, and growth, respectively (Allan *et al.*, 2006; Wolska *et al.*, 2007). The *Vibrio fischeri* biolumines-cence inhibition is the most common test; it is relatively simple to implement, and a large database of results for many chemicals and a standard (ISO 11348) protocol are available. One advantage of a commercially available system is that a supply of highly standardised organisms (usually lyophilised) is available. Standard protocols are also available for some green algae (e.g. *Selenastrum capricornutum* or *Pseudokirchneriella subcapitata*) that are used to measure the toxicity of herbicides that reduce photosyn-thetic activity and/or growth (Wadhia and Thompson, 2007). A further advantage of methods using microorganisms is that the assays can be miniaturised and automated. Other organisms that are commonly used include a number of crustaceans, such as *Daphnia magna* and *Gammarus pulex*, and the embryos of the mollusc *Crassostrea gigas*. Here the toxicological endpoints used include mortality, changes in feeding and locomotory activity (Gerhardt *et al.*, 2007).

Invertebrates are commonly used in chronic toxicity assays. Here endpoints include changes in growth rate or survival of crustaceans (e.g. *Hyalella azteca*, *Gammarus sp.* and daphnids), chironomid larvae (*Chironomus riparius*), oysters (*Crassostrea gigas*) and many other organisms under controlled conditions (Allan *et al.*, 2006b; De Lange *et al.*, 2005). Higher organisms such as fish have been used for risk assessment pur-poses in 96 hour exposure trials.

In order for these methods to become acceptable for use within the regulatory frame-work they need to be inexpensive, simple to implement, validated and used following standardised protocols. It is also necessary to demonstrate the relevance of the data provided by these assays to a wide range of species and at population levels. A couple of potentially useful applications would be the control of the toxicity of consented discharges, and monitoring changes in toxicity following accidental or illegal pollu-tion events. One disadvantage is that most of these methods rely on expensive spot sampling, though some of the microbiological assays have been miniaturised for use in the field. Some have been adapted for use *in situ* (e.g. a test using the alga *P. subcapitata* immobilised in alginate beads) and online (e.g. the *Daphnia* toximeter, the Multispecies Freshwater Biomonitor (MFB) based on *Gammarus pulex*, and the BBE Algae Toximeter). A recent promising approach in the Netherlands has used a battery of bioassays to measure the toxicity of raw effluents or concentrated extracts of surface waters, and has used the profile of responses of the various assays to indicate the nature of the toxicants present (Maas and van den Heuvel-Greve, 2004). One of the most important advantages is that these methods measure only the biologically relevant fraction of pollutants present in the water.

Field investigations into the structure of a population of an indicator species or the ecological status of a water body are time-consuming and expensive. They use a range of monitoring methods, and since the outcome of a survey depends heavily on the meth-ods used, some ISO and CEN standards are available to provide guidance on sampling protocols, especially for fish and invertebrates. Environmental quality is measured in terms of species diversity, or of population size and structure for key groups of organ-isms that are used as indicators of ecological health. The data from a survey site are compared with the expected status based on surveys at pristine reference sites (Logan,

2001). Several computer-based systems are available to carry out the comparisons between reference and survey data for various groups of organisms. Some use benthic invertebrates (RIVPACS, AQEM, ECOPROF, AusRivAs), fish (CITYFISH, FAME) or a combination of invertebrates, algae, fish and macrophyte (ECOFRAME and PAE-QANN) (Allan *et al.*, 2006b). Most of these tools are for freshwater systems. One major problem with this approach is that the flora and fauna in different European rivers and in different regions of one river can vary significantly. Hence it is difficult to achieve true comparability of data even within a river basin, and whilst the various computer-based tools use similar processes and statistical methods, they differ in the reference conditions used. The main advantage of this approach is that it provides direct measures of environmental health. The main problems are that the structure of populations in an ecosystem can fluctuate naturally with changes in weather and hydro-morphological factors, and there is a large natural variation involved. A disadvantage of this approach is that if problems are detected it is often difficult to attribute them to specific causes, since changes at the ecosystem level often take a long time to develop, and further, by the time the damage is detected it is often difficult and expensive to remediate. In some cases the time necessary for a system to recover can span years or decades.

These biologically-based methods have some important advantages over the current approach of sampling combined with chemical analysis. The most obvious of these is that the information provided is directly biologically relevant. A further advantage is that they are generally not focused on individual pollutants in a list, and take into account interactions between pollutants. They can provide early warnings of potential problems and allow preventative measures to be put in place before long-term damage is caused. These are true screening methods that can be used to focus expensive conventional monitoring where it is needed, and improve its cost-effectiveness by concentrating efforts in areas where there is a problem. Further, these methods can provide direct information on the effectiveness of remedial actions.

On the other hand, the biologically-based methods also have limitations, and one is associated with limited range of distribution of some species, so that their use is limited to a narrow range of external conditions. For instance, it is not possible to deploy freshwater organisms in mixohaline or marine environments, and some organisms are not able to tolerate a wide range of temperatures, as is found between different areas of Europe and within a region at different times of the year. Most organisms are not able to tolerate the high concentrations of contaminants, low dissolved oxygen concentrations and high temperatures associated with some industrial and domestic effluents, and it is not possible to place organisms in such environments. Where spot samples are taken to use in toxicological assays it is necessary to dilute them before introducing test organisms. A further drawback is that whilst these assays indicate the presence of a problem, they do not necessarily give guidance on the nature or source of the problem. Another important limitation is associated with the high cost of chronic exposure assays, since this dictates that routine assays are short-term (acute exposures) and last only 24–48 hours. Such assays miss the effects of prolonged exposure to low concentrations. Not all organisms are equally sensitive to any contaminant, and some of those used show tolerance to some important classes of pollutants. It is therefore difficult to find a single organism that can be used for all classes of pollutant, and which can provide representative information that will safeguard human health and all of the species present in the aquatic environment. Here the use of a battery of assays, as discussed above, may prove useful (Dalzell *et al.*, 2002; Maas and van den Heuvel-Greve, 2004). The data provided is qualitative, and is associated with large uncertainty because of the added factor of biological variation (between individuals and between species).

The cost of biological methods tends to increase with the complexity of the system used and the timescale of the assay. For instance, assays based on isolated biochemical systems, cells and tissues are rapid (minutes to hours) and relatively inexpensive; those based on individual organisms are more costly and take longer (days) to obtain the data. Measurements at population and ecosystem level are very costly and time-consuming. Whilst rapid assays provide early warnings of potential problems and permit timely remedial actions, when a problem is detected only at population or ecosystem level it is usually too late to take preventative action, and the cost of remediation is very high and the timescale for recovery very long. It also becomes increasingly difficult to associate biological effects with a specific causative agent as the complexity and timescale increase.

3.1.3   THE WAY FORWARD

Some of the methods discussed in this chapter could make useful contributions within the water monitoring programmes that are being developed across the European Union Member States. Potential roles have been identified in investigative monitoring that aims to identify causes when water bodies fail to reach the environmental objectives of the WFD, and in operational monitoring that aims to provide additional and essential data on water bodies at risk or that are failing to meet the necessary quality for identified reasons. However, there is still a reluctance to consider the use of emerging methods in surveillance monitoring that is focused on assessing long-term water quality changes and providing baseline data on river basins allowing the design and implementation of other types of monitoring. It is here that there is potential for these methods to make a valuable contribution by providing good-quality, temporally- and spatially-representative data at an affordable cost within national monitoring budgets. Much work is necessary to demonstrate the validity and cost-effectiveness of these methods before this can happen.

All of the emerging tools (in common with classical methods) have strengths and weaknesses. Some may be more cost-effective than others and this may become one of the most important factors in determining their application in support of WFD. However, in order for some of the emerging methods to be adopted for use in support of the legislation it will be necessary to develop new approaches to validation and QA, and to rethink the formulation of EQS. Whatever methods are selected for use in monitoring, they must yield reliable, consistent, representative fit-for-purpose data that are directly comparable between repeated samples, and most importantly between laboratories and EU Member States. Most of the reservations that monitoring authorities have concerning the adoption of new methodologies and technologies are associated with quality issues. The scientific community has not made much progress in defining the uncertainties associated with measurements made with some of the promising emerging methods, and much work in this area is urgently needed. In some applications a large uncertainty may be acceptable, but only if it is well defined. This field is now lagging behind the analytical chemistry area, where remarkable improvements were made through the development of appropriate QA and QC procedures, the establishment of widely accepted standards and the development of fit-for-purpose reference materials, which are essential to underpin this area (Quevauviller, 1999). For some methods, such as the use of immunoassays, a conventional approach to QA and QC may be tenable, and some preliminary efforts in this direction have been encouraging (Brunori *et al.*, 2007; Gonzalez *et al.*, 2007; Krämer *et al.*, 2007). However, in some cases, such as passive sampling, it is not possible to use the currently available certified reference material because of the large volumes (hundreds of litres) required and the associated expense. For such methods it will be necessary to develop some other form of reference materials, or carefully regulated reference sites.

Another factor that is holding back change is a failure to address the relevance or significance of what is being measured by the various methods, including bottle sampling of water combined with classical analytical methods. The data obtained are in some cases operationally defined, and their biological significance is not always taken into consideration. For instance, bottle sampling will provide different information depending on whether the sample is filtered, ultrafiltered or untreated before analysis. Further, the effects of the various pretreatments will vary depending on the physicochemical properties of the analyte. For some very nonpolar compounds, such pretreatments can significantly disturb the balance between bound and free analyte, and can reduce the concentration of freely dissolved material where it binds significantly to filter components or glassware (Lung *et al.*, 2000). In general, passive samplers measure only the freely dissolved fractions, and these are often regarded as the biologically available (significant) fractions. If emerging methods are to be used in a regulatory environment, EQS must be defined in terms of what those methods measure (e.g. concentration of freely dissolved pollutant, toxicity in specific assays). The current practice of using spot sampling and classical analysis as a sort of 'gold standard' is not helpful, particularly when comparing methods such as passive sampling with bottle sampling, since the latter can yield highly unrepresentative data when there are marked temporal fluctuations (Allan *et al.*, 2006), as are routinely found in sewage works, where inputs and hence outputs can vary on a diurnal basis. There is an urgent need for long-term (several years) comprehensive studies to establish the relationship between the data yielded by emerging methods and those provided by currently accepted methodologies (Booij *et al.*, 2006).

It is important to formulate a clear definition of what is a good measure of water quality, and to identify the most important variables that should be measured. Without this it is not possible to define a relevant error or estimate uncertainty. Whilst a concentration estimated by classical laboratory analysis of a sample of water may have a relatively low uncertainty associated with it, there may be a huge uncertainty associated with this estimate if it is meant to provide a measure of the TWA concentration. In some cases, where bias can be introduced during the various steps from sampling to analysis, it may be better to be roughly right than precisely wrong. It is important for the scientific community involved in developing the application of emerging methods to assess the relevance of monitoring activities and the data that these provide in terms of the aims of the WFD, where the key objectives are to improve and safeguard the quality and availability of water across all of the Member States. While the focus is on compliance with EQS it will be easy to lose sight of the long-term goals of the legislation and to fail to notice or measure compounds that are not on a priority list, since we do not generally find what we do not look for. However, in order for emerging monitoring methods to be useful in changing the approach to monitoring and to provide support for regulatory bodies, they need to be fully validated, and the uncertainty associated with the data they yield defined. This is a key area where much work is needed to carry these new technologies and methodologies forward from the laboratory to routine application in the field. Further, the nature and environmental relevance of the information provided by the various methods needs to be clearly understood. Without this, regulators will not have access to a well-stocked tool box of methods that will allow them to make a well-informed selection of the most appropriate tool for any particular monitoring task, and to obtain the representative information that they need at an affordable cost. Without good-quality, representative data, river basin management plans, associated risk assessments and planned remedial actions could be flawed. The cost of inappropriate actions based on inadequate information could be very high.

3.1.4   ACKNOWLEDGEMENTS

The authors are grateful for the European Union funding that supported the work upon which this chapter is based (STAMPS: EU 5th Framework contract no. EVK1-CT-2002-00119 and SWIFT-WFD: EU 6th Framework contract no. SSPI-CT-2003-502492). However, the views expressed in this chapter are those of the authors alone.

REFERENCES

Allan, I.J., Mills, G.A., Vrana, B., Knutsson, J., Holmberg, A., Guigues, N. *et al.* (2006a) Intelligent monitoring for the European Water Framework Directive, *Trends Anal. Chem.*, **25**, pp. 704–715.
Allan, I.J., Vrana, B., Greenwood, R., Mills, G.A., Roig, B. and Gonzalez, C. (2006b) A 'toolbox' for biological and chemical monitoring requirements for the European Union's Water Framework Directive, *Talanta*, **69**, pp. 302–322.
Blake, D.A., Blake, R.C. II, Khosraviani, M. and Pavlov, A.R. (1998) Immunoassays for metal ions, *Anal. Chim. Acta*, **376**, pp. 13–19.
Booij, K., Smedes, F., Van Weerlee, E.M. and Honkoop, P.J.C. (2006) Environmental monitoring of hydrophobic organic contaminants: the case of mussels versus semipermeable membrane devices, *Environ. Sci. Technol.*, **40**, pp. 3893–3900.
Booij, K., Vrana, B. and Huckins, J.N. (2007) Theory, modelling and calibration of passive samplers used in water monitoring, In: Greenwood, R., Mills, G.A. and Vrana, B. (eds) *Passive Sampling Techniques in Environmental Monitoring*, Comprehensive Analytical Chemistry Vol. 48, Elsevier, Amsterdam.
British Standards Institute (2006) Determination of priority pollutants in surface water using passive sampling, Publicly Available Specification PAS 61:2006.
Brunori, C., Morabito, R., Ipolyi, I., Pellegrino, C., Ricci, M., Bercaru, O. *et al.* (2007) The SWIFT-WFD proficiency testing campaigns in support of implementing the EU Water Framework Directive, *Trends Anal. Chem.*, **26**, pp. 993–1004.

Dalzell, D.J.B., Alte, S., Aspichueta, E., de la Sota, A., Etxebarria, J., Gutierrez, M. *et al.* (2002) A comparison of five rapid direct toxicity assessment methods to determine toxicity of pollutants to activated sludge, *Chemosphere*, **47**, pp. 535–545.

De Lange, H.J., De Haas, E.M., Maas, H. and Peeters, E.T.H.M. (2005) Contaminated sediments and bioassay responses of three macroinvertebrates, the midge larva *Chironomus riparius*, the water louse *Asellus aquaticus* and the mayfly nymph *Ephoron virgo*, *Chemosphere*, **61**, pp. 1700–1709.

Dixon, D.R., Pruski, A.M., Dixon, L.R.J. and Jha, A.N. (2002) Marine invertebrate eco-genotoxicology: a methodological overview, *Mutagenesis*, **17**, pp. 495–507.

Environment Canada (1995) *Effluent dispersion in the Fraser River from the Glenbrook Combined Sewer overflow at New Westminster, British Columbia*, DOE FRAP 1995-22, p. 48.

Farré, M., Brix, R. and Barcelo, D. (2005) Screening water for pollutants using biological techniques under European Funding during the last 10 years, *Trends Anal. Chem.*, **24**, pp. 532–545.

Galloway, T.S., Brown, R.J., Browne, M.A., Dissanayake, A., Lowe, D., Jones, M.B. and Depledge, M.H. (2004) A multibiomarker approach to environmental assessment, *Environ. Sci. Technol.*, **38**, pp. 1723–1731.

Gerhardt, A., Kienle, C., Allan, I., Greenwood, R., Guigues, N., Fouillac, A.-M. *et al.* (2007) Biomonitoring with Gammarus pulex at the Meuse (NL), Aller (GER) and Rhine (F) rivers with the online Multispecies Freshwater Biomonitor, *J. Environ. Monit.*, **9**, pp. 979–985.

Ghosh, D., Dutta, K., Bhattacharyay, D. and Sarkar, P. (2006) Amperometric detection of pesticides using polymer electrodes, *Environ. Monit. and Assess.*, **119**, pp. 481–489.

Gonzalez, C., Spinelli, S., Gille, J., Touraud, E. and Prichard, E. (2007) Validation procedure for existing and emerging screening methods, *Trends Anal. Chem.*, **26**, pp. 315–322.

Greenwood, R., Mills, G.A. and Vrana, B. (eds) (2007) *Passive Sampling Techniques in Environmental Monitoring*, Comprehensive Analytical Chemistry Vol. 48, Elsevier, Amsterdam.

Gunther, A.J., Davis, J.A., Hardin, D.D., Gold, J., Bell, D., Crick, J.R. *et al.* (1999) Long-term bioaccumulation monitoring with transplanted bivalves in the San Francisco Estuary, *Mar. Pollu. Bull.*, **38**, pp. 170–181.

Hanrahan, G., Patil, D.G. and Wang, J. (2004) Electrochemical sensors for environmental monitoring: design, development and applications, *J. Environ. Monit.*, **6**, pp. 657–664.

Honeychurch, K.C. and Hart, J.P. (2003) Screen-printed electrochemical sensors for monitoring metal pollutants, *Trends Anal. Chem.*, **22**, pp. 456–469.

Huckins, J.N., Petty, J.D. and Booij, K. (eds) (2006) *Monitors of Organic Chemicals in the Environment: Semipermeable Membrane Devices*, Springer, New York.

ISO 17025:2005 (2005) General requirements for the competence of testing and calibration laboratories.

Killard, A.J., Micheli, L., Grennan, K., Franek, M., Kolar, V., Moscone, D. *et al.* (2001) Amperometric separation-free immunosensor for real-time environmental monitoring, *Anal. Chim. Acta*, **427**, pp. 173–180.

Kim, S.J., Gobi, K.V., Harada, R., Shankaran, D.R. and Miura, N. (2006) Miniaturized portable surface plasmon resonance immunosensor applicable for on-site detection of low-molecular-weight analytes, *Sens. Actuators B*, **115**, pp. 349–356.

Krämer, P.M., Franke, A., Zherdev, A.V., Yanzynina, E.V. and Dzantiev, B.B. (2005) Comparison of two express immunotechniques with polyelectrolyte carriers, ELISA and FIIAA, for the analysis of atrazine, *Talanta*, **65**, pp. 324–330.

Krämer, P.M., Martens, D., Forster, S., Ipolyi, I., Brunori, C. and Morabito, R. (2007) How can immunochemical methods contribute to the implementation of the Water Framework Directive?, *Anal. Bioanal. Chem.*, **387**, pp. 1435–1448.

Livingstone, D.R., Chipman, J.K., Lowe, D.M., Minier, C. and Pipe, R.K. (2000) Development of biomarkers to detect the effects of organic pollution on aquatic invertebrates: recent molecular, genotoxic, cellular and immunological studies on the common mussel (*Mytilus edulis* L.), *Intern. J. Environ. Poll.*, **13**, pp. 56–91.

Logan, P. (2001) Ecological quality assessment of rivers and integrated catchment management in England and Wales, *J. Limnol.*, **60** (suppl. 1), pp. 25–32.

Louch, J., Allen, G., Erickson, C., Wilson, G. and Schmedding, D. (2003) Interpreting results from the field deployments of semipermeable membrane devices, *Environ. Sci. Technol.*, **37**, pp. 1202–1207.

Lung, S.-C., Yanagisawa, Y., Ford, T.E. and Spengler, J.D. (2000) Characteristics of sorption losses of polychlorinated biphenyl congeners onto glass surfaces, *Chemosphere*, **41**, pp. 1857–1864.

Maas, J.L. and van den Heuvel-Greve, M.J. (2004) Opportunities for bioanalysis in WFD chemical monitoring using bioassays, RIZA working document 2005.053X, Rijkswaterstaat RIZA, Lelystad, The Netherlands, p. 51.

Ouyang, G. and Pawliszyn, J. (2007) Configurations and calibration methods for passive sampling techniques, *J. Chromatogr. A*, **1168**, pp. 226–235.

Quevauviller, P. (1999) Reference materials: an inquiry into their use and prospects in Europe, *Trends Anal. Chem.*, **18**, pp. 76–85.

Roig, B., Valat, C., Berho, C., Allan, I.J., Guigues, N., Mills, G.A. *et al.* (2007) The use of field studies to establish the performance of a range of tools for monitoring water quality, *Trends Anal. Chem.*, **26**, pp. 274–282.

Salazar, M.H. and Salazar, S.M. (2006) Linking bioaccumulation and biological effects to chemicals in water and sediment: a conceptual framework for freshwater bivalve ecotoxicology, In: Farris, J.L. and van Hassel, J.H. (eds), *Freshwater Bivalve Ecotoxicology*, CRC Press, Boca Raton, FL, USA.

Seethapathy, S., Górecki, T. and Li, X. (2008) Passive sampling in environmental analysis, *J. Chromatogr. A*, **1184**, pp. 234–253.

Tschmelak, J., Proll, G. and Gauglitz, G. (2005) Optical biosensor for pharmaceuticals, antibiotics, hormones, endocrine disrupting chemicals and pesticides in water: assay optimization process for estrone as example, *Talanta*, **65**, pp. 313–323.

Van der Schalie, W.H., Shedd, T., Knechtges, P.L. and Widder, M.W. (2001) Using higher organisms in biological early warning systems for real-time toxicity detection, *Biosensors and Bioelectronics*, **16**, pp. 457–465.

Vasseur, P. and Cossu-Leguille, C. (2003) Biomarkers and community indices as complementary tools for environmental safety, *Environ. Intern.*, **28**, pp. 711–717.

Vrana, B., Mills, G.A., Dominiak, E. and Greenwood, R. (2006) Calibration of the Chemcatcher passive sampler for the monitoring of priority organic pollutants in water, *Environ. Poll.*, **142**, pp. 333–343.

Vrana, B., Mills, G.A., Kotterman, M., Leonards, P. and Greenwood, R. (2007) Modelling and field application of the Chemcatcher passive sampler calibration data for the monitoring of non-polar priority organic pollutants in water, *Environ. Poll.*, **145**, pp. 895–904.

Wadhia, K. and Thompson, K.C. (2007) Low-cost ecotoxicity testing of environmental samples using microbiotests for potential implementation of the Water Framework Directive, *Trends Anal. Chem.*, **26**, pp. 300–307.

Wolska, L., Sagajdakow, A., Kuczyńska, A. and Namieśnik, J. (2007) Application of ecotoxicological studies in integrated environmental monitoring: possibilities and problems, *Trends Anal. Chem.*, **26**, pp. 332–344.

Zhang, H. and Davison, W. (1999) Diffusional characteristics of hydrodrogels used in DGT and DET techniques, *Anal. Chim. Acta*, **398**, pp. 329–340.

3.2
Diagnostic Water Quality Instruments for Use in the European Water Framework Directive

**J.L. Maas, C.A. Schipper, R.A.E. Knoben,
M.J. van den Heuvel-Greve, P.J. den Besten and P.G.-J. de Maagd**

3.2.1   INTRODUCTION

The WFD is now the guiding policy framework for water quality management in Europe. The aim of the WFD is to realise protection and improvement of aquatic

The Water Framework Directive - Ecological and Chemical Status Monitoring   Edited by Philippe
Quevauviller, Ulrich Borchers, Clive Thompson and Tristan Simonart © 2008 John Wiley & Sons, Ltd

ecosystems and the sustainable use of water. It offers a framework and planning system for:

- Setting ecological targets for water bodies.

- Revealing problems and their causes.

- Deciding on measures.

- Monitoring quality to assess the effect of measures.

The ultimate goal is 'good ecological and chemical status' (GES and GCS) for all European waters by 2015. Under certain conditions, the deadline for the achievement of this goal may be extended to 2027 at the latest.

In this chapter the focus is laid on the ecological status. If a water system achieves a poor score on one or more of the ecological metrics or sub-metrics, the water manager must take action. In order to do so effectively and efficiently, insight is needed into why the water body in question has failed to achieve good ecological status. Often, several candidates – 'pressures' or 'stress factors', such as habitat change, the presence of chemicals or a change in predation pressure – may play a role.

To the authors' knowledge, no structured diagnostic system is available for identifying causes for a lack of ecological quality. The agencies responsible are forced to estimate the main problems encountered in their water bodies, and their causes, on the basis of the existing knowledge of burden and effect, coupled with their regional management expertise. To fill this gap, a Diagnostic Water Quality Instrument (DWQI) (Royal Haskoning, 2006) has been developed.

3.2.1.1   Position of the Diagnostic Instrument

The WFD requires three kinds of monitoring (Directive 2000/60/EC). First, *surveillance monitoring* (status and trend monitoring) assesses whether GES and GCS has been achieved. If this is not the case, *operational monitoring* is needed to assess the degree to which the actual status deviates from GES and GCS, and whether any measures taken have had an effect. If it is unclear what is causing the poor ecological status and what measures should be taken, the causes and appropriate measures must be identified by means of *investigative monitoring* (usually project-based).

For the surveillance and operational monitoring of surface water according to the European Water Framework Directive, a national guideline has been published in the Netherlands (Van Splunder *et al.*, 2006) based on the WFD CIS Guidance Document on Monitoring (WFD, 2003). Investigative monitoring is to be tailored to the specific case or problem under investigation. In some cases it will be more intensive in terms of monitoring frequency and will be focused on particular water bodies or parts of water bodies, and on relevant quality elements. The use of emerging tools, such as ecotoxicological monitoring, and assessment methods may in some cases be appropriate for investigative monitoring. The DWQI approach is linked to this last type (and phase) of monitoring. The DWQI enables water managers, through a step-by-step approach, to identify possible causes in cases where the reason for failing environmental objectives is unknown.

3.2.2 CONCEPTUAL FRAMEWORK OF THE DIAGNOSTIC INSTRUMENT

The DWQI focuses on the WFD system for monitoring and assessment of the ecological status of water bodies. The key issues are biological quality elements and their metrics. These quality elements define the quality of the water body in question. Within the WFD approach, physical-chemical and hydro-morphological variables can influence the ecological quality of a water system as well.

3.2.2.1 Conceptual Framework

The DWQI concept consists of a framework that encompasses the most important factors – or, rather, complex of factors – that influence the populations in a particular water body. The DWQI concept has been derived from the '5S system' (Verdonschot, 1995), which covers the most important factors or complexes of factors affecting the communities in brooks. This concept is also applicable to other types of water body if the factors structure and streaming are replaced by morphology and hydrology. In addition to this change in terminology, the concept has been further adapted to suit the purposes of the WFD. In the WFD, species and substances are the key factors, and physical-chemical, morphology and hydrology factors underpin the biological quality elements (Figure 3.2.1).

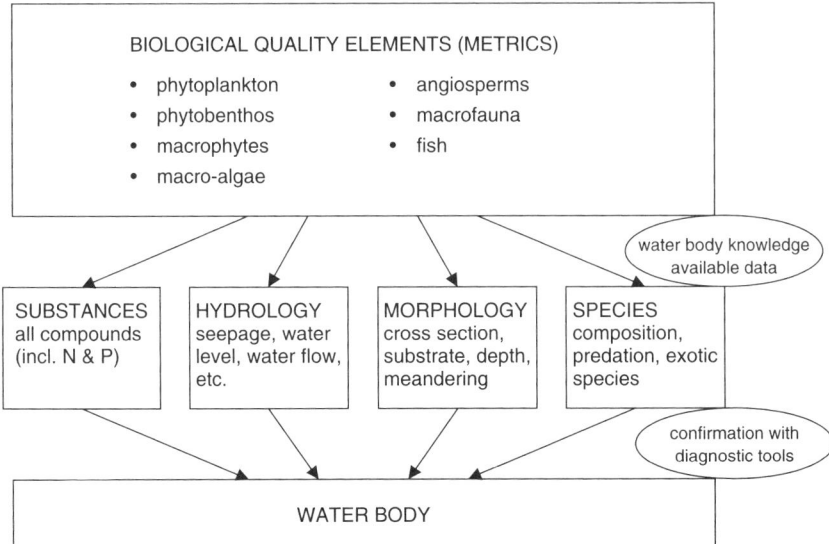

Figure 3.2.1 The conceptual framework: the DWQI starts with the biological quality elements of a water body which are causing its poor status. Pressures of the type of substances, hydro- and morphological aspects and species could affect the biological element in concern. Manipulation of the aspects mentioned will change the ecological quality of the water body. *Cross links between the different pressures have been omitted for reasons of simplicity*

Table 3.2.1 Biological quality requirements applicable to types in the natural waters category (Policy Summary on Monitoring, 2002)

Quality element	Rivers	Lakes	Transitional waters	Coastal waters
Phytoplankton	X	X	X	X
Phytobenthos	X	X		
Macrophytes	X	X		
Macroalgae			X	X
Angiosperms			X	X
Macrofauna	X	X	X	X
Fish	X	X	X	

Biological Quality Elements and Metrics

It is essential to WFD monitoring that ecological status be determined at the level of quality elements: biological, physical-chemical, hydrological and morphological. The overall biological quality requirements for natural waters are shown in Table 3.2.1. These quality elements determine the biological quality of a water body.

Each quality element is defined on the basis of a metric, most of which are built up of a number of sub-metrics. Biological quality elements, for example, include sub-metrics reflecting factors like species composition and abundance. The DWQI starts with the metric of the biological quality element of a water body which is causing its poor status.

Substances, Hydrology, Morphology and Species

Substances, hydrology, morphology and species constitute the second level of the diagnosis. The causes of poor ecological status in water bodies tend to be associated with these aspects. The DWQI gives an overview of these aspects, specified for different types of water body, which can be controlled or 'steered' to improve the ecological status of the water body concerned. Table 3.2.2 lists common 'steering variables' or pressures. Tools used for diagnosis and to determine what measures should be taken influence the relationships that are shown in Figure 3.2.1. Manipulation of the aspects mentioned will change the overall water body.

Besides the pressures mentioned in Table 3.2.2, aspects that cannot be steered directly (like climate change) and drivers (like shipping traffic, maintenance of flood control dams, fishery, etc.; see Policy Summary on Pressures and Impact Analysis (2002) also influence the ecological quality of water bodies. These aspects have been considered indirectly in the pressures corresponding to the Driver, Pressure, State, Impact and Response (DPSIR) framework mentioned in the WFD CIS Guidance document on analysis of pressures and impacts (WFD, 2002).

3.2.2.2   Description of the DWQI

The DWQI is an interactive PC program. Important words are highlighted in the tables and text. A click on these words takes the user further and further through the diagnostic instrument and the underlying information.

Table 3.2.2 Physical-chemical, hydrological and morphological pressures and species

	Pressures
Substances	Nutrients (N and P)
	Suspended solids (transparency)
	Oxygen
	Temperature
	Chloride (salinity)
	Macro-ions (including bicarbonate)
	Significant discharges
	Specific pollutants (toxicity)
Hydrology	Seepage
	Water level
	Water level dynamics
	Rate of flow
	Barriers
	Flood
	Temperature stratification
	Drainage dynamics
Morphology	Structure (sediment/sludge)
	Cross-section (embankment)
	Substrate (sediment)
	Flooded area
	Water flow
	Depth
	Shade
	Meandering
	Turbidity
	Covered area
	Spawn area
Species	Exotic species
	Predation
	Bank and water plants
	Maintenance of water system
	Fishery (cockles)
	Variation in salt marshes
	Fish population

A click on the metric (Table 3.2.1) displaying poor status forwards the user to the list of pressures for the type of water body concerned.

It is possible that more than one metric contributes to the poor status. In lakes it is common for quality elements of phytoplankton, macrophytes and fish to fall short of the quality criteria. In rivers, macrofauna and fish are likely to have inadequate scores. In both situations the steering factors form a complex. In the example of lakes, this is known as a eutrophication complex. In rivers it is caused by canalisation and normalisation. And, third, the complex causing an inadequate score for macrofauna and fish in the metric for transitional waters is called coastal defences.

The DWQI gives tables with a list of possible pressures per water body type. These pressures are arranged vertically in order of priority. Hydrological and morphological pressures tend to be the most important factors, followed by chemical factors and, finally, biological factors. Physical interventions will have the most impact on the ecology in the ecosystem.

Box 3.2.1 gives an example of an application of the DWQI. The particular metrics and pressures are highlighted in the tables. A click on a highlighted word gives access to underlying information and to the diagnostic tools that are available to gain more insight into the precise causes of the inadequate score.

Box 3.2.1 An example of the application of the DWQI: macrofauna in rivers

The composition of macrofauna in a river does not meet the criteria of a good ecological status. The user clicks on 'macrofauna' underneath 'rivers' in the table below (see shaded area) and ends up in a table with relevant pressures for macrofauna.

rivers	lakes	transitional waters	coastal waters
	phytoplankton	phytoplankton	phytoplankton
phytobenthos	phytobenthos		
macrophytes	macrophytes	macro-algae and angiosperms	macro-algae and angiosperms
macrofauna	macrofauna	macrofauna	macrofauna
fish	fish	fish	

On the basis of the table below, the user assesses whether the effect on macrofauna is caused by the absence of bank and water plants. Information about the pressure 'bank and water plants' is provided by a click on the word (shaded area).

Pressures on macrofauna in rivers			
hydrology	**morphology**	**substances**	**species**
canalisation and normalisation			
rate of flow			exotic species
	sediment/sludge	oxygen	maintenance
	substrate		bank and water plants
	flooded area		
eutrophication complex			
seepage (iron)		nutrients	
		micro pollutants	
		salinity	

Below the information a table is given, with information about the diagnostic tools and knowledge that can be used to examine this parameter, references, and examples of comparable areas and measures taken. Hyperlinks give the user access to this background information.

Bank and water plants

Bank and water plants are important as habitat for macrofauna and some fish species and as substrate for diatoms. Bank and water plants provide structure, protection and food. The quality of macrofauna, especially in lakes, will be influenced strongly by the occurrence of bank vegetation, underwater plants and the total of *overgrown area*. Bank vegetation can be important to fish as habitat, forage area (e.g. for pike, eel, perch) or as spawn area (e.g. for perch).

Water level, *depth* and *water level dynamics* are important 'steering factors' to stimulate the growth of bank vegetation.

tools	knowledge/models	references	examples
in situ tests	WFD-explorer	Vermaat, 2002	Lake Zuidlaarder
inundation tests	factsheets	Graveland and Coops, 1997	Friese Boezem
test for germination		Graveland, 1999	
mesocosms		Nagelkerke *et al.*, 1999	
fish stock		Boedeltje *et al.*, 2004b	

3.2.2.3   Diagnostic Tools

Diagnostic tools in the DWQI vary from laboratory and field experiments to literature reviews and a theoretical or model approach to the system. The DWQI gives an overview of possible tools and existing methods for different working fields. It also shares experiences from different projects and from other water managers. The DWQI does not give a solution for every problem in practice and does not aspire to be an encyclopaedia of all diagnostic tools that may exist for all problems in water bodies.

Experimental Approach

The laboratory and field experiments include both short- and long-term experiments. Examples of short-term experiments include tests to investigate the germinative capacity of seeds in substrate and standard laboratory experiments to measure the toxicity of the surface water (bioassays). Long-term experiments include compartmentalisation (enclosure) of a water body to prevent predation pressure from fish and the use of mesocosms (model ecosystem to scale) to assess the influence of tides on reed growth.

Theoretical and Model Approaches

Models for drawing up a mass balance and risk models like OMEGA (OMEGA, 2006), used to assess the possible adverse effects of substances on an ecosystem, are relatively simple. Examples of more complex and costly instruments include compartmentalisation, which can be used to determine how changes to soundings in a bank area reflect water quality or to assess the influence of streaming on oxygen content.

3.2.2.4   Measures

After diagnosis, measures should be selected by the water manager in order to improve the ecological quality. The entire range of possible measures is very extensive and is not the subject of the DWQI. Other projects in the Netherlands are considering measures and their relationship with effects on ecological elements.

3.2.3   THE DWQI DIAGNOSIS

The DWQI diagnosis takes the form of a decision tree. The diagnostic framework consists of a combination of decisions that follow from the flow chart or identification key, and of concrete tools. These tools vary from laboratory or field experiments to literature reviews and a theoretical or model approach to the system.

Analysing the various steps in the diagnosis allows us to pinpoint exactly where diagnostic tools might make a useful and efficient contribution. Every water system is however unique and complicated in its structure and functionality. Water managers' knowledge of their water system varies. If the DWQI cannot help solve the problem, a tailor-made solution will be needed.

The diagnosis and the input for the diagnostic tools are explained in the following steps (see Figure 3.2.2):

Step 1 (Identify Ecological Effect): The process begins at the moment when an effect is observed in a certain water body, causing that water body to be given a 'poor' score on one or more of the metrics or sub-metrics. This conclusion is drawn during surveillance monitoring and/or operational monitoring.

Step 2 (List Possible Causes): The next step is to list all possible causes of the effect observed. This requires knowledge of reference areas and similar areas where a similar effect may have been observed (see Table 3.2.2; Box 3.2.1).

Step 3 (Gather Available Data): To pinpoint the actual cause, access must be gained to all available information on:

- Auto-ecology (biological and ecological information on the organism concerned, not always available).

- The hydro-morphology of the area concerned (available through WFD monitoring).

- The chemical status of the water body in question (available via WFD monitoring and through risk models, which allows the substances most responsible for poor water quality to be identified).

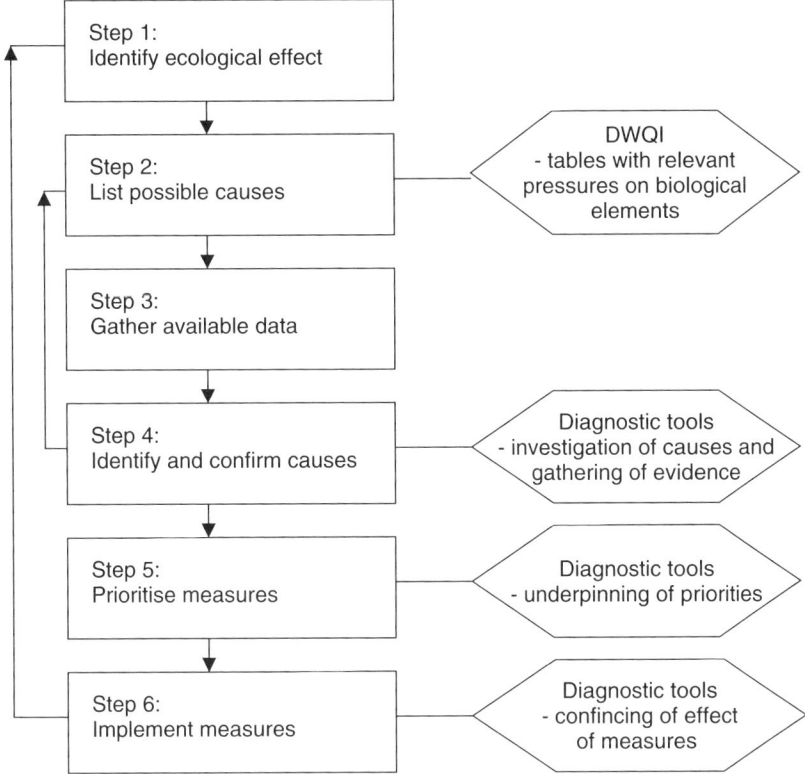

Figure 3.2.2 Decision tree of the diagnostic system for establishing cause in cases where poor ecological status is identified

- Problems in similar water bodies and the results of other projects. The more knowledge of this kind is available, the easier it will be to pinpoint the causes of poor status.

Step 4 (Identify and Confirm Causes): The shortlist of possible causes and the information gathered can be used to establish the most likely causes. Specific hypotheses can then be used to test whether these factors really are responsible for the low score. Even if the experts are certain of the cause after Step 3, this step can prove useful in demonstrating it to others (positive confirmation) or showing that other factors do not play a role (negative confirmation). This creates support for measures that could have far-reaching economical and social consequences.

Step 5 (Prioritise Measures): If several factors are found to be responsible for the observed effect, priorities have to be set. Considerations include the likely effectiveness and feasibility (including cost) of the planned measures. One must also investigate whether measures to improve one quality element are likely to adversely affect another (for instance, reducing the amount of nutrients in Lake IJsselmeer would lead to lower fish production).

Step 6 (Implement Measures): The final step is to implement the most suitable measures. The biological score will be reevaluated after a certain time, either because WFD testing is required or to obtain an interim assessment of the effect of the measures. If the situation has not improved despite the measures, the decision tree procedure can be repeated. The evaluation might also give cause for the deadline to be postponed or the targets adjusted. In that case the decision tree and the measures taken in response can be used to demonstrate that the Member State has done everything possible to remedy the situation.

3.2.4   CONCLUSIONS AND OUTLOOK

The DWQI offers water managers an instrument and tools that can assist in diagnosing water quality problems. It enables water managers to take more clearly defined and structured steps and will help to prioritise measures to improve water quality. The DWQI gives an overview of existing methods from different disciplines. It offers the opportunity to share knowledge and experiences from water managers and projects. The DWQI can adapt new information as a result of new experiences and developments. This will make the DWQI a 'living instrument'.

REFERENCES

Directive 2000/60/EC, Establishing a framework for community action in the field of water policy (the Water Framework Directive – WFD), Brussels, Belgium.

OMEGA (2006) Optimal Modelling in Ecotoxicological Assessment, version 6.0, RIZA, Lelystad, The Netherlands.

Policy Summary on Pressures and Impacts Analysis (2002) The key implementation requirements of the Water Framework Directive, Policy Summary to the Guidance Document, Brussels, Belgium.

Policy Summary on Monitoring (2002) The key implementation requirements of the Water Framework Directive, Policy Summary to the Guidance Document, Brussels, Belgium.

Royal Haskoning (2007) The diagnostic water quality instrument (guidance), version 2, 's Hertogenbosch, The Netherlands (in Dutch; to be translated).

Van Splunder, I., Pelsma, T.A.H.M. and Bak. A. (eds) (2006) Guidelines for monitoring surface water for the European Water Framework Directive, version 1.3, Lelystad, The Netherlands (in Dutch).

Verdonschot, P.F.M. (1995) Brooks flow: guidance for ecological recovery of brooks, 95-03 WEW-06.STOWA report 95-03/WEW-06, STOWA/Working Group Ecological Water Management Sub-group Recovery of Brooks, Utrecht, The Netherlands (in Dutch).

WFD (2002) Monitoring under the Water Framework Directive, CIS Guidance Document No. 7, Luxembourg, Luxembourg.

WFD (2003) Analysis of pressures and impacts, CIS Guidance Document No. 3, Luxembourg, Luxembourg.

Section 4
Modelling Tools in Support of WFD Monitoring

4.1

Joint Modelling and Monitoring of Aquatic Ecosystems

J.C. Refsgaard, L.F. Jørgensen, A.L. Højberg, C. Demetriou, G. Onorati and G. Brandt

The Water Framework Directive - Ecological and Chemical Status Monitoring   Edited by Philippe Quevauviller, Ulrich Borchers, Clive Thompson and Tristan Simonart © 2008 John Wiley & Sons, Ltd

4.1.1 INTRODUCTION

The big challenge for the European water and environmental managers these years is to implement the EU Water Framework Directive (WFD) (European Community, 2000). Successful implementation of the WFD calls for appropriate tools and models to support the management of the different technical and social aspects in different phases of the implementation (Rekolainen *et al.*, 2003; Wasson *et al.*, 2003). The use of models to address the technical challenges does, however, require observations of the physical system. Models may be thought of as sophisticated databases that provide an ordered way to store field data and define the relationships between data, e.g. the temporal development in streamflow as a response to a storm event. These relationships are generic formulations and models must therefore be adjusted to describe the site-specific conditions, which are commonly accomplished through calibration, where model parameter values are varied until the model reproduces the response of the reality within some predefined accuracy. If no site-specific data is available, the models cannot provide insight into the system of interest, but are limited to provide some general insight into the system behaviour, e.g. how different processes interact. Using such models to predict the state of a site-specific natural system and its response to stresses will be guesswork and monitoring data are therefore a prerequisite to site-specific modelling.

In the past, monitoring was traditionally considered an independent discipline, but within the last decades modelling has entered the arena as a supplementary tool to help extract the information retained in the observation data. In the research communities it is generally accepted that monitoring and modelling are interlinked activities (Holt *et al.*, 2000; Parr *et al.*, 2003). The need to support the EU environmental monitoring, specifically for the groundwater domain, with new knowledge and scientifically-based tools including an enhanced communication among scientists, stakeholders and policy makers is discussed by Quevauviller (2005). Descriptions of present practices, however, show that most often models are not considered an option when the monitoring obligation in the WFD is solved in practice (Arustiene *et al.*, 2005; Kamphorst *et al.*, 2005; Steenstra *et al.*, 2004). This may be due to a number of obstacles, such as lack of skill, lack of time, lack of awareness about what models can do, and also a lack of confidence in models by policy makers (Brugnach *et al.*, 2006). Furthermore, the WFD Guidance Document on Monitoring (European Community, 2003a) does, but only to a very limited extent, explicitly prescribe the use of modelling tools. To advance the use of models in practice there appears to be an imperative need to illustrate how models may improve the efficiency of monitoring data with respect to the specific monitoring obligations in the WFD.

The overall purpose of the monitoring programme is to establish a coherent and comprehensive overview of the water status within each river basin district, which forms the basis for assessing whether the chemical, qualitative and ecological status of a water body is in accordance with the predefined objectives. The specific objectives and requirements of monitoring in the WFD are described in a guidance document (European Community, 2003a). Here some general guidance is provided on where, what and when to monitor. These recommendations, however, do not specify the levels of precision and confidence required by the monitoring programmes, but instead state a key principle: '*the actual precision and confidence levels achieved should enable meaningful assessments of status in time and space to be made. Member States will have to quote these levels in River Basin Management Plans and will thus be open to scrutiny and comment by others*'. The acceptable level of precision and confidence is thus a subjective quantity that depends on the socioeconomic interests that are at stake and on the risk strategy of the decision makers. In this context it should be borne in mind that the costs of measures to improve the water status are orders of magnitude greater than the costs of monitoring. In general, the lower the desired risk of misclassification, the more monitoring (and hence cost) is required. Therefore, there should be a balance between the costs of monitoring and the risk of water bodies being misclassified.

Thus, although the WFD CIS Guidance Document on Monitoring (European Community, 2003a) does not explicitly emphasise modelling, the uncertainty and risk assessments cannot in practise be carried out without some modelling support. The above guiding principles based on uncertainty and risk assessments are therefore quite advanced, when considered as guidance to practitioners in water resources management. In full accordance with this line of thinking, the WFD CIS Guidance Document on Planning (European Community, 2003b) describes the benefits of using models to support the WFD implementation. The WFD therefore creates new challenges both for monitoring and modelling, and provides new incentives for improving the joint use of monitoring and modelling. One of the most often posed questions in relation to the design of monitoring systems is: 'Will the monitoring system be good enough to support the modelling activities that can be foreseen in subsequent WFD steps – can the monitoring programmes provide all the necessary data for modelling?'

The objectives of the present paper are: a) to illustrate for which tasks modelling can support the monitoring programmes and the WFD implementation; and b) to illustrate joint modelling and monitoring approaches in case studies.

4.1.2 WFD REQUIREMENTS

In the process of implementing the Water Framework Directive, the issue of data is extremely important. The WFD asks for ambitious goals for the aquatic environment, with the ultimate objective of obtaining good quantitative and qualitative status of all waters in Europe by 2015. Achieving this calls for the solution of a large number of technical tasks, all involving either using already-collected data or collecting new data: water bodies need to be delineated, classified and possibly grouped; water districts must be characterised; anthropogenic effects evaluated; trends identified; pressures and impacts assessed; programmes of measures established and evaluated, etc.

Four types of monitoring are described in the WFD CIS Guidance Document on Monitoring (European Community, 2003a):

- *Surveillance Monitoring*: To determine status; to supplement and validate the initial impact assessments; to assess long-term changes; to validate hypotheses and to assist in design of future monitoring programmes.

- *Operational Monitoring*: To assess status of water bodies at risk of failing to meet environmental objectives; to assess effects of programmes of measures. Operational monitoring must be carried out in all bodies receiving priority substances.

- *Investigative Monitoring (only surface water)*: Where the reason for any exceedance is unknown; where good status is not likely to be achieved and where operational monitoring has not already been established; to ascertain the magnitude and impacts of accidental pollution. Investigative monitoring only addresses the problem-defined variable(s).

- *Water Level Monitoring (only groundwater)*: For groundwater, quantitative monitoring is required. This should be designed to ensure that the available groundwater resource is not exceeded by the long-term annual average rate of abstraction; to ensure that abstractions and other anthropogenic alterations to groundwater levels do not adversely affect associated surface water bodies and terrestrial ecosystems that depend directly on groundwater with respect to their water needs; to ensure that anthropogenic alterations to flow direction have not caused, and are not likely to cause, saltwater or other intrusion.

4.1.3   FRAMEWORK FOR MODELLING SUPPORT WITHIN THE WFD

Modelling may be carried out for a number of different purposes. Each of these purposes may support various types of monitoring requirement, as outlined in Table 4.1.1. This is briefly outlined below, but described in more detail in Højberg *et al.* (2007a) and (2007b).

4.1.3.1   Quality Assurance of Monitoring Data

Quality assurance of data is, in the WFD CIS Guidance Document on Monitoring (European Community, 2003a), confined to sampling and laboratory procedures. While this is an important step, it is not sufficient to ensure good-quality data. It is a well known experience that when data are used for the first time in modelling using a mechanistic process-based model, some data often appear to be highly inconsistent with the model. This may be due to an incorrect model or it may be due to suspicious data quality. In any case it is a good quality assurance check. Quality assurance may in this way be seen as a natural spin off of any modelling activity, and in cases where apparent inconsistencies in data are subsequently validated in the field, it may as an additional benefit increase the water manager's confidence in models.

4.1.3.2   Interpolation and Extrapolation in Time and Space

Monitoring data are discrete in time and space, and interpolation and sometimes even extrapolation techniques are therefore required to achieve a continuous image of the conditions all over the time and space domain for which WFD decisions are made. Standard mathematical interpolation procedures such as kriging are constrained by their general assumptions, e.g. on second-order stationarity, and often in practice end up in linear interpolation. By use of models, it is possible to include data on nonstationarity in, for example, geological properties, and models can be seen as tools for making knowledge- or data-based interpolation.

4.1.3.3 Conceptual Model

A thorough conceptual understanding of the physical system is explicitly required in the WFD. It is vital to the design of all monitoring programmes and management strategies. If the conceptual understanding and the associated perceptions of dynamics and cause–effect relations are wrong, water bodies may be misclassified, monitoring programmes may be ineffective and programmes of measures may be non-optimal. Construction and test of a conceptual model is therefore a crucial but also often a very complicated task. A numerical model is based on the conceptual model. Test of the results of the numerical model against monitoring data is therefore also a test of the consistency of the conceptual model. If such tests reveal significant deviations between model results and field data, there may be a need to revise the conceptual model. If the deviations are within the acceptable accuracy, it causes an increased credibility of the conceptual model.

4.1.3.4 Assess Effects of Anthropogenic Activities

Monitoring data are used to identify trends in the ecological status and to assess whether an implemented programme of measures has had the expected effects. In this regard the natural climate variability induces variance in the variables to be analysed. Natural variability will, together with measurement errors, act as noise that may hide the signals from, and make it very difficult to identify the effects of, the anthropogenic activities. This is especially problematic in cases where the full effect of the intervention is slow, and the changes are modest initially and small compared to changes due to natural variations. Models may be useful here, if they are able to explain some of the natural variability and thereby enhance the signal from anthropogenic activities.

4.1.3.5 Design of Monitoring Programmes

There are many different objectives to design and operate a monitoring system. Because the operation of monitoring programmes is expensive, the question is how we can assess the balance between the information content and the cost of the monitoring programme. State variables (pressure, fluxes, concentration, etc.) show variability in space and time that may appear as noise when estimating their characteristics. Consequently, estimates of the characteristics are uncertain. In principle, the more variability in the state variables, the more measurements are required to arrive at the same level

Table 4.1.1 Overview of how modelling can support the different WFD monitoring requirements. Requirements are listed for monitoring related to groundwater (italic), surface water (bold) and both (italic and bold) (Højberg et al., 2007a)

	Quality assurance	Inter-polation	Conceptual model	Anthropogenic activities	Design of monitoring programmes
Level Monitoring network	G	G	*GW-SW interaction*	*Long term impact by abstraction*	
	e	e	*Characterisation & Risk assessment, quantitative status*	*GW abstractions impact on SW*	
	n	n		*Saltwater or other intrusions*	
	e	e			
Surveilliance Monitoring	r	r	*Characterisation & Risk assessment, qualitative status*	***Assessment of long-term changes: natural conditions and anthropogenic activities***	***Design of future monitoring programmes***
	a	a	*Status of GW bodies not at risk*		
	l	l			
Operational Monitoring	p	p	**Impact assessment**	***Evaluate effects of POMs***	
	u	u	***Status of water bodies at risk***	*Detection of pollution trends in GW*	
	r	r			
Investigative Monitoring	p	p	**Pressures unknown**		
	o	o			
	s	s	**WB at risk but no operational monitoring established**		
	e	e			
	s	s			

of uncertainty. For the design of a monitoring programme we have to assess the relationship between the measurement effort (number of locations and measurement frequency) and the reduction of the uncertainty. A model can be used to assess the uncertainty corresponding to various levels of measurement effort and in this way support an optimal design of a monitoring programme.

4.1.4 CASE STUDIES

4.1.4.1 Context and Aim of Workshop

In connection with the 3rd Harmoni-CA Forum and Conference in Osnabrück, April 2006, a workshop on Joint Use of Modelling and Monitoring for WFD Implementation was held. The workshop had 18 participants from 10 countries (BE, CY, DK, EE, IT, LT, LV, PL, UK and US), with backgrounds as policy makers, water managers, stakeholders, consultants and scientists. The objective of the workshop was to elaborate management strategies involving relevant use of monitoring and modelling in three real-life case studies. The participants were divided into three groups and each group worked on one case only. All groups were asked to prepare a flow diagram for the process of identifying and solving the problems at hand, and in this connection to pay particular attention to the role of monitoring and modelling. In addition, each group was asked to provide advice on local case problems. In the following, the key result from the three case studies is presented with a focus on the general flow diagrams. More details can be found in Jørgensen *et al.* (2006).

4.1.4.2 Case Study 1: Artificial Recharge of Tertiary Treated Wastewater, Cyprus

Description of the Case

Groundwater is used extensively in Cyprus for water supply. Treated sewage water can be used to artificially recharge aquifers, both to increase the available groundwater resources for irrigation and to mitigate the risk of seawater intrusion. The public has generally been very negative towards the idea of recharging aquifers with treated effluent, but in the area of the Ezousas aquifer, people accepted the idea with some scepticism.

The Ezousas aquifer is a confined river aquifer close to the sea at the western part of Cyprus, and it delivers groundwater to a number of private wells for irrigation. During the last few years, precipitation in Cyprus has in general decreased by around 20 %, resulting in reduced groundwater recharge. Besides, a dam was recently constructed upstream in the river. Because a) the aquifer is used intensively for irrigation during summer and is not recharged by the river as there is hardly any flow during the dry season due to the dam; b) the precipitation and groundwater recharge during the wet season is not sufficient to compensate for the abstraction; and c) private waste water recharge, which to some extent contributed to local groundwater recharge, has been closed and the waste water is now led to a public waste water treatment plant, the groundwater table is declining drastically, inducing water quality problems by seawater intrusion.

To reverse this development, tertiary treated wastewater is artificially recharged in ponds to the Ezousas aquifer to increase the amount of available groundwater for irrigation. The groundwater is mixed with water from the river upstream of the dam at a ratio of 1:20 and used for irrigation both locally and in a neighbouring region. The dominant crop in both areas is citrus trees. Domestic water supply is based on water from the reservoir and not on water from the Ezousas aquifer.

There is no previous tradition of using numerical models for studies of the Ezouras aquifer.

Conclusions from the Group

As the resulting flow chart (Figure 4.1.1) indicates, the group agreed that the starting point must be to identify the problem. In the Cyprus case the present problem comprises two aspects: to maximise the water available for irrigation and to prevent seawater intrusion. After this, potential solutions can be suggested, taking into account new problems and obstacles that may arise. For instance, local farmers are likely to be sceptical towards the use of recharged wastewater for irrigation purposes, because they fear it may affect the pricing of their crops.

The next step is to make sure that the system is well understood by both the managers and the stakeholders. Here monitoring data should be used to test the conceptual

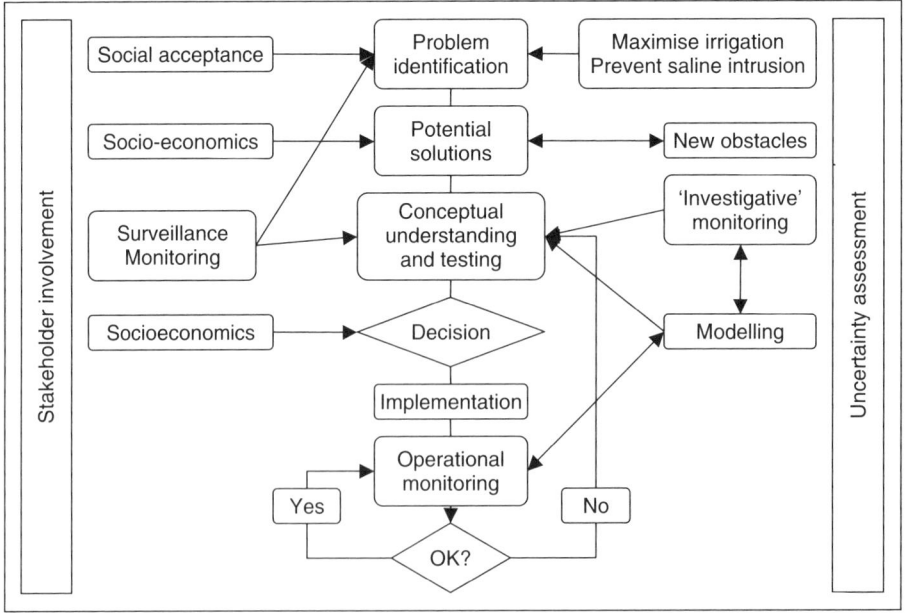

Figure 4.1.1 Flow diagram for the problem identification and solution process for the Ezousas Aquifer case (Cyprus). Modified from Jørgensen *et al.* (2006)

understanding. As a correct causal understanding is crucial in this case, the monitoring type is likely to be the most comprehensive one until the conceptual model is success-fully tested and accepted. The group denoted this 'investigative' monitoring, although they were aware that investigative monitoring according to the European Community (2003a) is only meant for surface water systems and not for groundwater systems. Models may also be useful for visualisation of the system dynamics and for making sure that all chemical, physical or other relations are understood. At this stage the effects of the different solutions can be assessed by modelling.

Now one or more solutions (programme of measures) must to be chosen, taking both the physical system behaviour and social and economic aspects into account. In implementing the chosen measure(s), an operational monitoring programme must be designed so that the effect of a given measure can be documented – again models can support in this process by being able to detect the effects more efficiently by filtering away some of the natural variations.

During the whole process it is important to involve stakeholders, to ensure that their interests and views are correctly understood in the problem identification and problem solution phases. They must also share the conceptual understanding of the system and have confidence in the scientific basis of the assessments of the pro-gramme of measures. Furthermore, it is important to evaluate the uncertainty in the different steps and tools used, for instance in the model simulations and in the data collected. Other factors such as social aspects, socioeconomics and restrictions in terms of economy, etc. have to be taken into account in the implementation plan. Stakeholder involvement and uncertainty assessments are not shown as individual steps in the flow chart in Figure 4.1.1, because they should be considered through the whole process.

4.1.4.3   Case Study 2: Nitrate in Groundwater, Campania, Italy

Description of the Case

The Campania region covers $13\,600\,km^2$ and has a population of 5.7 million. More than $90\,\%$ of the drinking water is abstracted from aquifers; hence the preservation of groundwater is crucial. Therefore, much groundwater quality data has been collected during the last years by the regional environmental agency, ARPAC, which started to regularly monitor the groundwater in Campania in 2002. This was preceded by a col-lection and analysis of 1996–2001 data, which allowed a preliminary characterisation of springs and wells in Campania and aided the design of a preliminary monitoring network with semi-annual monitoring campaigns. The network included 224 wells in 2006. Monitoring is carried out in both an upper and a lower aquifer. The collected data show that nitrate is the main overall pollutant. Locally, polycyclic aromatic hydro-carbons (PAHs), heavy metals and organic compounds are present in concentrations exceeding the allowed limits. The most polluted water bodies are located below the most permeable portion of the densely populated alluvial plains, while the less con-taminated aquifers stretch along the mountain areas of the Apennines, where human activities are sparse and protected wilderness is prevailing. More details on the case are provided in Onorati *et al.* (2005, 2006).

The group was particularly asked to consider the modelling needs for regionalisation of the well monitoring data, including questions on the relevance of using mass balance models and non-point source models for nitrate.

There is no previous tradition of using numerical models for studies of groundwater quality in the Campania area.

Conclusions from the Group

The group did not focus on problem identification, but went directly to the issue of conceptual understanding. Therefore the flow diagram in Figure 4.1.2 does not include surveillance monitoring.

A key problem in the Campania area is to assess how much of the groundwater contamination originates from non-point agricultural sources and how much comes from infiltrating sewage recharges from streams and from other sources. It is necessary to make a full analysis, taking all possible pressures into account. In order to quantify the effects of the various pressures, models can be useful in helping assess nitrogen mass balances. The first step is therefore to develop a conceptual model comprising a causal understanding of the origin of the problem (Figure 4.1.2, Step 1).

To obtain the necessary knowledge to establish credible causal relationships an investigative monitoring should be established (Figure 4.1.2, Step 2). This group also noted that although investigative monitoring as described in the WFD CIS Guidance Document on Monitoring (European Community, 2003a) only applies to surface water, it should here be extended to cover groundwater and in particular the interaction

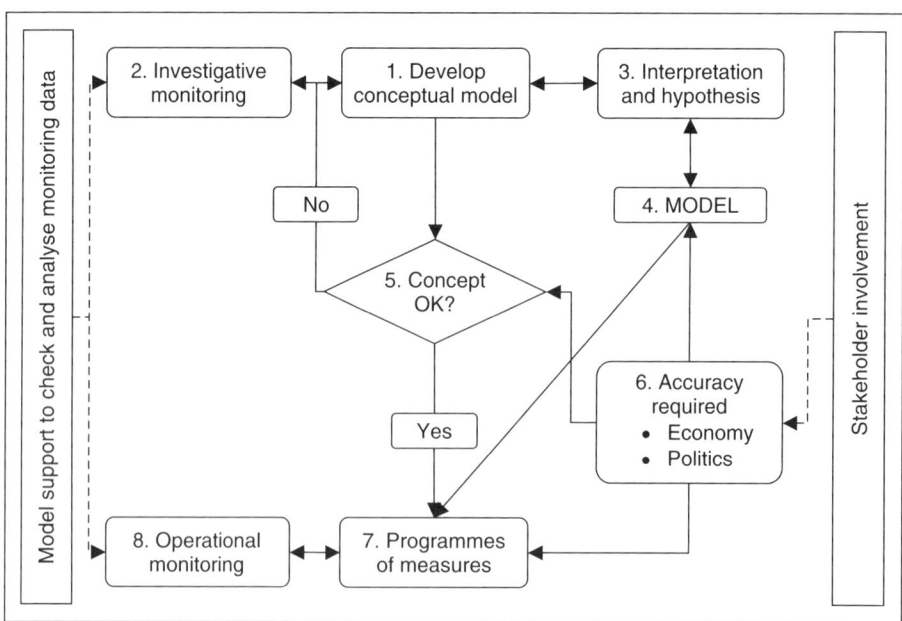

Figure 4.1.2 Flow diagram for the problem identification and solution process for the Campania case (Italy). Modified from Jørgensen *et al.* (2006)

between surface water and groundwater. As part of this monitoring, establishment of pilot areas with intensive monitoring specially designed to illustrate the most important processes, including the interaction between groundwater and surface water, should be considered. In this respect, models may be useful in helping test various hypotheses on the consistency of perceived process descriptions and on the origin of nitrate in groundwater (Figure 4.1.2, Steps 3 and 4). When it is generally accepted that a credible causal relationship is established (Figure 4.1.2, Step 5) the next steps on identification and assessment of programmes of measures (Figure 4.1.2, Step 7) and design and operation of an operational monitoring (Figure 4.1.2, Step 8) can be initiated.

Presently, vulnerability maps are produced on a $1\,km^2$ scale, where data from agricultural activities and wastewater plants are combined with the geological vulnerability. The maps depict vulnerable and non-vulnerable areas (non-vulnerable are where monitoring data shows values of nitrate below $50\,mg/L$). The group felt that there should be more than two classes of vulnerability, and at least one intermediate class. It is possible through numerical modelling to point out areas where data is insufficient for working out a vulnerability 'index' and also for actually compiling all data in order to 'map' an area's vulnerability. The group discussed whether it was justifiable to carry out solute transport groundwater modelling and concluded that such studies would in principle be useful. However, it should be evaluated whether the costs involved are worth the effort, as compared to using the same money to collect additional data.

The group emphasised that in order to properly design modelling studies it is important to define what is an acceptable level of uncertainty, taking into account what is at stake politically and economically. This requires involvement of stakeholders and policy makers (Figure 4.1.2, Step 6). Finally, it was noted that models may be useful tools for supporting the quality assurance and in analysing data from both the investigative and the operational monitoring.

4.1.4.4   Case Study 3: Groundwater Protection, Denmark

Description of the Case

Hørup, an area 225 km south-west of Copenhagen, is one of 56 well fields supplying the greater Copenhagen area with groundwater for drinking water production. The well field contains nine boreholes and the annual abstraction is 2.5 million m^3. The catchment area has been estimated at around $24\,km^2$, where the dominant land use is agriculture with small villages.

The groundwater is abstracted from a confined limestone aquifer that is overlaid by fluvioglacial sand, which again is covered with $15-30\,m$ of clayey till. In the catchment area there are many point sources (e.g. landfills) and non-point sources (e.g. roads, agriculture) that – actually or potentially – pose threats to the groundwater quality. A wellhead protection zone has been delineated around the wells, up to 1500 m from them. In this area an afforestation scheme has been initiated. When planted, the forest is expected to provide permanent groundwater protection. Several groundwater modelling studies have been conducted in this area during the past decade.

Conclusions from the Group

The management process would typically comprise the following six stages (Figure 4.1.3):

Stage 1 (Problem Identification): This comprises definition of the problem and suitable indicators that can help define monitoring activities and targets for modelling. The key problems in the present case are that groundwater abstraction may damage habitats and adversely affect the water quality and quantity in the surface water system, and that groundwater is at risk of being contaminated from point and non-point sources. The group assumed that the basis for the problem identification was data originating from some kind of surveillance monitoring, although this is not shown in the flow diagram in Figure 4.1.3.

Stage 2 (Collect Existing System Information): This includes information on habitat, geology, soil maps, topography, river system, etc. Here it may also be relevant to have model predictions of future development for aquifer boundaries and habitat changes.

Stage 3 (Assess Pressures): This includes collection and assessment of data on point sources, roads, abstraction wells, land use, historical climate data, groundwater abstractions and assessments of load due to point sources and non-point sources.

Stage 4 (Monitoring): This includes relevant monitoring of water quality in groundwater, surface water and drinking water (e.g. classical parameters, heavy metals, PAHs, pesticides, etc.); groundwater levels and discharges; habitats, etc. This will be operational monitoring.

Stage 5 (Building the Model): This involves the classical steps of defining the conceptual model, selecting model code, calibrating the model against field data and

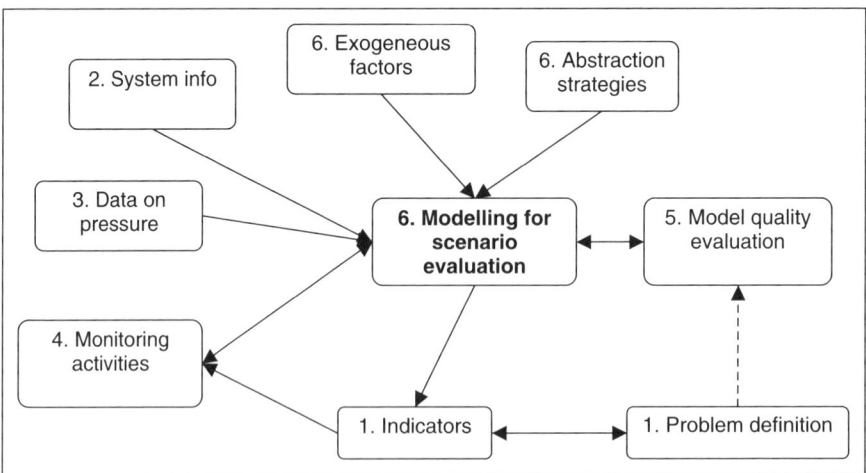

Figure 4.1.3 Flow diagram for the problem identification and solution process for the Hørup case (Denmark). Modified from Jørgensen *et al.* (2006)

evaluating/validating the model against independent data. These activities are heavily dependent on data from the monitoring programmes. The evaluation of when a model performance can be considered good enough will depend on the character of the problem.

Stage 6 (Scenario Building and Evaluation): When a model has been successfully tested it can be used to assess the effects of various scenarios. Key elements in the scenarios are the exogenous factors such as new legislation and policies (e.g. due to the EC's Common Agricultural Policy), new technologies and new potential pressures such as climate change. Stakeholder opinions must be taken into account when defining scenarios. The model should have documented capabilities to make predictions on water balance, groundwater flow, fate of pollutants, impacts on wetland and habitat. In addition, it should be able to support assessments of economic impacts and ecological and human health.

4.1.4.5 Lessons Learned from the Three Case Studies

The workshop cases were real-life with respect to problems, data availability, etc. A clear limitation of the outcome is that the three groups were put in the rather unrealistic situation where they did not have to consider political or economic constraints. This obviously made it easier to identify approaches that were acceptable to all participants in the groups. Nevertheless, the overall conclusion was that when water managers and researchers (or people from the monitoring and modelling communities) are brought together and given rather free hands, they are easily able to work together in a very inspiring and constructive way. Hence it is not a problem to consider monitoring and modelling jointly and to plan the use of these two disciplines in an effective manner.

The three participants who brought the cases to the workshop found that they had some interesting recommendations and considerations to bring back home.

The three cases were rather different in terms of problem, data availability and tradition for modelling. The three 'solutions', as represented in Figures 4.1.1, 4.1.2 and 4.1.3, appear at first glance rather different as well. However, a closer examination of the content and the underlying rationale reveals that there are many similarities in the identified solutions. In all cases there are several examples of recommended combinations of monitoring and modelling. The recommended use of models includes all five areas outlined in Table 4.1.1:

- Support for quality assurance of monitoring data (Case 1 and Case 2).

- A 'regionalisation' tool for interpolation/extrapolation of monitoring data to a larger area (Case 2).

- Support for development and testing of conceptual models (Case 1 and Case 2).

- A tool for evaluating the effects of possible management actions (programme of measures) in terms of environment and socioeconomy by using scenario approaches (Case 1, Case 2 and Case 3). This also included use of models to analyse whether the effects of implemented measures turned out to be according to plan (Case 1 and Case 2).

• Support for design of monitoring programmes (Case 1, Case 2 and Case 3).

The fact that not all groups included all five areas of joint monitoring and modelling should not necessarily be seen as a sign that they rejected some of these joint uses, but rather that the groups during their limited discussion time focused on what they felt were the most important issues in their respective cases.

Case 1 and Case 2 differ somewhat from Case 3, both in the ways the groups designed the flow charts shown in the three figures and in content. Case 1 and Case 2 have a large emphasis on establishment of a conceptual model with causal relationships for the key processes, and numerical models are in both cases seen as important tools towards that goal. Case 3 has more focus on using numerical models as tools in scenario evaluations. The main reason for this significant difference in focus is probably that the conceptual understanding in Case 3 is supposed to be well established already through a comprehensive data availability and, not least, through several previous modelling studies, where various conceptual models were already developed and tested.

In all situations where modelling was recommended for supporting the development and testing of a conceptual model or as a tool for design of a programme of measures, the data requirements for the modelling were assumed to be available through the associated monitoring programmes. Thus, the question 'are there enough data for modelling?' was not posed. Instead, the monitoring programme was supposed to be designed in such a way that it could provide the necessary data.

In Case 1 and Case 2, investigative monitoring was recommended for gaining more insight into process descriptions and causal relationships in groundwater and in the interaction between groundwater and surface water. This use of investigative monitoring is slightly beyond the recommendations in the WFD CIS Guidance Document on Monitoring (European Community, 2003a), where investigative monitoring is only envisaged for surface water systems.

4.1.5   CONCLUSIONS

Modelling and monitoring activities can support each other in many ways. On the one hand, useful modelling is not possible without appropriate data, which have to come from monitoring. On the other hand, modelling can support monitoring activities through different activities such as a) quality assurance of monitoring data; b) knowledge-based interpolation and extrapolation in time and space; c) establishment of a conceptual model with causal relationships; d) assessment of effects of anthropogenic activities, e.g. in relation to a programme of measures; and e) design of monitoring systems.

Results from a workshop with participants from policy makers, water managers, consultants and scientists illustrate how modelling and monitoring jointly can support problem identification and solution in three real-life cases that differed significantly in terms of type of problem, data availability and tradition of modelling. The specific recommendations differed between the three cases, but the overall conclusion was that much can be gained by considering the monitoring and modelling disciplines in an integrated manner, both in cases with and cases without previous traditions of modelling.

4.1.6   ACKNOWLEDGEMENT

The present work was carried out within the Concerted Action Harmoni-CA, which is funded under the European Commission's 5th Framework Programme (Contract EVK1-CT2001-00192).

REFERENCES

Arustiene, J., Vaitiekūnienė, J. and Jørgensen, L.F. (eds) (2005) Joint use of monitoring and modelling when implementing the WFD, 2nd Workshop: Potential Modelling Support, Workshop report, Harmoni-CA – Harmonised Modelling Tools for Integrated River Basin Management, WP 2: Toolbox and WP 4: Joint use of monitoring and modelling, http://www.harmoni-ca.info.

Brugnach, M., Tagg, A., Keil, F. and de Lange, W.J. (2006) *Water Res. Manag.*, http://dx.doi.10.1007/s11269-006-9099-y.

European Community (2000) Directive 2000/60/EC of the European Parliament and of the Council of October 23 2000, establishing a framework for Community action in the field of water policy, *Official Journal of the European Communities*, **L327/1–L327/72**.

European Community (2003a) Water Framework Directive, Common Implementation Strategy, Guidance Document No. 7, Monitoring, http://forum.europa.eu.int/Public/irc/env/wfd/library.

European Community (2003b) Water Framework Directive, Common Implementation Strategy, Guidance Document No. 9, Planning processes, http://forum.europa.eu.int/Public/irc/env/wfd/library.

Holt, M.S., Fox, K., Griessbach, E., Johnsen, S., Kimnunen, J., Lecloux, A. *et al.* (2000) *Chemosphere*, **41**, p. 1799.

Højberg, A.L., Refsgaard, J.C., van Geer, F., Jørgensen, L.F. and Zsuffa, I. (2007a) *Water Res. Manag.*, http://dx.doi.org/10.1007/s11269-006-9119-y.

Højberg, A.L. Jørgensen, L.F. and Refsgaard, J.C. (2007b) Joint use of monitoring and modelling, Harmoni-CA Guidance Document, Report under preparation – will become available on http://www.harmoni-ca.info/.

Jørgensen, L.F., Brandt, G. and Vanderberghe, V. (2006) Joint use of modelling and monitoring for implementation of the Water Framework Directive, 3rd Workshop: Case Studies, Workshop Report, Harmoni-CA – Harmonised Modelling Tools for Integrated River Basin Management, WP 2: Toolbox and WP 4: Joint use of monitoring and modelling, http://www.harmoni-ca.info/.

Kamphorst, E., Jørgensen, L.F., van Griensven, A. and Vanrolleghem, P.P. (eds) (2005) Joint use of modelling and monitoring for implementing the Water Framework Directive, workshop report, 1st Workshop: State of the art on existing monitoring programmes around Europe, Harmoni-CA, http://www.harmoni-ca.info/.

Onorati, G., Di Meo, T. and Mottola, A. (2005) The approach of Campania Region to groundwater quality monitoring, Abstracts from Aquifer Vulnerability and Risk, 2nd International Workshop, 4th Congress on the Protection and Management of Groundwater, Parma.

Onorati, G., Di Meo, T., Bussettini, M., Fabiani, C., Farrace, M.G., Fava, A. *et al.* (2006) *Physics and Chemistry of the Earth*, Parts A/B/C, **31**(17), pp. 1004–1014.

Parr, T.W., Sier, A.R.J., Batterbee, R.W., Mackay, A. and Burges, J. (2003) *Sci. Total Environ.*, **310**, p. 1.

Quevauviller, P. (2005) *J. Environ. Monitor.*, **7**, p. 89.

Rekolainen, S., Kämäri, J. and Hiltunen, M. (2003) *Int. J. River Bas. Manag.*, **1**, p. 347.

Steenstra, M., Troch, P., Santbergen, L. and Jørgensen, L.F. (eds) (2004) Remote sensing and data assimilation techniques, Workshop Report, Harmoni-CA, http://www.harmoni-ca.info/.

Wasson, J.G., Tusseau-Vuillemin, M.H., Andréassian, V., Perrin, C., Fauer, J.B., Barreteau O. *et al.* (2003) *Int. J. River Bas. Manag.*, **1**, p. 125.

4.2

Integrated River Basin Management: Harmonised Modelling Tools and Decision-making Process

Zbigniew W. Kundzewicz and Fred F. Hattermann

4.2.1   INTRODUCTION

The Water Framework Directive (WFD), formally the Directive 2000/60/EC of the European Parliament and of the Council of 23 October 2000 (European Community,

The Water Framework Directive - Ecological and Chemical Status Monitoring   Edited by Philippe Quevauviller, Ulrich Borchers, Clive Thompson and Tristan Simonart © 2008 John Wiley & Sons, Ltd

2000), guiding the European Community (EC) action in the field of water policy, creates a legal obligation for the EC Member States to achieve a 'good water status' for all waters by 2015. The Directive establishes a framework for the protection of all waters (including inland surface waters, transitional waters, coastal waters and ground-water). It aims at preventing further deterioration, reducing pollution, and protecting and enhancing the status of water resources and the aquatic environment. It also pro-motes sustainable water use and contributes to mitigating the effects of hydrological extremes – floods and droughts.

The Directive constitutes a considerable challenge to EC nations and their water sectors. It is particularly difficult for the new Member States (access dates: 1 May 2004 for ten countries and 1 January 2007 for two countries), in many of which the water quality problems are serious and widespread. It has to be implemented on the river basin scale, under a variety of existing conditions: water quality, legislations, standards, priorities, traditions, institutions, environmental awareness, etc. Perspec-tives of different stakeholders and sectors, and the projections of future changes in socioeconomic, terrestrial and climatic systems, have to be taken into account.

Since the Directive regulates management of water resources within each river basin district, the concept of integrated water resources management is implemented at the catchment scale (integrated river basin management).

In order to support the coherent and harmonious implementation of the Directive, several research and development projects have been launched within the Sixth Frame-work Programme of the European Union (EU). A cluster of projects, under the name of CATCHMOD (CATCHment MODelling), was funded by the EU. One such project was a concerted action, 'Harmonised Modelling Tools for Integrated Basin Management (Harmoni-CA)', launched in 2002, with the aim of bringing together the knowledge obtained in the different projects and bridging the areas of science/research and imple-mentation/practice. The objectives of Harmoni-CA include:

- Harmonisation of current scientific expertise in use and development of information and communication tools (ICT) relevant to the implementation of the WFD.

- Promotion of model-supported water management and the use of ICT tools in well-documented and monitored basins.

- Preparation of widely-accepted guidance documents for potential end users of ICT tools.

- Stimulation of broad availability of and support for ICT tools.

- Promotion of a forum for unambiguous communication and information exchange.

4.2.2 THE MODEL-SUPPORTED PLANNING FRAMEWORK

In order to fulfil the objectives of the Directive, it is necessary to manage water, land and related resources in an integrated manner. The need for integrated river basin man-agement has arisen because managing processes independently (without integration) is not likely to be sufficient and may not lead to optimal solutions for increasingly complex problems. The notion of integration, referring to the water resources management in the Water Framework Directive, is very broad and embraces integration of:

(1) Quality, quantity and environmental objectives.

(2) All water resources (fresh surface water and groundwater).

(3) All water uses, functions and values.

(4) Disciplines (hydrology, hydraulics, ecology, chemistry, soil sciences, agronomy, forestry, technology, engineering and economics).

(5) Water legislation in different sectors and levels.

(6) Stakeholders and the civil society in decision making, at different levels (local, regional and national).

(7) Water management between Member States sharing river basins.

Integrated water resources management is a complex task, so there are not many successful examples of model-supported management of water resources related to the WFD implementation (see http://www.euroharp.no). Among the possible reasons for failures are: inadequate data support; lack of experience of model users; overselling of models when applying for projects; deficits in model structure, spatial distribution and process description; and insufficient communication between modellers and users.

In order to facilitate model support in the process of implementation of the WFD by water managers, it is of crucial importance to structure the process in an appropriate way. Figure 4.2.1 presents the WFD's implementation timetable, indicating the role of modelling. References to articles of the Directive are also given.

Mathematical models can considerably help at several stages of implementation of the Water Framework Directive, as illustrated in Figure 4.2.1. Models are indispensable for impact prediction and what-if analysis (in planning, design and management of water resources). Models can help us understand and optimise resource use efficiency. However, models are also useful in other stages of the planning process, such as: assessment of the current status and preliminary gap analysis; establishment of monitoring programmes; gap analysis; setting up of the programme of measures; development of river basin management plans; implementation of the programme of measures; and informing of and consultation with the public, with active involvement of all the interested parties.

Four main management tasks can be identified where models could play a particularly important role in supporting the implementation process of the WFD, providing additional information about the chemical and/or ecological status of the specific water bodies and potential impacts of management options (Refsgaard and Henriksen, 2004):

- Identification – assessment of the current status and setup of monitoring programmes (should be completed now; the deadline was 22 December 2006).

- Design – setup of programmes of measures and river basin management plans (until 22 December 2009).

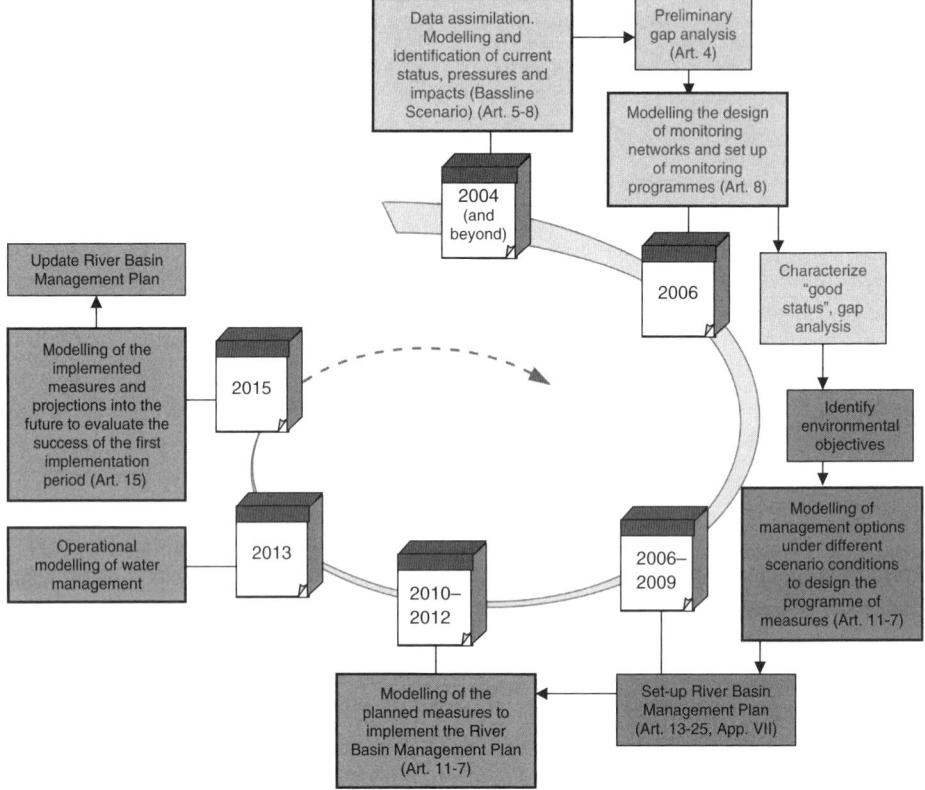

Figure 4.2.1 Timetable of implementation of the Water Framework Directive and the role of modelling to support different management tasks: identification of pressures and impacts (blue), design of programmes of measures (red), implementation of measures (green) and evaluation of the results (purple) (see Plate 3)

- Implementation – bringing the programmes of measures into action (until 22 December 2012).

- Evaluation – review of the programmes of measures and improvements achieved (22 December 2015).

At the stages of *identification* and *characterisation* of the individual river basin, including assessment of the current ecological status, impacts and pressures, as well as establishment of environmental objectives, modelling may be useful to support the definition of the reference conditions and to assess the possible pressures.

The evaluation of the susceptibility of water status to pressures can be achieved using both *monitoring* and *modelling*. Modelling can be useful, because in the initial phase available data are typically insufficient, or else the data records are incomplete in time and space, or simply erroneous. In combination with monitoring information (from both remote sensing and ground truth stations) and expert judgment, models can be used in assessing the impacts of the various pressures, but also in design of adequate monitoring networks. At the stage of *designing river basin management plans and programmes of measures*, modelling is a useful tool in supporting the assessment and quantification of the effects and costs of various measures under consideration. Further, at the stage of *implementation of the measures*, real-time modelling is used in some cases to support operational decision making, for example in reservoir operations, flood protection (forecasting and warning) and urban drainage systems. Finally, at the stage of *evaluation of the effects* of the planned or implemented measures on the environment, modelling may support the monitoring in order to extract maximum information from the monitoring data, e.g. by indicating errors and inadequacies in the data and by filtering out the effects of climate variability.

Monitoring and *modelling* are the backbones of the integrated river basin management. Monitoring enables us to identify progress in the pursuit of the policy long-term objectives, and to justify the needed actions. A comprehensive understanding of the water systems, with the help of a model representation of the relevant processes and variables, in combination with monitoring data, helps characterise the situation and identify pressures (Jørgensen *et al.*, 2007).

Monitoring and observations are indispensable for developing the models (in the stages of model identification and validation) and for using models aimed at simulating and/or forecasting the performance of the system. Models can be revised and updated as new data and information become available. Equally, modelling can clearly benefit monitoring. Models may indicate errors and weaknesses in the monitoring network. The joint use of data and information from different sources (field monitoring, remote sensing, modelling), with different accuracies and different spatial and temporal resolutions, and the continuous update of the model using these data, allows one to get comprehensive (as complete as possible) information about the system.

A possible way to overcome the problem of unknown (or changing) boundary conditions is to integrate and describe possible changes in climate, water and land use management in the form of *scenarios*, where readily validated models can be used to evaluate the possible impacts on the water quantity and quality under scenario conditions.

The additional value of model systems, of relevance to the management problem, is that they facilitate investigation of the complex processes in a river basin and possible impacts of planned measures in a cost-effective way, in short time and without the need for an active experiment (which would not be possible anyway). Therefore, they can be used, along with expert judgments and stakeholder dialogues, to describe processes, to identify and characterise water bodies, to evaluate impacts of planned measures in a river basin, to support the implementation of monitoring networks, and more generally to run decision support systems in the planning process and in the operational mode.

Figure 4.2.2 illustrates the main tasks which have to be considered in order to integrate participatory river basin planning and modelling in the process of the WFD implementation. Fundamental works to develop the scheme have been carried out in the German GLOWA-Elbe Project (Wenzel, 2005) and by Soncini-Sessa *et al.* (2007a, 2007b). Figure 4.2.2 demonstrates that the use of models is potentially beneficial in a number of stages of the framework (blue colour in boxes in Figure 4.2.2). Red colour in boxes indicates the need to involve stakeholders' participation (see also Hattermann and Kundzewicz, 2007).

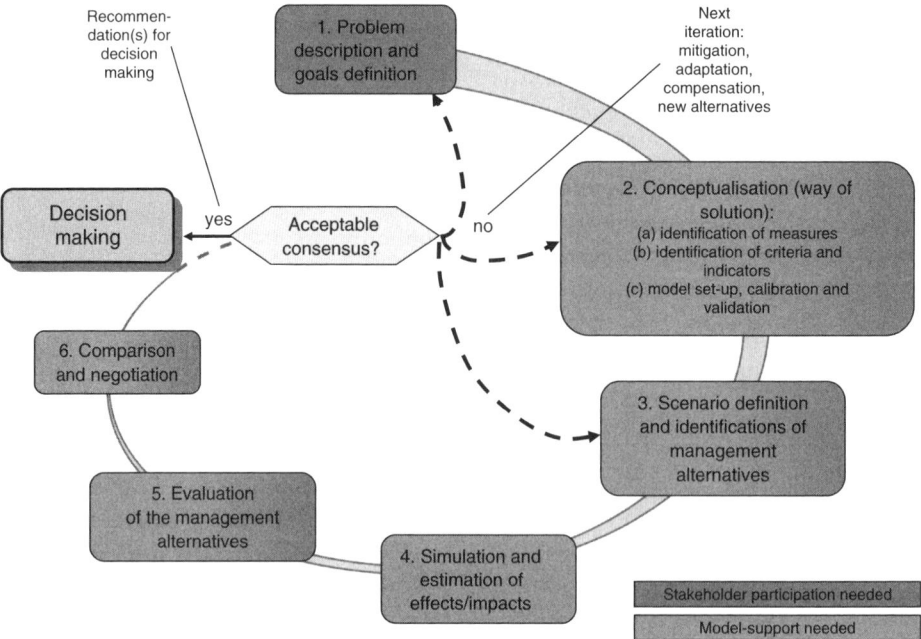

Figure 4.2.2 Framework for model-supported participatory planning of measures and integrated river basin management (planning framework, Hattermann and Kundzewicz, 2007; Wenzel, 2005) (see Plate 4)

Following Pahl-Wostl *et al.* (2004), three types of stakeholder and public participation can be distinguished:

- Information (the lowest form of participation).

- Consultation (different stakeholder groups and the public at large are asked to give their opinion on a management plan/scenario).

- Active involvement of stakeholders and the public in river basin management.

The last option is preferable, and it can lead to a proactive management regime via a co-production of knowledge and of co-decision making, and, in the end, can help to prevent undesirable impacts of unsuitable management measures.

The planning framework shown in Figure 4.2.2 describes the main tasks to be considered for integrating participatory river basin planning and modelling along the first main stages of the WFD implementation. It can be applied to manage and structure the entire implementation process, but also to organise individual modelling tasks. A very detailed methodology for structuring the modelling process as such has been compiled by the EU HarmoniQuA Project (MoST – Modelling Support Tool, see http://www.wise-rtd.info). For more detailed information about the stepwise approach of the planning framework see Hattermann and Kundzewicz (2007) and Soncini-Sessa *et al.* (2007a, 2007b).

4.2.2.1 Step 1: Problem Description and Goal Definition: 'Setting the Scene'

A clear definition of the management problem should be available at the beginning of the study, in order to have a cost-effective project performance and allocation of available resources. The aim of this step is to develop the working concept of the project. Starting from the problem description and goal definition, the 'means of solution' must be conceived and designed. This phase aims at defining the *system* of interest (river basin or river basin district), describes the problem to be solved (pressures and impacts) and the *goals* of the plan, including '*good water status*', as required by the WFD. The state of the system has to be characterised, including the relevant drivers and pressures. Looking at the entire WFD implementation process, this phase corresponds to the first three implementation steps according to the European Community (2003), namely:

- Assessing the current status and preliminary gap analysis.
- Setting up the environmental objectives.
- Establishing monitoring programmes.

The preliminary gap analysis consists of collecting the background information to define and describe the modelling problem ('what to model'); and the setting up of the environmental objectives is the input for developing a model study plan ('how to model'). For that purpose it is important to define the boundaries of the system (climate and management) and the time and space domains, as well as the *authorities*, *stakeholders* and *sectors* involved, and the institutional and legal framework of the plan. Stakeholder participation helps to identify major challenges and renders the management process more proactive.

4.2.2.2 Step 2: Conceptualisation

While the previous step aimed at 'setting the scene', this step intends to conceptualise the management and modelling problem.

Step 2a: Identification of Measures

In consultation with experts and stakeholders, possible measures to overcome the gaps found in Step 1 are identified. Examples of possible measures are the implementation of reservoirs or sewage treatment plants and changes in water management in general, but also changes in land use management (watershed management), reestablishment of wetlands and creation of riparian buffer zones. These measures should increase the buffering capacity and resilience of the hydrological system.

A coherent set of measures (following one storyline) represents a management *alternative*, whereby part of the action identification should be to clearly state '*who* is doing *what* and *when*'.

Step 2b: Criteria and Indicator Identification

A very sensitive task is to define criteria and indicators which can be used to describe the water-related problems and to measure the consequences of implementing new management measures.

In addition to the criteria defined by the WFD, and the existing legislation (including national or regional regulations), stakeholders must define a set of *evaluation criteria* reflecting the values that underlie their judgments (Soncini-Sessa *et al.*, 2007a, 2007b). These are the 'goal criteria', such as 'threshold' values of water levels, discharges, or water quality and ecological criteria not to be exceeded in order to reach a good status of the water body. They can be structured as a hierarchy, starting off with the goal of the plan and iteratively refining it.

Indicators are required to evaluate and compare how the hydrological system behaves without any action taken and under the effect of the management alternative(s), i.e. measure(s). Examples of such indicators are water levels, discharges, water quality characteristics, e.g concentrations of N and P, BOD, biomass, etc. These indicators also specify the type of model to be applied to investigate the state of the hydrological system.

Step 2c: Model Setup, Calibration and Validation

In order to simulate the system response to changes in management or climate and the trajectories needed to quantify the indicators, a site-specific model description of the entire system is required. Three sub-steps can be distinguished in building the model:

(1) model setup

(2) model calibration

(3) validation.

For a comprehensive introduction to issues of modelling in support of the implementation of the WFD, see the EU project HarmoniQuA (http://harmoniqua.wau.nl) and Refsgaard *et al.* (2005).

First of all, the decision must be made whether and where models are to be applied and what types of model (e.g., detailed, parsimonious) could be used. The most important selection criterion is the required accuracy of the results: if there is demand for very accurate and detailed model results, a more sophisticated model has to be applied, and relevant data have to be collected accordingly (Højberg *et al.*, 2006). Important aspects should be uncertainty assessment and quality assurance.

Figure 4.2.3 illustrates a more complex case of integrated modelling in support of the Directive – for the Elbe River Basin, embracing a large network of rivers, reservoirs, balance points and users. It is an international river, whose basin is shared by the Czech Republic and Germany. Much of the basin area consists of agricultural lands, but there are also large cities in the basin, such as Prague, Berlin, Leipzig, Dresden and Hamburg. Hence the low present per capita water availability ($680 \, m^3$) and projections of increasing stress (due to climatic changes – increase in temperature and decrease of summer precipitation) are reasons for concern (Wechsung *et al.*, 2005).

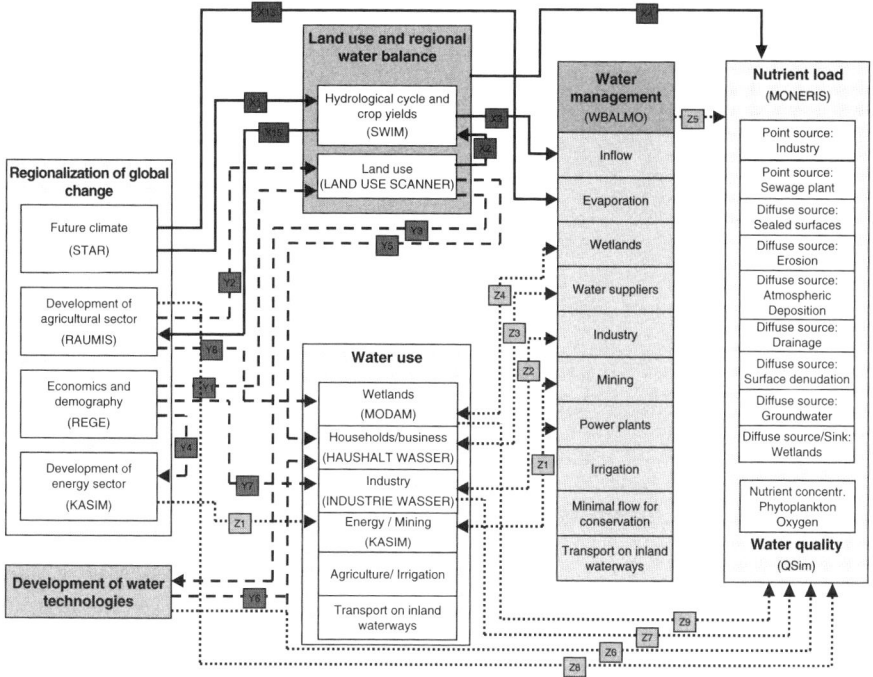

Figure 4.2.3 Conceptualisation of the domains to be included in the model setup to facilitate integrated river basin modelling; example of the GLOWA-Elbe project (Wechsung *et al.*, 2005)

In multi-scale studies, the larger-scale study can be based on a simpler (parsimonious) model, while in the smaller-scale study a more detailed model is required.

4.2.2.3   Step 3: Scenario Definition and Action Alternative

Projections into the future are required in order to define the boundary conditions under which the programme of measures will be implemented. First, the driving forces unaffected by local management, describing alternative external conditions, have to be considered, e.g. climate change and globalisation. Their trajectories are called *scenarios*, since they describe the background scene on which the alternatives act. Both the scenarios and the management alternatives (the second input) have to be quantitative and the scope of this step is defined by determining them.

Step 3a: Choice of Scenario

The scenario(s) may be chosen by experts or obtained by running models that describe the processes producing the driving forces; e.g. the future scenario of rainfall can be produced by a climate model, while the future scenario of land use is often proposed by an expert (Soncini-Sessa *et al.*, 2007a, 2007b).

The length of the time horizon defining the scenario must be sufficient to observe all the types of possible significant event in the system (e.g. 20 or 50 years). There may be different scenarios for designing alternatives (*design scenarios*) and for estimating impacts.

Step 3b: Design of the Programme of Measures (Action Alternatives)

The term *measure* or '*action alternative*' is used to emphasise that system modelling serves the investigations of 'alternative measures'. Measures are often suggested by the stakeholders and local water managers. This is a convenient starting point, but one should consider all alternative measures obtained by all combinations of actions identified in Step 2 as leading to achievement of the environmental objectives ('*good water status*' proposed by the WFD). Generally, the resulting number of alternatives can be fairly large. Therefore it is necessary to screen them in such a way that optimal combinations in terms of benefits for the different stakeholders are selected (the so-called *Pareto optimality*, where *subordinate alternatives* are removed; see Soncini-Sessa *et al.*, 2007a, 2007b).

4.2.2.4 Step 4: Simulation and Estimation of Effects/Impacts

The *impacts* resulting from each scenario and the advantages and disadvantages of each measure (alternative) have to be estimated by computing the values the indicators take for each alternative. The estimation is done using the model calibrated during the previous step. At the end of this step an *impact matrix* is produced whose elements are the indicator values for each scenario and alternative (see Figure 4.2.4). This serves evaluation and comparison of results.

4.2.2.5 Step 5: Evaluation of the Alternatives

Given the impact matrix, the goal of this step is to determine the '*value*' that each sector assigns to each alternative. A number of *evaluation* techniques exist which can assist the analyst in reaching this goal (e.g. multi-attribute value theory and analytic hierarchy process; see Soncini-Sessa *et al.*, 2007a, 2007b).

In cases where only one decision maker and one stakeholder exist, the *optimal alternative* can easily be found by ranking the alternatives with respect to their values compared with the adopted goal criteria and indicators, after which the procedure terminates (Soncini-Sessa *et al.*, 2007a, 2007b). If this is not the case, a different ranking of alternatives can be found by each stakeholder or decision maker at the end of this step; the decision process is not yet completed and it is necessary to proceed with the next step.

4.2.2.6 Step 6: Comparison and Negotiation

The aim and optimum result of this step is identification of a set of measures (pro-gramme of measures) which are perceived as a fair *trade-off* between different stake-holders' interests, without encountering anyone's opposition and taking Articles 4.5–7 of the WFD into account (see Hattermann and Kundzewicz, 2007). If such an alter-native cannot be found, another possibility is to identify the alternatives that live

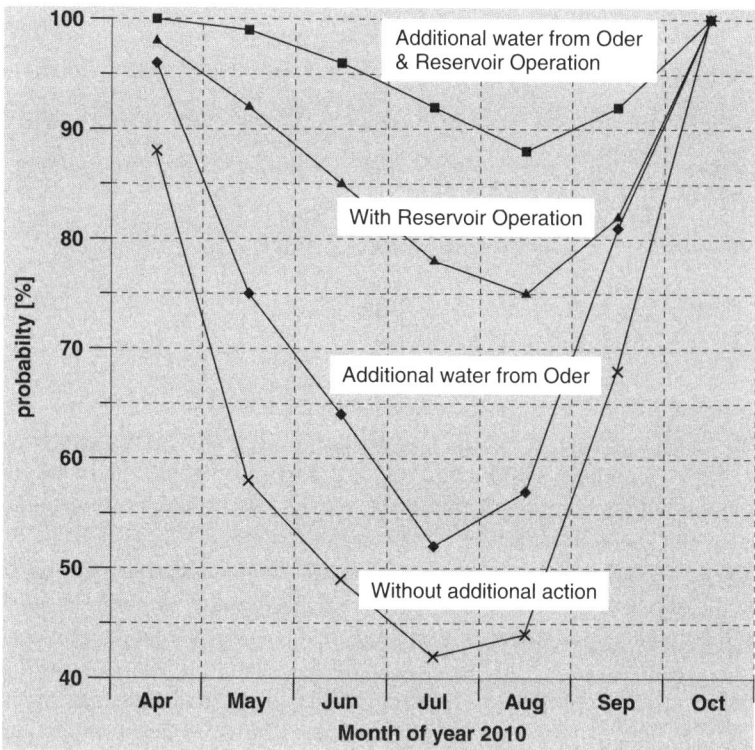

Figure 4.2.4 Reliability of having enough water flow in the River Spree (Berlin, $8\,m^3\,s^{-1}$) under scenario conditions. The impacts of a set of measures are shown (Becker and Kaden, 2007)

up to the environmental objectives (*'good water status'*) and gather a broad (yet incomplete) consensus amongst the stakeholders. Supporters and opponents of each of these alternatives must be identified (Soncini-Sessa *et al.*, 2007a, 2007b). The process begins by briefing each stakeholder on the other stakeholders' points of view. This includes a briefing on the negative effects that the actions preferred by the stakeholder in question will have on the other stakeholders and the environment (Article 14 of the WFD). Once this information has been shared, the core of this step is the *negotiation* amongst stakeholders to reach a compromise (Soncini-Sessa *et al.*, 2007a, 2007b). The result of this step is a set of management instruments that form the programme of measures to be included in the River Basin Management Plan.

4.2.2.7 Iterative Process of Arriving at the Decision

Normally a best compromise cannot be found in the first iteration of the planning framework ('no' in Figure 4.2.2). An iterative application of the planning process should be started in which, depending on the result, different alternatives for proceeding can be considered further.

The following alternatives may primarily be taken into account:

- Mitigation of negative impacts by technical measures (e.g. dykes, floodplains).

- Compensation by external (political or economic) measures (e.g. by subsidies).

- Introduction of new alternatives of process control (e.g. reservoir building, extending wastewater treatment capacities, etc.).

- Adaptation through a social learning process with the stakeholders.

4.2.3 WISE WEB PORTAL

One of the principal means through which Harmoni-CA aims to reach the objective of knowledge transfer is the WISE-RTD web portal (Figure 4.2.5), (http://www.wise-rtd.info/), which contains information on the experiences gained in pilot projects involved in the implementation of the WFD, as well as information on available tools, technologies and methodologies (Willems and de Lange, 2007).

This web portal has links to web sites containing information relevant for the implementation of the Directive. The linked web sites contain a wide range of information, such as (CIS) guidance documents, selection of ICT tools, methodologies and results of research projects.

The guidance material available via WISE-RTD covers more general topics such as guidance on the use of tools (e.g. model selection, model linking, model calibration and validation, model sensitivity and uncertainty analysis, etc.), the monitoring process, the stakeholder participation, etc. Such technical guidances are set up by various RTD

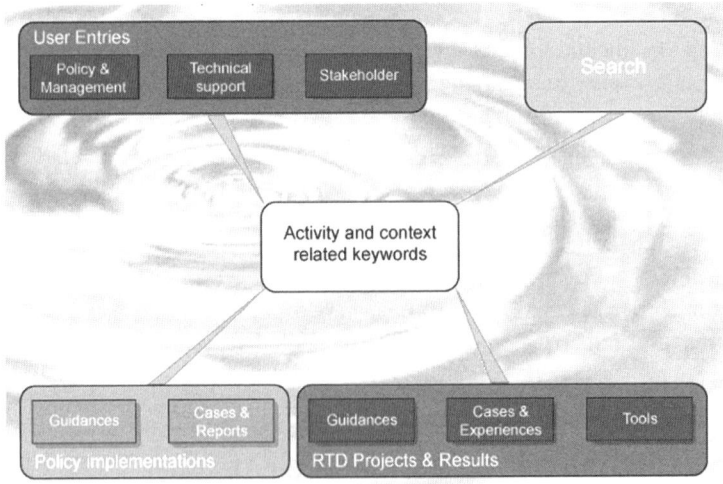

Figure 4.2.5 The scheme of the WISE-RTD web portal (Water Information System Europe – Research, Technology and Development; see http://www.wise-rtd.info/)

projects (e.g. projects of the CATCHMOD cluster of the EC), complemented and generalised with technical guidances by Harmoni-CA.

The WISE web portal has entries addressed to different user groups. It contains information targeted to such categories of users as those dealing with policy and management, technical support and stakeholders.

Experts implementing policy are interested in information about the possible types of technique and tool that can be used for the various policy implementation tasks, and in information on the data needs and on costs/benefits. There are 17 volumes of CIS Guidance Documents (GDs), being examples of such guidances applied to the WFD and its implementation tasks. These GDs are classified as per type of WFD tasks or per water-related discipline (economic analysis, analysis of pressures and impacts, monitoring, etc.). One of the CIS Guidance Documents (GD11) focuses on the planning in the WFD implementation (European Community, 2003).

It is useful for both scientists and practitioners to consult examples of existing policy implementations and the results of the use of RTD tools available via the WISE-RTD web portal. These examples complement the guidance material and can contribute to the dissemination of experience among different players in the area. This dissemination step is necessary to narrow the science–policy gap.

The list of RTD results and related tools and instruments that have been produced and developed by the scientific community and software developers is lengthy. In order to enable their analysis and reporting on their applicability to policy implementation, classifications have been proposed by discipline, by policy implementation task, by input and output variables, and by physical processes considered (e.g. Rekolainen *et al.*, 2004).

4.2.4 CONCLUDING REMARKS

It is broadly accepted that mathematical models play a considerable role in the various stages of implementation of the Water Framework Directive, and the present chapter provides ample illustrations. In order to improve the chances of successful model-supported implementation of the WFD, one should pay attention to the following essential items:

(1) Clear definition of goals.

(2) Focusing on a small number of clearly defined indicators, e.g. concentrations of substances, river flow indices, etc.

(3) Precision of the model study plan.

(4) Allowance for adaptive management at the stage of project design.

(5) Definition of goals, indicators, scenarios and management options.

(6) Effective communication between model developers and users at every stage of the process.

4.2.5 ACKNOWLEDGEMENTS

This is a contribution to the Harmoni-CA project, sponsored by the European Commission within its Sixth Framework Programme.

REFERENCES

Becker, A. and Kaden, S. (2007) In: Hattermann, F.F. and Kundzewicz, Z.W. (eds) *Model-supported Implementation of the Water Framework Directive: A Water Manager's Guide*, http://www.wise-rtd.info/html_docs/planning_guidance/planning_guidance.html, forthcoming.

European Community (2000) Establishing a framework for community action in the field of water policy, Directive 2000/60/EC of the European Parliament and of the Council of 23 October 2000, *Official Journal of the European Communities*, Brussels.

European Community (2003) Guidance Document No 11: Common implementation strategy for the water framework directive (2000/60/EC), Office for Official Publications of the European Communities, Luxembourg.

Hattermann, F.F. and Kundzewicz, Z.W. (eds) (2007) *Model-supported Implementation of the Water Framework Directive: A Water Manager's Guide*, http://www.harmoni-ca.info/products/, forthcoming with John Wiley & Sons, Ltd.

Højberg, A.L., Refsgaard, J.C. and Jørgensen, L.F. (2006) In: Refsgaard J.C. and Højberg A.L. (eds) (2006) *Nordic Water 2006, Proc. XXIV Nordic Hydrological Conference 2006: Experiences and Challenges in Implementation of the EU Water Framework Directive*, NHP Report No. 49.

Jørgensen, L.F., Refsgaard, J.C. and Højberg, A.L. (2007) *Environ. Monit.*, **9**, p. 931.

Pahl-Wostl, C., Schmidt, S. and Jakeman, T. (eds) (2004) The implications of complexity for integrated resources management, *iEMSs 2004 International Congress: 'Complexity and Integrated Resources Management'*, International Environmental Modelling and Software Society, Osnabrück, Germany.

Refsgaard, J.C. and Henriksen, H.J. (2004) *Advances in Water Resources*, **27**, p. 71.

Refsgaard, J.C., Henriksen, H.J., Harrar, W.G., Scholten, H. and Kassahun, A. (2005) *Environmental Modelling & Software*, **20**(10), p. 1201.

Rekolainen, S., Kämäri, J. and Hiltunen, M. (2004) *Int. J. River Basin Manag*, **1**(4), p. 347.

Soncini-Sessa, R., Castelletti, A. and Weber, E. (2007a) *Integrated and Participatory Water Resources Management: Theory*, Elsevier, Amsterdam, The Netherlands.

Soncini-Sessa, R., Cellina, F., Pianosi, F. and Weber, E. (2007b) *Integrated and Participatory Water Resources Management: Practice*, Elsevier, Amsterdam, The Netherlands.

Wechsung, F., Becker, A. and Gräfe, P. (eds) (2005) *Auswirkungen des globalen Wandels auf Wasser, Umwelt und Gesellschaft im Elbegebiet*, Weißensee Verlag Ökologie, Potsdam, Germany.

Wenzel, V. (2005) Der Integrative methodische Ansatz im stringenten Sprachkalkül, In: Wechsung, F., Becker, A. and Gräfe, P. (eds) *Auswirkungen des globalen Wandels auf Wasser, Umwelt und Gesellschaft im Elbegebiet*, Weißensee Verlag Ökologie, Potsdam, Germany.

Willems, P. and de Lange, W.J. (2007) *Environ. Sci. Pol.*, **10**(5), p. 464.

Section 5
Hydrogeological Components and Groundwater Status

5.1

Groundwater Quality Monitoring: The Overriding Importance of Hydrogeologic Typology (and Need for 4D Thinking)

Didier Pennequin and Stephen Foster

5.1.1   INTRODUCTION

Compared to the monitoring of all other natural waters, the satisfactory collection and interpretation of groundwater quality data requires an essentially 4D (four-dimensional) approach, because the large volume and the complexity of many aquifer systems (or hydrogeologic typologies) result in flow paths with residence times of very variable timescale, ranging from months to millennia (and usually counted in decades or

The Water Framework Directive - Ecological and Chemical Status Monitoring   Edited by Philippe Quevauviller, Ulrich Borchers, Clive Thompson and Tristan Simonart © 2008 John Wiley & Sons, Ltd

centuries), which often leads to major variations in groundwater quality with depth below the land surface, in addition to the normal areal variations and short-term temporal fluctuations also characteristic of other water bodies.

Reliable data on groundwater quality are essential to guide policy for sustainable resource management and effective resource protection. Whilst appropriate and standardised analytical and sampling procedures are required for this purpose, they on their own are not enough, since it is even more critical that the groundwater samples collected are:

- Representative of the 'compartment' of the groundwater body under consideration.

- Compatible with the specific objectives that are being pursued (e.g. protection against diffuse or point source pollution, evaluation of the effects of groundwater abstraction, setting up of programmes of measures for groundwater management, etc.).

Groundwater quality monitoring networks must therefore be designed with regard to:

- *Groundwater System Characteristics or Hydrogeologic Typology*: Subsurface environments, as mentioned above, are often complex, with groundwater flow paths and residence times being determined by such complexity. It is therefore essential to acquire a basic understanding of the subsurface geometry of aquifers and their associated groundwater flow system and hydrogeochemical regime before designing and installing a groundwater quality monitoring network. Indeed, a necessary condition for a groundwater monitoring network to be fully effective is that all significant processes controlling subsurface flow and chemical reactions are understood. This is particularly true if monitoring networks are to satisfy EC-WFD (European Commission – Water Framework Directive) requirements; given that working scales most of the time are large and network densities are generally low, network optimisation in this case is mandatory.

- *Clearly-defined Objectives*: In the case of the EC-WFD, the objectives set down for groundwater quality monitoring are very clear, and require monitoring networks variously sufficient to a) assess the overall chemical status of groundwater bodies, b) evaluate long-term trends in contaminant concentrations and c) determine the effectiveness of pollution control measures (and provide evidence of contaminant 'trend reversal').

- *Prevailing Pressures on the Groundwater Resource*: To meet these objectives it is necessary to determine the existing and potential future socioeconomic pressures (such as the nature, type and distribution of industrial activities, agricultural practice and urbanisation) that are generating or could generate a contaminant load on the subsurface and on the groundwater body concerned, since human activity can introduce contaminants into soils and surface water bodies that often will eventually infiltrate to groundwater.

This chapter will focus on the first of these three considerations – the importance to groundwater quality monitoring of taking into careful consideration groundwater system characteristics or hydrogeologic typology, *which exhibit wide geographical variations across the EU (and within many individual EC nations – see Figure 5.2.6 in Chapter 5.2 of this book)*, since this:

- Is of overriding importance when it comes to effective collection and interpretation of groundwater quality data.

- Tends most often to be overlooked or neglected.

5.1.2 GROUNDWATER SYSTEM CHARACTERISTICS

Five basic facets of groundwater systems need to be appreciated in order to design and implement adequate groundwater quality monitoring networks and to lay the groundwork for scientifically sound interpretation of groundwater quality variations:

- Groundwater flow regime geometry (from recharge to discharge areas).

- Groundwater flow velocities and residence times (which control 'aquifer system inertia').

- Natural groundwater quality controls.

- Mechanisms of subsurface contamination (and role of the unsaturated zone storage).

- Aquifer pollution vulnerability.

Each of these is discussed in some detail in the following sections.

5.1.2.1 Groundwater Flow Regime Geometry

Groundwater flows through aquifers which consist of permeable geologic layers. Most aquifers are formed from either the primary pore space of unconsolidated sediments or from fractured and fissured rocks, in which the secondary discontinuities usually result from tectonic and/or weathering processes. In many limestone formations the primary rock matrix can also be partly dissolved by chemical solutions, producing irregular galleries or karstic features. In porous media, groundwater flows through the connected intergranular pore space, while in fissured, fractured and karstic media groundwater flow can occur in two modes; mainly through the secondary open space, but also through the residual intergranular pore space.

In relation to groundwater monitoring, the following simple division is often used:

- Shallow aquifers occur close to the ground surface and their groundwater is said to be unconfined, being in direct continuity with the atmosphere and often in hydraulic connection with surface water bodies (rivers and lakes). They are made up of two distinct zones: a) the 'unsaturated zone', closest to the ground surface, where vertical seepage predominates but some water is often held in the finer pore space, and, below this, b) the 'saturated zone', where all pores, fissures or fractures are completely full of water and most groundwater flow occurs.

- Deeper aquifers, in contrast, usually have hydrostatic pressures exceeding atmospheric and their groundwater is said to be confined. They are fully saturated with water from the bottom to the top. They are often connected vertically or laterally 'up hydraulic gradient' to shallow aquifers and thus can then be regarded together as part of a larger 'aquifer system'.

Indeed, often many shallow and deeper aquifers are separated by less permeable geological horizons, known as aquitards, which retard, reduce and deflect subsurface water flow; in these cases, hydraulic connections do exist between adjacent aquifers, and significant water exchange can take place between them. Sets of hydraulically connected aquifers form aquifer systems, and this is today often the unit taken or the scale used to study, protect and manage groundwater resources. Aquifer systems are in turn separated by aquicludes, made of highly impermeable material, which do not allow water to get across.

Groundwater flow is therefore complex and must be regarded in terms of both micro and macro processes; *overall it is highly dependent on the nature and the micro- and macro-geometries of the aquifer system*.

Groundwater flows from recharge to discharge areas (in effect from areas of high to low hydrostatic head or pressure). Recharge areas are often underlain by the shallower parts of the aquifer system, which receive continuous or regular water inputs from the biosphere (such as infiltration fronts from major rainfall events). Discharge areas are the parts of groundwater systems through which water returns to the biosphere (such as springheads and wetlands), and in part can be ephemeral. In reality the recharge, transfer (intermediate between recharge and discharge areas) and discharge areas of groundwater systems often present rather complicated geometries as a result of the interaction of numerous geological, geomorphological and climatic factors. The geometry of the aquifer system itself (micro and macro) along with its hydraulic properties (permeability and storativity) determine the groundwater flow pattern, the flow lines, flow fields and flow paths, and the scale of the groundwater resource involved. Overall, the groundwater flow fields tend to be subhorizontal in aquifers and subvertical in aquitards (see Figure 5.1.1). In reality, the actual flow pattern can show a high degree of complexity due to the presence of geologic heterogeneities or geometric discontinuities; indeed, the downstream travel of water is highly influenced by the various difficulties encountered along its path, which may result in very tortuous flow lines and flow paths (see Figure 5.1.2). The driving mechanisms for groundwater flow are the three-dimensional hydraulic gradients which are established as a result of the interaction between the input of recharge water, the aquifer geometry and its hydraulic properties.

Groundwater flow-fields often undergo continuous natural evolution, reflected by the groundwater level fluctuations observed in monitoring piezometers/boreholes, which often appear to be in a perpetually transient state (see Figure 5.1.3). This is mostly due to temporal variations in the recharge inflow. In many cases, this natural evolution leads to periodic modifications in the geometry of the flow lines and the flow paths, which may sometimes become significant and prompt flow reorientation in some portions

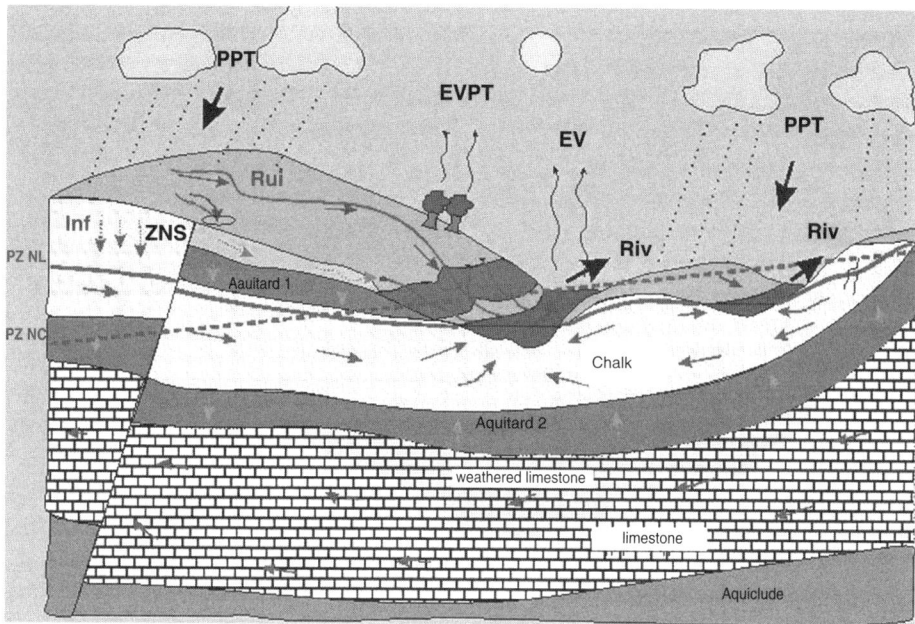

Figure 5.1.1 Typical groundwater flow in a lowland sedimentary aquifer system. Basically groundwater flows in a subhorizontal direction in aquifers (here a chalk and a limestone) and sub-vertically through the unsaturated zone (= infiltration between the ground surface and the water table - PZ NL) and the aquitards. The hydraulic head difference between the upper unconfined chalk (PZ NL) and the deeper confined limestone (PZ NC) drives and orientates the flow exchange between the two aquifers (PPT: precipitations; EVTP: evapotranspiration; EV: evaporation; INF: infiltration; ZNS: unsaturated zone)

of the aquifer system. In some rare cases flow reversal may even occur. In addition, pumping water supply wells (especially those of large yield in areas of high demand) can also radically disturb natural groundwater flow fields on a local- and sometimes larger-scale basis (see Figure 5.1.4).

5.1.2.2 Groundwater Flow Velocities and Residence Times

It is important from the outset to distinguish two independent but related concepts: aquifer pressure transfer and groundwater flow. When a stress is applied on a ground-water system (for example by an infiltration front prompted by a rainfall event in the recharge area), it reacts with a groundwater level rise, which will be propagated through the system. This 'pressure transfer phenomenon' happens relatively quickly – but the groundwater level rise down hydraulic gradient does not mean that recently-infiltrated water from the rainfall event has reached the downstream point of pressure response. Groundwater pressure transfer in aquifer systems is usually very much faster than actual groundwater flow, which in many cases will take from several months or many years, to decades and even thousands of years in large aquifer systems.

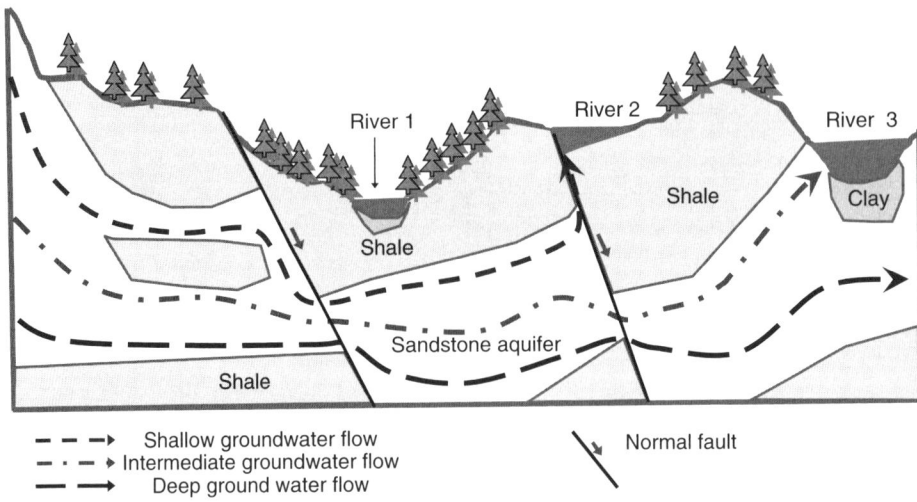

Figure 5.1.2 Groundwater flow is often strongly influenced by geologic heterogeneities and discontinuities. This example from a mountain area shows tortuous flow lines, which tend to adapt their orientation in order to travel preferentially through areas of greater permeability. This process can bring shallow, intermediate and deeper groundwater to have different flow paths and discharge areas, as shown in the figure. In this case, shallow groundwater discharges into River 2 after having seeped upward through a permeable normal fault, while intermediate groundwater feeds River 3 further eastward, and deep groundwater continues its subsurface flow to the east beyond River 3

The groundwater flow velocity and aquifer residence time will depend both on the aquifer permeability characteristics and on the hydraulic gradient. Groundwater flow velocities are very much slower than river flow velocities, ranging for the most part from cm/d to m/d – but may exceptionally reach km/d in some karstic aquifer systems. Although groundwater flow is generally a slow process, substantial variations may occur spatially within the same aquifer system according to existing heterogeneities, and these variations may further fluctuate with time as the system undergoes its natural evolution. *Groundwater velocities in an aquifer system therefore often vary both in space and in time.*

From the above discussion, it will readily be seen that the *residence time of groundwater in aquifer systems is generally large* – that is, *aquifers are most often 'high inertia systems'* with respect to water transfer, even if the scale may vary from one groundwater system to another, or even from one part of a given system to another. Variations in the inertia or in the residence time of water in large aquifers or aquifer systems most often stem from either or both velocity differences or path lines of different lengths (see Figure 5.1.5).

5.1.2.3   Natural Groundwater Quality Controls

Groundwater is never 'pure' but acquires a chemistry that reflects water–rock interactions which occur during its slow passage through the subsurface. It should thus

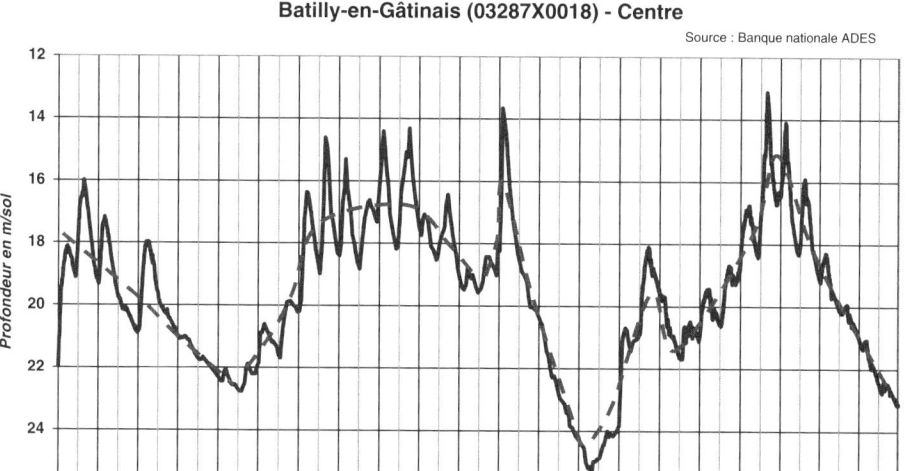

Figure 5.1.3 Groundwater level fluctuations displayed in a piezometer monitoring the Beauce limestone aquifer system in the Gatinais region (south of Paris). Different levels of fluctuations can be observed; seasonal variations (blue line) occur at the scale of the seasons and result mostly from winter recharge and summer discharge processes (most groundwater recharge takes place during the winter months, while discharge mechanisms from the groundwater system are predominant in summer). In addition, inter-annual groundwater level variations (red line) also occur due to fluctuations in the recharge volumes available from one year to the next. Sometimes, as it is the case here, irrigation may to some extent affect the natural water level fluctuations. (Source: modified from http://www.ades.eaufrance.fr)

be noted that 'natural background quality' may occasionally include concentrations of certain substances in excess of EC Drinking Water Guidelines (e.g. fluoride or arsenic).

As groundwater progresses through the subsurface, a series of hydrogeochemical reactions take place, commencing in the soil and continuing in the unsaturated and saturated zones. During this process, groundwater becomes enriched in desorbed and dissolved substances (consuming its dissolved oxygen and carbon dioxide), until it eventually reaches some equilibrium with the host rock of the aquifer. Then down the flow path it can cross into a different host rock or hydrogeochemical environment (say when dissolved oxygen is no longer present) and be subject to further reactions until a new equilibrium is again eventually reached; and later enter a third rock type or geochemical environment, and so on... all the way to the discharge area.

The chemical composition of groundwater resulting from natural subsurface flow and reaction processes is said to be its *'background quality'* or *'baseline quality'* – which can exhibit significant spatial variation, from upstream to downstream as seen above, but also laterally and with depth, as the path lines may cross out different geological

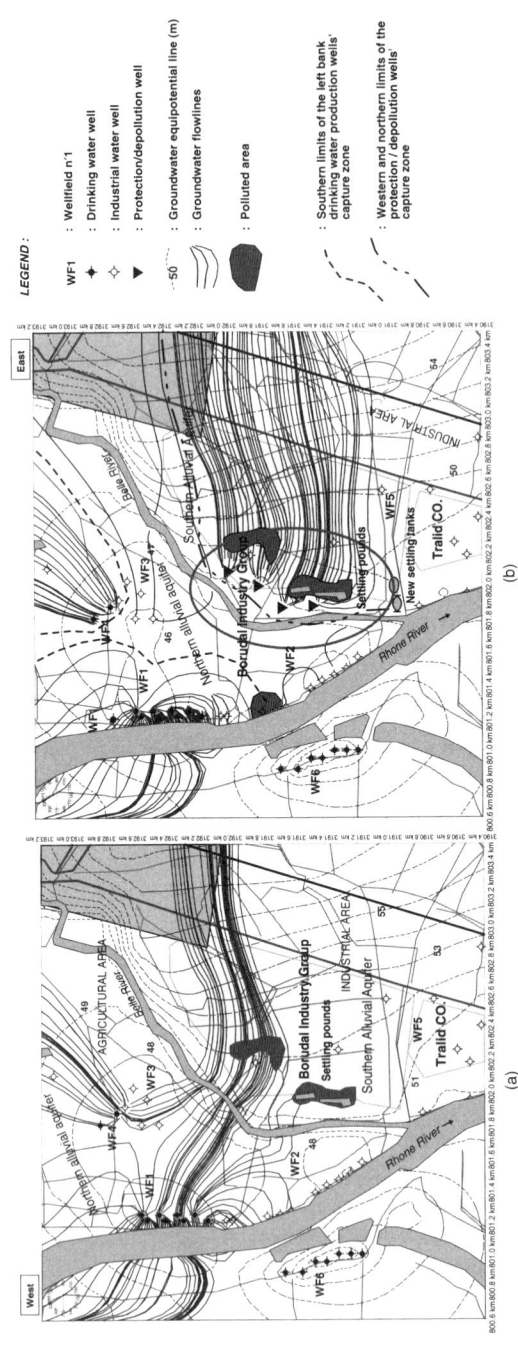

Figure 5.1.4 Example of the effect of pumping well fields on groundwater flow. In this area natural groundwater flow generally occurs in a south-westward direction in the left bank of the Rhone, and toward the south-east in the right bank. When active drinking water well fields were created on the left bank (a), significant modifications of this natural flow field took place, especially in the area located between the Rhone and the Belle River, where north-west groundwater flow started to predominate (replacing the natural south-west flow). An additional well field used for depollution was later installed (b – inside the red ellipse), which has further modified groundwater flow in the same area, this time inducing a significantly reduced divergent flow (Pennequin, 2000; Pennequin *et al.*, 2003)

204

Pessisjåkka

Abiskojokk (A-ätno)

Kaitumälven
Muddusälven
Lansån
Alep Uttjajåkkå

Sangisälven
Rakkurjaurbäcken
Skellefte älv
Storbäcken (Ostträsk)
Bjurbäcken

Fiskonbäcken
Röströmsälven (Korpån)

Anjanån
Semlan

Vindelälven
Sämsjöån
Moälven
Kvarnån
Ammerån
Kvarnån
Viskansbäcken

Ljusnan

Framsängsån (Ulvsjön)

Hångelån

Fämtan

V. Dalälven
Strömarån
Vistebyån (Sävjaån)

Klarälven
Mansån
Trösälven

Töftedalsån

Kagghamraån

Ålbergaån
Tivedalsbäcken
Svedån
Häragsbäcken

Anråsälven

Pipbäcken

Domneån

Västergarnsån

Färgeån

Nyrebäcken
Björkeredsbäcken

Norrhultsbäcken

Klingavälsån

Dammån

Skärån
Verkaån
Tolångaån

Plate 1   (See Figure 2.1.5)

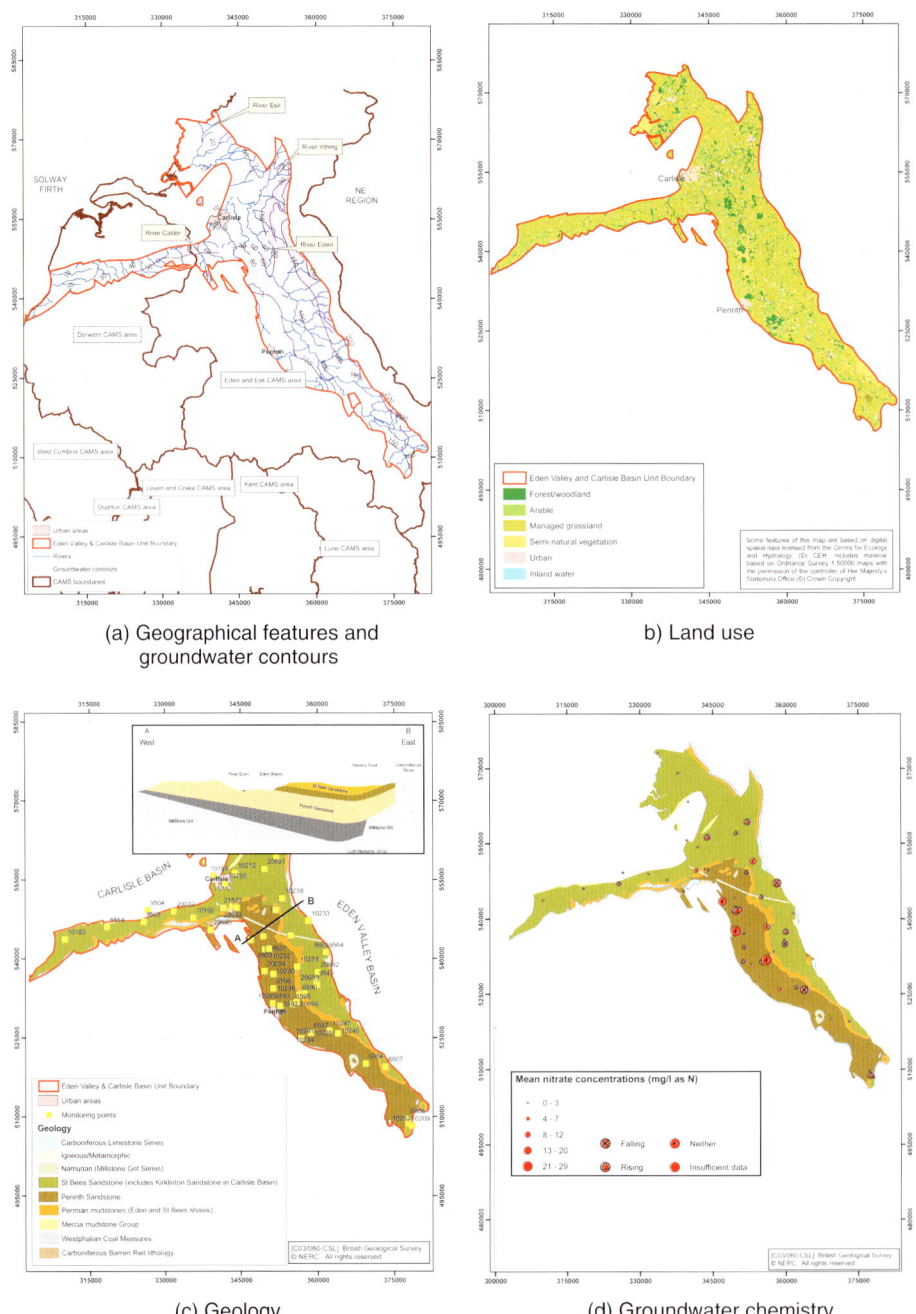

(a) Geographical features and
groundwater contours

b) Land use

(c) Geology

(d) Groundwater chemistry

Plate 2 (See Figure 2.3.2)

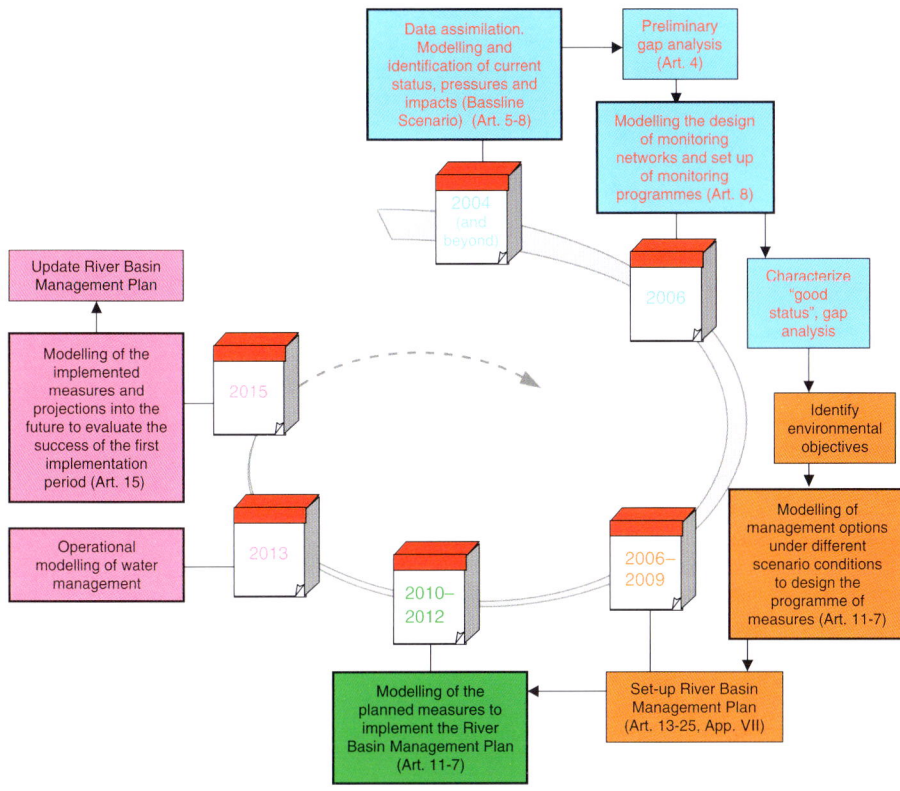

Plate 3   (See Figure 4.2.1)

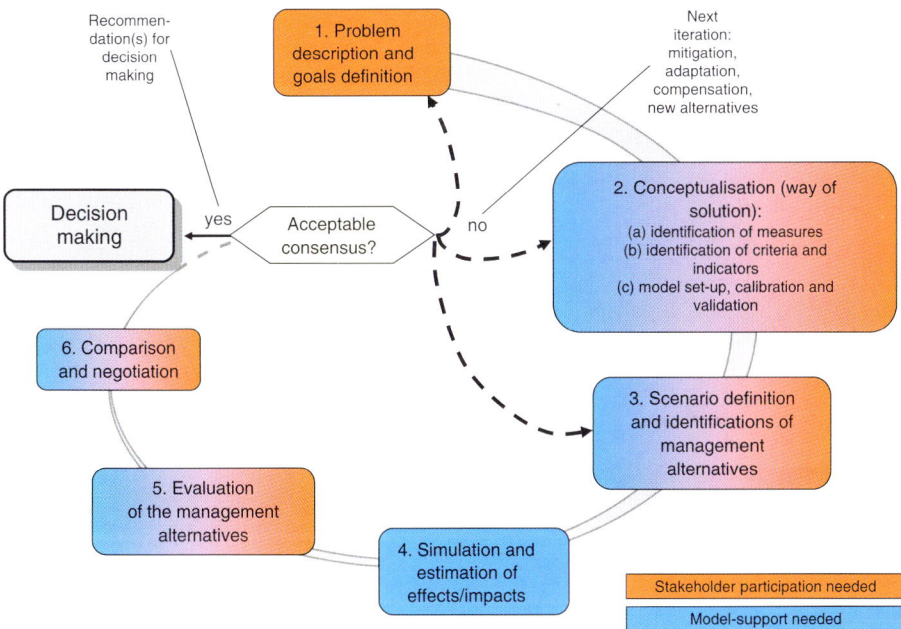

Plate 4   (See Figure 4.2.2)

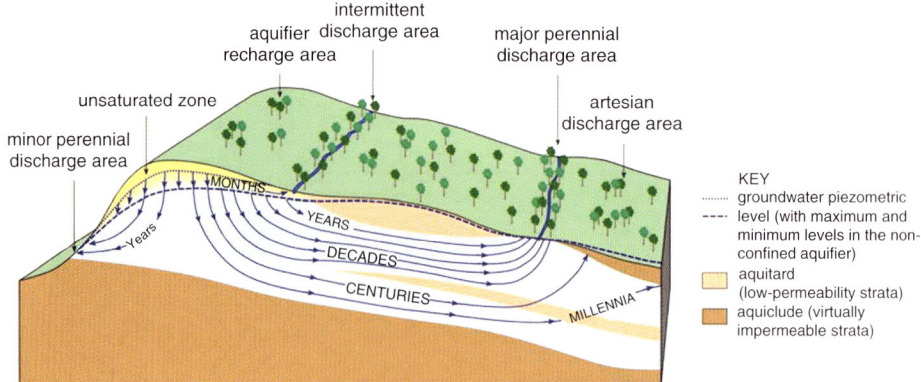

Plate 5 (See Figure 5.1.5)

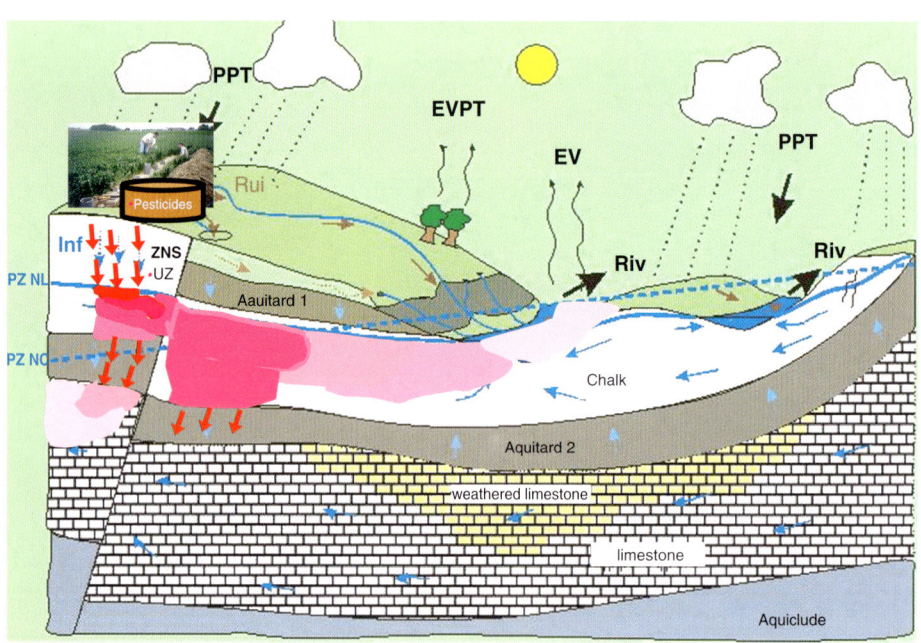

Plate 6 (See Figure 5.1.6)

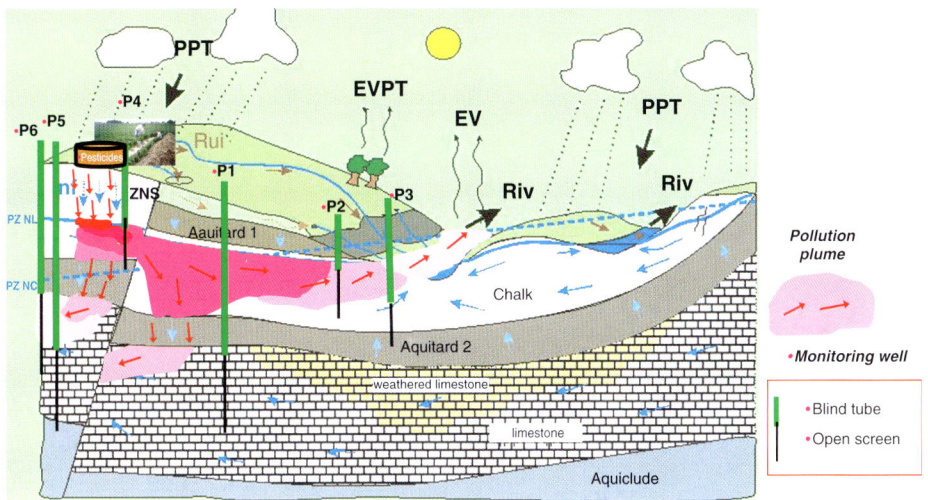

Plate 7 (See Figure 5.1.7)

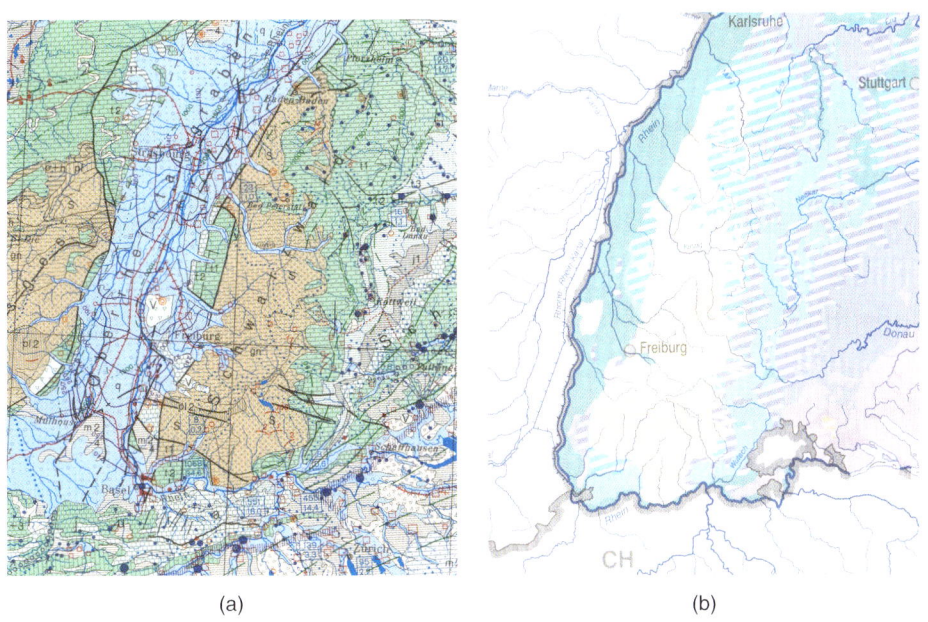

(a)   (b)

Plate 8 (See Figure 5.2.2)

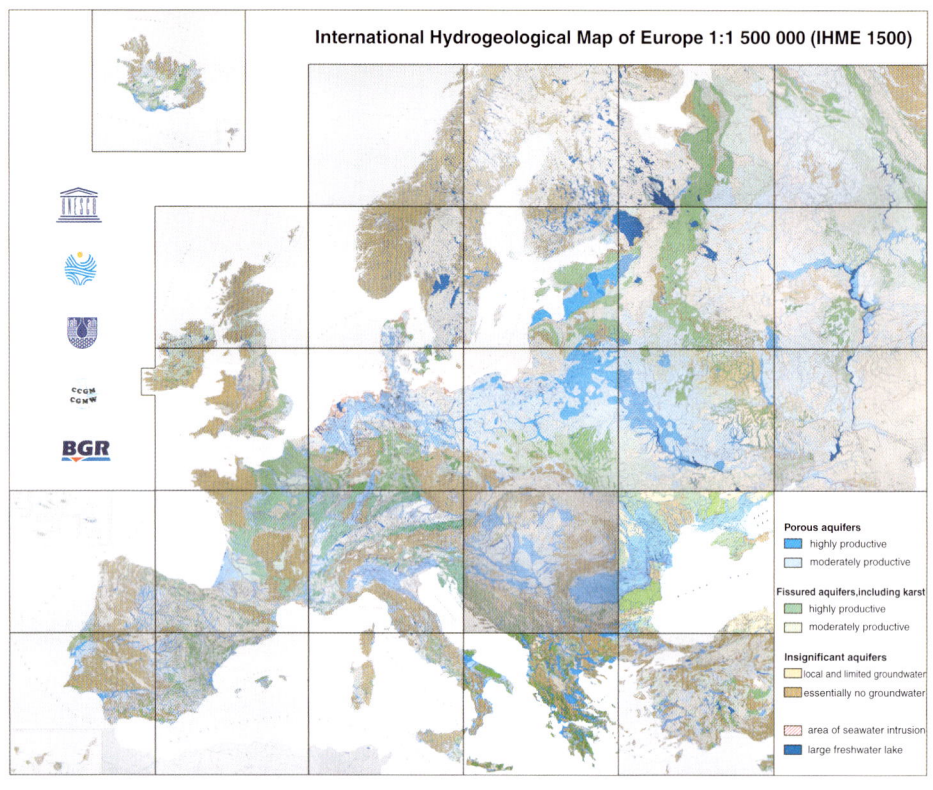

Plate 9   (See Figure 5.2.3)

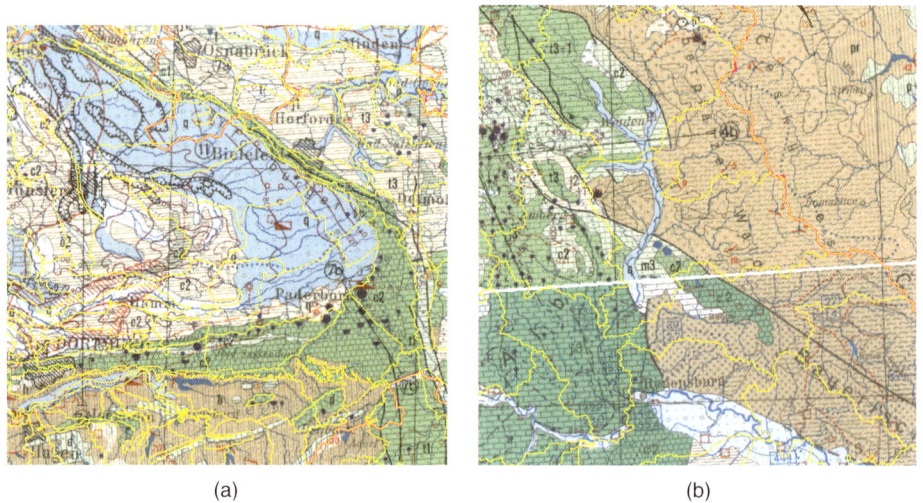

(a)   (b)

Plate 10   (See Figure 5.2.4)

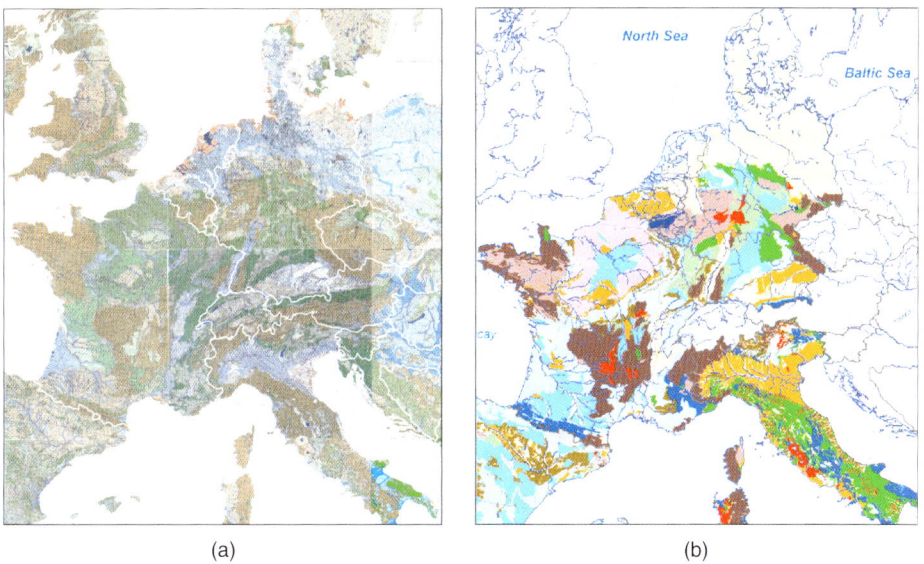

(a)   (b)

Plate 11   (See Figure 5.2.6)

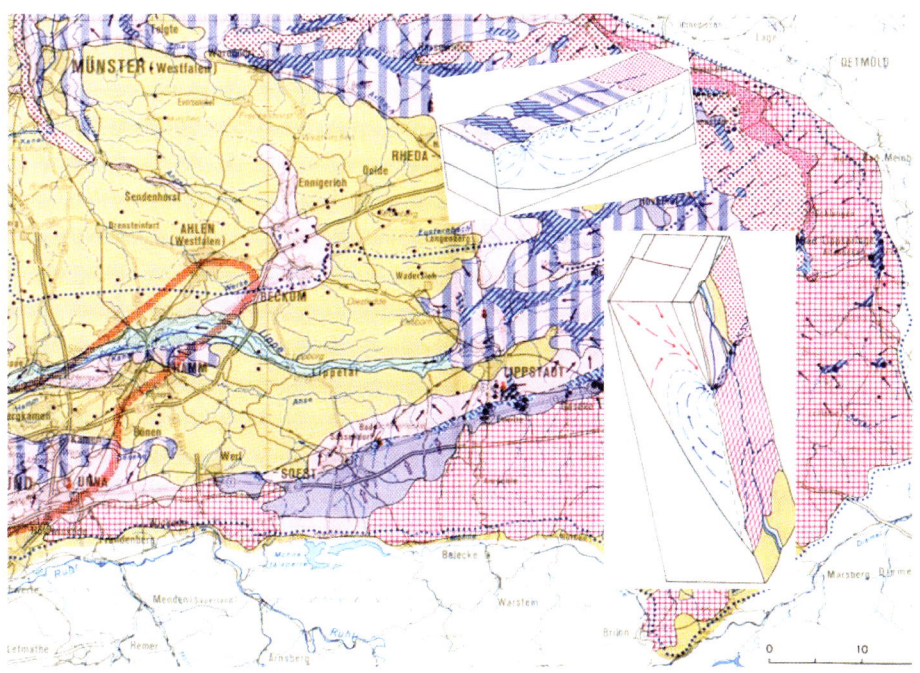

Plate 12   (See Figure 5.2.7)

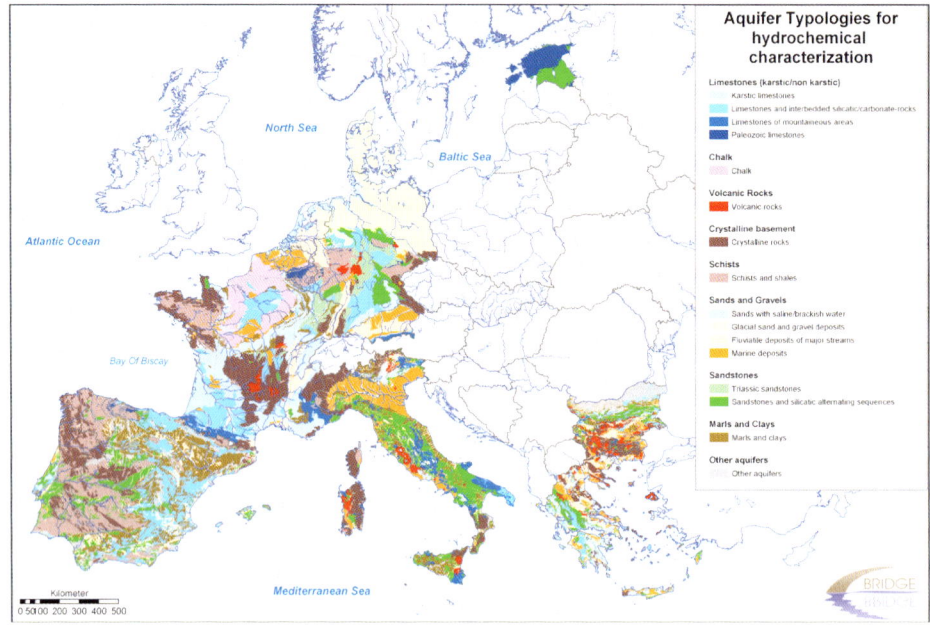

Plate 13   (See Figure 5.3.2)

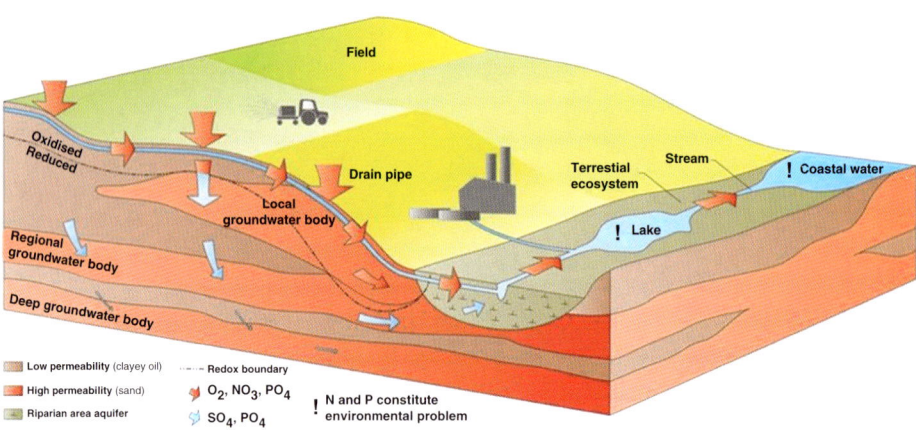

Plate 14   (See Figure 5.3.4)

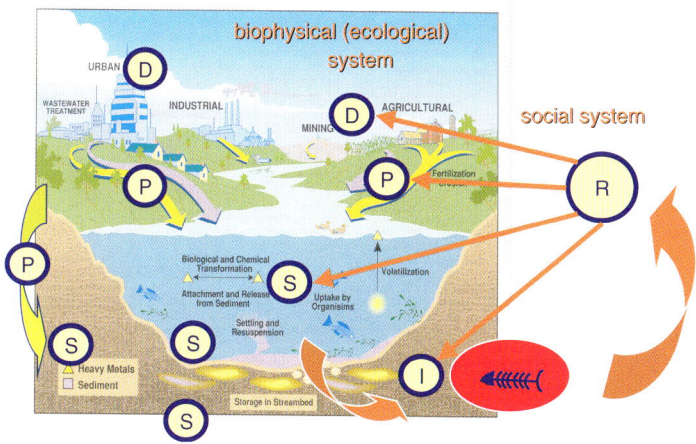

Plate 15   (See Figure 7.1.1)

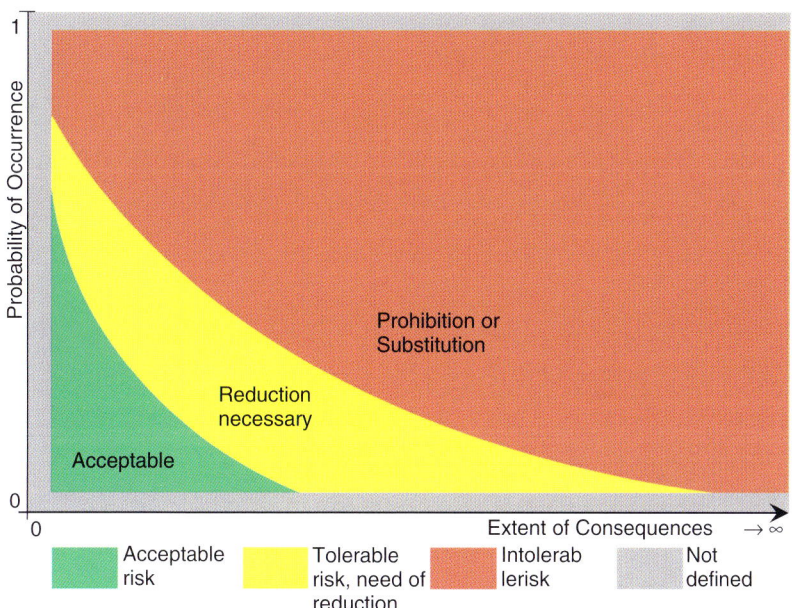

Plate 16   (See Figure 7.1.2)

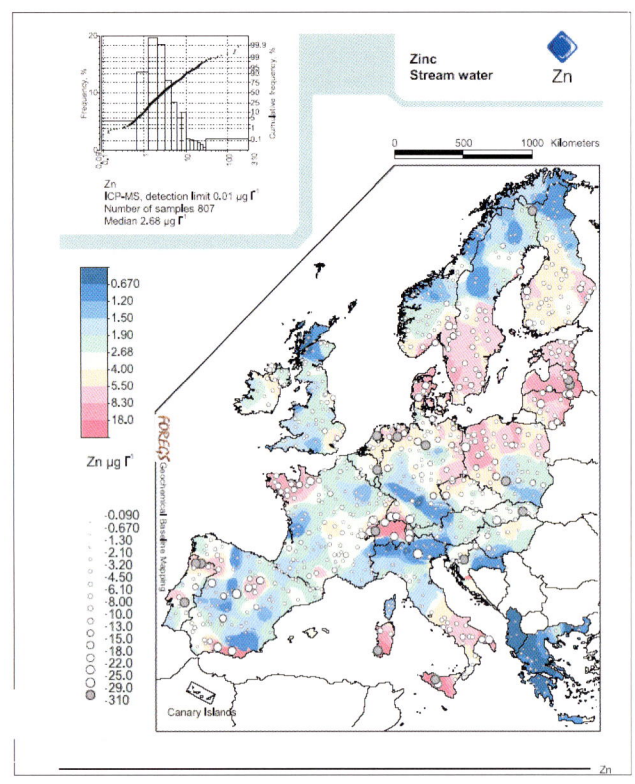

Plate 17 (See Figure 7.3.1)

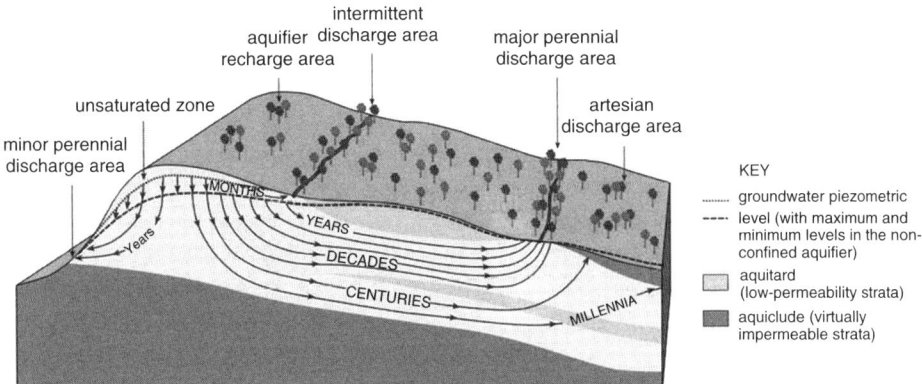

Figure 5.1.5 In large aquifer systems, shallow, intermediate and deep groundwater flow paths may vary significantly. Close to the ground surface, groundwater travel from recharge to the discharge areas may take only several years, while decades, centuries or more may be needed for deep groundwater circulation. This often results in different inertias and residence times for shallow, intermediate and deep groundwater. In addition, water quality may significantly differ from the top to the bottom of the aquifer system, due to flow paths crossing different geologic layers and to varying degrees of water–rock interaction (increased water–rock interaction often result in more mineralized water at depth) (Foster and Gomez, 1990) (see Plate 5)

facies or as the residence time of water, and thus the contact time with the host rock, is longer.

In addition, there may also be variation of background water quality with time, for example linked to a rising or falling groundwater table, or related to evolution of physicochemical conditions engendered by intensive groundwater pumping.

5.1.2.4 Mechanisms of Subsurface Contamination (and Role of the Unsaturated Zone Storage)

A wide range of urban, industrial and agricultural practices can, if not properly managed and controlled, generate significant pollutant pressures, and if the natural pollution attenuation capacity of the shallow subsurface is exceeded, groundwater pollution often results.

Subsurface contaminant migration is controlled primarily by the flow field, the properties of the contaminant(s) and the attenuation characteristics of the host rock involved.

The groundwater flow field (in the saturated and unsaturated zones) lays out the main pathways for contaminant underground transport. Basically, pollutant migration from ground surface to the groundwater table roughly occurs in a downward direction, carried along by rainfall events or other fluid discharges, since significant downward flow in the unsaturated zone can only take place when moisture content is close to saturation. Once pollutants reach the water table they start to travel along the groundwater flow lines, often in a subhorizontal direction.

However, a series of additional mechanisms are superimposed on the general scheme above, which can significantly control, impede or enhance subsurface pollutant transport:

- *Hydrodynamic Dispersion*: Due to the tortuousity of groundwater flowlines, which is linked to the irregular geometry of both primary and secondary aquifer pore space, the pollution plume tends to spread out and dilute in the three-dimensional space as it moves downstream. This dispersion can contribute either to acceleration or delay of pollution transfer.

- *Molecular Diffusion*: A micro-scale phenomenon occurring at slow groundwater flow velocities, which tends to spread dissolved pollutants into the finer pore space of the aquifer matrix and thereby greatly reduce their rate of migration, but greatly increase their persistence.

- *Density Effect*: Resulting from density contrasts between natural water and pollutants; immiscible organic compounds and highly saline brines often have lighter or heavier densities than water and as a result they behave differently and tend not to follow exactly the general flow field as they concentrate either close to the surface or in the deeper layers according to their pollutant:water density ratio. This often hampers their migration within the aquifer system.

- *Adsorption/Desorption*: While some contaminants move freely down the flow path at a velocity close to that of the groundwater itself, some tend to enter into complex adsorption/desorption reactions on the surfaces of soil and rock minerals. This often greatly retards their transport and increases their persistence into the aquifer system. Many organic compounds and heavy metals are prone to adsorption/desorption mechanisms.

- *Chemical Reactions and Transformations*: These are operative to varying degree in the soils, unsaturated and saturated zones. They can eliminate certain contaminants or create others.

- *Biodegradation*: One of the most complex series of mechanisms involving microorganisms, which remains poorly understood. Biodegradation can result in the complete breakdown (neutralisation) of some pollutants or in the transformation of others into daughter compounds (e.g. the pesticide atrazine to diethyl-atrazine) whose toxicity and mobility in the subsurface environment can be enhanced or decreased.

Most aquifer systems exhibit very high inertia, and such behaviour is even more pronounced in relation to pollutants, whose transport is generally slower than that of the groundwater. *The capacity of the unsaturated zone in effect to receive and store pollutants is especially significant*, since here groundwater movement is slowest and provides ample opportunity for molecular diffusion of pollutants from macropores into the finest micropores. *This zone can then act as a 'secondary source' of pollution*, in which pollutants can be retained for long periods of time and released periodically – this is the explanation of why 'pollution spikes' are often detected in groundwater downstream of an industrial site of point source pollution long after corrective measures have been implemented to prevent any further pollutant input into the subsurface environment.

Two distinctive types of occurrence of groundwater pollution are generally recognised:

- *Point Source Pollution*: Associated with clearly defined 'hot points' of pollutant discharge to the subsurface due to accidental chemical spills/leakages or improperly disposed wastes, generating high contaminant concentrations over small areas, affecting the soil, the immediately underlying unsaturated zone and the downstream groundwater, sometimes over great distances.

- *Diffuse Source Pollution*: Associated with widespread multiple sources, such as those derived from many agricultural practices (e.g. excessive or inappropriate use of fertilisers and/or pesticides), often generating pollutants in moderate to somewhat elevated concentrations in the soils and in the unsaturated zone over large areas, which eventually migrate down to the water table with infiltration fronts during or after rainfall events.

Overall subsurface pollutant transport advances very slowly, and as pollutants migrate down hydraulic gradient they tend to spread out and become diluted; but although they are diluted and sometimes undergo degradation to harmless byproducts, they more often *result in unacceptable groundwater chemical status, threaten groundwater supply for human consumption and remain in the aquifer system for very long periods of time*.

Furthermore, pollutants are not restricted to shallow aquifers only; if the pollution source is persistent or is widespread, pollutants in shallow aquifers can eventually manage to reach and spread into deeper aquifers, as with time they often end up crossing through low-permeability aquitards in areas affected by downward hydraulic gradients. Figure 5.1.6 schematically shows the geometry of subsurface pollutant transport.

5.1.2.5 Aquifer Pollution Vulnerability

Most pollutants are generated at the ground surface and in many cases they can infiltrate groundwater directly in the vicinity of the source area, or indirectly farther downstream through downward seepage from surface water bodies (after prior transport by overland flow to these surface water bodies).

The vulnerability of an aquifer system to pollution from the land surface is primarily determined by the thickness and characteristics of the unsaturated zone and/or of the capping strata which overlie it (soil layers included). Basically, the lower the permeability and the finer the grain size of the unsaturated zone and capping strata, the greater the natural protection of the aquifers against surface pollutants, and the lower their vulnerability to pollution. On the other hand, overlying coarse-grained materials or fissures, fractures and geometric discontinuities in the capping layers generally tend to increase their vulnerability to surface pollution.

In general, shallow aquifers are more vulnerable than deeper aquifers (but much less vulnerable than surface waters) to pollution. Deep aquifers are indeed generally better

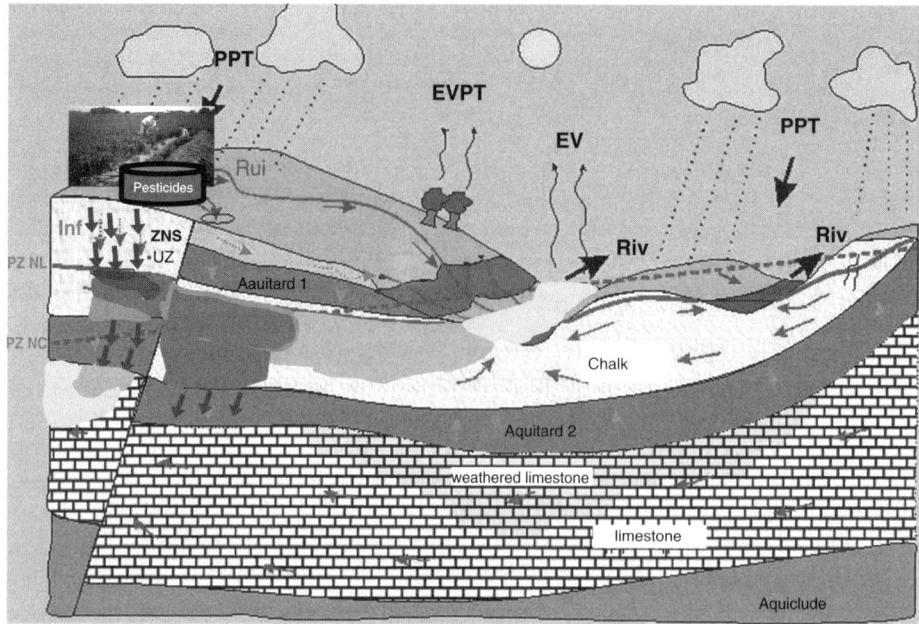

Figure 5.1.6 The main driving mechanism for pollutant transport below the ground surface is groundwater flow. Basically, contaminants migrate downward through the unsaturated zone to reach the water table (the groundwater surface). Once they penetrate into the shallow aquifer, their transport to the discharge areas is subhorizontal. Subvertical seepage through the aquitards may further occur according to hydraulic gradients and flow direction. If a pollution source is persistent, pollutants can eventually cross through most aquitards and contaminate the down-flow areas of deeper aquifers (see Plate 6)

protected by the presence of overlying aquitards, although as mentioned above, any persistent pollutants from the ground surface can eventually also contaminate them, provided that the vertical hydraulic gradients allow it to happen. More precisely, in situations of downward hydraulic gradients through the capping aquitard (e.g. water flows from the shallow to the deeper aquifer), the deep aquifer vulnerability to surface pollution increases, and the contrary is true when the direction of the gradient is upward (water flows from the deep to the shallow aquifer).

In the special case of rivers flowing across shallow aquifers, groundwater vulnerability to percolation of pollutants carried out by rivers is determined by two factors:

- The nature and the thickness of river bed deposits, which can range from coarse-grained and highly permeable to fine-grained with a very low permeability in the case of some perched or slow-flowing rivers.

- The direction of the vertical hydraulic gradients across the river bed, which control water exchange (same principle as for aquitards: downward gradients bring the river to feed the aquifer and upward gradients do the reverse).

Experience, however, shows that most of the river beds are permeable to at least some extent, and that the actual vulnerability of the underlying aquifers depends more often on the direction of the vertical hydraulic gradients (greater vulnerability when the vertical gradients are downward). However, both the nature and the thickness of the river beds and the vertical hydraulic gradients usually vary with time, with periods of floods and droughts, and as a result, the permeability of the river beds can increase and decrease, and so can the vertical gradients. The latter, in fact, can even become temporarily reversed at times, thereby modifying the direction of the water exchange (e.g. an aquifer normally feeding a river can sometimes switch to be fed by the same river and vice versa). The vulnerability of aquifer systems to infiltrations of contaminated river water is therefore a tricky concept, which can often fluctuate through time.

In the case of lakes overlying shallow aquifers, the groundwater vulnerability to polluted lake water is determined in a similar way as for rivers; often, however, the vertical hydraulic gradients tend to display a greater stability through time in these cases.

Aquifer systems thus naturally show a widely varying vulnerability to pollution from the land surface, which may even fluctuate with time, as in the vicinity of rivers.

5.1.3 DESIGNING GROUNDWATER QUALITY MONITORING NETWORKS

To fully meet EC-WFD objectives, *groundwater quality monitoring networks should be designed and installed following a systematic assessment of the groundwater flow regime, the flow dynamics and the pollution vulnerability of the aquifer system, together with the land use pressures and the location and type of drinking water sources and groundwater-dependent ecosystems that need to be protected.*

In practice, adequate and cost-effective selection of monitoring wells requires some prior basic knowledge of the main characteristics of the aquifer system under consideration, including *at minimum* some information on:

- *The Geometry and the Geology of the Aquifer System*: Necessary in order to install the monitoring wells in representative sections of each aquifer, to specify the depth (vertical position) of the well screen and to determine the sections of the annular space that need to be sealed in order to prevent unnatural mixing of water from adjacent shallower or deeper aquifers, which would totally confuse data interpretation. Particular attention must be paid to geologic properties and heterogeneities in the aquifer system and to the geometric discontinuities in the capping layers, as they usually have a significant impact on, respectively, groundwater velocities and flow direction, and the vulnerability of the aquifer system to surface pollution. In addition, information on geology is essential to establish the system potential for geochemical reactions with pollutants and to establish natural background water quality (necessary later for data interpretation).

- *The 3D Configuration of Overall Groundwater Flow Lines*: Necessary to establish the general pathway for pollutant transport. This requires general delineation of recharge, transfer and discharge areas, together with some basic data on groundwater hydraulic gradients to determine flow direction in the aquifers, and on the presence and properties of any significant aquitards within the overall aquifer system, to assess natural water exchange capacities between shallow and deeper aquifers. Without such knowledge, the area of influence or the capture zone of monitoring wells cannot be identified and thus what the observed groundwater quality data in fact represent cannot be assessed.

- *The General Groundwater Flow System Dynamics and Inertia*: Including basic information on groundwater level fluctuations, flow velocities and residence times, which are needed, for example, to locate monitoring wells according to the objectives that are being pursued (e.g. early detection of potential threats to groundwater quality, overall pollutant trend analysis, etc.), or to estimate migration speed of mobile contaminants (those which can move at the speed of groundwater) and help set up the basis for sound data interpretation.

- *The Vulnerability of the Aquifer System to Surface Pollution*: As it determines the areas through which pollutants could travel downward to reach groundwater and contaminate it. This information is essential when setting up water quality monitoring networks, as *the interaction between aquifer vulnerability and overlying land use is the normal approach to making reconnaissance assessments of the pollution risks to groundwater which are necessary to guide quality monitoring strategies.*

- *The Type and Spatial Distribution of Polluting or Potentially Polluting Activities*: Which provide useful indications of the groundwater pollutants most likely to be found in the aquifer system. This data, along with the above information (vulnerability, flow field, geology, etc.), is necessary to determine the pollutants' possible entrance points into the aquifer system and their potential subsurface migration, attenuation and transformation, as well as to guide sampling priorities and protocols.

- *The Geographic Distribution of the Potentially Impacted or Impactable Receptors*: Such as drinking water sources and groundwater-dependent ecosystems, to determine, along with the above data, appropriate locations for monitoring wells to aid their protection.

This list does not pretend to be exhaustive, but shows clearly how many different parameters must be considered in order to set up groundwater quality monitoring networks that are both functionally adequate and cost-effective. It will also be evident that network design is a specialist work, which will rely heavily on all available information about the aquifer system (including existing water quality data) and on expert judgements if the assigned objectives are to be fulfilled. Ideally, optimisation of monitoring networks should be undertaken by specialists with the use of mathematical or numerical modelling.

In many situations, however, sufficient information for adequate network design may simply not be available initially, and periodic reinterpretation of groundwater flow and hydrogeochemical regimes will be required to guide network improvement and consolidation. Figure 5.1.7 shows some examples of commonly found situations where water quality monitoring wells do not always fulfil their role.

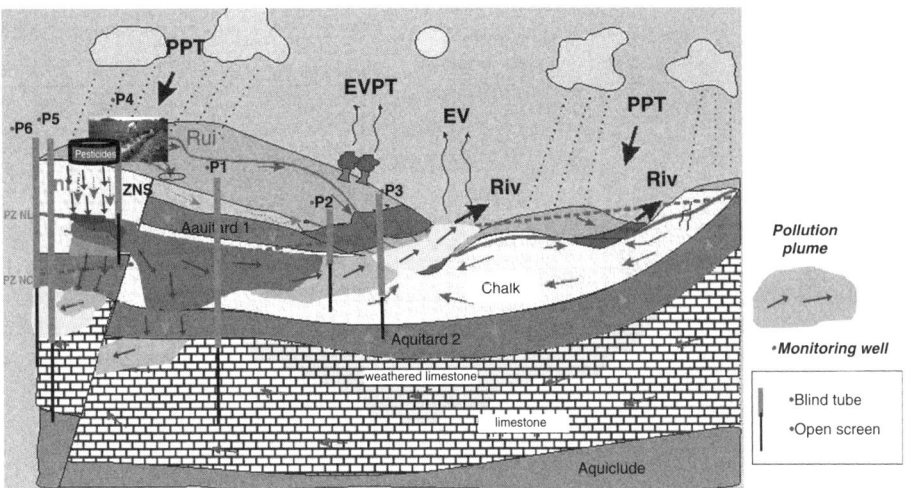

Figure 5.1.7 Insufficient information often leads to designing inefficient monitoring networks. A minimum understanding of the geology, the groundwater flow field, the system inertia, its vulnerability to pollution, as well as a good knowledge of potential pollution sources and receptors to be protected in the area, is required in order to install cost-efficient monitoring wells and to optimise the monitoring network. Here, in the case of the shallow aquifer, a network including monitoring wells P4 or P2 would detect the upstream pollution source, whereas a network with P3 simply would not, because the open screen is placed below the path of most pollution plumes, which is determined by the upward flow field leading to the river bed. P4 is, however, most efficient if it has to detect a new pollution very early in the recharge area, which would allow prompt reaction with corrective measures at a reasonable cost (by the time it reaches P2, the pollution has already had plenty of time to seep its way through the deeper aquifer, and it becomes much more difficult and costly to solve the problem). The same reasoning can be made for the deep aquifer, where a monitoring network including P1 and P5 would simply not detect the pollution (P1 is placed upstream, and the screen of P5 is outside the part of the flow field which carries out the pollution plumes). Only P6 is placed correctly in this case (see Plate 7)

5.1.4   INTERPRETING GROUNDWATER QUALITY DATA

An understanding of how groundwater flow systems function and what their character-istics are is equally necessary for the proper interpretation of groundwater quality data generated by a monitoring network. Misinterpretation of monitoring results is com-monplace without such understanding. Furthermore, this needs to be complemented with knowledge of the natural background groundwater quality, which is the starting point for any groundwater quality assessment.

The distribution and the nature of past, existing and potential pollution sources have also to be thoroughly established and assessed in light of the pollution vulnerabil-ity of the aquifer system concerned. A general appreciation of the reactive processes which may take place during contaminant transport, in all sub-surface compartments (soils, cappings layers, unsaturated and saturated zones in aquifers, river beds, . . .) is required too.

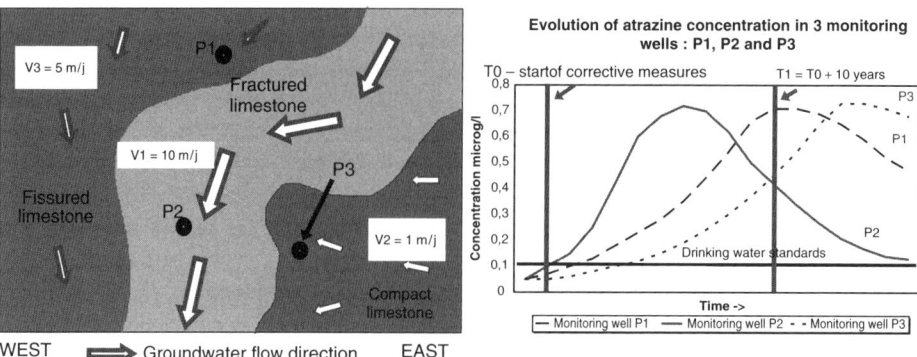

Figure 5.1.8 Example of the overriding influence of hydrogeologic typologies in response to measures applied to correct a diffuse pollution situation. Increasing concentration of atrazine in a heterogeneous chalk aquifer watershed led to a stop on using this pesticide in the early 1990s. Ten years after stopping atrazine applications, one monitoring well located in a low groundwater flow velocity aquifer compartment, P3, still displays a strong upward concentration trend for this pesticide, whereas P1, situated in moderate flow velocity materials, seems to suggest a possible future improvement, in the sense that the rising concentration curve starts to progressively level off. P2, in a high-velocity compartment, clearly shows the positive effects of the measures, with a downward concentration trend. Aquifer property differences within an aquifer system often lead to varying response times to corrective measures

All this information constitutes the necessary minimum background for the correct interpretation of groundwater quality data from a monitoring network. Water quality data analysis must then be performed both in the 3D space and in time. The 3D working basis stems from the geometry of the flow field and of the contaminant path lines. Taking into account of the time factor, or the 4th dimension, is required as groundwater systems inertia is almost always high and the processes leading to groundwater quality are very slow. Indeed, the introduction of corrective management or pollution control measures (e.g. to reduce nitrogen or pesticides inputs to groundwater) may often take years or even decades to start showing positive effects on groundwater quality (e.g. trend reversal), even in relatively shallow groundwater systems. At depth, this often takes even longer periods of time. Furthermore, the benefits of such measures may accrue at very different rates in different sections of a groundwater system, according to the prevailing local hydrodynamic and hydrogeochemical conditions. Figure 5.1.8 shows a typical example of this, where 10 years after stopping atrazine applications, two monitoring wells in low and moderate groundwater flow velocity aquifer compartments, P3 and P1, still display respectively a strong and a slight upward atrazine concentration trend, whereas P2, located in a high-velocity compartment, clearly shows a decreasing trend, meaning that removal of pollutants is more rapid here, but not necessarily (although it is possible) that the efficiency of the measures is definitely higher (their effect may indeed simply take a longer time to appear in the P3 and P1 compartments due to more constraining aquifer properties).

Pollutant concentration trend analysis is therefore a difficult task to undertake, as similar trend geometries may result from many different, independent causes. Indeed, a rising trend in pollutant concentration in a group of monitoring piezometers or boreholes may either simply reflect a general increase in direct input of the pollutant into the aquifer system or, equally, the arrival of a past pollution front at these observation points, or even a remobilisation of past trapped pollution (secondary pollution source) from the unsaturated zone below the recharge area. Likewise, a decreasing trend in pollutant concentration may reflect an actual reduction in pollutant input, but without additional information one cannot rule out the possibility that upstream pollution sources (primary or secondary) are still active, with pollutants underway that have not yet reached the observation points concerned. *Without the proper level of understanding of the 4D processes which operate to affect groundwater quality, no definite conclusions on the nature of the problems, nor on the efficiency of possible undergoing programmes of measures can be drawn.*

5.1.5 CONCLUDING REMARKS

(1) Groundwater flow and hydrogeochemical regimes often exhibit a significant degree of complexity, which must be assessed and understood (at least to a minimum extent) before attempting to design and install a groundwater quality monitoring network.

(2) In essence, groundwater quality monitoring network design and data interpretation must be undertaken from a 4D standpoint, which embraces an appreciation of the mechanisms affecting the three-dimensional spatial distribution of contaminant, and of the 4th dimension, which is the great inertia of most aquifer systems, and the very long time scales potentially involved in groundwater and pollutant transport.

(3) Goundwater monitoring (sampling and analytical) strategy needs to be guided by knowledge of potential present and past sources of pollution, which indicate pollutant pressures on the subsurface environment and, when combined with maps of aquifer pollution vulnerability, the probable nature and scale of the main groundwater pollution threats to drinking water sources and groundwater-dependent ecosystems.

(4) The problem of diffuse-source agricultural and urbano-industrial pollution should command special attention, in part because of the complex role played by the unsaturated zone, which can encompass a mix of both contaminant attenuation and contaminant accumulation, with periodic release into the aquifer system. River beds can play a similar role in the case of many industrial pollutions.

(5) The general philosophy advocated in this paper on groundwater quality monitoring is of special relevance for the implementation of the EC-WFD, where investments to optimise groundwater quality monitoring networks and representative data collection are mandatory to allow scientifically sound assessment of contaminant trends, so as to establish effective programmes of measures in line with the objective of achieving good groundwater chemical status.

(6) In order to fully embrace the objectives of the EC-WFD, a tri-component monitoring network for groundwater quality will ideally need to be developed with separate and adapted monitoring piezometers/boreholes, focusing on:

– *The Recharge Zone*: For early-warning (or impact) monitoring, to evaluate and validate the risk to groundwater bodies and to assess the effectiveness of corrective measures.

– *The Main Groundwater Body (or Transfer Zone)*: For strategic surveillance monitoring, to assess general groundwater body chemical status and trends in water quality.

– *The 'Discharge Receptors'*: For compliance monitoring, to guarantee their protection according to EC drinking water and environmental guidelines.

REFERENCES

Chilton, P.J. and Foster, S.S.D. (1997) Monitoring for groundwater quality assessment: current constraints and future strategies, *RIZA International Conference on Monitoring*, Tailor-Made II, Nunspeet, The Netherlands, pp. 53–64.

Foster, S.S.D. (2006) Monitoring and managing diffuse agricultural impacts on groundwater: key challenge for the new EC Water Directives, IGME Publication *Serie Hidrogeologia y Aguas Subterraneas*, **21**, pp. 3–16.

Foster, S.S.D. and Candela, L. (2008) Diffuse groundwater quality impacts from agricultural land-use: management and policy implications of scientific realities, In: Quevauviller, P. (ed.) *Groundwater Science and Policy: an International Overview*, RSC Publishing, Cambridge, UK, pp. 454–470.

Foster, S.S.D. and Gomes, D.C. (1990) Groundwater quality monitoring: an appraisal of practices and costs, WHO-PAHO-CEPIS Publication, Lima, Peru.

Foster, S.S.D., Hirata, R., Gomes, D., D'Elia, M. and Paris, M. (2002) Groundwater quality protection: a guide for water utilities, municipal authorities and environment agencies, World Bank Publication, Washington DC, USA.

Parker, J.M. and Foster, S.S.D. (1986) Groundwater monitoring for early warning of diffuse pollution, *IAHS Publication*, **157**, pp. 37–46.

Pennequin, D. (2000) Combined field and mathematical approach to protect two drinking water production wellfields from contaminant plume migration, *1st World Water Congress of the International Water Association (IWA)*, pp. 417–426.

Pennequin, D. (2002) Fonctionnement des systèmes aquifères, *Annales du colloque "les Entretiens de l'Environnement"*, APESA, PAU.

Pennequin, D. (2008) Les eaux souterraines, ressource vulnérable à préserver, *Pour la Science*, Dossier Janvier–Mars 2008, pp. 60–61.

Pennequin, D., Poitrinal, D., Pointet, T. and Machard de Grammont, H. (2003) Techniques d'optimisation environnemento-économique appliquées à la gestion intégrée des ressources en eau, *La Houille Blanche*, **3**, pp. 77–85.

5.2

Contribution of Hydrogeological Mapping to Water Monitoring Programmes

Wilhelm F. Struckmeier

The Water Framework Directive - Ecological and Chemical Status Monitoring   Edited by Philippe
Quevauviller, Ulrich Borchers, Clive Thompson and Tristan Simonart © 2008 John Wiley & Sons, Ltd

5.2.1 SURFACE WATER AND GROUNDWATER WITHIN THE WATER CYCLE

5.2.1.1 Introduction

The concept of the coherent water cycle is the basis for our understanding of all water flows in the atmosphere, as well as on and beneath the land surface. Rain or snow from clouds, snow and ice cover, water bodies in rivers and lakes are the visible parts of the water cycle, but infiltrating soil water and deeper groundwater in aquifers, although less spectacular, are of key importance for water supply in many countries in the world. Groundwater is the invisible part of the water cycle, and its turnover times are much slower than those of the atmospheric components. Hence it can be considered as a buffer of climatic extremes, as it can store storm waters in the unsaturated zones of aquifers, and it contributes substantially to river runoff during drought periods, through the discharge of base flow from groundwater stored in aquifers.

A number of important interactions between the surface water and the groundwater are shown in Figure 5.2.1, from a groundwater systems conceptual perspective. Without this conceptual understanding we cannot explain correctly the water quantity and quality data measured in monitoring programmes.

5.2.1.2 Hydrogeological Systems and Nested Groundwater Bodies

Aquifer layers in the underground are built up of unconsolidated formations, such as gravel, sand or silt, or semi-consolidated or fissured hard rock formations that are permeable enough to store and transmit noticeable quantities of groundwater. Therefore, the geological conditions, in particular the lithological composition, the bedding

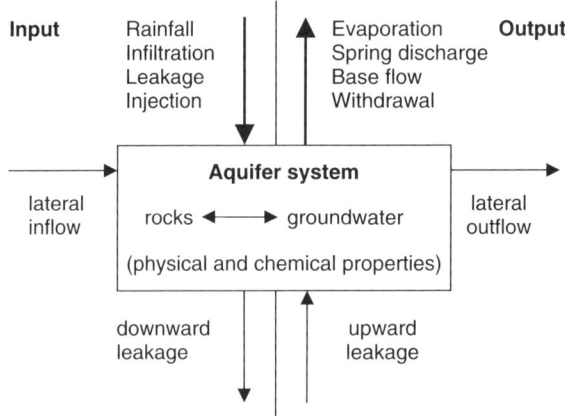

Figure 5.2.1 Conceptual framework of the complex system of groundwater and aquifer formations

structure and the deformation history (folding, faulting, fissuring) play a major role for the storage and movement of groundwater in the upper crust of the earth.

Groundwater bodies cannot be dissociated from the rock bodies in which they are nested. The minerals building up the hydrogeological formations interact closely with the groundwater contained in the openings of the pores and fissures. Therefore, an aquifer layer is usually associated with a certain characteristic, natural chemical composition of its groundwater.

The rock formations making up aquifers and less permeable layers separating aquifer layers from each other in the vertical succession are hosting groundwater flow systems and groundwater bodies. As such, they are much more persistent in their geographical location, composition and structural setting than the groundwater itself, being subjected to the driving forces of the water cycle, which changed significantly during the evolution of the earth.

An aquifer formation may be more or less filled with groundwater, depending on the climate and recharge conditions. It will stay in its place for thousands and millions of years, except if the rock body is removed, e.g. by quarrying sand, gravel or other building material. Hence, the mapping of hydrogeological units has become a basic tool in the assessment of groundwater flow systems, their quantity and quality aspects.

5.2.1.3 Coherence of Lithological, Hydrogeological and Hydrochemical Conditions

Hydrogeology is a complex science studying the interrelations between the lithological conditions of rock formations (consolidated and unconsolidated), the structural and tectonic setting of aquifers, aquitards and aquicludes, the input to and output from an aquifer in terms of fluxes, the water–rock interactions, as well as the chemical composition and evolution of the groundwater.

Aquifer formations and the overlying soil cover form one coherent system, since soils are usually the decomposition products of the underlying rock formations, and their physical and chemical characteristics usually mirror the properties of their parent formations. As the groundwater is recharged from rain and surface water percolating through the unsaturated soil zone, the physical properties of the soil layer are extremely important for the recharge rates, but also for capillary rise from the aquifer to the root zone of vegetation. In addition to this, the soil zone and the unsaturated upper part of an aquifer layer can be crucial for holding back contaminations from the surface. This transition zone is also known for its cleaning capabilities through absorption and/or slow biodegradation of chemical impacts from the surface.

The hydrogeological setting is a decisive factor for the groundwater flow in the subsurface. As the fresh, useable groundwater resources are of prime importance for water supply and management, only the upper freshwater horizons generally overlying the deeper parts of the earth's crust, filled with saline water or brines, are considered. This freshwater zone may range from almost close to surface to a depth of a thousand metres and more, depending on the regional hydrogeological conditions. In many sedimentary basins, we find a vertical succession of aquifers and nonaquifers (aquitards, aquicludes), and the individual aquifers might be quite distinct in their hydraulic properties (hydraulic conductivity, water levels and pressure heads) as well as their hydrochemical composition. This layering (hydrostratigraphy) and vertical hydrogeochemical zoning is a particular challenge for hydrogeological mapping and requires a three-dimensional approach of describing the underground conditions correctly.

Owing to the intensive interaction between the aquifer body and the groundwater nested therein, the chemical composition of groundwater is very specific. A comparison between a hydrochemical map of the German Hydrological Atlas and the International Hydrogeological Map of Europe shows a clear correlation in the areal extent of hydrochemical groundwater types and mapped hydrogeological units (HAD, 2002; Struckmeier and Winter, 2003).

To a similar extent, the characteristics of rivers and streams at the surface are clearly influenced by the underlying geological conditions. Throughout Germany, these features of river courses have been mapped by the German Environment Agency, and it was identified that e.g. river beds in the Plain of northern Germany underlain by quaternary aquifers deposited during the last glaciation periods are very different from river beds in Triassic sandstone regions of southern Germany, which in turn are different from rivers underlain by karstic limestone. The relationship between river structure, base flow in rivers, groundwater and geology has also been confirmed by the BRIDGE project throughout Europe (Pauwels *et al.*, 2007).

A large geochemical survey of sediments in river beds, together with a hydrogeochemical survey of base flow river water, has proven the same relationship, as published in the Geochemical Atlas of Germany (BGR, 1985).

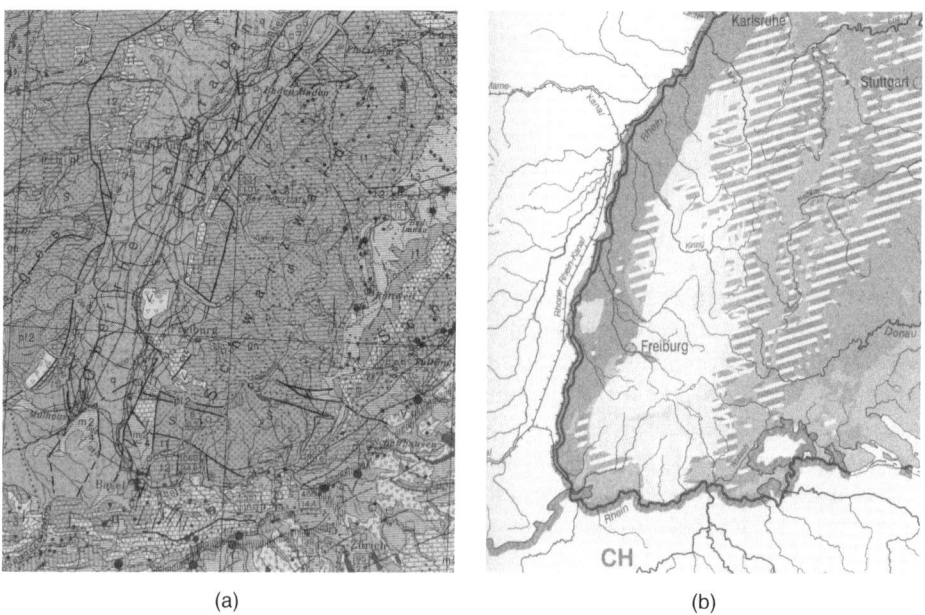

(a)   (b)

Figure 5.2.2   Comparison of hydrogeochemical and hydrogeological maps in south-eastern Germany: a) International Hydrogeological Map of Europe; b) Map of hydrogeochemical units, part of the Hydrological Atlas of Germany (HAD, 2002) (see Plate 8)

Biota and microbiological life in rivers waters also reflect to a certain degree the close relationship with the hydrogeochemical zonation, based on the geological setting.

5.2.1.4 Divergence of Political and Administrative Boundaries vs. Aquifers and Groundwater Bodies

In contrast to the coherence in the natural systems related to water and groundwater, political and administrative boundaries rarely coincide with natural system delineations. As river basins or groundwater bodies may extend over several territories or administration areas, the trans-boundary aspect of such bodies has to be taken into account. This is also a key rationale for the European Water Framework Directive (WFD) ultimately requiring a coherent water management in river basin districts (RDB). However, beyond the RDB there might be groundwater flow systems and groundwater bodies crossing surface water bodies or even extending beyond river basin districts. A striking example is the Dinaric karst area in SE Europe, where surface catchment boundaries differ greatly from the boundaries of groundwater flow systems in the underground.

The issue of trans-boundary aquifers (TBAs) and aquifer systems has been dealt with recently by UN-ECE (UN-ECE, 1999, 2007). However, the 1999 inventory yielded merely incomplete results, because the inventory was done on the basis of a questionnaire and several countries did not reply or did not recognise that they had trans-boundary aquifers with their neighbours. In the study published in 2007, the south-eastern European and the western Asian regions were investigated in conjunction with surveys executed by scientists mapping the hydrogeological conditions. The results clearly showed that coherent hydrogeological structures and nested transboundary groundwater systems can be identified by international hydrogeological mapping programmes and may provide important contributions for a better knowledge of the trans-boundary groundwater flow systems.

5.2.2 HYDROGEOLOGICAL MAPPING IN EUROPE

5.2.2.1 Introduction

Hydrogeological mapping in Europe usually started in the second half of the twentieth century, when the industrialisation, economic growth and welfare caused surface water degradation on the one hand, and on the other hand groundwater came into focus as a high yielding local resource in many places, appropriate to meet the increasing water demands and provide security of water supplies for domestic, industrial and agricultural purposes.

With the ongoing development, both the scales and detail of hydrogeological maps increased, and the mapping tools underwent a drastic change into the new digital world. Therefore, databases are used instead of filed forms, lists and tables, and modern geo-information systems now handle the geo-referenced digital data from filed surveys and monitoring programmes.

5.2.2.2 Hydrogeological Mapping at National Level

From the 1950s, a wealth of national hydrogeological maps was produced, chiefly in the western and central European countries, as part of national reconnaissance surveys to identify the existence and potential of natural resources for national development. However, these groundwater-related maps differed substantially in their scales, format, thematic content and mode of cartographic representation. It was therefore impossible to recognise, from this patchwork of national hydrogeological maps, a regional picture of the European groundwater features, most of which were supposed to extend beyond national borders.

The heterogeneity and graphic diversity of the national hydrogeological maps was clearly demonstrated in an international water conference in Helsinki in 1960, where several tens of groundwater-related maps were exhibited. This gave rise to a common endeavour of the groundwater community to harmonise and standardise the thematic maps related to groundwater, by developing and agreeing common standards and harmonised legends for such maps.

5.2.2.3 Maps at International Level

In the early 1960s the International Association of Hydrogeologists (IAH), UNESCO, the Commission for the Geological Map of the World (CGMW) and the Geological Surveys in Europe initiated a project for the preparation of an International Hydrogeological Map of Europe at a scale of 1:1 500 000 (IHME, 1500). The general purpose of this map was to provide a simplified representation of the groundwater setting in Europe and adjacent areas, as related to the geological situation. This small-scale map provides a general picture and is therefore used primarily for information, teaching purposes, planning and scientific work. Its main objective is to show the location, geographic extent and characterisation of the major hydrogeological units or groundwater bodies, classified according to the main types of aquifers (Gilbrich *et al.*, 2001).

The European hydrogeological map was based on a set of colours, symbols and graphic elements harmonised throughout Europe. This led to a general legend consisting of geologic-petrographic ornaments that characterise the lithology of hydrogeological units in porous, fissured and karstified rocks, in combination with three wash colours: blue was used for unconsolidated, porous aquifers, green for fissured and karstic aquifers, while brown symbolised poor aquifers and nonaquifers, both consolidated and unconsolidated. The aquifer productivity was shown by shades of these colours. In addition, a wealth of overlay information about groundwater and springs, surface water and water usage was printed on the map. The mapping standards and the general legend were developed and tested in Europe in a variety of climatic and environmental settings. In the following years they were recommended for international use and published in various publications of IAH, UNESCO and CGMW. The success of this map and its legend principles inspired

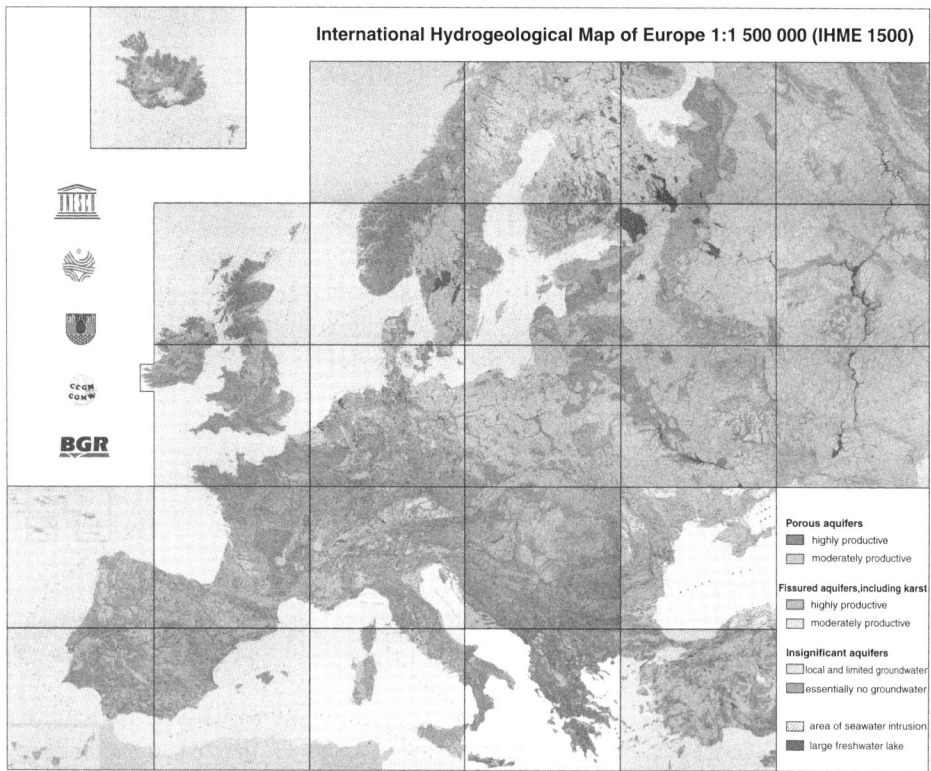

Figure 5.2.3 Mosaic of hitherto published and presently edited map sheets of the series of the International Hydrogeological Map of Europe at the scale of 1:1 500000 (IHME 1500); status: end of 2007 (see Plate 9)

hydrogeologists in other regions and continents to adopt this model and to draw maps according to their requirements, the availability of data and the size of their continents.

Today, practically the whole globe has been mapped, for the benefit of scientists, educators, planners and politicians. The maps deliver the tools for water management, water sharing and eco-hydrological management of the groundwater resources, in order to ensure safe groundwater for mankind on a sustainable basis for the regeneration of disturbed ecosystems.

The map series of the IHME at the scale of 1:1 500000 presents a unique, harmonised hydrogeological representation of Europe and adjacent areas to the east. Figure 5.2.3 shows a mosaic derived from scans of the 22 printed individual map sheets and 3 draft maps still under preparation.

The scanned map sheets are geo-referenced at the moment and are made available on a CD-ROM and as a web mapping service in 2008, to underpin the Water Information System for Europe (WISE).

5.2.3 USING HYDROGEOLOGICAL MAPS FOR GROUNDWATER MONITORING

5.2.3.1 Introduction

Hydrogeological maps should be used to a much greater extent for groundwater monitoring programmes, as foreseen in the groundwater directive of the European Water Framework Directive (WFD). The dependency of groundwater flow systems and groundwater bodies from the underlying hydrogeological setting has been clearly demonstrated (Kunkel *et al.*, 2004), particularly by the EU BRIDGE project (Pauwels *et al.*, 2007). However, it is still common practice to define groundwater bodies on the basis of surface water catchment boundaries, to match the surface water management areas with the underlying groundwater bodies in a way that both delineations are made congruent by a vertical stamping of groundwater bodies according to the catchment subdivision.

This is a very pragmatic procedure of 'harmonising' the management areas of surface water and groundwater. However, it may lead to a segmentation of coherent groundwater flow systems into sub-units that are closely related to the sub-units 'upstream' or 'downstream' of them. This segmentation may only be compensated for by grouping all coherent groundwater sub-units into relevant 'groups of groundwater bodies', as foreseen in the WFD.

The reality in certain regions of Europe unfortunately only partially copes with this hydrogeological conceptual thinking, since many water bodies delineated for the WFD do not really take into account the underlying groundwater situation.

5.2.3.2 Comparison of Groundwater Bodies Delineated in Different Areas of Germany

Two examples may highlight the different approaches chosen for delineating water bodies; both examples show an overlay of the actual delineation of water body management areas according to Article 5 of the WFD against the background of the International Hydrogeological Map of Europe.

Figure 5.2.4a highlights a region of eastern Northrhine-Westfalia and south-western Lower Saxony at the southern edge of the Northern German glacial plain, in the transition zone to the Mesozoic hilly mountain land. Here, most of the water bodies are rather small and are clearly linked with the underlying hydrogeological units. Their boundaries are more or less identical with the boundaries on the hydrogeological map. It is therefore expected that groundwater monitoring programmes will capture coherent entities and the data will be consistent.

Several large groundwater bodies outlined in Figure 5.2.4b are however composed of different hydrogeological terrains, e.g. basement (brown colour), Mesozoic limestones (green) and Quaternary sand and gravel (blue). Lumping these different aquifers into one groundwater body unit makes coherent groundwater monitoring almost impossible, because of the risk of mixing various groundwater quality patterns together.

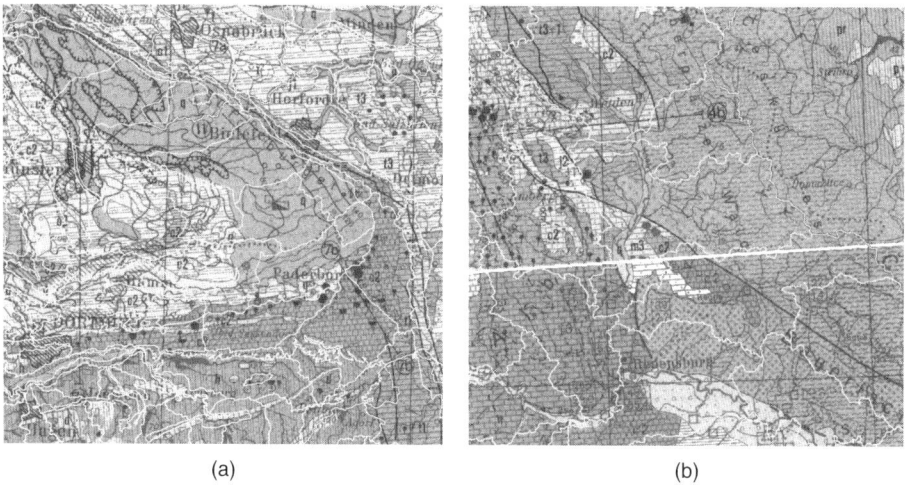

(a)   (b)

Figure 5.2.4 Two regions in Germany showing an overlay of the groundwater body delineations (in yellow) on the background of the International Hydrogeological Map of Europe 1:1 500 000; a) Northrhine-Westfalia and Lower Saxony; b) Bavaria (see Plate 10)

From the principles outlined in Section 5.2.1.3 and evidence from Bavaria, it can be seen that groundwaters in different hydrogeological entities have their own characteristic fingerprints (Wagner *et al.*, 2003). For instance, an evaluation of several hundred groundwater samples in three hydrolithological units of Bavaria shows clearly the different characteristic patterns of their 90 percentiles of the major cations and anions (see Figure 5.2.5). Hence, coupling different types of groundwater quality of distinct hydrolithological entities will greatly complicate the evaluation of hydrochemical monitoring.

In addition, a vertical stratification of hydrochemical layers is often recognised, requiring a clear identification of the depth from which a groundwater sample stems. Moreover, there may be an evolution in the chemical composition of groundwater along the flow path, particularly in large groundwater flow systems where the flow from the recharge to the discharge area takes centuries or millennia. This may lead to a certain variation of absolute concentrations, though the typical geochemical patterns will be maintained.

5.2.3.3 Mapping of Hydrolithological Typologies in Europe

An important step towards a better identification of the natural geochemical characteristics was achieved within an EU-funded research project under the acronym BRIDGE (Background cRiteria for the Identification of Groundwater treEsholds), in which a methodology for separating natural and anthropogenic chemical constituents in groundwaters in many European countries was tested (Pauwels *et al.*, 2007). To cope with variations in the natural background levels of aquifer units, a European aquifer typology was suggested (see Figure 5.2.6b). It coincides nicely with the aquifer types

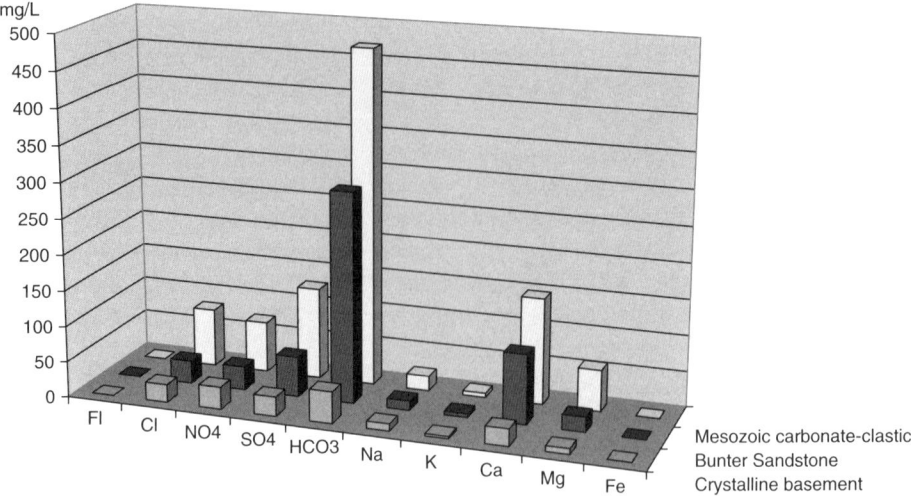

Figure 5.2.5   Average concentrations of important cations and anions in hydrolithological units in Bavaria (90 percentile of several hundreds of samples) (see Wagner *et al*., 2003)

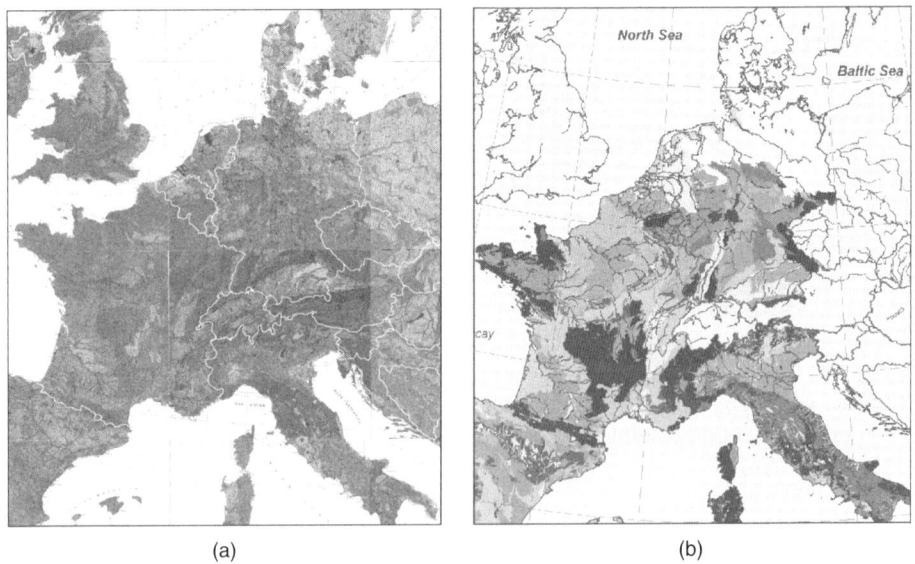

Figure 5.2.6   Comparison between the International Hydrogeological Map of Europe (a) and the European aquifer typology derived from the BRIDGE project (b) (see Plate 11)

outlined in the International Hydrogeological Map of Europe (IHME, 1500). Since only a smaller number of European countries have joined this European aquifer typology so far, this useful exercise should be completed on the EU level, making use of the IHME map by extracting relevant hydrolithological information.

5.2.3.4 Challenges for Hydrogeological Mapping in Relation to Monitoring

Knowledge of the underground conditions close to the surface has progressed continuously over the past decades and is at present good enough for a sound delineation of groundwater bodies and trans-boundary aquifer systems. However, solid data about deeper aquifer structures and their hydrodynamic and hydrochemical conditions is as yet rather weak in many places in Europe, and the interaction between regional and local groundwater flow systems is also poorly known.

The hydrodynamic position within an aquifer system or a groundwater body is essential information for the assessment and evaluation of pollution impacts into a groundwater system. Whilst pollution sources in recharge areas frequently endanger the whole groundwater flow system in the long run, the same sources would stay confined to a small area in a groundwater discharge area, though they might present a real threat in the short term to rivers or groundwater dependent wetlands. Therefore, hydrogeological entities such as groundwater flow systems and groundwater bodies should be described and mapped with respect to the flow of the groundwater, i.e. into recharge, lateral flow or discharge areas. One example of a hydrodynamic map outlining regional, sub-regional and local groundwater systems and their classification into recharge, transit and discharge areas has been prepared and tested in the Münsterland Basin, a well-known hydrogeological structure of more than $10\,000\,km^2$ in northwestern Germany (Struckmeier, 1989). Figure 5.2.7 shows the south-eastern region of that basin, where a karst aquifer and an overlying sand aquifer are recharged over vast regions. Two overlying perspective diagrams indicate conceptually the flow within the groundwater system and the chemical type of the groundwater.

5.2.4 CONCLUSIONS

Hydrogeological maps are useful tools for groundwater monitoring programmes, because they chiefly provide geo-referenced information about the position and size of groundwater bodies, and characteristics of hydrogeological entities, based on their hydrolithological compositions. In addition, a wealth of complementary information, e.g. on the relationship between surface water and groundwater, groundwater flow directions, spring discharge and groundwater quality, may also be displayed on such maps.

A large number of hydrogeological maps have been prepared in many European countries in the past decades on the basis of systematic hydrogeological surveys. In most countries the information relevant to hydrogeology and groundwater is now held in geo-information systems (GIS), allowing up-to-date retrieval and visualisation of the data. However, there is a lack as yet of standardised, harmonised and interoperable groundwater data on the European level. A few coherent hydrogeological data sets that cover part of the region of the European Union exist at the EU Joint Research Centre or in the institutions that have cooperated within the BRIDGE project.

However, the only coherent set of hydrogeological maps of the whole of Europe in the geographic sense, up to the Urals, is the series of the International Hydrogeological

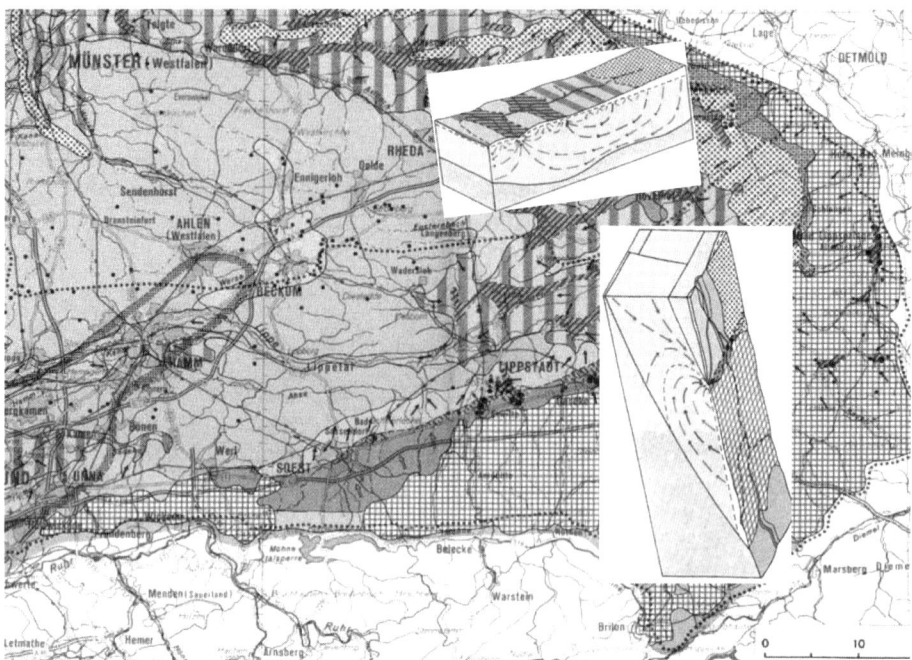

Figure 5.2.7   Hydrodynamic map of the Münster Basin (in the back) and perspective diagrams of groundwater flow systems (after Struckmeier 1989) (see Plate 12)

Map of Europe at the scale of 1:1 500 000 (IHME), to which more than 40 geological surveys and national geoscientific institutions on the European continent have contributed. These maps are scanned and geo-referenced, to make this unique set of groundwater-related information fit for European tools, such as the Water Information Systems of Europe (WISE), and to provide a reliable, harmonised background picture for the future.

REFERENCES

Bundesanstalt für Geowissenschaften und Rohstoffe (BGR) (1985) Geochemischer Atlas von Deutschland (numerous maps and explanations), Hannover.
BGR, EuroGeoSurveys, CGMW, IAH and UNESCO/IHP (2008) International Hydrogeological Map of Europe, reduced to the scale of 1: 5 000 000, Hannover.
Gilbrich, W., Krampe, K. and Winter, P. (2001) Internationale Hydrogeologische Karte von Europa 1:1 500 000, Bemerkungen zum Inhalt und Stand der Bearbeitung, *Hydrologie und Wasserbewirtschaftung*, **45**(3), pp. 122–125, Koblenz.
Hydrologischer Atlas von Deutschland (HAD) (2002) Hydrological Atlas of Germany (numerous maps and explanations), Freiburg.
Kunkel, R., Wendland, F., Voigt, H.-J. and Hannappel, S. (2004) Die natürliche, ubiquitär überprägte Grundwasserbeschaffenheit in Deutschland, *Schriften FZ Jülich, Reihe Umwelt*, **47**, p. 204.

Pauwels, H., Muller, D., Griffioen, J., Hinsby, K., Melo, T. and Brower, R. (2007) BRIDGE – Background cRiteria for the IDentification of Groundwater thrEsholds, Publishable final activity report.

Struckmeier, W. (1989) Wasserhaushalt und Hydrologische Systemanalyse des Münsterländer Beckens, *LWA Schriftenreihe*, **45**, p. 72.

Struckmeier, W. and Winter, P. (2003) The International Hydrogeological Map of Europe, Scale 1:1 500 000: a coherent hydrogeological representation for Europe and adjacent areas to the East, EAGS meeting, Hannover.

UN-ECE (1999) Inventory of transboundary groundwaters, Task Force on Monitoring and Assessment, Lelystad.

UN-ECE (2007) Our waters: joining hands across borders, New York/Geneva, available at http://www.unece.org/env/water/publications/pub76.htm.

Wagner, B., Töpfner, C., Lischeid, G., Scholz, M., Klinger, R. and Klaas, P. (2003) Hydrogeochemische Hintergrundwerte der Grundwässer Bayerns, *GLA Fachberichte*, **21**, p. 250.

Walter, T., Beer, A., Dreher, T., Elbracht, J., Fritsche, H.-G., Hübschmann, M. *et al.* (2006) Ermittlung und Darstellung der natürlichen Hintergrundwerte der Grundwässer in Deutschland, In: Voigt, H.-J., Kaufmann-Knoke, R., Jahnke, C., Herd, R. (eds) *Indikatoren im Grundwasser: Kurzfassungen der Vorträge und Poster*, Tagung der FH-DGG, Cottbus.

5.3
Establishing Environmental Groundwater Quality Standards

Dietmar Müller[1]

5.3.1   INTRODUCTION

Developed under Article 17 of the Water Framework Directive (WFD), the new European Groundwater Directive (GWD) sets out criteria for assessing the chemical status of groundwater bodies. In accordance with an analysis of pressures and

---

[1] The views expressed in this chapter are purely those of the author and may not in any circumstances be regarded as stating an official position of the European Commission

---

The Water Framework Directive - Ecological and Chemical Status Monitoring   Edited by Philippe Quevauviller, Ulrich Borchers, Clive Thompson and Tristan Simonart © 2008 John Wiley & Sons, Ltd

impacts, Member States are required to identify pollutants that are representative for groundwater bodies found as being at risk and for which threshold values should be established as (environmental) quality standards at the most appropriate scale (national, river basin district or groundwater body).

BRIDGE as a policy support-oriented research project under the 6th Framework Programme of Research and Technological Developments, aimed to contribute to the WFD Common Implementation Strategy (CIS) by developing a scientifically sound but also practicable common methodology for deriving groundwater thresholds. To support sustainable groundwater management, the methodology shall ensure that the derivation process for groundwater thresholds will be based on common criteria, and comparability of status classification for groundwater bodies across Europe can be assumed.

5.3.2 ENVIRONMENTAL OBJECTIVES FOR GROUNDWATER MANAGEMENT

Groundwater is a key environmental resource and factor supporting sustainable regional development. This is recognised by the WFD (European Community, 2000) and the new GWD (European Community, 2006a), which represent a major move in water policy towards an integrated management of groundwater resources and provide a new regulatory setting for the protection of groundwater quality. The 'prevent and limit' concept of the Groundwater Directive 80/68/EEC (to be repealed under the WFD by 2013), aiming at a control of point sources and at protecting groundwater quality at local scale, has been complemented by the concept of managing the environmental status of groundwater resources, aiming at the control of widespread pollution by diffuse sources and at preserving groundwater quality at regional scales. According to the environmental objectives defined by the WFD, the final goal of groundwater management at larger scales is to preserve good status (quantitative and chemical) to maintain groundwater in its functions to support aquatic and terrestrial ecosystems. In particular, this recognition of groundwater as being a major ecological factor asks for a new understanding, which goes beyond the general perspective on groundwater as a resource for human needs, like drinking water or irrigation. Finally, the status assessment objectives are complemented by trend assessments and the objective of avoiding and reversing significant and sustaining upward trends in contaminant concentrations.

5.3.3 FRAMEWORK FOR THRESHOLD VALUES

Status assessment of groundwater bodies asks for Environmental Quality Standards (EQS). With respect to the variability of natural hydrogeological settings and the resulting hydrogeochemical composition of groundwater, the GWD indicates that groundwater threshold values are quality standards that are not set at a European level but by Member States at the most appropriate scale.

Thus a general framework to structure the derivation process of threshold values is needed. Any methodology for deriving environmental thresholds for groundwater bodies has to be based on an integrated characterisation process built on three pillars:

- Characterisation of potential pollutants and any parameters indicative of pollution, including description of the properties which influence their fate and transport, e.g. transport through and out of the aquifer, including transport in unsaturated zone, the behaviour of hydrogeochemical environments, ecotoxicology and toxicology, and possible impacts on ecosystems.

- Characterisation of groundwater bodies, including a description of the hydrogeo-chemical setting of aquifers, the background quality (natural and anthropogenically influenced) and any dependencies of water quality on quantitative aspects (like the variability of water levels due to the hydrological cycles during the year, groundwater to surface water interactions, or the water balance in the long term).

- Characterisation of receptors, including aquatic ecosystems, dependent terrestrial ecosystems and groundwater.

Obviously, the methodology has to refer to the definitions provided by the WFD, which are generally focused on possible impacts on ecological receptors (associated surface waters, dependent terrestrial ecosystems) and a qualification of the significance of these impacts (e.g. significant diminution of the ecological or chemical quality of surface water bodies or any significant damage to terrestrial ecosystems which depend directly on the groundwater body).

Consequently, any environmental threshold to indicate the turn between good and poor chemical status has to be derived with a strong orientation to a risk-based approach

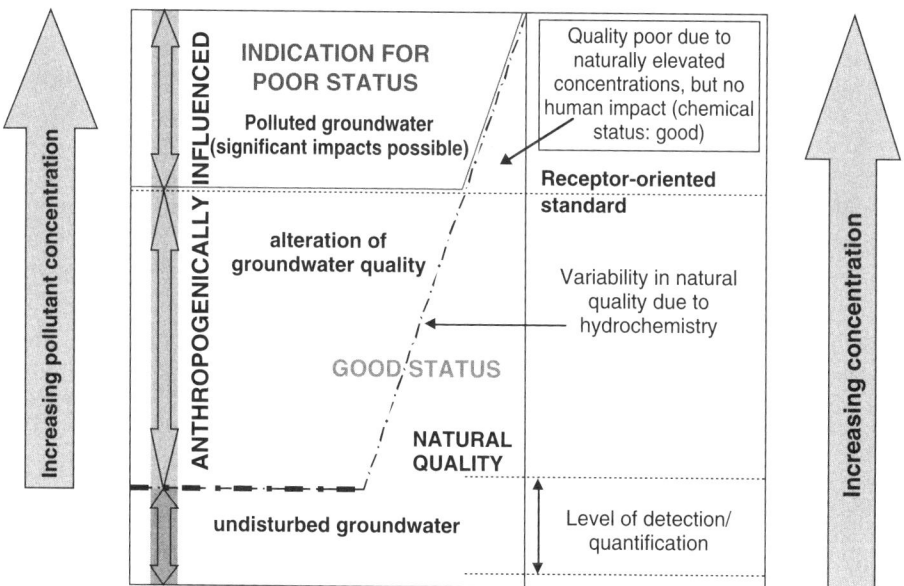

Figure 5.3.1 Groundwater quality and status – general relationship under the WFD

and the likelihood that receptors are or might be harmed. Furthermore, the differences between synthetic substances and anthropogenically introduced but naturally occurring substances also need to be taken into account, together with natural variations in quality both within and between groundwater bodies. Moreover, some substances, though considered as pollutants, may be present in naturally elevated concentrations. Such water may then be considered to have a poor quality (with regard to possible uses, e.g. drinking water abstraction) but still represents a good chemical status (see also Figure 5.3.1).

Environmental thresholds requested under the regime of the new GWD aim to assess the 'overall health' of groundwater bodies and therefore are not to be compared to any quality standard under use to control local pollution of groundwater caused by point sources (prevent and limit). Groundwater status and its assessment refers to specific pollutants, defined by a risk characterisation process, and reasonably large management units at ranges up to more than $10\,000\,km^2$, whereas quality assessment refers to small scales (e.g. wellfields), taking account of any possible negative impact.

5.3.4 CRITERIA TO ASSESS QUALITY AND STATUS

Groundwater quality assessment is a routine, which may aim at various possible objectives, such as monitoring human impacts at local scales, controlling drinking water supply or characterising pristine conditions. In contrast, groundwater status assessment, as requested by the WFD, is a new task, lacking experiences and defined procedures. As for developing groundwater protection towards an integrated management of groundwater resources, criteria applied for quality and status assessment should be in correspondence and interlinked.

To assess groundwater quality at local scales it is quite common to make use of *natural background levels* and *generic reference values* according to possible receptors, which are in general ecosystems and human uses. As status assessment can be understood as being an integrated assessment at larger scales of groundwater bodies, these two categories of criteria have to be included. Moreover, status assessment has to take into account the fact that the general chemical quality of groundwater, as well as fate and transport of contaminants, is determined by a variety of factors, where petrographic properties of rocks in the vadose and groundwater saturated zone, regional hydrological and hydrodynamic conditions and hydrogeochemical processes controlling the behaviour of natural and anthropogenic substances are of major importance. Thus status assessment would need to go beyond quality assessment and consider *attenuation criteria* like dilution, diffusion, retardation and degradation, which might be described referring to properties of contaminants, hydrogeological units, where specific hydrogeochemical processes govern, and the interaction with surface waters.

5.3.4.1 Natural Background Levels

Depending on data availability, natural background levels (NBLs) can be defined following a hierarchy of possible options. To unify the starting point for groundwater status assessment, BRIDGE has recommended a European aquifer typology, classifying

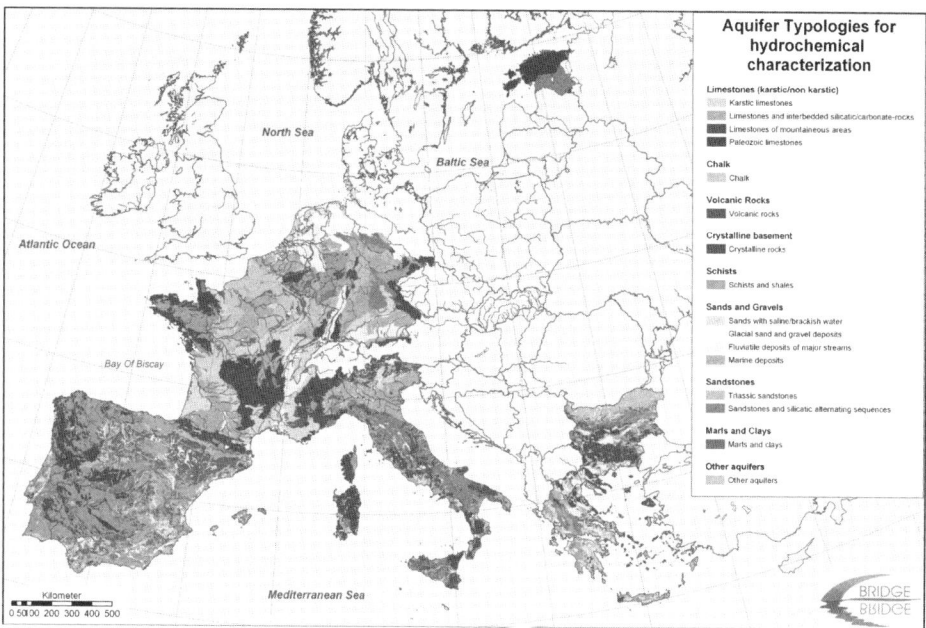

Figure 5.3.2 European aquifer typology for hydrochemical characterisation of groundwater (Wendland *et al.*, 2007) (see Plate 13)

16 types (Wendland *et al.*, 2007; see also Figure 5.3.2), and also has referenced NBLs from specific national studies accordingly. These NBLs might be used within the procedure to derive threshold values if no appropriate groundwater quality data are available, but only the hydrogeological units of a specific groundwater body can be described.

Given a groundwater body where a limited set of quality data is available, a second option with a simplified and practical approach to determine NBLs based on a preselection method can be employed. As a prerequisite for applying a simplified preselection method, common minimum requirements for groundwater quality data (e.g. deviation of the ion balance <10 %) and appropriate preselection criteria to identify groundwater samples showing no significant anthropogenic impact (e.g. nitrate <10 mg/l) have been defined. Finally, given a groundwater body where a broad set of quality data is available, the third option for estimating NBLs is to apply scientifically sound methods (e.g. hydrochemical simulations, component separation by concentration separation analysis) which have already been established at national or international levels.

Table 5.3.1 shows a comparison of natural background levels derived on the basis of the recommended preselection method for a groundwater body in Austria, against the results of a project (GEOHINT) which assembled groundwater data across the country.

Comparing both approaches for natural background levels, the resulting concentrations of major ions and trace elements were found at quite similar ranges. The national approach generally proved to be less conservative for metals but more sensitive for parameters describing the general hydrogeochemical composition. Regarding metals,

Table 5.3.1 Groundwater body 'Southern Vienna Basin' – natural background levels and reference values

Parameter		NBL (BRIDGE)	NBL (GEOHINT)	DWS	AA-EQS	MPA (national)
calcium	mg/l	123	134	n.a.	n.a.	n.a.
magnesium	mg/l	39.4	40.8	n.a.	n.a.	n.a.
chloride	mg/l	41.2	32	200 (Ind.)	n.a.	n.a.
sulphate	mg/l	164	80	250 (Ind.)	n.a.	n.a.
phosphate	mg/l	0.08	0.14	n.a.	n.a.	n.a.
boron	mg/l	0.041	0.04	1.0	n.a.	n.a.
chromium	μg/l	0.9	4.0	50	n.a.	8.5
arsenic	μg/l	1.0	4.0	10	10	24
cadmium	μg/l	0.16	0.2	5	0.08–0.25	n.a.
nickel	μg/l	1.9	4.0	20	20	n.a.

NBL: natural background level; DWS: drinking water standard; AA-EQS: annual average EQS (EU); MPA: max. permissible addition (national): Ind. indicator parameter.

the natural composition of groundwater resources in Austria generally does not exceed reference values for human use or aquatic ecosystems (see Table 5.3.1).

It has to be noted that if long-term series of monitoring data are available, as in Austria, these may already indicate reasonable trends for any preselection parameters. In particular, increasing trends for nitrates may significantly reduce the number of sampling sites that can be considered within the preselection period. Thus it was concluded that besides nitrate, the use of further preselection parameters indicating anthropogenic impacts should be considered and the choice of time periods could reasonably influence the derivation process for natural background levels.

5.3.4.2 Generic Reference Values

Besides NBLs, generic reference values concerning receptors which might be harmed by groundwater contaminants are used for groundwater quality assessment. Giving a focus on ecosystems and human uses means that Environmental Quality Standards (EQS) for surface waters or drinking water standards (DWS) are to be transferred and linked into groundwater status assessment too. The mentioned reference values might be defined at European level or at national level. With respect to the variety of possible substances contaminating groundwater, it is also likely that for some substances no reference values are available at all. Again, as a hierarchy of options, it can be envisaged that unified European standards such as the EQS for priority substances set out by the proposal of the European Commission (European Commission, 2006b) are preferable to and should overrule national standards. If no agreed European standards exist, national reference values can be used. Finally, for substances without established receptor-orientated reference values at European or national scale, a survey and evaluation of human toxicity or ecotoxicity data will be necessary. Depending on data availability, the evaluation of these toxicity data should again be based either on agreed European procedures or on national agreements. As for ecotoxicity, there are few data available concerning groundwater organisms, but test results from aquatic organisms are generally recognised as comparable. Considering long residence times and the relatively slow dynamics of groundwater, emphasis is to be given to ecotoxicity data on chronic effects.

Regarding surface water as a receptor, it has to be noted that the existing European groundwater quality standards set out by the WFD for nitrate and pesticides may not always be protective for ecological receptors. Particularly for nitrates, it is likely that protecting surface waters from eutrophication might demand lower quality standards for some groundwater bodies (see also Section 5.3.6). Still missing is a specific legal and scientific background on appropriate status objectives for dependent terrestrial ecosystems. Thus, as long as a sound evidence base is missing, the general assumption could be that terrestrial ecosystems need to adapt to similar conditions to neighbouring aquatic ecosystems. However, if damages to dependent terrestrial ecosystems are monitored, specific investigations by ecologists will need to be conducted.

5.3.4.3 Attenuation Criteria

Pollutant attenuation may occur along the flow path of groundwater, at the interface between groundwater and surface water (the hyporheic zone; Smith, 2005), and at the surface water itself. General consideration might be given to dilution, dispersion, diffusion, volatilisation, sorption, and chemical and biological degradation. The physical, chemical and biological processes which occur in aquifers, and which may act to naturally attenuate pollutants, are well known and widely reviewed for the purpose of assessing point source pollution. The same processes can also act over much larger scales, e.g. on groundwater bodies. Although rapid assessment methodologies (e.g. Smith and Lerner, 2007) have been developed during the last few years, it still needs to be pointed out that a real in-depth conceptual understanding of the groundwater and receptor system is needed, and the demand on specific data consequently increases. Currently we have insufficient knowledge on the surface water interface, but science is on its way.

In comparison, attenuation at the receptor surface water might easily be described, as in general dilution will be the major attenuating process. The estimation of quantity relationships of groundwater flow against surface water flow can be made rather easily by a variety of methods for estimating the base flow (e.g. by hydrograph separation, temperature or water quality surveys, tracer analysis). In comparison, contaminant attenuation processes are highly spatially heterogeneous and often occur in localised areas (in particular at environmental interfaces), such as the soil layer, the unsaturated zone/saturated zone, and the riparian/hyporheic areas (McClain *et al.*, 2003). Therefore, a pragmatic approach is to consider dilution as separate, generic criteria, whereas the description of all other attenuating processes would need in-depth investigations, which might be necessary for a final status assessment but are hardly to be considered for the derivation of threshold values. Also, the case studies conducted within BRIDGE proved that it is generally easier to establish dilution factors. Nevertheless, situations of groundwater–surface water interaction were found where no significant dilution was given but degradation could be proved, and accordingly an attenuation factor was established (see Section 5.3.6).

5.3.5   THRESHOLD VALUES BY A TIERED APPROACH

To combine the described criteria within a receptor-oriented status assessment, a tiered approach for deriving appropriate threshold values for groundwater bodies is recommended. Tiers may provide intermediate levels based on an increasingly detailed understanding of a groundwater body. Therefore, each tier will involve increasingly sophisticated levels of data collection and analysis. An initial analysis (Tiers 1 and 2) can use conservative and rather simple criteria (e.g. Tier 1: check of monitoring data against natural background levels; Tier 2: check against environmental quality standards defined for associated surface waters), whereas further steps of detailed analysis (Tiers 3 and 4) would mean a thorough evaluation of specific groundwater body characteristics (e.g. Tier 3: back calculation for an associated surface water body, taking into account the contribution of the groundwater to the total pollutant burden; Tier 4: taking into account further attenuating capabilities of the subsurface environment, specifically for the aquifer, the hyporeic and the riparian zone). As a result of the detailed analysis, the final separation between good and poor status by a refined threshold is identified (see Figure 5.3.3). With respect to the heterogeneity of groundwater bodies on the one hand and the limited data availability in practice on the other, it seems likely that the detailed analysis will often be limited to a rather simple

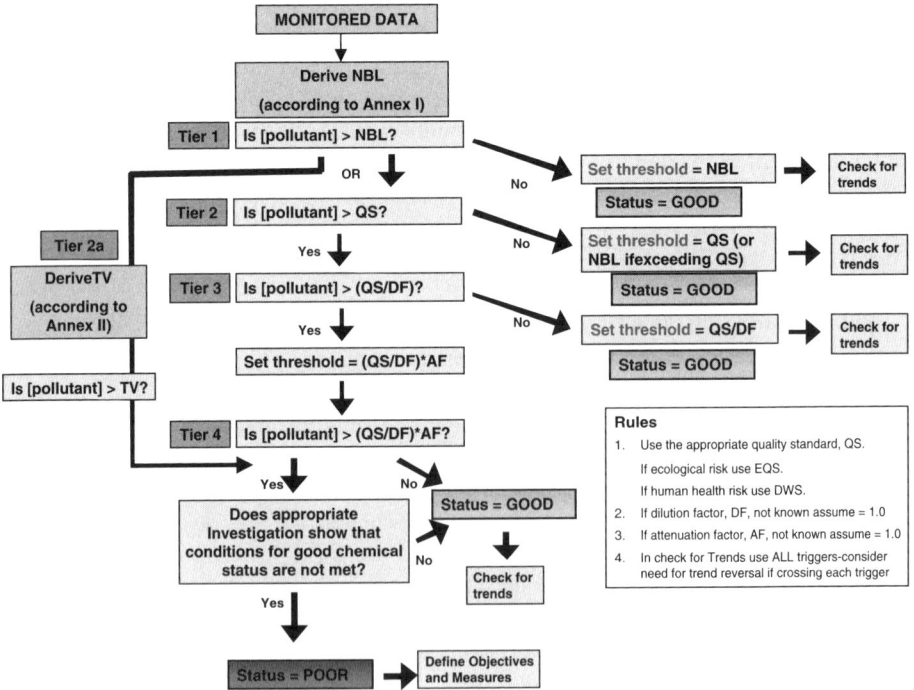

Figure 5.3.3   Flow chart for status assessment and identification of a groundwater threshold (BRIDGE report D 18)

Tier 3 by describing groundwater and surface water interaction in terms of quantity relationships.

5.3.6 HOW TO DETERMINE A THRESHOLD VALUE

Groundwater thresholds are generally only required if a groundwater body has been characterised as being at risk for a specific pollutant. Relevant risks have already been identified by Member States during the process of characterisation, as set out by the Water Framework Directive (Article 5). Given the situation of a groundwater body which may have an impact on an associated surface water, the threshold value for groundwater should consider natural background levels, Environmental Quality Standards and attenuation (dilution) as described. This means that status assessment may follow a maximum of four possible tiers and the different tiers may refer to the following sequence of considerations:

- *Tier 1: Natural Background Level*: Pollutant concentrations within the range of natural background levels (NBL) would cause the groundwater body to be determined as having good status. For naturally occurring substances, background levels have to be considered as concentration ranges and a consequently derived threshold has to be set above this range to allow variation. For substances without a natural origin, the connected thresholds would be zero, but for practicality have to be referenced by a distinct factor to the limit of detection (LOD).

- *Tier 2: Quality Standards*: The groundwater monitoring data are compared to the quality standard established for the associated surface water body. If the quality of groundwater flowing into the receptor is below the receptor quality standard then the risk to that receptor must be derived from some other source and groundwater is to be determined as having good status.

- *Tier 3: Considering Dilution*: The groundwater provides only a proportion of the pollutant flow to the receptor. The concentration of a substance in the groundwater is compared to the quality standard multiplied by a dilution factor (DF). Where the dilution factor is not known it is assumed to be 1.

- *Tier 4: Considering Attenuation*: The pollutant might be attenuated on its path to the receptor. If further information is known about the attenuation, a relating factor could be introduced. This is the ultimate threshold in this scheme. Failing this will classify the groundwater body as having poor status, whereas failing the earlier tiers just leads to more investigations.

One of the few BRIDGE case studies where data availability supported a process up to Tier 4 was the Odense river basin (Hinsby *et al.*, 2007). The characterisation of this area ($\sim$1050 km^2; see Figure 5.3.4) shows pressures from agricultural land use (68 % of the surface) and surface waters (lake Arresov and the estuary of the Odense Fjord) impacted by eutrophication due to severe nutrient loads (nitrogen and phosphorus).

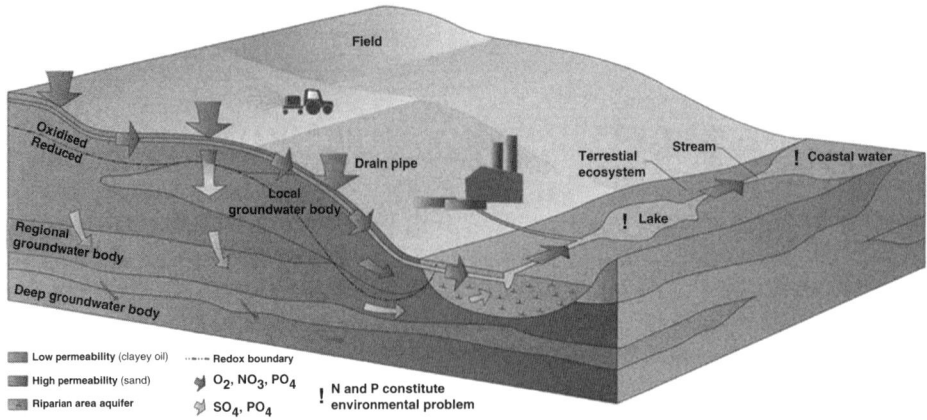

Figure 5.3.4   Conceptual model of the groundwater system at the Odense river basin, with dependent terrestrial and aquatic ecosystems (Hinsby *et al.*, 2007) (see Plate 14)

As groundwater discharge is almost the only sustainable source of surface waters, dilution cannot be accounted for, but due to nitrate degrading processes at different interfaces an attenuation factor of 0.5 was established. However, despite this natural breakdown it was proved that, specifically at the Odense river basin, the generic Environmental Quality Standard for nitrates of 50 mg/l is not protective to aquatic ecosystems. Specifically for this groundwater body and its interactions with surface water, a nitrate threshold value less than 20 mg/l would be necessary to decrease the nutrient load to an acceptable level.

5.3.7   COMPLIANCE REGIME

A standard may have widely different impacts and implications, depending on the compliance regime selected. Therefore, a clear definition is at least necessary for:

- A summary statistic (such as a mean, 95 percentile or maximum value).

- A time period over which compliance is assessed.

- Area over which the criteria are applied and interpreted via the monitoring network.

Following the concept of a tiered and receptor-orientated approach, the compliance regime may vary widely, e.g.:

- Data aggregation over time only for monitoring stations at Tier 1, and data aggregation over all monitoring stations and time within a defined area (groundwater body) at Tier 2.

- Data aggregation over monitoring stations with an associated surface water system.

It has to be recognised that any quality standard (limit values, thresholds, indicators) must be associated with explicit compliance regimes that enable a straightforward assessment of compliance via the monitoring data obtained from the operational monitoring network. Therefore, the continuation of discussions between scientists, regulators and stakeholders after agreeing upon the threshold methodology is a fundamental prerequisite to developing a consistent approach for groundwater status assessment. The coordination of practice and the ongoing implementation of the Water Framework Directive by the Member States is another challenge and chance in developing a common and comparable groundwater management which respects a Europe of regions and can contribute to enhanced sustainability.

5.3.8 CONCLUDING REMARKS

This chapter highlights some of the main findings of the BRIDGE project, which provide a scientific background for agreeing on a common methodology for the establishment of threshold values in the light of the requirements of the new Groundwater Directive. It does not prejudge concerning the final methodology which will be agreed to and adopted by the Member States in the framework of the Working Group on Groundwater of the Common Implementation Strategy of the WFD (Grath *et al.*, 2007).

5.3.9 ACKNOWLEDGEMENTS

BRIDGE is funded by the European Commission, DG Research within the 6th Framework Programme under Priority 8 [Contract No. 006538 (SSPI) – Scientific Support to Policies].

Disclaimer: The views expressed in this paper are purely those of the authors and may not in any circumstances be regarded as stating an official position of the European Commission.

REFERENCES

BRIDGE reports, available at http://www.wfd-bridge.net.
European Community (1979) Council Directive 80/68/EEC of 17 December 1979 on the protection of groundwater against pollution, *Official Journal of the European Communities*, **L20**, p. 43.
European Community (2000) Directive 2000/60/EC of the European Parliament and of the Council of 23 October 2000 establishing a framework for Community action in the field of water policy, *Official Journal of the European Communities*, **L327**, p. 1.
European Community (2006a) Directive 2006/118/EC of the European Parliament and of the Council of 12 December 2006 on the protection of groundwater against pollution and deterioration, *Official Journal of the European Communities*, **L327**, p. 19.
European Community (2006b) Proposal for a directive of the European Parliament and of the Council on environmental quality standards in the field of water policy and amending Directive 2000/60/EC, COM (2006) 397 final.

Grath, J., Ward, R., Legrand, H., Blum, A., and Broers, H.P. (2007) Draft guidance on groundwater chemical status and threshold values, Working Group C Groundwater, Activity WGC-2, 'Status compliance & trends', Groundwater chemical status and threshold values, version 2.0.

Hinsby, K., Condesso de Melo M. T. and Dahl, M. (2008). European case studies supporting the derivation of natural background levels and groundwater threshold values for the protection of dependent ecosystems and human health, *Sci Total Environment*, **401**, Issues 1-3, pp. 1–20.

McClain, M.E., Boyer, E.W., Dent, C.L., Gergel, S.E., Grimm, N.B., Groffman, P.M. *et al.* (2003) Biogeochemical hot spots and hot moments at the interface of terrestrial and aquatic ecosystems, *Ecosystems*, **6**, pp. 301–312.

Smith, J.W.N. (2005) Groundwater-surface water interactions in the hyporheic zone, Environment Agency, Science Report SC030155/SR1.

Smith, J.W.N. and Lerner, D.N. (2007) A framework for rapidly assessing the pollutant retardation capacity of aquifers and sediments, *Quarterly Journal of Engineering Geology & Hydrogeology*, **40**, pp. 137–146.

Wendland, F., Blum, A., Coetsiers, M., Goroya, R., Griffioen, J., Grima, J. *et al.* (2007) European aquifer typology: a practical framework for an overview of major groundwater composition at European scale, *Environmental Geology*, DOI 10.1007/s00254-007-0966-5.

Section 6
Sediment Monitoring

6.1

Sediment Dynamics and their Influence on the Design of Monitoring Programmes

Sue White

6.1.1   INTRODUCTION

The supply and transfer of sediment to and through the aquatic system is a highly spatially and temporally dynamic process, which has been described as a 'jerky conveyor belt' (Ferguson, 1981). The processes involved in sediment supply, transport and deposition are generally not linearly dependent on the causal factors, nor do they relate to one another in a linear manner. Rather, at any scale from the field to a large river basin there will be a cycle of supply–transport–deposition (Figure 6.1.1), which will repeat in both space and time. Thus, sediments that are deposited at a point in a river basin may become part of the supply chain for sites further downstream or may remain deposited for long time periods. This sediment erosion–transport–deposition cycle is summarised in the concept of a sediment delivery ratio (Walling, 1983, 1988),

The Water Framework Directive - Ecological and Chemical Status Monitoring   Edited by Philippe Quevauviller, Ulrich Borchers, Clive Thompson and Tristan Simonart © 2008 John Wiley & Sons, Ltd

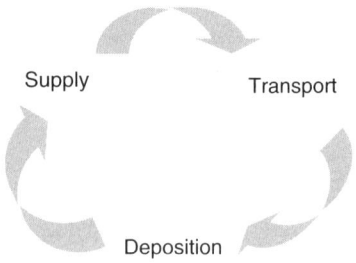

Figure 6.1.1 The sediment cycle

which is the proportion of detached sediment across a river basin (the supply) which actually leaves the basin.

The initiation of sediment movement depends on a supply of energy from raindrop impact, overland or river flow, or a mass movement of soil. Similarly, deposition occurs when sufficient energy is not available to maintain sediment particles in motion. The energy thresholds for initiation and cessation of sediment movement are not the same, being influenced differently by a range of conditions such as flow depth, velocity and turbulence, and sediment characteristics such as cohesion, particle shape and particle density.

There are many equations which allow calculation of sediment transport rate within a water body, or sediment flux (see for example Task Committee of Computational Modeling of Sediment Transport Processes, 2004 for a review). However, these equations tend to be for a uniform sediment distribution, which is far from the variable source supply of material seen in events when the majority of sediment is moving. It is also generally considered that a particular flow has a maximum capacity to transport sediment, although the concentration this relates to depends again on sediment characteristics. Hence there are examples in China where sediment concentrations can reach several tens of thousands of parts per million for very fine particles, whereas a flow may become saturated with sand-sized particles at far lower concentrations. Rivers are often considered to be either capacity- or supply-limited in terms of their sediment transporting dynamics. However, in practice for most rivers, most of the time, sediment transport is limited by a complex and dynamic pattern of sediment supply.

The supply of sediment is controlled by a mix of factors relating to geology, rainfall distribution, land use and management, soil type and distribution, topography, landscape complexity and antecedent conditions, and the relationship between them in space and time. Approaches to estimating sediment supply from the landscape tend to be limited to the small scale (e.g. Flanagan *et al.*, 2007; Morgan *et al.*, 1988; Wischemeier and Smith, 1978) although many of these have now been applied across a landscape or river basin (e.g. Brath *et al.*, 2002; De Roo, 1998; Lewis *et al.*, 2005). For supply of sediment from river banks, the controlling conditions are related to the material of the river bank and its variability with depth, sequences of changes in flow, access and damage to the river bank (e.g. by cattle damage), and antecedent conditions, particularly the wetness of river bank material (see e.g. Lawler, 1992). In addition, changes in flow characteristics which affect the channel-forming flows, generally considered to be 1.5-year return period events (Leopold *et al.*, 1964), will cause remobilisation of river channels through the alluvial deposits in the river valley. This remobilised material will in part be relocated at other points in the river system, but is of importance, as alluvial deposits may be thousands of years old and may contain a range of historical contaminants, which will also be remobilised. This is likely to be an important consequence of climate change. The key message is that sediment supply will vary both in space and time, and that sediments being transported through the river system today may have started their journey through the catchment recently or thousands of years ago.

Large time lags may be built into the sediment supply–transport–deposition system through manmade structures such as dams, weirs or estuarine barrages, although natural barriers such as hedgerows, topographic depressions or alluvial sediment deposits may also store sediments for periods of months to millennia. Sediment release from such stores may happen in extreme or catastrophic events, or may be a result of human intervention. The removal of weirs and in-river structures as part of the programmes of measures to meet WFD targets are examples of such interventions (see e.g. Bednarek, 2001). Similar sediment release occurs as a result of mass movements or river bank collapses, which may be associated with very wet conditions or occurrences such as earthquakes. Such sudden high inputs of sediment would mean that supply is likely to exceed the capacity of water to move these sediments either locally or further through the system, at least over the short term.

Generally one expects to find sediment deposits fining with downstream distance; this is because rivers tend to be wider and less steep, and velocities slower towards the sea, such that coarser material is not often transported. This fine material is, however, easier to remobilise. Therefore one might expect a decrease in sediment residence time (the amount of time sediment stays deposited in a location) with distance downstream.

The tendency for sediment supply to be concentrated in high rainfall-runoff events, which coincide with high river flow and thus high sediment transport capacity, means that sediment transport (flux) is concentrated in high-flow events, which also have the highest transporting capacity. A study by White *et al.* (2005) for several European rivers showed that typically 70 % of suspended sediment transported moves in the top 20 % of flows (Figure 6.1.2). This skew in sediment flux is due to higher concentrations of sediment and higher flows coinciding. Peak high-flow sediment concentrations are seen in steep mountain rivers, and, as a broad generalisation, peak concentrations decrease as rivers become less steep. Groundwater-fed rivers behave somewhat differently because flow is not dominantly supplied by overland routes, which would deliver sediment, and thus there is not such a marked increase in sediment concentrations at high-flows.

The exact percentage moved in the top flows differs from river to river and from year to year, depending on a wide range of factors relating to both land management and rainfall flow regime. For example, a study on the River Tees in Northern England (White *et al.*, 2003) gave sediment flux percentages of 76.7 % to 94 % moving in the top 20 % of flows, over a monitoring period of 3.5 years (Figure 6.1.3).

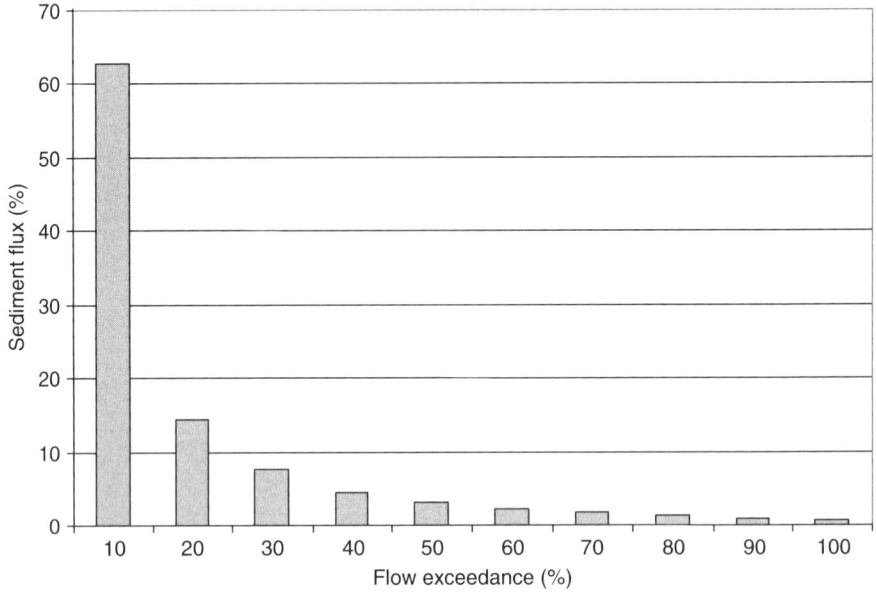

Figure 6.1.2 Percentage of sediment flux by flow exceedance class for major European rivers (from Becvár, 2005)

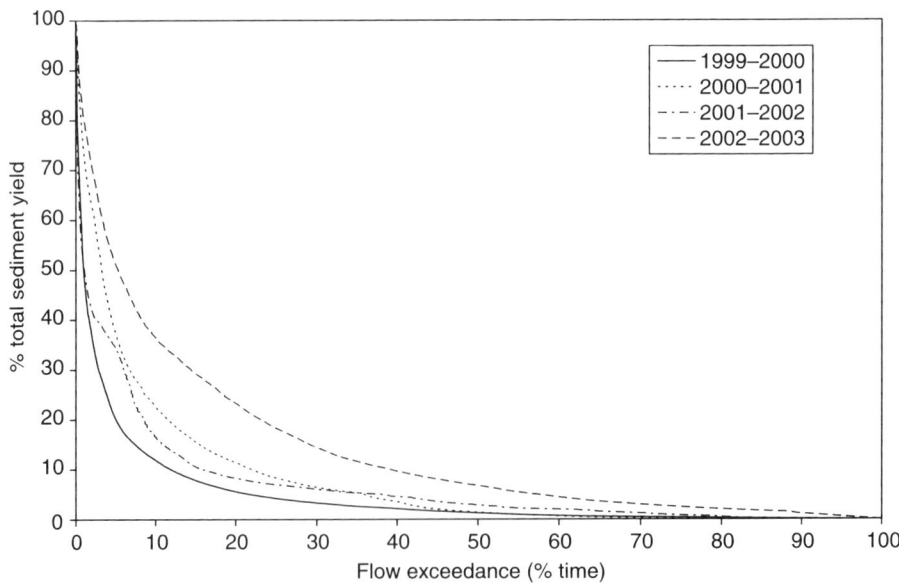

Figure 6.1.3 Sediment exceedance as a function of flow exceedance for the River Tees, UK

6.1.2 SEDIMENT TRANSPORT AND MONITORING METHODS

Within a river, sediments move in different ways depending on the sediment characteristics (size, shape, density) and the flow rate. Normally we would consider three sediment transport modes:

- *Wash load,* which is typically taken to consist of all particles of less than 63 µm. Here sediment concentration measured at any point in the river cross-section should be constant. In practice, the size of particles being transported as wash load may vary with flow velocity.

- *Suspended bed material load,* which consists of coarser particles which are intermittently moved from the river bed into the flow profile. Typically concentrations vary logarithmically with depth, with mean concentration for a vertical profile being seen at 0.6 × depth, measured from the flow surface. Concentrations also vary across the river section, being highest in the centre, where flow velocity is highest, and lowest at the river banks.

- *Bed load,* which consists of coarse material moving intermittently, always in contact with the river bed.

Clearly the measurement of these three transport modes requires different techniques. Wash load and suspended bed material load tend to be monitored together, either directly through sampling of water and sediment from the flow profile or indirectly through monitoring of turbidity.

Sampling from the profile is normally done as a grab sample by lowering a container into the flow and allowing it to fill. Alternatively, pumped sampling has been used, normally from a fixed sampling point in the flow profile, with sample timing being controlled by depth or flow or rate of change in either. These two methods provide information about sediment concentration at one point and time. Other methods allow collection of sediment at a point over time (e.g. Schulze *et al.*, 2007). However, because sediment concentration varies across the channel and through depth, in order to assess sediment flux through a river section, sampling on several profiles is required, concurrent with velocity measurement. Samples and velocity data are collected on a number of vertical profiles across the river (as for the velocity area method of flow gauging), with sediment flux being calculated either by mean-section or mid-section methods. An alternative approach to sampling at several points on the vertical is the use of depth-integrating samplers, where a sample tube is lowered and then raised at constant velocity through the flow vertical, sampling water and sediment throughout its travel time (see ASCE (2006) for an overview of sediment monitoring techniques). Any of the point, pumped or depth-integrated sampling techniques that involve water and sediment being removed from the river involve a further step of laboratory analysis to determine the sediment concentration.

The indirect turbidity approach is best applied in tandem with a programme of sediment sampling. This is because the techniques used for monitoring of turbidity show a variable response with sediment colour and particle size and shape. Thus the

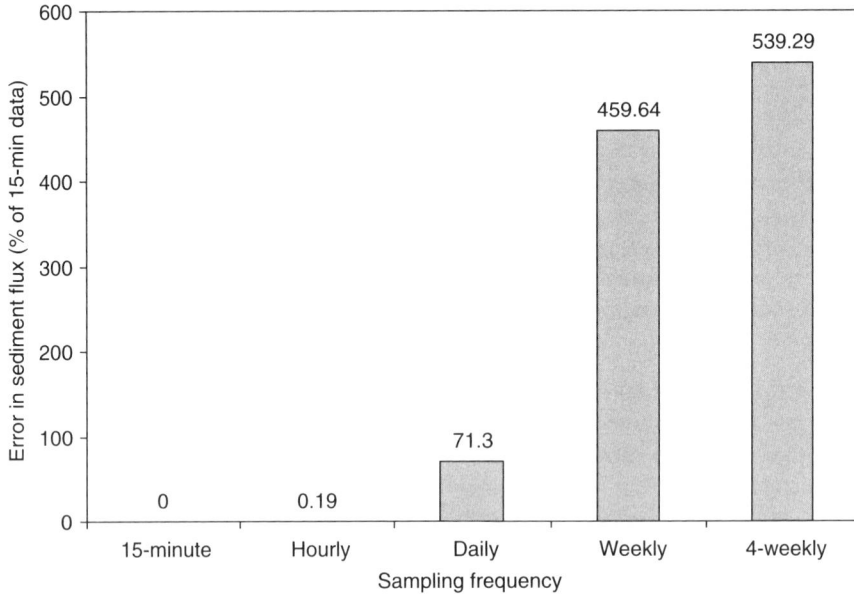

Figure 6.1.4 Error in sediment flux relative to flux calculated from 15-minute data, River Tees, 1999–2003 (the numbers above the bars are the percentage difference from the 15-minute data)

relationship between turbidity and sediment concentration is not constant and needs readjustment, particularly during high-flow events.

For the River Tees study referred to above, 15-minute turbidity data, corrected to sediment concentration values, are available for 3.5 years. Using this data set, it is possible to calculate the error in sediment flux that would accrue if samples had only been taken at hourly, daily, weekly or 4-weekly intervals. Figure 6.1.4 shows how error increases as sampling frequency decreases. This general pattern will be the same for most rivers, although the absolute percentage error, and the way it increases with sampling frequency, will depend on the supply of sediment and the flashiness of the river regime.

Bedload measurements are more complex and offer no single standard approach. Small upstream channels are frequently the focus of bedload measurement (e.g. Leeks and Marks, 1997; Newson and Leeks, 1985; Reid and Frostick, 1986; Richards and McCaig, 1985), but there are few reported studies for downstream sites (e.g. Habersack *et al*, 2001; Meigh, 1987). More recent developments relate to acoustic or impact devices (e.g. Reid *et al.*, 2006; Rickenmann and McArdell, 2007) and the recently founded Bedload Research International Cooperative (http://www.nced.umn.edu/bricportal.html) to study novel measurement technologies.

6.1.3 SEDIMENT RATING CURVES

In many situations, sediment flux is estimated through use of a sediment rating curve (see e.g. Julien, 1998), where a relationship is derived between sampled sediment concentration, C, and river discharge, Q. Relationships are normally in the form of a power law, where:

$$C = aQ^b$$

and a and b are constants for a given sampling location on a river.

The capacity of a river to move sediment is related to the size distribution of sediment particles, particle shape, sediment density, river flow velocity and river depth which will all change over time.

The concepts of limited sediment supply to rivers and an upper limit to the capacity of a particular flow rate to transport sediment of a particular particle size distribution and density explain the phenomenon of hysteresis, where sediment concentration for a given flow rate differs on the rising and falling limbs of a flow event. There is thus no simple relationship between flow and concentration, and a sediment rating curve will inherently have wide error bars.

6.1.4 SEDIMENT MONITORING IN EUROPE TODAY

For the majority of European countries, sediment transport, if monitored at all, is measured as part of a routine river monitoring programme at intervals of between one and four weeks, and is normally restricted to the material in suspension (i.e. wash load + suspended bed material load). Where there is particular concern, or sometimes during high-flow events, sampling frequency may be increased, but it rarely exceeds once per day except in research studies. Sampling is done as either a grab or an integrated sample, sometimes from an unspecified point in the flow profile, or at a fixed point above the river bed. Samples are then analysed in a variety of ways to give one or more of:

- Total sediment concentration (often referred to as total suspended solids): samples may be prefiltered to remove fine particles. Filter size (i.e. size of sediment particle removed) can vary from place to place.

- Fine (wash load) concentration: determined from that portion of the sediment passing through a filter. There are a range of analysis techniques which can give variation in concentration data.

- Organic material concentration: determined from the difference in dry sediment weight before and after heating to burn off organic material. The time and temperature of heating needs to be to a standard to provide a consistent data set.

- Inorganic sediment concentration: that sediment remaining after heating.

There are often inconsistencies between sampling protocols and/or analysis techniques within a country, where sediment sampling may be carried out for different purposes, and thus different outputs may be required. It is often difficult to determine the underlying sampling or analysis technique for particular data, and thus difficult to produce reliable and consistent data sets to analyse, for example, trends. Data availability also varies between countries, with some being available online, some available on request and some very difficult to acquire.

Where rivers cross national or regional boundaries, differences in monitoring protocols can lead to difficulties in looking at sediment supply and transfer through a river system. Such an understanding is necessary if sediment is to be understood and managed at a river basin scale.

6.1.5 HOW SHOULD WE MONITOR SEDIMENT FLUX IN EUROPE?

Although the requirement to monitor and assess sediment is not explicitly mentioned in the Water Framework Directive, it is clear that sediment can be both an ecological threat in its own right (due to both lack and excess of material) and a vector for transfer of other contaminants (see e.g. Brils *et al.*, 2007; van der Meulen *et al.*, 2007). It is clear that sediment is an issue of importance for the delivery of good ecological status in many, if not all, European rivers. Whilst sediment is related to hydromorphological aspects considered in the WFD, its importance and influence extends beyond this, to the sorbtion, release and transfer of contaminants, the effect of turbidity, changing river bed sediment texture, deposition on or erosion from flood plains, and so on. Therefore, in order to make sensible management decisions as part of the programmes of measures at river basin scale, data need to be collected appropriately and consistently. Better information on sediment flux through a river system could help us better identify sediment sources and deposition zones, estimate sustainable sediment extraction rates, allow estimation of dredging requirements, better estimate reservoir and lake sedimentation rates, and better quantify and understand downstream pollution transfers.

It is widely accepted that there is a need for more evidence (system understanding) to support sediment management decisions in relation to both WFD and other legislative drivers, such as the Birds and Habitats Directive. Currently the only suspended sediment guideline in legislation is from the Fisheries Directive, which sets a target value of less than 25 ppm mean annual suspended sediment concentration, which may be sufficient to protect fisheries but is unlikely to address all water body ecosystem requirements.

Current monitoring protocols fail for various reasons:

(1) Inadequate temporal scale of data: routine monitoring is at regular time intervals, whilst bulk sediment transport is highly skewed to high flows.

(2) Inadequate spatial resolution of data: data may not be collected at places most relevant for increasing our system understanding.

(3) A need for new guideline values in terms of whole ecosystem functioning, i.e. based on impacts rather than concentrations.

If any quantitative or qualitative target for sediment is to be set, this will need to take into account the highly dynamic nature of rivers. Furthermore, there is a need to differentiate by river type, to be inclusive of the whole river ecosystem and to look at the continuum of sediment quantity gradient and biological response. There is a requirement for scientists to improve our sediment system understanding, i.e. understanding not only of how sediment is moving, but of how it impacts on ecosystem function, at river basin scale, so that more robust, evidence-based monitoring can be implemented.

There are various sediment-related factors that may be of importance for assessing impacts on ecological status. Those that could be directly monitored are:

- *Sediment Flux/Budget:* Measured consistently and coherently across rivers to allow better understanding of how sediment flux and dynamics vary between river types. Evidence from many European rivers, excluding those that are dominantly ground-water fed, suggests that monitoring should be focused in the top 20–30 % of flows. Continuous turbidity monitoring with sampling (possibly using flow rate-controlled automated pump samplers) during high-flow events may offer a way forward, although this requires more manpower (and hence cost) than currently involved in four-week spot sampling.

- *Particle Size Distribution:* This can currently only be operationally assessed through analysis of sediment samples ex situ. Such samples would be provided through the automated pump sampling suggested above. One disadvantage of using such samples is that they tend to be small (typically a maximum of 1 litre) and therefore may contain few particles for analysis. Collection of larger sediment volumes requires more sophisticated pumping and sampling setups. Novel approaches, using spectral interferometry, which allow particle size distribution to be determined in situ will help to improve system understanding.

However, other issues such as sediment-borne contaminants, residence times of sediment at a location in the river network, remobilisation of historical sediment deposits (and contaminants), possibly through remobilisation of river channels as climate change increases sediment supply, channel-forming flows and flood frequency, and changing land use and sediment supply patterns may mean that a reappraisal of sediment monitoring and of the impacts of sediment on ecosystem functioning is required as we move forward into the next phase of European environmental legislation (Apitz *et al.*, 2006). It is unlikely that we will ever have sufficient data to completely characterise sediment flux throughout all river systems. Approaches to management, whilst based on improved system understanding, will need to be based on assessments of risk in future (Heise, 2007).

REFERENCES

ASCE (2006) *Sedimentation Engineering: Theory, Measurements, Modeling and Practice*, ASCE, Reston, Virginia.

Apitz, S.E., Elliott, M., Fountain, M. & Galloway, T.S. (2006) European environmental management: moving to an ecosystem approach. Integrated Environmental Assessment and Management, **2**(1), pp. 1–6.

Becvár, M. (2005) Estimating typical sediment concentration probability density functions for European rivers, Unpublished MSc thesis, Cranfield University, UK.

Bednarek, A.T. (2001) Undamming rivers: a review of the ecological impacts of dam removal, *Environmental Management*, **27**(6), pp. 803–814.

Brath, A., Castellarin, A. and Montanari, A. (2002) Assessing the effects of land-use changes on annual average gross erosion, *HESS*, **6**(2), pp. 255–265.

Brils, J., Brauch, H.J., Liska, I., Miloradov, M., Nachtnebel, H.P., Pirker, O., Rast, G., Almodovar, M., Dias, E.B., Gomes, F.V., Heise, S., Portela, L., Vale, C., Forstner, U., Gabriel, T., Heininger, P., Netzband, A., Sassen, K., Schulz, S., Brien, J., Morris, R., Owens, P., White, S., Whitehead, P., Winn, P., den Besten, P., Della Sala, S., Eisma, M., Hauge, A., Keller, M. & Slob, A. (2007) Sediment management: An essential element of river basin management plans. *J. Soils and Sediments,* **7**(2), pp. 117–132

De Roo, A.P.J. (1998) Modelling runoff and sediment transport in catchments using GIS, *Hydrol. Proc.*, **12**(6), pp. 905–922.

Ferguson, R.I. (1981) Channel form and channel changes, In: Lewin, J. (ed.), *British Rivers*, Allen and Unwin, pp. 90–125.

Flanagan, D.C., Gilley, J.E. and Franti, T.G. (2007) Water Erosion Prediction Project (WEPP): development history, model capabilities, and future enhancements, *Trans ASABE*, **50**(5), pp. 1603–1612.

Habersack, H.M., Nachtnebel, H.P. and Laronne, J.B. (2001) The continuous measurement of bedload discharge in a large alpine gravel bed river, *J. Hydraul. Res.*, **39**(2), pp. 125–133.

Jaeggi, M.N.R. (1995) Sediment transport in mountain rivers: a review, *Proc. Int. Sabo Symp.*, Tokyo, Japan.

Julien, P.Y. (1998) *Erosion and Sedimentation*, Cambridge University Press, Cambridge, UK.

Lawler, D.M. (1992) Process dominance in bank erosion systems, In: Carling, P.A. and Petts, G.E. (eds) *Lowland Floodplain Rivers: Geomorphological Perspectives*, John Wiley and Sons, Ltd, Chichester, UK, pp. 117–143.

Leeks, G.J.L. and Marks, S.D. (1997) Dynamics of river sediments in forested headwater streams: plynlimon, *Hydrol. Earth System Sci.*, **1**(3), pp. 483–497.

Leopold, L.B, Wolman, M.G and Miller, J.P. (1964) *Fluvial Processes in Geomorphology*, W.H. Freeman, San Francisco.

Lewis, L.A., Verstraeten, G. and Zhu, H.L. (2005) RUSLE applied in a GIS framework: calculating the LS factor and deriving homogeneous patches for estimating soil loss, *Intl J of GIS*, **19**(7), pp. 809–829.

Meigh, J. (1987) Bedload transport in a gravel-bed river, PhD thesis, University of East Anglia, UK.

Morgan, R.P.C., Quinton, J.N., Smith, R.E., Govers, G., Poesen, J.W.A., Auerswald, K. *et al.* (1988) The European Soil Erosion Model (EUROSEM): a dynamic approach for predicting sediment transport from fields and small catchments, *Earth Surface Processes Landf*, **23**(6), pp. 527–544.

Newson, M.D. and Leeks, G.J. (1985) Mountain bedload yields in the United Kingdom: further information from undisturbed fluvial environments, *Earth Surface Processes Landf.*, **10**, pp. 413–416.

Reid, S.C., Lane, S.N., Berney, J.M. and Holden, J. (2006) The timing and magnitude of coarse sediment transport events within an upland, temperate gravel-bed river, *Geomorphology*, **83**(1–2), pp. 152–182.

Reid, I. and Frostick, L.E. (1986) Dynamics of bedload transport in Turkey Brook, a coarse-grained alluvial channel, *Earth Surface Processes Landf.*, **11**, pp. 143–155.

Richards, K. and McCaig, M. (1985) A medium-term estimate of bedload yield in Allt A'Mhuillin, Ben Nevis, Scotland, *Earth Surface Processes Landf.*, **10**, pp. 407–411.

Rickenmann, D. and McArdell, B.W. (2007) Continuous measurement of sediment transport in the Erlenbach stream using piezoelectric bedload impact sensors, *Earth Surface Processes Landf.*, **32**, pp. 1362–1378.

Schulze, T., Ricking, M., Schoter-Kermani, C., Korner, A., Denner, H.D., Weinfurtner, K., Winkler, A., Pekdeger, A. (2007) The German Environmental Specimen Bank – Sampling, processing, and archiving sediment and suspended particulate matter. *J. Soils and Sediments,* **7**, pp. 361–367.

Schumm, S.A. (1977) *The Fluvial System*, John Wiley & Sons, Ltd, New York, USA.

Task Committee of Computational Modeling of Sediment Transport Processes (2004) Computational modeling of sediment transport processes, *J. Hydr. Eng.*, **130**(7), pp. 597–598

van der Meulen, M.J., van der Spek, A.J.F., de Lange, G., Gruijters, S.H.L.L., van Gessel, S.F., Nguycn, B.L., Maljers, D., Schokker, J., Mulder, J.P.M. & Krogt, R.A.A.D. (2007) Regional sediment deficits in the Dutch lowlands: Implications for long-term land-use options. *J. Soils and Sediments,* **7**(1), pp. 9–16.

Wass, P.D. and Leeks, G.J.L. (1999) Suspended sediment fluxes in the Humber catchment, UK, *Hydrol. Processes*, **13**, pp. 935–953.

Walling, D.E. (1983) The sediment delivery problem, *Journal of Hydrology*, **65**, pp. 209–237.

Walling, D.E. (1988) Erosion and sediment yield research: some recent perspectives, *Journal of Hydrology*, **100**, pp. 113–141.

White, S.M., Fredenham, E. and Worrall, F. (2005) Sediments in GREAT-ER: Estimating sediment concentration ranges for European Rivers, Report to the European Chemicals Industries Council (CEFIC).

Wischmeier, W.H. and Smith, D.D. (1978) Predicting Rainfall Erosion Losses. A guide to conservation planning. Agriculture Handbook No. 537. USDA-SEA, US. Govt. Printing Office, Washington, DC. 58pp.

White, S.M., Worral, F. and Pender, G. (2003) The long term sustainability of estuarine impoundments, Final report to the Engineering and Physical Sciences Research Council, UK.

White, W.R., Meyer, P.E., Rottner, J., Engelund, F., Ackers, P. Milli, H. *et al.* (1975) Sediment transport theories, *ICE Proceedings*, **59**(2,2), pp. 265–292.

Yang, C.T. (2003) *Sediment Transport: Theory and Practice*, Krieger Publishing Co.

6.2

Monitoring Sediment Quality Using Toxicity Tests as Primary Tools for any Risk Assessment

Wolfgang Ahlf, Ute Feiler, Peter Heininger and Susanne Heise

The Water Framework Directive - Ecological and Chemical Status Monitoring   Edited by Philippe Quevauviller, Ulrich Borchers, Clive Thompson and Tristan Simonart © 2008 John Wiley & Sons, Ltd

6.2.1 INTRODUCTION

The Water Framework Directive (WFD) is the first EU legislative instrument which requires a systematic monitoring of biological, chemical and quantitative parameters in European waters at a wide geographical scale. The implementation of the WFD is changing the scope of water management from the local scale to the more complex scale of a river basin. Thus, in managing river water quality, not only distinct point pollution sources but also diffuse pollution sources from groundwater, land erosion and especially contaminated sediment have great influences on water quality and should be considered (Heise and Ahlf, 2002). The WFD is concerned with protection of the environment and the achievement of a good ecological status. The central operational instrument of the WFD in the control and reduction pollution is a combined approach using emission standards on the source side and quality standards on the effects side. Satisfactory monitoring concepts are needed to establish a rational and wide-ranging overview of water status within each river basin district. The WFD approach encompasses measures of ecological, hydrological and hydrogeological systems, including targets reflecting the ecological integrity of the water body. With respect to this objective, the aquatic sediments have special significance as habitats of species-rich biocoenoses, e.g. of the macrozoobenthos as one of the indicator groups. Because of their high potential for accumulation of contaminants, sediments are particularly sensitive to anthropogenic impacts, which may disturb the natural state of waters. The continuous exchange between sediment and water phase during settlement and resuspension of contaminated particles could impact less contaminated areas (Förstner *et al.*, 2004). Therefore the risk of bound contaminants being spread in the river basin due to different hydrological conditions is growing with the increasing amount of sediment trapped in a river basin (Salomons, 2005).

To evaluate sediment quality, a triad approach with chemical, ecotoxicological and ecological parameters is often used (Chapman, 1996). Since the basic question is whether biological impairments are caused by contaminated sediments, the biological assessment facilities are the basis for the development of a regulatory scheme. We think that understanding the bioavailability of pollutants is the key issue for assessing sediment quality, which is increasingly being realised as the primary issue for risk management (Ahlf and Förstner, 2001). The recommendations from a risk assessment are directed to environmental authorities, which need:

- To classify hot spots, rank contaminated sites using chemical and biological methods.

- To make decisions for more detailed studies on the site-specific damage of aquatic communities.

- To trigger regulatory action and establishment of target remediation objectives, e.g. for dredged material.

Bioassays provide a more direct measure of environmentally relevant toxicity than chemical analyses do, since they integrate environmental variables and contaminants (Keddy *et al.*, 1995). The value of toxicity tests within an integrated sediment assessment system is given in Table 6.2.1.

Table 6.2.1 Integrated sediment assessment system: a hierarchical strategy

Regulatory Concerns	Assessment Tools	Problems	Interpretation
Contamination	Set of chemical analyses	Single contaminants	Quality criteria
Toxicity	Toxicity tests	Single effects	Quality criteria
Hazard	Set of toxicity tests	Cause unknown	Multivariate statistic approach
Management	Toxicity identification evaluation	Single pollutants	From case to case
Ecological quality	Analysis of benthic structure, geochemical habitat	Reference of unpolluted areas	From case to case
Risk assessment	All above	No standardised set of methods	Unknown, no general model

Table 6.2.2 List of presently available ISO standardised biological tests with whole organisms for the detection of environmental pollutants in water and sediments

Norm	Title	Exposure Phase
ISO 10712:1995	*Pseudomonas putida* growth inhibition test (pseudomonas cell multiplication inhibition test)	Water
ISO 11348-1:1998	Determination of the inhibitory effect of water samples on the light emission of *Vibrio fischeri* (luminescent bacteria test) Part 1: Method using freshly prepared bacteria	Water
ISO 11348-2:1998	Determination of the inhibitory effect of water samples on the light emission of *Vibrio fischeri* (luminescent bacteria test) Part 2: Method using liquid-dried bacteria	Water
ISO 11348-3:1998	Determination of the inhibitory effect of water samples on the light emission of *Vibrio fischeri* (luminescent bacteria test) Part 3: Method using freeze-dried bacteria	Water
ISO 8692:1989	Fresh water algal growth inhibition test with *Scenedesmus subspicatus* and *Selenastrum capricornutum*	Water
ISO 14442:1999	Guidelines for algal growth inhibition tests with poorly soluble materials, volatile compounds, metals and waste water	Water
ISO 10253:1995	Marine algal growth inhibition test with *Skeletonema costatum* and *Phaeodactylum tricornutum*	Water
ISO 6341:1996	Determination of the inhibition of the mobility of *Daphnia magna* Straus (Cladocera, Crustacea) Acute toxicity test	Water
ISO 14669:1999	Determination of acute lethal toxicity to marine copepods (Copepoda, Crustacea)	Water
ISO 20079:2005	Lemna test	Water
ISO 16712:2005	Determination of acute toxicity of marine or estuarine sediments to amphipods	Sediment

A number of tests have been developed to evaluate the biological significance of sediment contamination. These tests may be as simple as short-term (acute) bioassays involving a single contaminant using a single species, or as complex as microcosm studies in which the long-term (chronic) effects of mixtures of contaminants on ecosystem dynamics are investigated. In addition, tests may be designed to assess the toxicity of whole sediments (solid phase), suspended sediments, elutriates, sediment extracts or pore water. The organisms that are usually tested include microorganisms, algae, invertebrates and fish. In principle a suite of bioassays is needed for an ecotoxicological approach with organisms' exposure to bulk sediments, detecting acute, chronic and subcellular endpoints like growth, reproduction, genotoxic, immunotoxic and endocrine effects.

Although sediment toxicity tests are widely applied in monitoring and assessment studies, the methods standardised are almost exclusively toxicity tests using the water phase (Table 6.2.2). Biological testing is required to provide reliable information regarding the toxicity of bed sediments. That means requirements for quality control/assurance are most stringent when deriving data for regulatory purposes.

6.2.2 SEDIMENT ECOTOXICOLOGICAL DATA QUALITY

For environmental testing, bioassays provide an integrated picture of the overall toxicity of pore water, sediment elutriate or sediment from a contaminated site. Various aquatic organisms, such as vertebrates, invertebrates, protozoa, algae, macrophytes and bacteria are used to test environmental samples. The idea behind these toxicity tests is that the test organisms will react in a predictable way to various types of environmental contaminants.

In principle, the influencing factors affecting the test results and the uncertainty are the same as for chemical analyses:

- Human factors: operator carrying out the studies must be competent in the field of work under study and have practical experience related to the work to be able to make appropriate decisions from the observations made as the study progresses.

- Environmental factors: control of oxygen conditions, geochemical composition.

- Instrumental and technical factors: equipment within specification, working correctly, properly calibrated, procedures established for operational control and calibration, traceability of measurement to the criteria for test validity.

In addition, quality assurance/quality control (QA/QC) requirements for the biological tests are highly specific, because a calibration using reference material does not cover all aspects of biological variability (Simpson *et al.*, 2005).

6.2.2.1 Quality Objectives

Accuracy criteria are not applicable to toxicity testing endpoints, because there are no standard organism responses against which to compare test results. In place of

Table 6.2.3 Data acceptability criteria for toxicity testing samples

Test Organism (Protocol – MPSL SOP)	Acceptability Criteria
Ceriodaphnia dubia (chronic 7 day) (US EPA 1994 – MPSL SOP 2.3)	a) Survival in the controls $\geq 80\%$ b) Surviving females: average ≥ 15 neonates c) Surviving females: 60% have 3 or more broods d) All performance criteria outlined in SOP are met
Holmesimysis costata (7 day) (US EPA 1995 – MPSL SOP 2.6)	a) Survival in the controls $\geq 75\%$ b) $>0.4\,\mu g$ average dry weight in controls. c) Survival and growth NOECs $\leq 100\,\mu g/L$ Zn d) $<40\%$ MSD for survival and $<50\,\mu g$ MSD for growth e) All performance criteria outlined in SOP are met
Hyalella azteca (10 and 28 day) (US EPA 2000 – MPSL SOP 2.7)	a) Survival in the controls must be $\geq 80\%$ b) Measurable growth in controls c) All performance criteria outlined in SOP are met
Mytilus galloprovincialis (48 hour) (US EPA 1995 – MPSL SOP 2.9)	a) Control normal survival $\geq 70\%$ (or with two endpoints: survival $\geq 50\%$ and normal development $\geq 90\%$) b) $<25\%$ MSD c) All performance criteria outlined in SOP are met
Pimephales promelas (chronic 7 day) (US EPA 1994 – MPSL SOP 2.10)	a) Survival in the controls $\geq 80\%$ b) $>0.25\,mg$ average weight of control larvae c) All performance criteria outlined in SOP are met

an absolute measurement of accuracy for toxicity tests, reference toxicant tests are performed to determine whether organism response is within prescribed acceptability criteria. One of the advantages of norms is that acceptability criteria are described and test results have to be checked for being within the range of given limits. Table 6.2.3 presents some acceptability criteria for toxicity tests standardised from the US EPA.

QA/QC requirements for sediment toxicity tests generally deal with ensuring that test conditions remain within control limits during the tests and do not contribute to observed effects and thereby confound interpretations regarding the toxicity of the sediments. For sediment toxicity tests, there are control limits for temperature, dissolved oxygen, salinity and pH. Monitoring of sulphides and ammonia in the test chambers may be appropriate for sediments where either of these chemicals is suspected as being a problem, and may be useful for interpreting test results. The sediment toxicity test protocols also require the testing of control samples as negative controls, positive controls and reference sediments (see SEKT paragraph). The criteria for determining test validity are an essential component of all standardised bioassays and are specific for each toxicity test species. The test results of control samples have to be compared to performance standards, which are used to validate acceptability limits, like mortality in a control sediment. All generated data should be presented in a report or in a standard operating procedure (SOP).

The SOPs should cover all aspects of the assay from the time the sample is collected and reaches the laboratory until the results of the bioassay are reported. A description of experiments concerning the validation conducted to determine variability, limit of quantification and the quality controls should be documented for data audit and inspection; the traceability is a requirement for good analytical practice. Any deviations from SOPs should be documented with justifications for deviations.

6.2.2.2   Quality Control Procedures

Laboratory quality control procedures for sediment bioassays are listed in Ecology (2003). Here we will give a brief overview of how control procedures ensure the quality of ecotoxicological tests.

Control and Reference Sediments

All solid phase tests measure toxicity relative to a negative control or reference sediment (ASTM, 2003). For bulk sediment tests a negative control will be a sediment that is essentially free of contaminants. Such a negative control provides evidence of test organism health, which is in most acute tests defined as mortality lower than 10 %, and for chronic tests a survival of more than 80 % is sufficient. Control sediments can be provided from field-collected sediments or from artificial or formulated sediments. The physicochemical properties such as grain size, TOC and background levels of contaminants should be determined. However, these properties could be different to those of the area studied.

Although sediment test organisms should tolerate a wide range of physicochemical sediment conditions, the contribution of those confounding factors may be assessed using a reference sediment as a parallel test only. Reference sediments are ideally collected from sites near the contaminated site, representing the same sediment conditions exclusive of contaminants. The impact of the reference sediment has to be determined, inclusive of the site-specific variability. Whereas the control sediment provides a reference point for interpreting effects from the test, the reference sediment can help estimate the relative contributions of natural and anthropogenic stress.

Positive Controls

A positive control uses a reference toxicant that affects the test organism in a reproducible manner. Reference toxicants provide a general measure of the precision of a toxicity test method over time. Acceptability limits are in general a ± 2 standard deviation of the EC50. The criteria for determining the sensitivity of the test organisms is an important component of good quality assurance.

Reference toxicants recommended by Environment Canada (1995) are copper and fluoranthene. Both chemicals are reference toxicants used preferentially in spiked control sediments for chronic whole-sediment tests. If the aim of the positive control is to measure the sensitivity of the test organisms in acute tests, water-only exposures may be used.

6.2.2.3 Data Management Procedures

The project proponent is responsible for the quality assurance review of data generated in any sediment investigation. There are two levels of quality assurance review applicable for sediment data. On the first level a review of bioassay data covers field and reporting elements and evaluates the acceptability of test results for positive controls, negative controls, reference sediment, replicates and experimental conditions (temperature, salinity, pH, dissolved oxygen). Detailed guidance on review procedures is available from Ecology (2003).

The second level represents a more vigorous level of quality assurance review, and is appropriate for sediment data that are to be used for the development of numerical chemical criteria or the derivation of effect classes (Ahlf and Heise, 2005). Such a review is also recommended in cases where the data may be used in litigation. We expect a more complex environmental scene investigation in future, due to the fact that point sources are less important than diffuse ones, and all lines of evidence have to be used to characterise the environmental impact (Wenning *et al.*, 2004).

6.2.2.4 Uncertainty of Laboratory Toxicity Tests

Uncertainties of laboratory toxicity tests have been regarded as falling into two categories: 1) uncertainties related to the phase tested, and 2) uncertainties related to the selection of endpoints measured in toxicity tests (Ingersoll *et al.*, 1997). While those parameters that need to be controlled and monitored before and during a test have been mentioned above, this sub-chapter will address intrinsic properties of tests which cannot easily be overcome but have to be known and evaluated in order to interpret the results well. Among these, specific properties of the environmental samples themselves influence the degree of uncertainty of a toxicity measurement, such as the heterogeneity of sediments, which has to be considered in sediment contact tests:

Test-specific Uncertainties Related to the Tested Phase

Toxicity investigations can comprise (from Ingersoll *et al.*, 1997): a) whole-sediment tests that are carried out incubating test organisms in direct contact with sediment and which should reflect the effects of in-place (e.g. adsorbed or absorbed) pollutants; b) pore water test systems, assuming that the concentration in the pore water is in equilibrium with the sediment and the main exposure pathway is through contact with the pore water; c) toxicity tests with organic extracts, simulating a worst-case scenario in which even strongly bound contaminants may become available; and d) tests with elutriate samples and/or suspended solids to simulate resuspension events. For evaluation of toxicity data, information is needed on a number of aspects, which are influenced by the matrix. Among these are precision of the test system, standardisation, sensitivity and interference of the sediment matrix.

Results from round-robin tests and inspection of variability of positive controls over time have indicated that laboratory precision (i.e. precision not related to sampling collection, handling and storage) seemed to be good for whole-sediment tests and in the same range as elutriate, organic extracts and pore water tests. Intra- and inter-laboratory variability were low (Ahlf and Heise, 2005; Burton *et al.*, 1996; Mearns *et al.*, 1986). Only tests on suspended matter showed high uncertainty, due to the low standardisation of this testing method (ASTM, 1995).

Sensitivity of biotests in terms of uncertainty considerations refers to the potential of a test system to indicate or to predict correctly the effects of contaminants in the sediment, thereby minimising false positive (nontoxic sample incorrectly classified as toxic) and false negative (toxic sample incorrectly classified as nontoxic) results. Ingersoll *et al.* (1997) pointed out that whole-sediment tests with benthic organisms and acute measurement endpoints showed a high degree of certainty in this respect. This has also been shown by Rönnpagel *et al.* (1998) for a bacteria contact test which correlated well with the autochthonous microbial activity in experiments with spiked sediments.

Interference of sediment compounds with measurement endpoints has been described for organic extracts in the luminescence bacteria test (Greene *et al.*, 1992), when extracted pigments interfere with the luminescence, for the algae growth inhibition in cases of shading effects of elutriated compounds (Cleuvers *et al.*, 2002), and for bacterial contact tests, where sediment properties reduced the measurable endpoint due to adsorption of organisms (solid phase Microtox®; Ringwood *et al.*, 1997) or of the indicator substance (bacterial contact assay; Heise and Ahlf, 2005).

6.2.3 THE SeKT JOINT RESEARCH PROJECT: EVALUATION OF A BIOTEST BATTERY PROPOSED FOR SEDIMENT MONITORING AND RISK ASSESSMENT

The SeKT joint research project was initiated with the aim of comparing recently developed sediment contact assays by addressing reference conditions, control sediments and toxicity thresholds for their application in limnic sediment toxicity assessment (Feiler *et al.*, 2005).

Sediment contact tests are biological methods for the determination of toxic effects induced by whole sediments in direct contact with test organisms, taking into account all possible pathways of contaminant uptake (particle contact, food, pore water). Sediment contact tests are highly relevant in order for an ecosystem approach to consider the actual bioavailability of contaminants sufficiently.

The complexity of the sediment matrix (particles and water) places high requirements on the biological test methods. The bioassays must be able to distinguish between anthropogenic contamination and the influences of natural factors (e.g. the grain-size spectrum), thus making risk assessments possible. Over the past ten years, fruitful efforts were invested in developing suitable sediment contact tests, standardising them and validating them for routine application (Egeler *et al.*, 2005; Feiler *et al.*, 2004; Hollert *et al.*, 2003; Liss and Ahlf, 1997; Neumann-Hensel and Melbye, 2006; Traunspurger *et al.*, 1997; Weber *et al.*, 2006).

However, the above-described fundamental problems (e.g. environmental factors, controls, test-phases; see Section 6.2.2.) cannot be resolved by further developments of individual tests, but require a comprehensive comparative examination. Such a study should comprise the definition of reference conditions and standardised reference sediments (controls) as the basis for the comparability of several sediment contact tests that are used within one test battery. Further, toxicity thresholds should be defined to assure assessment methods. Meaningful answers to these fundamental questions of sediment risk assessment and classification can be found by pooling research efforts in a joint research project. Such a project is the here described SeKT joint research project (SeKT: Sediment Kontakt Test).

The studies of SeKT have the objective of testing the applicability of sediment contact tests with a possibly wide range of different sediments. This means primarily determining the variability of the test results with the aim of improving the reliability of the tests. The project plan comprises i) the application of the sediment contact tests with different sediments in order to identify the influences of natural sediment properties on the test systems; ii) the definition of reference conditions, including the standardisation of negative controls; iii) the determination of toxicity thresholds for the individual sediment contact tests. Further, iv) the test systems should be validated with contaminated natural sediments and by means of dose–effect relations with sediment samples that were spiked with selected contaminants. The results obtained within the project should serve as a data base for improved interpretation and evaluation of ecotoxicological sediment analyses.

6.2.3.1 Selection Criteria for and Choice of Sediment Contact Tests

The project uses test systems that were developed in Europe and have a good record of application in a wide range of questions. Seven different test systems were included in the SeKT battery. The selected test organisms represent three trophic levels. Moreover, several uptake pathways were considered and compared (see Table 6.2.4).

6.2.3.2 Definition of Reference and/or Control Sediments

As already described (see Section 6.2.2.3), the choice of a suitable control is one of the crucial steps for obtaining reliable results in a sediment contact test. In a first step of the SeKT project, chemically unpolluted natural and artificial sediments were studied with the sediment contact test battery. The influences of natural sediment properties on the individual test systems were investigated. If possible, a control sediment (negative control) should be defined suitable for all test organisms. For this purpose, twelve chemically non- or minor polluted natural sediments and five artificial sediments were tested. Criteria for selecting the natural sediments were the degree of anthropogenic impact, their grain-size composition, their content of organic carbon (TOC) and the type of the water body. The variability of the test results from all natural sediment investigations was determined to improve their reliability. In addition,

Table 6.2.4   Choice of sediment contact tests for the SeKT battery

Test	Organism	Organisation Level	Trophic Level	Pathway	Endpoint
Macrophyte contact test	Myriophyllum aquaticum	Plant	Producer	Pore water (pw), particle contact	Growth
Nematode test	Caenorhabditis elegans	Invertebrate	Consumer	Pw, particle contact, particle ingestion	Growth, reproduction
Oligochaetes test	Lumbriculus variegatus	Invertebrate	Consumer	Pw, particle contact, particle ingestion	Growth, reproduction, bioaccumulation
Fish-egg test	Danio rerio	Vertebrate	Consumer	Pw, particle contact	Mortality
Bacteria contact test	Arthrobacter globiformis	Bacteria	Decomposer	Pw, particle contact	Metabolism
Yeast contact test	Saccharomyces cerevisiae	Yeast	Decomposer	Pw, particle contact	Metabolism

a range of artificial sediments were tested, generally used in the particular sediment contact tests (e.g. the OECD-218) as control sediments. Two natural sediments and one artificial sediment were found appropriate as control sediments for the entire sediment contact test battery. One of the natural sediments and the artificial one were taken for spiking experiments in order to obtain information on the sensitivity of the test systems.

6.2.3.3   Variability and Sensitivity of the Sediment Contact Tests

After finding suitable control sediments (see above), these sediments were spiked with two different mixtures of pollutants (heavy metals and organic substances), in order to obtain information on the sensitivity of each test system. A wide range of EC50 values between the different test systems and applications was observed, indicating different exposure pathways and sensitivities for the various test organisms. The different toxicities of metals and organic substances in artificial compared with natural sediments in all test systems suggests different bioavailability of the toxicants in the respective sediment, probably due to a specific binding capacity of the diverse pollutants. The results indicate that the proposed sediment contact tests complement one another, representing all possible exposure pathways and trophic levels of a benthic habitat.

6.2.3.4 Toxicity Comparison of the Sediment Contact Tests in Natural Polluted Sediments

The SeKT battery was applied to polluted sediments from several German river systems. As control/reference sediments, the artificial sediment and the two natural sediments were used in all test systems. A wide variety of toxic effects were obtained. The inhibition values ranged between no effect and almost 100 %, depending on the test system and the sediment studied. Thus, an allocation of the sediment toxicity was possible for the range of sediments, based on the multiple toxic effects of the six test systems performed.

A sediment contact test battery as applied in the SEKT project can therefore be used as a tool for sediment monitoring.

6.2.4 SEDIMENT RISK ASSESSMENT

Qualitative sediment monitoring programmes should address the risk coming from contaminated sediments and the temporal and spatial changes in sediment quality. The current state of the science in ecological risk assessment is the use of a weight of evidence approach similar to that used in effects-based sediment toxicity testing. In fact, sediment toxicity testing and ecological risk assessment have been described as complementary components of a sediment assessment framework. However, several sediment assessment methodologies have evolved using a variety of approaches with wide ranges of scientific uncertainty and predictability. Here, emphasis is given to the application of a sediment toxicity testing approach, leaving aside ecological risk assessment as a separate theme.

One approach to increase certainty of data has been suggested e.g. by Suter (1983): the use of several lines of evidence in order to make a best-judgement weight of evidence decision. In this context, the application of different biotests with different endpoints, exposure routes and sensitivities towards contaminants could be regarded as different lines of evidence. If such a biotest battery, however, comprises e.g. sediment contact tests as well as elutriate tests, additional information could be drawn from the results on whether potential risks of contaminants are sediment-focused or could also affect organisms in the water column upon resuspension.

Another parameter mentioned above is the precision of biotest data depending on a variety of factors. These factors comprise the heterogeneity of the matrix and the preparation procedure, but also the number of organisms in the test system and the robustness of the bioassay. Precision can be estimated by reproducibility of e.g. positive controls or replicability of environmental samples.

The integration of ecotoxicological data into risk assessment schemes for sediments has been frequently demanded by scientists and seldom carried out by stakeholders. While biotests have been part of governmental assessment schemes in the US for a long time (US EPA, Canadian EPA), their integration into European regulations is very slow. In an overview of European sediment risk assessments, only Belgium, the Netherlands and Germany have biotest batteries that relate to sediment management practices included in their regulations – for determining either in situ risks (in situ BEBA) or ex situ risks, if the fate of dredged material has to be considered (ex situ BEBA) (den Besten *et al.*, 2003). There are in general two ways to include ecotoxicological tests in hazard assessment: First, in the form of sediment quality guidelines. These have been defined as numerical chemical concentrations intended to be protective of biological resources, predictive of adverse effects to those resources, or both (Wenning *et al.*, 2004). There are a variety of SQGs, like the EqP approach (DiToro *et al.*, 2004), effects range low (ERL) and effects range median (ERM) approaches (Long *et al.*, 1995), threshold effects level (TEL) and probable effects level (PEL) approaches, PEC/PNEC concept, and more, which are all based on comparison of the ambient chemical concentration to concentrations that have been estimated from ecotoxicological effects in spiked material (water or sediment). A second possibility is the direct application of biotests to sediments and the interpretation of these data as one line of evidence. Applying biotests directly to sediments (or other environmental samples) makes use of the greatest advantage of biotests: the integration of effects of mixtures of contaminants and reaction to those contaminants that have not been detected chemically.

The disadvantage, however, is the difficult interpretation. Chemical SQG serve the need of stakeholders for yes or no answers, as they are either exceeded or not. Biotest results as such, especially if they are part of a biotest battery, do not provide one single number but a range of values, which have to be interpreted carefully and with the characteristics of the different biotests in mind. How a biotest battery has to be constituted has been explained in detail elsewhere (Ahlf *et al.*, 2002), but it should be emphasised that a test battery should comprise organisms with different exposure routes and different sensitivities.

In order to simplify the complex interpretation of ecotoxicological data, mostly just one result out of a battery of tests is considered, which is usually the most sensitive one, in order to have false positive rather than false negative results (e.g. BfG, 1999). However, there have also been attempts to integrate results from various biotests into one single number: simple addition of inhibition values, comparative ranking by cluster analyses (Ahlf and Wild-Metzko, 1992), the Sed-Tox Index (Bombardier and Bermingham, 1999), Hasse Diagramm Technique (Brüggemann and Halfon, 1997), the toxicological risk ranking model (Hartwell, 1999) and others. Some of these methods have already been compared in Hollert *et al.* (2002).

Basing decisions in risk management on only one of a number of applied biotests means losing valuable information that could be used to characterise the risk. The information from different test systems in a biotest set is considered to be complementary. Various biotests of a battery are not applied to validate each other but to describe the system more thoroughly.

In order to get the highest amount of information out of a biotest battery, it should fulfil the following demands:

- There should be no persistent correlation between any of the tests.

- All tests should contribute significant information to the overall result.

- No redundancies should occur, in order to keep the test system cost-efficient.

With test systems that are independent of each other, a weight of evidence approach can be applied: the more evidence is given by the test systems for a certain conclusion (e.g. three instead of one tests showing high toxicity), the higher the certainty that this conclusion is correct. Currently, concepts are being discussed and approaches tested for using the response pattern of a biotest battery to characterise environmental hazards and to quantify the potential risk by developing integrated ecotoxicological classification systems. If this system is established, the implementation into regulation for sediment management will proceed.

6.2.5 LIST OF ABBREVIATIONS

BEBA   Biological Effect-based Assessment
EqP approach   Equilibrium Partitioning Approach
MPSL   Marine Pollution Studies Laboratory
MSD   Minimum Significant Difference
NOEC   No Observed Effect Concentration
QA/QC   Quality Asssurance/Quality Control
SeKT   Sediment Kontakt Test joint research project
SOP   Standard Operating Procedure
SQG   Sediment Quality Guideline
TIE   Toxicity Identification Evaluation
TOC   Total Organic Carbon
WFD   Water Framework Directive

ACKNOWLEDGEMENT

We acknowledge the financial support which has been provided by BMBF, PT Forschungszentrum Karlsruhe, No. 02WU0598

REFERENCES

Ahlf, W., Braunbeck, T., Heise, S. and Hollert, H. (2002) Sediment and soil quality criteria, In: Burden, F.R., McKelvie, I., Förstner, U. and Günther, A. (eds), *Environmental Monitoring Handbook: 17.11–17.18*, McGraw-Hill, New York, USA.

Ahlf, W. and Förstner, U. (2001) Managing contaminated sediments, I: improving chemical and biological criteria, *JSS – J. Soils & Sediments*, **1**, pp. 30–36.

Ahlf, W. and Heise, S. (2005) Sediment toxicity assessment: rationale for effect classes, *JSS – J. Soils & Sediments*, **5**(1), pp. 16–20.

Ahlf, W. and Wild-Metzko, S. (1992) Bioassay responses to sediment elutriates and multivariate data analysis for hazard assessment of sediment-bound chemicals, *Hydrobiologia*, **235**, pp. 415–418.

ASTM (1995) Standard guide for developing conceptual site models for contaminated sites, American Society for Testing and Materials. West Conshohocken, PA, USA, **E**, pp. 1689–95.

ASTM (2003) Standard guide for designing biological tests for sediments, American Society for Testing and Materials. West Conshohocken, PA, USA, **E**, pp. 1367–1403.

BfG (1999) Handlungsanweisung für den Umgang mit Baggergut im Küstenbereich (HABAK-WSV), Bundesanstalt für Gewässerkunde.

Bombardier, M. and Bermingham, N. (1999) The sed-tox index: toxicity-directed management tool to assess and rank sediments based on their hazard-concept and application, *Environmental Toxicology and Chemistry*, **18**(4), pp. 685–698.

Brüggemann, R. and Halfon, E. (1997) Comparative analysis of nearshore contaminated sites in Lake Ontario: ranking for environmental hazard, *J. Environ. Sci. Health*, **32**(1), pp. 277–292.

Burton, G.A., Norberg-King, T.J., Ingersoll, C.G., Ankley, G.T., Winger, P.V., Kubitz, J. *et al.* (1996) Interlaboratory study of precision: Hyalella azteca and Chironomus tentans freshwater sediment toxicity assays, *Environ. Toxicol. Chem.*, **15**, pp. 1335–1343.

Chapman, P.M. (1996) Presentation and interpretation of sediment quality triad data, *Ecotoxicology*, **5**, pp. 327–339.

Cleuvers, M., Altenburger, R. and Ratte, H.T. (2002) Combination effect of light and toxicity in algal tests, *J. Environ. Qual.*, **31**, pp. 539–547.

Den Besten, P.J., Deckere, E., Babut, M.P., Power, B., DelValls, T.A., Zago, C. *et al.* (2003) Biological effects-based sediment quality in ecological risk assessment for European waters, *JSS – J. Soils & Sediments*, **3**(3), pp. 144–162.

DiToro, D.M., Mahony, J.D., Hansen, D.J., Scott, K.J., Carlson, A.R. and Ankley, G.T. (2004) Acid volatile sulfide predicts the acute toxicity of cadmium and nickel in sediments, *Environ. Sci. Technol.*, **26**, pp. 96–101.

Ecology (2003) Sediment sampling and analysis plan appendix, Washington State Department of Ecology, Olympia, WA, USA, Publication No. 03-09-043.

Egeler, P., Meller, M., Schallnaß, H.-J. and Gilberg, D. (2005) Validation of sediment toxicity test with the endobenthic aquatic oligocheate Lumbriculus variegates by an international ring test, UBA report.

Environment Canada (1995) Guidance document on measurement of toxicity test precision using control sediment spiked with a reference toxicant, Environment Canada Environmental Protection Series Report EPS 1/Rm/30, Ottawa, ON, Canada.

Feiler, U., Ahlf, W., Hoess, S., Hollert, H., Neumann-Hensel, H., Meller, M. *et al.* (2005) The SeKT joint research project: definition of reference conditions, control sediments and toxicity thresholds for limnic sediment contact tests, *ESPR – Environ Sci. Pollut. Res.*, **12**(5), pp. 257–258.

Feiler, U., Kirchesch, I. and Heininger, P. (2004) A new plant-based bioassay for aquatic sediments, *JSS – J. Soils & Sediments*, **4**(4), pp. 261–266.

Förstner, U., Heise, S., Schwartz, R., Westrich, B. and Ahlf, W. (2004) Historical contaminated sediments and soils at the river basin scale: examples from the Elbe River catchment area, *JSS – J. Soils & Sediments*, **4**(4), pp. 247–260.

Greene, M.W., Bulich, A.A. and Underwood, S.R. (1992) Measurement of soil and sediment toxicity to bioluminescent bacteria when in direct contact for a fixed time period, *Proc. 65th Annual Conference and Exposition of the Water Environment Federation*, 20–24 September, New Orleans, LA, USA, pp. 53–63.

Hartwell, S.I. (1999) Empirical assessment of an ambient toxicity risk ranking model's ability to differentiate clean and contaminated sites, *Environ. Toxicol. Chem.*, **18**(6), pp. 1298–1303.

Heise, S. and Ahlf, W. (2002) The need for new concepts in risk management of sediments, *JSS – J. Soils & Sediments*, **3**, pp. 4–8.

Heise, S. and Ahlf, W. (2005) A new microbial contact assay for marine sediments, *JSS – J. Soils & Sediments*, **5**(1), pp. 9–15.

Hollert, H., Heise, S., Pudenz, S., Brüggemann, R., Ahlf, W. and Braunbeck, T. (2002) Application of a sediment quality triad and different statistical approaches (Hasse diagrams and fuzzy logic) for the comparative evaluation of small streams, *Ecotoxicology*, **11**, pp. 311–321.

Hollert, H., Keiter, S., König, N., Rudolf, M., Ulrich, M. and Braunbeck, T. (2003) A new sediment contact assay to assess particle-bound pollutants using zebrafish (Danio rerio) embryos, *JSS – J. Soils & Sediments*, **3**, pp. 197–207.

Ingersoll, C.G., Ankley, G.T., Baudo, R., Burton, G.A., Lick, W., Luoma, S.N. *et al.* (1997) Work-group summary report on uncertainty evaluation of measurement endpoints used in sediment ecological risk assessment, In: Ingersoll, C.G., Dillon, T. and Biddinger, G.R. (eds) 297 SETAC, Pensacola, FL.

Keddy, C.J., Greene, J.C. and Bonnell, M.A. (1995) Review of whole-organism bioassays: soil, freshwater sediment, and freshwater assessment in Canada, *Ecotoxicol. Environ. Safety*, **30**, pp. 221–251.

Liss, W. and Ahlf, W. (1997) Evidence from whole-sediment, porewater, and elutriate testing in toxicity assessment of contaminated sediments, *Ecotoxicol. Environ. Safety*, **36**, pp. 140–147.

Long, E.R., MacDonald, D.D., Smith, S.L. and Calder, F.D. (1995) Incidence of adverse biological effects within ranges of chemical concentrations in marine and estuarine sediments, *Environmental Management*, **19**, pp. 81–97.

Mahony, J.D., Hansen, D.J., Scott, K.J., Hicks, M.B., Mayr, S.M. and Redmond, M.S. (1990) Toxicity of cadmium in sediments: the role of acid volatile sulphide, *Environ. Toxicol. Chem.*, **9**, pp. 1487–1502.

Mearns, A.J., Swartz, R.C., Cummins, J.M., Dinnel, P.A., Plesha, P. and Chapman, P.M. (1986) Inter-laboratory comparison of a sediment toxicity test using the marine amphipod, Rheposynius abronius, *Mar. Environ. Res.*, **18**, pp. 13–37.

Neumann-Hensel, H. and Melbye, K. (2006) Optimisation of the solid-contact test with Arthrobacter globiformis, *JSS – J. Soils & Sediments*, **6**(4), pp. 201–207.

Ringwood, A.H., DeLorenzo, M.E., Ross, P.E. and Holland, A.F. (1997) Interpretation of Microtox® solid-phase toxicity tests: the effects of sediment composition, *Environ. Toxicol. Chem.*, **16**, pp.1135–1140.

Rönnpagel, K., Jansen, E. and Ahlf, W. (1998) Asking for the indicator function of bioassays evaluating soil contamination: are bioassay results reasonable surrogates of effects on soil microflora? *Chemosphere*, **6**, pp. 1291–1304.

Simpson, St.L., Bateley, G.E., Chariton, A.A., Stauber, J.L., King, C.K., Chapman, J.C. *et al.* (2005) *Handbook for Sediment Quality Assessment*, CSIRO, Bangar, NSW, p. 126.

Salomons, W. (2005) Sediments in the catchment-coast continuum, *JSS – J. Soils & Sediments*, **5**(1), pp. 2–8.

Suter, G.A. (1993) *Ecological Risk Assessment*, Lewis, Boca Raton, Florida, USA.

Traunspurger, W., Haitzer, M., Hoss, S., Beier, S., Ahlf, W. and Steinberg, C. (1997) Ecotoxicological assessment of aquatic sediments with Caenorhabditis elegans (nematoda): a method for testing liquid medium and whole sediment samples, *Environ. Toxicol. Chem.*, **16**, pp. 245–250.

Weber, J., Kreutzmann, J., Plantikow, A., Pfitzner, S., Claus, E., Manz, W. and Heiniger, P. (2006) A novel particle contact assay with the yeast Saccharomyces cerevisiae for ecotoxicological assessment of freshwater sediments, *JSS – J. Soils & Sediments*, **6**(2), pp. 84–91.

Wenning, R.J., Batley, G.E., Ingersoll, C.G. and Moore, D.W. (2004) Use of sediment quality guidelines and related tools for the assessment of contaminated sediments, Society of Environmental Toxicology and Chemistry (SETAC).

Section 7
Risk Assessment Linked to Monitoring

7.1

River Basin Risk Assessment Linked to Monitoring and Management

Jos Brils, Damia Barceló, Winfried E.H. Blum, Werner Brack, Bob Harris, Dietmar Müller, Philippe Négrel, Vala Ragnarsdottir, Wim Salomons, Thomas Track and Joop Vegter

The Water Framework Directive - Ecological and Chemical Status Monitoring   Edited by Philippe
Quevauviller, Ulrich Borchers, Clive Thompson and Tristan Simonart © 2008 John Wiley & Sons, Ltd

7.1.1 INTRODUCTION

7.1.1.1 Managing Rivers at the Basin Scale

Approaches to the management of water, water bodies and the wider environment across Europe have been radically altered with the introduction of the European Water Framework Directive (European Commission, 2000) and the subsidiary Groundwater Daughter Directive (European Commission, 2006). The WFD promotes the integrated management of water resources based on the natural geographical and hydrological unit of the river basin rather than administrative or political boundaries. Whereas previous approaches would generally assess the chemical quality of a stretch of river or a water body such as a lake or an aquifer, the WFD and GWD have stipulated that environmental quality should be assessed holistically on a larger scale (Griffiths, 2002). Consequently, in order to achieve 'good ecological status' and 'good chemical status' for all waters and develop strategies for achieving these goals through river basin management plans, it is now necessary to assess the whole biophysical river-groundwater-soil-sediment system in an integrated way (Chapman *et al.*, 2008).

The recently published proposals for an EU soil protection policy (European Commision, 2006b) and a Soil Framework Directive (European Commission, 2006c) are continuing the trend in environmental policy towards holistic planning approaches. Although soil protection in the EU is significantly lagging behind water management, it is important to note that most threats to soil identified in the Thematic Strategy on Soil Protection (European Commission, 2002) have strong relationships with the way water is managed. Therefore, in the managing of surface waters, ground waters and sediments to achieve an ecologically satisfactory state, it is essential to manage and protect soils as well.

For the successful implementation of the WFD and future regulations on soil protection (land use), a holistic understanding of both the biophysical (river-sediment-soil-groundwater) system and the social system (management, policy making, society and the economy) and how they all interrelate is required (Heinz, 2006). In addition, the future functioning of the system as a whole, including the impacts of changing climate conditions, land use practices and pollution needs to be understood and integrated in order to adequately establish river basin management plans (RBMP) as well as groundwater quality control plans that will continue to be relevant throughout their lifespan (six years in the case of RBMPs).

7.1.1.2 Rivers Basins at Risk

River basins throughout the world are under pressure from economic activities. These affect the chemical and ecological status of rivers, lakes and groundwater and deplete available soil-sediment-water resources. The wide range of economic activities and the hydrological complexity of many river basins, both in terms of the functioning of the soil-sediment-water system and the links between water quantity, quality and economic activities, make the integrated management of river basins both complex and highly challenging (Chapman *et al.*, 2008).

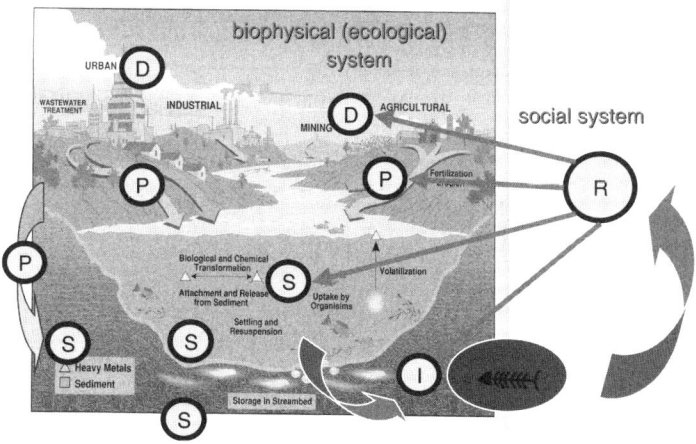

Figure 7.1.1 The Drivers-Pressures-State-Impact-Response (DPSIR) approach (base sketch provided by Garbarino *et al.*, 1995) (see Plate 15)

A useful conceptual framework for environmental processes and the links between human activities and their impact on the ecosystem functioning is provided by the Drivers-Pressures-State-Impact-Response (DPSIR) model (Figure 7.1.1). It treats the environmental management process as a feedback loop controlling a cycle consisting of five stages (European Environment Agency, 2000): Driving forces (D), Pressures (P), State (S), Impacts (I) and Responses (R). Economic activities (Driving forces) such as industry, agriculture, tourism, etc., lead to increasing Pressures on the natural environment, as these activities result in the use of natural resources and/or emissions (accidental or controlled) of waste to (ground) water (e.g. European Environment Agency, 2003), soil (e.g. Blum *et al.*, 2004a, 2004b; Van-Camp *et al.*, 2004) and sediment (e.g. Brils, 2005; Salomons and Brils, 2004).

The use of resources and/or emissions will change the State of these environments in quantity and/or quality. Sediment, water and soil resources are depleted (erosion), and/or they are loaded (contaminated) with hazardous substances originating from the economic activities. Above a certain level of depletion and/or contamination there will be significant further Impacts on the environment, e.g. loss of biodiversity, vulnerability to floods and landslides, decreased chemical and/or ecological water, soil or sediment quality and/or a shortage of these resources. Several Response measures, implemented at any of the DPS or I phases, could prevent this from happening or mitigate the impacts to a level deemed acceptable or tolerable by society. Moreover, the DPSIR framework also aims at bridging between science on one side and decision making and politics on the other side, bringing information from those who have it to those who need it.

In Europe, pollution from agriculture and hydraulic engineering (e.g. for navigation, water supply, hydroelectricity and flood control) are seen as the two main factors inhibiting the achievement of good ecological status of European river basins (Menedez *et al.*, 2006). In addition, water is both an input to many industrial production processes and a sink for their pollutants and wastewater, while households also consume water and cause pollution with hazardous substances. Furthermore, other economic sectors such as navigation and hydroelectric power rely on minimum water levels for their functioning. Moreover, the river basin ecology is damaged by shortages of water and water pollution. Water prices are generally low and water-efficient technologies and practices are not yet fully implemented in many sectors. Additional factors such as population growth, economic growth and possible effects of climate change on river flow are expected to increase existing pressures on river basins (Chapman *et al.*, 2008).

7.1.2   RISK ASSESSMENT AND RISK (-BASED) MANAGEMENT

7.1.2.1   Risk Assessment

There are many types of risk assessment. It is very important to select the type according to which policy question has to be answered or what kind of decision is expected to be made. Although risk has become a central concept in environmental policy and practice, this does not mean that it is an easy concept to define well. From the early 1980s there has been an ongoing debate in most developed countries between those who assess risk scientifically and technologically on the one hand, and social scientists and psychologists on the other, about the measurability or predictability of risk. Most of this debate, although relevant for the development of risk assessment procedures and risk management decisions for land and water resources, has not addressed river basin management specifically. The debate can, to a large extent, be characterised by two contrasting points of view: the scientific approach (formal risk assessment) versus the risks as perceived by individuals or the general public (intuitive risk assessment). The need for better integration of these objective and subjective components is becoming increasingly recognised in most fields of risk assessment.

7.1.2.2   Risk Management

There is a distinction between risk assessment (the objective, scientific part) and risk management (the policy-driven decisions about risks). Risk assessment generally starts with a respective risk agent or source. It identifies both the potential damage scenarios and their probabilities, and then models the potential consequences over time and space. In contrast, risk management addresses a much larger area of potential interventions and may alter human demands or needs (so that the agent is not even created or continued). It can suggest substitutes or alternatives for the same need. Risk management can relocate or isolate activities so that exposure is prevented, or it can make risk targets less vulnerable to potential harm. While risk assessment is mainly a scientific-technical approach, risk management is mainly based on political decisions, which depend on specific time and economic conditions. Therefore, it is very important to distinguish between both approaches.

Risk assessment and management are therefore not symmetrically opposed. Risk management encompasses a much larger domain and in the real world the risk assessment stage is often omitted or delayed, since management decisions are often based on considerations that are influenced by other factors (e.g. economic, political). In more general terms, risk management refers to the creation and evaluation of options for initiating or changing human activities or (natural and artificial) structures, with the objective being to increase the net benefit to human society and prevent harm to humans and what they value. The identification of these options and their evaluation is guided by systematic and experimental knowledge gained and prepared for this purpose by experts and stakeholders. A major proportion of that relevant knowledge comprises the results of risk assessments. However, risk managers also need to act in situations of 'non-knowledge' or insufficient knowledge about potential outcomes of human actions or activities. Hence they have to manage uncertainty.

Uncertainties are perhaps the most misunderstood concept in the field of risk analysis, even though the uncertainty analysis is an important component. It can indicate technical errors (standard deviation, variability or random error), lack of knowledge or data/information, or even more subjective aspects (such as expert judgment, quality of the problem formulation). There are many scientific and technical uncertainties in decision making and the CARACAS project discussed a number of these in relation to risk assessment (Ferguson *et al.*, 1998). Uncertainties will always be there, and the river basin management should provide ways to deal with them in a systematically and explicitly. Apart from the scientific uncertainties, there may be uncertainties in the needs of society, particularly in the future. This is not just a question of knowing whether the land use may change or not. The way land is used in the future may be very different from the forms of land use known today. These uncertainties translate into land management problems, with regard to both environment and spatial planning. Managing the unknown potential needs of society is more difficult, and to a large extent will always be a political choice, based on priorities for the current generation as much as for the future. It has to be based on the general principles of sustainable development (e.g. Brantley *et al.*, 2007).

7.1.2.3 Risk-based Management

Risk-based management or decision making is only better if the uncertainties associated with the decision are addressed in a transparent way. As a consequence, decision making can be more efficient in terms of the information needs. The basic question is: 'do we have enough information and/or are we certain enough to make a decision?' Risk-based decision making usually consists of a tiered approach. If there is not enough information/knowledge/certainty to make a decision, additional investigations are carried out.

The most critical part of handling risks relates to the understanding, description and justification of 'tolerable' or 'acceptable' risks. The term 'acceptable' refers to an activity or situation where the remaining risks are so low that additional efforts for risk reduction are not seen as necessary. The term 'tolerable' refers to an activity or situation that is seen as worth pursuing (for the benefit it carries) yet requires additional efforts for risk reduction within reasonable limits. The distinction between tolerability and acceptability can be applied to a large array of risk sources.

Zones of tolerability and acceptability can be identified in a risk diagram (where probabilities are plotted on the y-axis and consequences on the x-axis). The concept can be illustrated by the well-known traffic-light model (Figure 7.1.2; International Risk Governance Council, 2005). The red zone represents intolerable risk, the yellow zone indicates tolerable risks in need of further management actions (in accordance with the ALARP – 'as low as reasonably possible' – principle and the BAT 'best available technique' – approaches) and the green zone shows acceptable or even negligible risk. The task within risk-based management is to prohibit intolerable risks and to reduce tolerable risks by suitable measures, while ensuring that the current or planned uses and functions of soil, sediment and water resources can continue.

At the European scale, the Common Forum (http://www.commonforum.eu) and its initiatives CARACAS (Umweltbundesamt, 1998) and CLARINET (Vegter *et al.*, 2003) were the first to endorse and promote the concept of risk-based management, specifically for contaminated land. They named it risk-based land management (RBLM). RBLM is primarily a framework for the integration of two key decisions for the remediation of contaminated land:

(1) *The Time Frame*: This requires an assessment of risks and priorities, but also the consideration of the longer-term effects of particular choices.

(2) *The Choice of Solution*: This requires an assessment of overall benefits, costs and environmental side effects, value and circumstances of the land, community views and other issues.

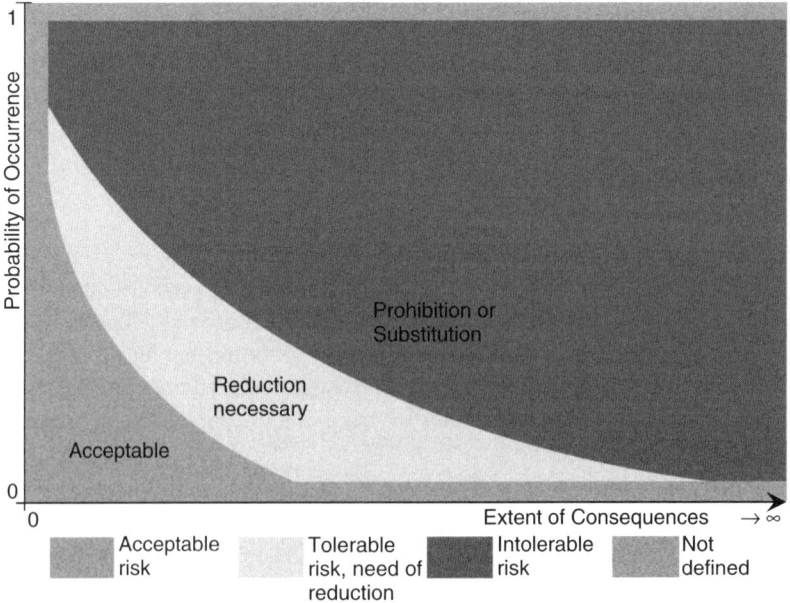

Figure 7.1.2 Acceptable, tolerable and intolerable risks (traffic-light model) (International Risk Governance Council, 2005) (see Plate 16)

These two decisions have to be taken at both an individual site level and at a strategic level, especially as the impact of contaminated land on the environment can not only have a large-scale regional dimension but also potentially wide-ranging long-term impacts. The decision making process needs to consider three main components which form the core of the RBLM concept: 1) fitness for use, 2) protection of the environment and 3) long-term care. The first two components describe goals for the safe use of land, including the prevention of harm and resource protection. The third allows for a more rigorous assessment of the manner in which these goals can be achieved sustainably. The three components need to be in balance with each other to achieve an appropriate solution (Vegter, 2004).

7.1.3   TOWARDS RISK-BASED MANAGEMENT OF RIVER BASINS (RISKBASE)

7.1.3.1   The Need for a Coordination Action

In the last decade several EC RTD Framework Programme (FP) projects (FP4–FP6) and other major research initiatives have addressed and promoted issues related to risk assessment-based management. Most of these initiatives focus on quality and management aspects of one specific compartment, such as (ground)water (BRIDGE), soil and groundwater (INCORE, JOINT, WELCOME), sediment (SEDNET) or contaminated land (CARACAS, CLARINET, NICOLE). However, all these initiatives stressed the importance of an integrative approach for understanding and managing the multi-compartment water-sediment-soil system at the river basin scale. In order to improve the scientific basis for the integrated management of this soil-sediment-water system a number of EU RTD and SSA initiatives have been funded, for example HARMONICA in FP5, the Integrated Projects AQUATERRA and MODELKEY, and the Specific Support Action (SA) SoilCritZone in FP6. Furthermore, Europe is implementing the Environmental Technology Action Plan (ETAP) on the level of mitigation technologies, because Europe needs to increase investment in more innovative environmental protection methods while boosting competitiveness (European Commission, 2004).

In relation to risk-based management, the JOINT project concluded that 'accompanying activities are necessary that help to network within the scientific and end-users communities, to evaluate the results and up-date the state of the art, and to focus on the main needs, strategies and future perspectives'. Furthermore, 'an integrated management approach is needed, which considers soil and waters as one system, interacting with other compartments of the environment and with the socio-economic world through users and functions' (JOINT, 2005).

Hence, at the end of EC FP6, there was a clear need to bring all soil-water RTD efforts together through a coordination action (CA) as a stepping stone towards further policy development and implementation, whilst addressing the RTD topics highlighted in the FP7 programme. This CA is the three-year FP6 project RISKBASE, full title: Coordination Action on Risk-based Management of River Basins (GOCE 036938, start date 1 September 2006).

7.1.3.2   RISKBASE

In RISKBASE, leading European scientists and representatives of major European stakeholder groups review and synthesise the outcome of EC RTD Framework Program projects and other major initiatives related to integrated risk assessment-based management of the water-sediment-soil system at the river basin scale. The synthesis will lead to the development of integrated risk assessment-based management approaches enabling the prevention and/or reduction of impacts caused by human activities on that system.

RISKBASE aims to deliver: 1) an overarching concept, generic approach and guiding principles to integrated risk-based management of river basins; 2) recommendations towards evolution and implementation of risk-based management operations in national and community policies and towards implementation in management; and 3) a proposal for the European research agenda related to risk-based management.

To be able to achieve this, RISKBASE organises several workshops dedicated to specific issues related to risk-based management at the river basin scale. Furthermore, RISKBASE annually organises a general assembly (GA) and makes use of EUGRIS (http://www.eugris.info) as its web-based information exchange structure. The workshops, GA and the web site (http://www.riskbase.info) are open to all who are interested and willing to contribute to achieving the RISKBASE goals and objectives.

7.1.3.3   The Risk Objective in RISKBASE

Logically, risk is always connected to an object or area of concern. Within RISKBASE this risk object is defined as the goods and services provided by the biophysical soil-sediment-water system (ecological system), with a specific focus on the resilience of the system. Furthermore, RISKBASE focuses on the goods and services that are directly (or immediately indirectly) affected by rivers, lakes and groundwater.

7.1.3.4   Ecosystem Goods and Services

Societies (present and future generations) depend for their well-being on the goods and services provided by ecosystems. Such goods comprise, amongst other things, (drinking) water, food, fuel, medicines and building materials, whilst services are the benefits people obtain from ecosystems, e.g. life support (biodiversity, fishery, fertile soils for agriculture, water supply, protection against natural hazards, etc.), regenerative services (cycling of nutrients) and cleansing services (clean water). The well-being that society derives from a healthy ecosystem is also such a service. Unlike goods bought and sold in markets, many ecosystem services are not traded in markets for readily observable prices. This means that the importance of natural processes for the well-being of humans is still ignored by financial markets (Hawken *et al.*, 2000; http://www.greenfacts.org), except for carbon sequestration. Biodiversity is popularly seen as a metaphor for the health of ecosystems and thus is of great importance for the functioning of natural processes. Hence, according to the EC Commissioner for the

Environment, Stavros Dimas, biodiversity should be pushed to the top of the political agenda: 'While climate change takes most media attention, there is one fundamental way in which biodiversity loss is more important – it cannot be undone' (opening session Green Week 2006).

Although there are large regional differences, two thirds of ecosystem services are in decline across the world. This is evidenced by collapsing fish stocks, widespread loss of soil fertility, crashes in pollinator populations (Hawken *et al.*, 2000) and reduced water retention capacity of our river systems through the loss of floodplains and wetlands by uninformed 'improvements' to the land. Ecosystem services are further compromised by overuse and loss of the species richness which ensures their stability. Two key drivers that underlie these pressures are our increasing technological abilities to efficiently consume natural resources and the combination of population growth and growing individual consumption. More specific pressures in Europe are the demands for housing and transport infrastructure (Crutzen, 2002). Added to that is the effect of climate change, which already has an observable effect on biodiversity (changing distribution, migration and reproductive patterns). Therefore, 'to conserve and restore biodiversity and ecosystem services' is seen as one of the main EU environmental policy objectives, with the Birds and Habitats Directive and the Natura 2000 network as the bases for action (European Commission, 2006a). Biodiversity reflects the number, variety and variability of living organisms. It includes diversity within species, between species and among ecosystems. The concept also covers how this diversity changes from one location to another and over time. Indicators such as the number of species in a given area can help in monitoring certain aspects of biodiversity (http://www.greenfacts.org).

7.1.3.5 Resilience

The Resilience Alliance network (http://www.resalliance.org) defines resilience as 'the amount of change a system can undergo (its capacity to absorb disturbance) and remain within the same regime and retain the same function, structure, and feedbacks'. Within resilience thinking, social and ecological systems are regarded as one closely interlinked and (very) dynamic 'social-ecological system'. The metaphor of 'adaptive cycles' describes this system's dynamic character, i.e. the progression through various phases (rapid growth → conservation → release → reorganisation → rapid growth → etc.) of organisation and function (Walker and Salt, 2006).

According to the Stockholm Resilience Center (the leading institute in the Resilience Alliance network), 'resilience refers to the capacity of a social-ecological system both to withstand perturbations from e.g. climate or economic shocks and to rebuild and renew itself afterwards. Loss of resilience can cause loss of valuable ecosystem services, and may even lead to rapid transitions or shifts into qualitatively different situations and configurations, described for e.g. people, ecosystems, knowledge systems, or whole cultures. The resilience lens provides a new framework for analysing social-ecological systems in a changing world facing many uncertainties and challenges. It represents an area of explorative research under rapid development with major policy implications for sustainable development.'

Hence, at its core, resilience is about risk and complexity. Social-ecological systems can be at risk as there are limits to the inherent resilience of these systems to absorb shocks and disturbances. Linking this to the river system ecosystem services concept raises the following scenario: the overuse of natural resources (sediment, soil, water), together with other pressures, such as changes in river hydromorphology and contamination, leads to a degradation of ecosystem functioning. Once the loss of species passes a certain threshold, these river ecosystems are often very difficult or impossible to restore. Those who want to restore these systems to a previous state have therefore to work against the resilience of the new (degraded) system. A good example is the dramatic change in aquatic ecosystems due to excessive nutrient load. Drastic lowering of nutrient loads, even below 'before pollution' levels, does not bring the system back to its previous state.

A risk-based river basin management approach should therefore be helpful in: 1) identifying the thresholds (or early warning indicators for these) and 2) defining measures for increasing the resilience (or decreasing the vulnerability) of social-ecological river systems. Measures should be aimed at decreasing the likelihood that these thresholds will be crossed and hence preventing the system moving to a different regime or state.

7.1.4 RISKBASE DRAFT CONCEPTUAL FRAMEWORK FOR RISK-BASED MANAGEMENT

Within RISKBASE a first (draft) version of a conceptual framework for risk-based management of river basins was developed (Figure 7.1.3). At the heart of the framework is the 'sources-pathways-receptors' (SPR) paradigm. In any river basin there are numerous sources of risk and pathways where these risks can be propagated through the basin towards the receptors that will be affected. As described above, the risk receptors in RISKBASE are the goods and services provided by the river ecosystem. Furthermore, what is important is that RISKBASE primarily focuses on those risks that are or can be propagated at the scale of a river basin. However, this does not mean that RISKBASE does not want to learn from site-specific risk-based management experiences.

Central to the RISKBASE management framework (Figure 7.1.3) are:

(1) The improved understanding of the functioning of the biophysical (or ecological) soil-sediment-water system (SPR pathways, affected by socioeconomic and global changes).

(2) The effectiveness, efficacy and efficiency of measures taken by the social system that aims to increase the resilience of that system. In this framework, the social-ecological system is regarded as a single system. Measures taken (in addition to the background effects of socioeconomic and global change) have an effect on the ecological system, but the resulting changes in the state of the ecological system will have a reverse effect on the social system.

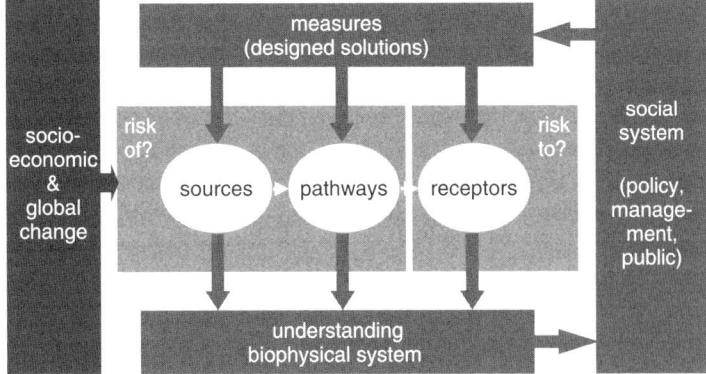

Figure 7.1.3 RISKBASE (draft) concept for a risk-based management framework of river basins. Within RISKBASE the 'receptor' is defined as the goods and services provided by the biophysical soil-sediment-water system (ecological system), with a specific focus on the system's resilience

Improved social-ecological system understanding will inevitably lead to new, additional or adapted measures, etc. Thus an adaptive management approach is proposed in which continuous learning (improved social-ecological system understanding) takes a central role. Because no fundamental, river system-orientated policies have been develop in Europe so far, existing river basin management policies are seen as supportive to this adaptive management framework, rather than this framework being supportive to the strict implementation of existing policies.

This draft framework is further developed and tested in practice through river basin case studies, in a series of workshops, general assemblies and other activities (co)organised by RISKBASE. Bringing this holistic and abstract framework to an operational level is a big challenge. This may be done by identifying indicators for the 'resilience state' of the system and by identifying or (further) developing tools and methods for assessment (measure/monitor and valuate) of the indicators. If this is successful, the insight can then be used to identify and develop practical measures aimed to increase the resilience of river systems and their ecosystem services. Measures are practical if they are: 1) technically feasible, 2) cost-effective (work with nature...), 3) socially/politically acceptable (now or in the near future) and 4) show a clear cause–response relationship.

REFERENCES

Blum, W.E.H., Büsing, J. and Montanarella, L. (2004a) Research needs in support of the European thematic strategy for soil protection, *Trends in Analytical Chemistry*, **23**, pp. 680–685.

Blum, W.E.H., Barcelo, D., Büsing, J., Ertel, T., Imeson, A. and Vegter, J. (2004b) Scientific basis for the management of European soil resources, Research Agenda, ISBN 3-900782-47-4.

Brantley, S.L., Goldhaber, M.B. and Ragnarsdottir, K.V. (2007) Crossing disciplines and scales to understand the critical zone, *Elements*, **3**, pp. 307–314.

Brils, J.M. (2005) Commission will continue its efforts to overcome the lack of knowledge on sediment quality in the EU, *Journal of Soils and Sediments*, **1**, pp. 48–49.

Chapman, A., Brils, J., Ansink, E., Herivaux, C. and Strosser, P. (2008) Conceptual models in river basin management, In: Quevauviller, P. (ed.) *Groundwater Science & Policy: An International Overview*, RSC Publishing.

Crutzen, P.J. (2002) Geology of mankind, *Nature*, **415**, p. 23.

European Commission (2000) Directive 2000/60/EC of the European Parliament and of the Council establishing a framework for the Community action in the field of water policy, Brussels.

European Commission (2002) Communication from the Commission to the Council, the European Parliament, the Economic and Social Committee and the Committee of the Regions: towards a thematic strategy for soil protection, COM (2002) 179 final, Brussels, 16 April 2002.

European Commission (2004) Communication from the Commission to the Council and the European Parliament: stimulating technologies for sustainable development: an environmental technologies action plan for the European Union, COM (2004) 38 final, Brussels, 28 January 2004.

European Commission (2006a) Communication from the Commission: halting the loss of biodiversity by 2010 – and beyond, sustaining ecosystem services for human well-being, COM (2006) 216 final, Brussels, 22 May 2006.

European Commission (2006b) Communication from the Commission to the Council, the European Parliament, the European Economic and Social Committee and the Committee of the Regions: thematic strategy for soil protection, COM (2006) 231 final, Brussels, 22 September 2006.

European Commission (2006c) Proposal for a Directive of the European Parliament and of the Council establishing a framework for the protection of soil and amending Directive 2004/35/EC, COM (2006) 232 final, 2006/0086 (COD), Brussels, 22 September 2006.

European Commission (2006d) Directive 2006/.../EC of the European Parliament and of the Council on the protection of groundwater against pollution and deterioration, PE-CONS 3658/06, Brussels, 22 November 2006.

European Environment Agency (2003) Europe's water: an indicator-based assessment, Summary, Office for Official Publications of the European Communities, Luxembourg.

Ferguson, C., Darmendrail, D., Freier, K., Jensen, B.K., Jensen, J., Kasamas, H. *et al.* (1998) Risk assessment for contaminated sites in Europe, Volume 1, Scientific Basis, LQM Press, Nottingham.

Garbarino, J.R., Hayes, H.C., Roth, D.A., Antweiler, R.C., Brinton, T.I. and Taylor, H.E. (1995) Heavy metals in the Mississippi River, In: Meade, R.H. (ed.) *Contaminants in the Mississippi River, 1987 – 92*, US Geological Survey Circular 1133, pp. 52–71.

Griffiths, M. (2002) The European Water Framework Directive: an approach to integrated river basin management, European Water Association, available at http://www.ewaonline.de/journal/2002_05.pdf.

Hawken, P., Lovins, A. and Hunter Lovins, L. (2000) *Natural Capitalism: Creating the Next Industrial Revolution*, Earthscan Publications.

Heinz, I. (2006) The economic value of water, Presentation from the International Workshop on Hydro-economic Modelling and Tools for the Implementation of the European Water Framework Directive, Valencia, Spain, 30–31 January 2006.

International Risk Governance Council (2005) Risk governance: towards an integrative approach, White Paper No. 1, International Risk Governance Council, Geneva.

JOINT (2005) Risk based management of contamination and protection of the soil system in urban environments, JOINT research Agenda 2005, Umweltwirtschaft GmbH, Stuttgart, Germany.

Menedez, M., de Rooy, M., Broseliske, G. and Mol, S. (2006) Key issues and research needs under the Water Framework Directive: final document, comprising Phase 1 and Phase 2, December 2005, Issue date: 26 January 2006.

Salomons, W. and Brils, J.M. (eds) (2004) Contaminated sediments in European river basins, SedNet booklet as final report for the EC FP5 Thematic Network Project SedNet (EVK1-CT-2001-20002).

Umweltbundesamt (1998) Final report on risk assessment models and risk management for contaminated sites, Concerted Action on Risk Assessment for Contaminated Sites in the European Union (CARACAS), Berlin.

Van Camp, L., Bujarrabal, B., Gentile, A.-R., Jones, R.J.A., Montanarella, L., Olazabal, C. and Selvaradjou, S.-K. (2004) Reports of the technical working groups established under the Thematic Strategy for Soil Protection, EUR 21319 EN/1, Office for Official Publications of the European Communities, Luxembourg.

Vegter, J.J. (2004) Risk based land management: status and perspectives for policy, In: Grotenhuis, T. and Tabak, H. (eds) *Soil and Sediment Remediation*, P. Lens.

Vegter, J.J., Lowe, J. and Kasamas, H. (eds) (2003) Sustainable management of contaminated land: an overview, Austrian Federal Environment Agency on behalf of CLARINET.

Walker, B. and Salt, D. (2006) *Resilience Thinking: Sustaining Ecosystems and People in a Changing World: How can Landscapes and Communities Absorb Disturbance and Maintain Function?*, Island Press.

7.2

Emerging Contaminants in the Water-sediment System: Case Studies of Pharmaceuticals and Brominated Flame Retardants in the Ebro River Basin

Mira Petrovic, Ethel Eljarrat, Meritxell Gros, Agustina de la Cal and Damià Barceló

The Water Framework Directive - Ecological and Chemical Status Monitoring   Edited by Philippe Quevauviller, Ulrich Borchers, Clive Thompson and Tristan Simonart © 2008 John Wiley & Sons, Ltd

7.2.1   INTRODUCTION

A wide range of manmade chemicals designed for use in industry, agriculture and consumer goods, and chemicals unintentionally formed or produced as by-products of industrial processes or combustion, are potentially of environmental concern. Beside recognised pollutants, numerous new chemicals are synthesised each year and released into the environment with unforeseen consequences. This group is mainly composed of products used in everyday life, such as surfactants and surfactant residues, pharmaceuticals, personal care products, gasoline additives, fire retardants and plasticisers, etc. The characteristic of these group contaminants is that they do not need to be persistent in the environment to cause negative effect since their high transformation/removal rates can be compensated by their continuous introduction into the environment.

Emerging contaminants enter the river system through various pathways.

Point sources are identifiable points that are (fairly) steady in flow and quality (over the timescale of years). The magnitude of pollution is not influenced by the magnitude of meteorological factors. Major point sources under this definition include municipal wastewater effluents and industrial wastewater effluents.

Diffuse sources are highly dynamic, spread out pollution sources and their magnitude is closely related to meteorological factors such as precipitation. Major diffuse sources under this definition include surface runoff (load from atmospheric deposition), groundwater, erosion (load from eroded material), diffuse loads of paved urban areas (atmospheric deposition, traffic, corrosion) including combined sewer overflows, since these events occur discontinuously over time and are closely related to precipitation (it has to be pointed out that emissions from urban areas are also partly involved in the point source term, so these discharges are not constant in reality). Both point and diffuse sources contribute to the total contaminant load of rivers.

This chapter describes two monitoring studies performed with the objective of determining the contribution of point sources, such as sewage treatment plants (STP) and specific industrial plants, in the contamination of the Ebro River by two groups of emerging contaminants: pharmaceutical residues and brominated flame retardants.

7.2.2   CASE STUDY 1: SEWAGE TREATMENT PLANTS (STP) AS A SOURCE OF PHARMACEUTICALS IN THE EBRO RIVER BASIN

Pharmaceutically active substances are a class of new so-called 'emerging' contaminants that have raised great concern in the last years (Daughton and Ternes, 1999). Pharmaceuticals in their native form or as metabolites are continuously introduced to sewage waters, mainly through excreta, disposal of unused or expired drugs, or directly from pharmaceutical discharges. Their significance as trace environmental pollutants in waterways, and on land to which treated sewage sludge or wastewater has been applied, is largely unknown. They deserve special attention for the following reasons: i) they are continuously introduced via effluents from sewage treatment facilities and from septic tank systems (pharmaceuticals are referred to as 'pseudo' persistent contaminants, i.e. their high transformation/removal rates are compensated by their continuous introduction into the environment); ii) they are developed with the intention of performing a biological effect; iii) they often have the same type of physico-chemical behaviour as other harmful xenobiotics (persistence, in order to avoid the substance becoming inactive before having a curative effect, and lipophilicity, in order to be able to pass membranes); iv) they are used by man in rather large quantities (i.e. similar to the widespread use of many pesticides). The most important issue of concern about their presence in the aquatic environment, and the main reason why they need to be included in monitoring programmes as environmental contaminants, is the ecotoxicological effects that they may cause.

In 2005, Spain was the eighth largest pharmaceutical market in the world (IMS Health 2006). Such high consumption leads to the presence of pharmaceutical residues in sewage effluents; however, the data on STP efficiency and possible contamination of aquatic system are sparse. Generally, sewage treatment processes in STP in Spain consist of a primary and a secondary treatment. For the latter, biological filters and activated sludge are the most frequently used processes. Tertiary treatments are seldom applied.

Low removal rates for some pharmaceuticals mean that STPs are main contributors, and therefore hot spots, of environmental contamination, at least when dealing with pharmaceuticals used by man.

As the target compounds included in this survey are such pharmaceuticals, the screening was focused on the analysis of the river waters located close to the main cities along the basin (Figure 7.2.1 and Table 7.2.1).

Total loads ranged from 2 to 5 g/day/1000 inhabitants in influent wastewaters and from 0.5 to 1.5 g/day/1000 inhabitants in effluent wastewaters. The average removal of ketoprofen, naproxen, ibuprofen, diclofenac and mefenamic acid in STPs was higher than 67 %, reaching values of >95 % in some cases. Lipid regulators, antihistaminics, antibiotics and β-blockers were not so efficiently eliminated (average removal ranged

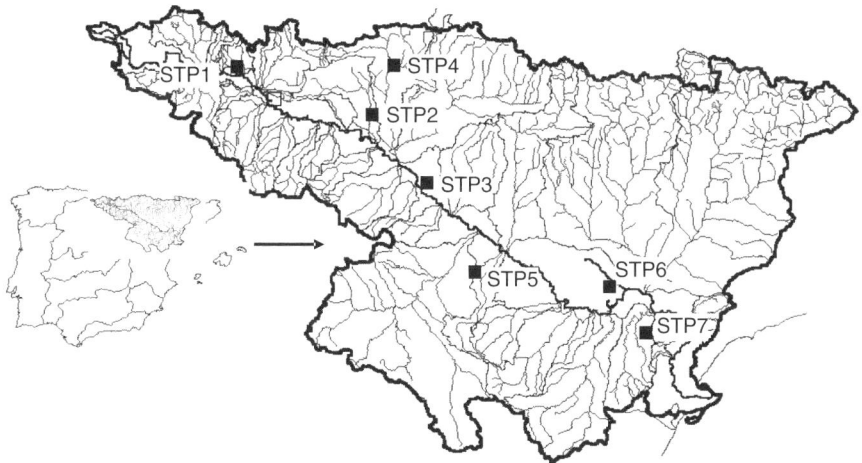

Figure 7.2.1 Map of the sampling sites (STP effluents and river water downstream of each effluent discharge)

Table 7.2.1 Characteristics of the STP monitored and the receiving waters where their effluents are discharged

STP	Population equivalent	Flow (m³/h)	Receiving river water	Type of wastewater treated	Hydraulic retention time (h)	Primary treatment	Secondary treatment
STP1	52 700	533	Vallas	Urban	32	???	Activated sludge
STP2	90 000	833	Ebro	Urban	18	Primary settling	Biological filters
STP3	466 560	2500	Iregua	Urban and industrial	8	Primary settling	Activated sludge
STP4	773 312	4313	Arga	Urban and industrial	25	Primary settling	Activated sludge
STP5	835 000	6833	Ebro	Urban	10	Primary settling	Activated sludge
STP6	18 000	2917	Segre	Urban	6–10	Primary settling	Activated sludge
STP7	36 625	305	Ebro	Urban	33	???	Activated sludge

from 30 to 62 %), but still showed high rates (over 80 %) in some plants. On the other hand, the antiepileptic carbamazepine was the most recalcitrant pharmaceutical, showing very low removal, or in some cases no removal at all. Similarly, several other compounds such as azythromycin, trimethropirm, sotalol and propranolol showed poor or no removal, leading to the conclusion that a wide range of pharmaceuticals is not undergoing the transformation in STPs, so the active drug component reaches the environment.

Figure 7.2.2 shows the most common pharmaceuticals of each therapeutic group detected in the receiving river waters. These results show that STP effluents represent the main source of pharmaceuticals in river water, and lower concentrations detected in river water are mainly due to dilution of the STP effluents. Even though total concentrations in river water were rather low, ranging from 100 to 600 ng/L, a wide spectrum of pharmaceuticals was detected. Highest levels were attributed to Pamplona and the area around Zaragoza, which are two of the most populated cities. Effluents from STP2, STP5 and STP7 are discharged in the Ebro River, whereas effluents from STP1, STP6, STP4 and STP3 go to tributaries which consist in the Vallas, Segre, Arga and Iregua, respectively. The dilution factor in the Ebro River is controlled, and averages 70 in Zaragoza, the area surrounding STP5 (1.9 m³/s of STP effluent is mixed with 150 m³/s of river flow), and around 1500 in Tortosa, STP7 (0.085 m³/s of STP effluent is mixed with 150 m³/s of river flow). A similar situation occurs in Lleida (STP6), as 0.8 m³/s of effluent wastewater is mixed with 50 m³/s of river water, showing a dilution factor of approximately 63. Nevertheless, in the Arga River, which receives the effluent from STP4, the dilution factor was much lower, averaging a value of 10. Such factors could not be estimated for Tudela (STP2), Miranda de Ebro (STP1) and Logroño (STP3), since no data referring to the river flows in these points were reported in the database of the Confederacion Hidrologica del Ebro (CHE, Ebro River

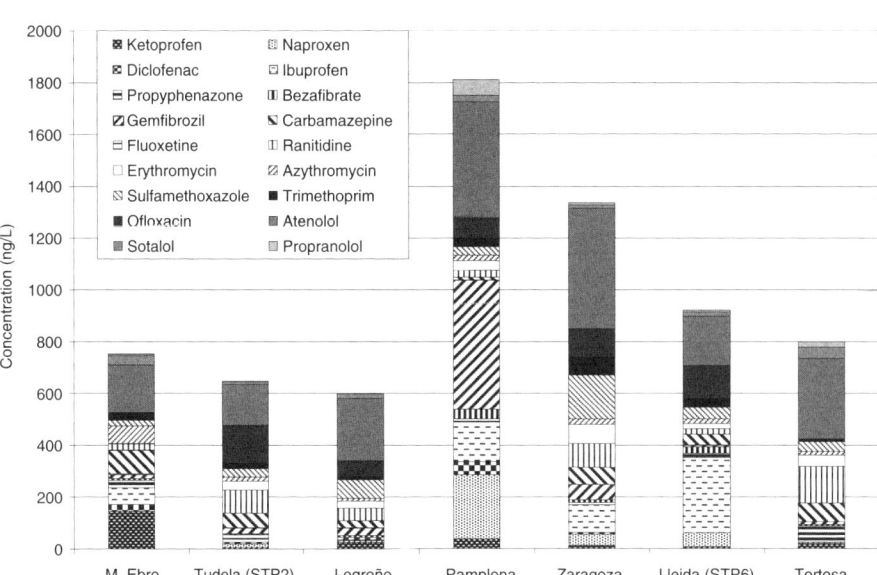

Figure 7.2.2 Concentrations (ng/L) of the most ubiquitous pharmaceutical residues detected in river water downstream of the STP monitored

basin management). Even though pharmaceuticals present in effluent wastewaters are generally diluted when entering river waters, the same spectrum of compounds found in the effluents is also detected in the river water, but generally at one order of magnitude lower than in effluent wastewaters, typically at a wide range of ng/L levels. These results show that STP effluents represent the main source of pharmaceuticals in river water, and lower concentrations detected in river water are mainly due to simple dilution of the STP effluents.

The potential ecological effects associated with the presence of pharmaceuticals in the environment have been largely ignored, and their toxicity (especially long-term (chronic) effects) to organisms and impact on the trophic chain is still not well documented. By comparing the environmental levels detected in both surface and waste waters in the Ebro River basin with the concentrations showing ecotoxicological effects, according to the information found in the scientific literature available (Fent *et al.*, 2006), it could be concluded that acute toxicity to aquatic organisms is unlikely to occur at measured concentrations, as levels reported to cause such effects are from 100 to 1000 times higher than the residues found in the surveyed sampling sites.

7.2.3 CASE STUDY 2: OCCURRENCE, FATE AND BEHAVIOUR OF BROMINATED FLAME RETARDANTS IN THE EBRO RIVER BASIN

Within a monitoring programme at different risk zones in this river, two highly contaminated areas were detected. The first one was located along the Cinca River, a tributary of the Ebro River, downstream of a heavily industrialised town (Monzón), and data showed a high hexabromocyclododecane (HBCD) contamination in this area. The second was located along the Vero River, a tributary of the Cinca River, downstream of an industrial park in Barbastro. In this case, high contamination of decabromodiphenyl ether (deca-BDE-209) was found. Our work included the analysis of sediments and biota, with special attention on aspects such as temporal trends, bioavailability and bioaccumulation of these contaminants. Moreover, an attempt at identification of source contamination was carried out, with the analysis of industrial effluents. In both cases, the industry responsible for the contamination was identified.

7.2.3.1   Area of Study

The study area is located in the north-east of Spain, along the Cinca and Vero rivers, in the Ebro River basin (Figure 7.2.3). Four different sampling stations were selected

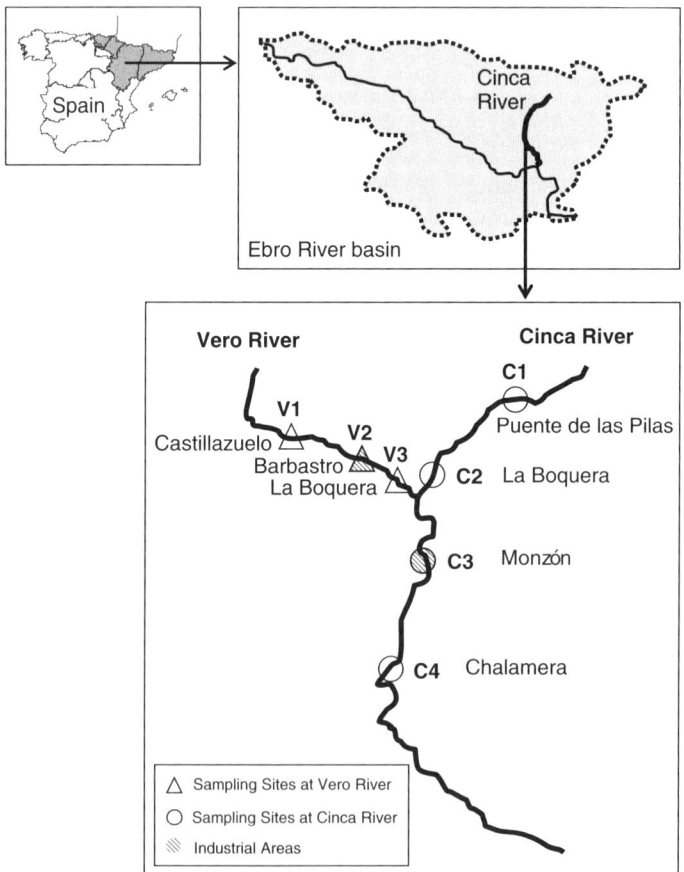

Figure 7.2.3   Geographical location of the area of study and sampling stations along the Cinca and Vero rivers

at the Cinca River: sites C1 (Puente de las Pilas) and C2 (La Boquera), 20 and 12 km respectively upstream from Monzón; site C3, just downstream from Monzón, a heavily industrialised town with a very important chemical industry; and site C4 (Chalamera), 30 km downstream of site C3. Moreover, three sampling stations were selected at the Vero River: site V1 (Castillazuelo), 11 km upstream from an industrial park; site V2 (Barbastro), just 1 km downstream from the industrial park; and site V3 (La Boquera), 4 km downstream of site V2.

Different sampling campaigns were carried out from 2002 to 2005. Surface sediments (0–2 cm) were collected at each selected site. Attempts were made to collect several fish from the same locations as the sediment samples. Different fish species were sampled: barbels (*Barbus graellsii*), bleaks (*Alburnus, alburnus*), south-western nases (*Chondrostoma toxostoma*) and carps (*Cyprinius carpio*). The whole fish was analysed for bleaks and south-western nases, whereas muscle and liver tissues were analysed for barbels and carps.

In order to determine the sources of contamination, some water samples, including wastewaters and effluents from different industries, were also taken. Along the Cinca River and in the industrial area of Monzón, industrial effluents from two different industries were selected: the first one produced EPS (expandable polystyrene) treated with flame retardants and ABS (acrylonitrile-butadiene-styrene), and the second one produced PVC (polyvinyl chloride). As regards the Vero River, three industries were sampled: the first one was a textile industry which produced polyester fibres treated with flame retardants, the second produced epoxy resins, and the third focused on polyamide polymerisation.

7.2.3.2 Results: Sediment Samples

PBDEs were detected in all the sediment samples collected along the Cinca River, at concentrations ranging from 2 to 131 ng/g dw (Table 7.2.2). HBCD was detected only in samples collected downstream of Monzón, with levels ranging between 90 and 1613 ng/g dw. In these samples, HBCD contamination was greater than that observed for PBDEs. Site C3 was found to be the most contaminated zone, followed by site C4 > site C1 = site C2. As expected, decabromodiphenyl ether (PBDE) and HBCD levels were greater near the site of industrial impact.

Table 7.2.3 shows the concentrations of BDE congeners from the different selected sites along the Vero River during the two sampling campaigns. PBDEs were detected in all the sediment samples at concentrations ranging from 11 to 14 400 ng/g dw. Site V2 was found to be the most contaminated zone, followed by site V3 > site V1. Sediments affected by the source contamination (sites V2 and V3) showed a congener pattern clearly dominated by the BDE-209, which was present at concentrations varying from 1910 to 12 500 ng/g dw. The highest BDE-209 level was found at the sampling site closest to the source of contamination (V2–November 2005, located five metres downstream from the effluent discharge), and contamination decreased with distance from the industrial area. High levels of BDE-154 (844 ng/g dw) were also detected in this sediment sample. Moreover, some octa- and nona-BDEs were also present at relatively high levels: 37–39 ng/g dw for octa-BDEs, and 169–375 ng/g dw for nona-BDEs. The

Table 7.2.2 PBDE results for sediment samples (ng/g dw) and biota samples (ng/g ww) obtained for samples collected along the Cinca River

				Site C1	Site C2	Site C3	Site C4
October 2002	Sediment	Total PBDEs		2.4	2.6	42	40
		HBCD		nd	nd	514	90
	Barbel[1]	Total	Muscle	1.3	4.5	281	96
		PBDEs	Liver	0.2	0.4	237	75
		HBCD	Muscle	nd	nd	530	90
			Liver	nd	nd	554	432
	Bleak[1]	Total PBDEs		–	4.7	555	232
		HBCD		–	nd	1501	760
November 2004	Sediment	Total PBDEs		14	38	131	87
		HBCD		nd	nd	1613	866
	South-western nase[1]	Total PBDEs		2.0	38	–	520
		HBCD		nd	nd	–	4863

[1]Mean values

Table 7.2.3 PBDE results for sediment samples (ng/g dw) and biota samples (ng/g lw) obtained for samples collected along the Vero River

			Site V1	Site V2	Site V3
November 2004	Sediment	BDE-209	7.5	5395	1911
		Total PBDEs	11	5531	1930
	Barbel[1]	BDE-209	nd	–	67
		Total PBDEs	54	–	791
	Carp[1]	BDE-209	–	–	80
		Total PBDEs	–	–	1560
November 2005	Sediment	BDE-209	27	12 459	7454
		Total PBDEs	30	14 395	7767
	Barbel[1]	BDE-209	nd	–	195
		Total PBDEs	83	–	1007

[1]Mean values

distribution of octa- and nona-BDE congeners in this sediment sample correlates well with the pattern observed in the technical deca-BDE formulation containing 97–98 % of BDE-209 and up to 3 % nona-BDEs.

7.2.3.3 Results: Biota Samples

PBDEs were detected in all the biota samples collected along the Cinca River, at concentrations ranging from 0.2 to 555 ng/g wet weight (ww) (Table 7.2.2). HBCD was detected only in samples corresponding to sites downstream of Monzón. In these samples, HBCD contamination was similar to or greater than that observed for PBDEs. HBCD levels ranged from 90 to 4863 ng/g ww. Similar to our findings in sediment samples, site C3 was found to be the most contaminated zone, followed by site C4 > site C1 = site C2. Comparison between levels found in muscle tissues and livers showed that similar contamination was detected in both matrices. However, slightly higher BFR levels were found in muscle tissues.

PBDEs were also detected in all the fish muscle samples collected along the Vero River, at concentrations ranging from 28 to 2092 ng/g lw (Table 7.2.3). Similar to our findings in sediment samples, samples from the site downstream of the industrial park (V3) were found to be much more contaminated than those collected at the site upstream of the park (V1). The comparison between the two sampling campaigns showed that levels during 2005 were slightly higher than those found in 2004, probably reflecting the same situation observed for sediment samples, with an increase of contamination with time.

7.2.3.4 Bioavailability and Bioaccumulation

When sediment concentrations were compared to those in biota collected at the same sites along the Cinca River, high fish-to-sediment ratios were seen. Specially, levels of BDE-47, BDE-153 and BDE-154 in fish were high compared with levels in sediment (13–21-fold), indicating the high bioavailability of these congeners. This is also supported by the fact that for some BDE congeners detectable levels were found in fish but not in corresponding sediments: 17 different PBDEs were detected in fish samples, whereas only 8 PBDE congeners were found in sediments. Only BDE-209, the main BDE congener found in sediments from the area, was not found in biota samples. The main reason for its absence in biota seems to be its relatively low bioaccumulation potential. The fact that HBCD is found in fish indicates that it is also bioavailable. However, fish-to-sediment ratio is low (below 1). Previous studies showed that PBDE levels increased with the age of the fish, indicating bioaccumulation. Fish length is

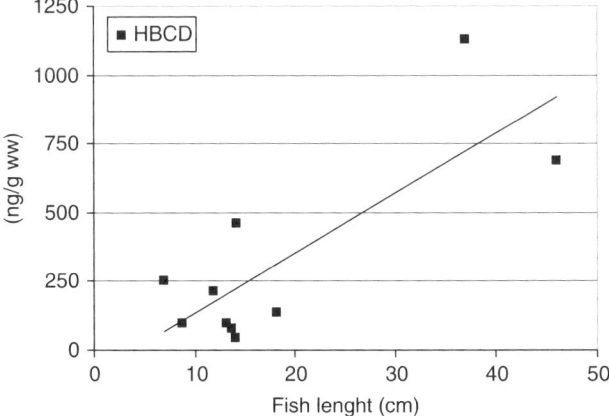

Figure 7.2.4 Length of Cinca River barbel (sites C3 and C4) versus concentrations of HBCD in the muscle. Line is linear regression on the data (Reprinted with permission from Environ. Scl. Technol. Copyright American Chemical Society, 2004)

directly related to fish age. Figure 7.2.4 shows the measured concentrations of HBCD as a function of fish length at sites C3 and C4. The correlation coefficients indicate acceptable correlations ($R^2 = 0.71$).

As regards results obtained along the Vero River, 12 different PBDEs were detected, ranging from tri- to deca-brominated compounds. The most relevant finding is the presence of significant concentrations of BDE-209 in fish samples collected downstream of the industrial park: 14 out of 15 biota samples (barbels and carps) collected downstream from the source of contamination showed BDE-209 levels ranging between 20 and 707 ng/g lw. BDE-209 values in fish collected during the second sampling campaign, with a mean value of 195 ng/g lw and a median value of 86 ng/g lw, were higher than those found in barbels collected during November 2004 (mean value of 67 ng/g lw and median value of 32 ng/g lw). These results are in accordance with the higher values found in sediment samples collected at the same site (V3). It is interesting to note the different congener distribution found in fish collected up- and downstream from the source of contamination (Figure 7.2.5). For samples at the control site (V1), the distribution showed a clear predominance of the tetra-BDE-47 (69 % and 77 % for the first and the second sampling, respectively). However, in fish from the contaminated area (V3), the BDE-47 contribution to the total PBDE burden decreased to 29 %. In contrast, samples from V3 contained a high contribution of hexa-BDE-154, with values between 16 % (sampling of 2004) and 20 % (sampling of 2005). The contribution of this hexa-BDE congener was only 5 % and 7 % for the first and the second sampling, respectively, at site V1. The higher predominance of BDE-154 in the contaminated

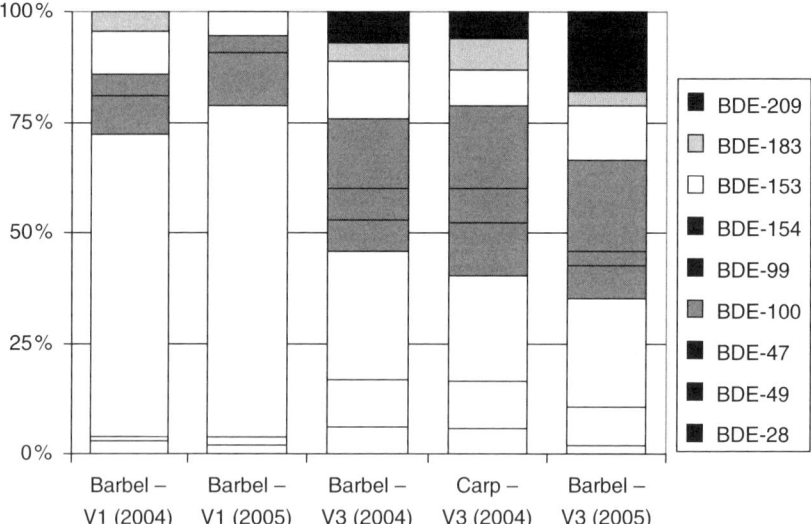

Figure 7.2.5 Percentage contribution of various congeners to the total PBDEs detected in fish samples from the control site (V1) and contaminated area (V3) during the two sampling campaigns (2004 and 2005) (Reprinted from Chemosphere, 69, Elijarret, Labandeira, Marsh, Raldua and Barcelo, *Decabrominated diphenyl ether in river fish and sediment samples collected downstream an industrial park*, 9, Copyright 2007, with permission from Elsevier)

area could be related to high contamination of this congener in the sediment of this area. But it could also be attributed to biotransformation of deca-BDE-209. Stapleton *et al.* (2004) studied the debromination of BDE-209 by caged carp following dietary exposure. They concluded that debromination of BDE-209 occurs in carp. Although they did not detect any accumulation of BDE-47 and BDE-99, accumulation of other PBDEs, such as BDE-154, occurs.

These results indicated that BDE-209 is bioavailable, but the bioavailability potential is lower than that observed for other BDE congeners, with fish-to-sediment values much lower (0.0 013 and 0.0 011 for the 2004 and 2005 sampling campaigns, respectively). It should be pointed out that an important factor affecting the ratios may be that the system is not in equilibrium because PBDEs are released directly to the river in industrial park effluents, and new inputs are ongoing. The low fish-to-sediment ratios obtained could be an indication of a recent release of BDE-209 from the industrial park that has contaminated the sediment but not yet been taken up by barbels or carps.

7.2.3.5 Sources of Contamination

Two industries were suspected to have caused the HBCD contamination along the Cinca River: the first one produced EPS treated with flame retardants and ABS, and the second one produced PVC. Analysis of industrial effluents of each industry revealed that the one which produced EPS and ABS was responsible for the HBCD contamination, with concentration levels around 5 µg/L. Three industries were suspected to have caused the BDE-209 contamination along the Vero River: the first one was a textile industry which produced polyester fibres treated with flame retardants, the second one produced epoxy resins, and the third one focused on polyamide polymerisation. Analysis of industrial effluents of each industry revealed that the industry focused on polyamide polymerisation was the main cause of the BDE-209 contamination, with concentration levels around 2600 ng/L. Nevertheless, the two other industries also contributed in some way to the total contamination.

7.2.4 CONCLUSIONS

STPs are found to be the major contributor of pharmaceuticals in the Ebro River water. Compounds more frequently detected in the Ebro River basin were analgesics (diclofenac, naproxen, ibuprofen), lipid regulators (gemfibrozil, bezafibrate), antibiotics (azythromycin, trimethoprim, erythromycin, sulfamethoxazole), the antiepileptic carbamazepine, the antihistaminic ranitidine, and the ß-blockers atenolol and sotalol, which are the ones of major consumption in Spain as well as the ones excreted at higher percentages as parent drugs. Concentrations detected in both waste and surface waters are from 100 to 1000 times lower than the levels reported to cause acute toxicity. However, with respect to chronic effects, for some of the most ubiquitous compounds the margin of safety is narrow. As a wide spectrum of pharmaceuticals has been detected in natural waters, effects of mixtures should also be taken into account, hence the overall toxicity could be the result of the combination of individual concentrations or the interaction of different compounds, causing effects at levels below the no observed effect concentration (NOEC) of individual substances.

As regards brominated flame retardants, and due to industrial activities, two highly polluted areas were detected. High concentration levels of HBCD were detected in sediment, but also in biota samples. This fact indicates that HBCD is bioavailable. Moreover, HBCD concentrations are correlated with fish length, indicating the bioaccumulation of this contaminant. Our study also shows that BDE-209 is bioavailable from sediment to fish and that bioaccumulation of highly brominated congeners is possible in highly polluted aquatic environments. Moreover, BDE-209 is probably a source material for lower brominated products, i.e. hexa-BDE-154, that are more readily accumulated, to a greater extent and with a greater degree of persistency. The high BDE-209 concentrations in sediments emphasise the need for a thorough understanding of the fate of this compound. Finally, the BDE-209 releases from some industries, such as textile, epoxy resin and polyamide polymerisation, merit special scrutiny.

REFERENCES

CHE (Confederacion Hidrografica del Ebro), http://www.chebro.es/.

Daughton, C.G. and Ternes, T.A. (1999) Pharmaceuticals and personal care products in the environment: agents of subtle change? *Environmental Health Perspectives*, **107**, pp. 907–938.

IMS Health 2006, http://www.imshealth.com.

Fent, K., Weston, A.A. and Caminada, D. (2006) Ecotoxicology of human pharmaceuticals, *Aquatic Toxicology*, **76**, pp. 122–159.

Stapleton, H.M., Alaee, M., Letcher, R.J. and Baker, J.E. (2004) Debromination of the flame retardant decabromodiphenyl ether by juvenile carp (Cyprinus carpio) following dietary exposure, *Environ. Sci. Technol.*, **38**, pp. 112–119.

7.3

Assessment of Metal Bioavailability and Natural Background Levels – WFD Monitoring from the Perspective of Metals Industry

Patrick Van Sprang, Katrien Delbeke, Lidia Regoli, Hugo Waeterschoot, Frank Van Assche, William Adams, Delphine Haesaerts, Claire Mattelet, Andy Bush, Lynette Chung and Violaine Verougstraete

7.3.1   INTRODUCTION

In the European Union (EU), the Water Framework Directive (WFD) (European Community, 2000) will provide the basis for EU water legislation for many years to come and will determine the way in which concentrations of chemicals in the aquatic environment will be monitored and regulated (Crane, 2003). The WFD aims to move Europe

The Water Framework Directive - Ecological and Chemical Status Monitoring   Edited by Philippe Quevauviller, Ulrich Borchers, Clive Thompson and Tristan Simonart © 2008 John Wiley & Sons, Ltd

towards an ecosystem-based management on the basis of sound scientific information and toxicological principles, thereby accounting for biological hazard while at the same time avoiding overestimation of ecological risks (Crane, 2003).

Quite often, these differences in bioavailability have not been properly addressed in currently used methodologies to derive Environmental Quality Standards (EQS) for metals. Although many EQS for metals are still based on total concentrations, regulatory authorities recognise the principle that mobility, bioavailability and toxicity of metals are a function of metal speciation. Furthermore, authorities recognise that metal speciation depends on site-specific seasonal and spatial variations existing in a particular ecosystem (Landner and Reuther, 2004). Extensive evidence has shown that neither total nor dissolved aqueous metal concentrations are good predictors of metal bioavailability and toxicity (Bergman *et al.*, 1997; Campbell, 1995; Janssen *et al.*, 2000) and the importance of explicitly considering bioavailability in risk assessment of metals in general and development of EQS in particular has been demonstrated scientifically (Allen and Hansen, 1996; Ankley *et al.*, 1996; Di Toro *et al.*, 1991; Janssen *et al.*, 2000). These concepts and the scientific methods to assess bioavailability of metals were proposed conceptually over a decade ago (Bergman and Dorward-King, 1996) and are gaining increasing recognition by regulatory authorities.

The Biotic Ligand Model (BLM) concept, which has recently been applied successfully to predict metal bioavailability and acute toxicity for several metals, i.e. Cu (De Schamphelaere *et al.*, 2002a, 2002b; Santore *et al.*, 2001), Ni (Hoang *et al.*, 2004; Keithly *et al.*, 2004), Ag (Paquin *et al.*, 1999) and Zn (Heijerick *et al.*, 2002a, 2002b) in surface waters, is currently the most developed practical applicable technique to assess the ecotoxicity of metals on a site-specific basis. The BLM provides a mechanistic understanding of the interactions of metals with organic, inorganic and biotic ligands at the biological membrane, and allows for the quantification of the bioavailability under given abiotic conditions of the water. Recently, the BLM concept has been extended successfully for predicting chronic toxicity of metals, i.e. copper, nickel and zinc, for different trophic levels (Deleebeeck *et al.*, 2007; De Schamphelaere and Janssen, 2004; De Schamphelaere *et al.*, 2003). Excellent reviews with regard to the developed BLM models have been written by Paquin *et al.* (2002) and Niyogi and Wood (2004). For cadmium, an acute BLM for fish is available (Hydroqual, 2003), as well as a hardness correction model (US-EPA, 2001), while acute/chronic BLMs are in development for the metals Ag, Al, Mo, Pb and Co.

Accurately assessing the ecological risks of metals requires the ability to account not only for metal-specific properties such as bioavailability, but also for their natural occurrence. Consequently, natural background concentrations and the exposure due to these background concentrations should be taken into account in the setting of EQS or during risk assessment. Due to the ubiquitous presence of metals in the natural environment, organisms have become adapted during the course of evolution to natural background concentrations and have developed the capacity to cope with natural variations (e.g. seasonal changes or fluctuations in river flow rates or different background metal concentrations). Specific detoxification mechanisms such as binding to metallothioneins, formation of metal granules or increased excretion mechanisms have therefore been developed in organisms as a means of providing protection from toxicity. Furthermore, species have the ability to adapt or acclimatise to short-term and moderate changes in metal (background) concentrations, at least to a certain extent. For these reasons, the natural background level represents the theoretical lower limit of the EQS, a threshold concentration below which risk to ecosystems is not expected.

The above concepts are applicable for all metals and are even more crucial for essential elements (EE) because the sensitivity of organisms to EE is determined to a large extent by the bioavailable EE level that the organism experienced before testing. The sensitivity of organisms to EE is determined to a large extent by the EE concentrations present in the natural environment. In general, organisms acclimatised at low metal concentrations become more sensitive to stress, including exposure to metals. Conversely, organisms acclimatised to media with elevated metal concentrations (e.g. natural waters or contaminated waters) may become less sensitive (Muyssen and Janssen, 2002a, 2002b). Organisms have the ability to regulate EE (i.e. uptake, storage or excretion) and can tolerate a wider range of exposures than for nonessential elements.

Moreover, the availability of metals for uptake by organisms under field will vary from site to site and is highly dependent on the speciation of the metal and the water chemistry (e.g. pH, Dissolved Organic Carbon [DOC]). Hence it is of utmost importance to take the speciation and/or bioavailability into account.

7.3.2 HOW TO ACCOUNT FOR NATURAL BACKGROUND LEVELS?

In Europe, with exception for some remote and unpopulated areas, unaltered natural background concentrations are infrequently observed in surface waters as a result of both historical and current anthropogenic input. Therefore, the term 'baseline concentration' is often used to express the concentration in the present or past corresponding to very low anthropogenic pressure. The baseline concentration is the sum of the natural background and the fraction of metal that has been historically introduced (or removed) to the environment by man over time (decades to centuries). The added fraction is often referred to as historical contamination. In many cases this historical contamination cannot be distinguished from the natural background concentration. Moreover, one should be aware that natural background concentrations within an environmental compartment may vary markedly by several orders of magnitude between geologically disparate areas, and are determined by various factors like the site/regional-specific bedrock composition, effects of climate on the degree of weathering, etc. Also, due to natural dynamic processes like weathering, introduction of metal-containing organic material (leaves) in the autumn, and uptake of metals by plants during spring and summer, natural background concentrations may show annual cycles. This means that it is impossible to attribute single values to natural background concentrations of specific metals within a certain compartment. It should be noted that under natural conditions in certain regions, clearly elevated natural background concentrations can be encountered. When assessing the natural background concentration within a certain area, these 'outliers' should not be used or included in the calculation of the standard background concentrations as they would give a nonrepresentative picture thereof.

Several methods are available for determining natural/baseline background concentrations. Measured metal levels at selected sites considered to be undisturbed by human activities must be preferably used. An important data set containing recent, reliable baseline concentrations in different environmental compartments (stream water, stream sediment, floodplain sediment, soil and humus) has been developed from the results of the FOREGS Geochemical Baseline Programme (FGBP) (http://www.gsf.fi/publ/foregsatlas/index.php). The main objectives of this European survey were: 1) to apply standardised methods of sampling, chemical analysis and data management to prepare a geochemical baseline across Europe; and 2) to use this reference network to level national baseline data sets. Samples of stream water (but also stream sediment and three types of soil (organic top layer, minerogenic top and sub-soil)) have been collected at 900 stations, each representing a catchment area of $100\,km^2$, corresponding to a sampling density of about one sample per $4700\,km^2$. Running stream water was collected from the small, second order, drainage basins ($<100\,km^2$) at the same site as the active stream sediment. This database represents the current best way forward to assess the variability of metal baseline concentrations at a regional scale. An example of the baseline concentrations across different regions in the EU is presented for Zn in Figure 7.3.1.

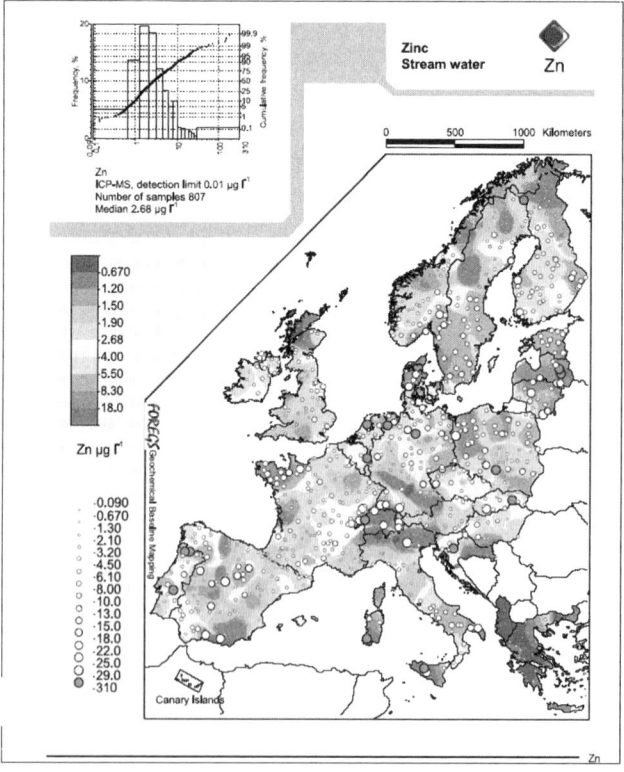

Figure 7.3.1   Baseline Zn concentrations in EU surface waters (see Plate 17)

Apart from the straightforward method of measuring metal levels at selected sites that are considered to be undisturbed by human activities, several additional methods are available for the estimation of the natural background concentration in surface waters, such as geochemical modelling, calculation based on natural background concentration in sediment and the equilibrium partition coefficient, and assessment of the metal concentrations in the deeper ground water. It must be verified that the well water or groundwater samples are free of current or historic pollution. Moreover, due to their contact with deeper, mineral rocks, metal background concentrations in these waters can be higher than those in surface waters, where there is an additional dilution with rain water. An overview of total and dissolved background concentrations in freshwater surface waters as estimated using the geochemical modelling approach is presented by Zuurdeeg *et al.* (1992).

Historical contamination of the environment can have long-lasting influence on metal concentrations in the water because of continuing emissions (e.g. from contaminated soils). Because of their large capacity to 'hold' metals, this is even more relevant for sediments which have been characterised as 'sinks'. Historical contamination for sediment monitoring data therefore reflects both the effect of current metal emissions and historical emissions over long periods of time. Therefore, other processes, such as resuspension of sediments, can also influence the concentration of metals in the water column. Moreover, the contribution of historical pollution to the measured baseline concentration could become important when enclosed water bodies with low turnover (e.g. lakes, reservoirs) that have been affected by important anthropogenic inputs in the past are considered. These types of water body should therefore only be used for the determination of natural background levels when there is no indication that metal levels have been affected by anthropogenic contributions in the past.

In order to deal with the role of natural background concentrations, different concepts have been developed, such as the added risk approach (ARA) (Crommentuijn *et al.*, 1997) and the total risk approach (TRA) concepts. The TRA assumes that 'exposure' and 'effects' should be compared on both the fraction compiling the natural background and the added anthropogenic concentrations. Both approaches have been used in the EU risk assessments context for metals, e.g. the ARA has been used in the EU Zn risk assessment and the TRA has been used in the EU Cd RAR. In general, it can be argued that the total risk approach is the most scientifically defensible option in the context of a risk assessment. Unlike the TRA, the ARA assumes that the baseline concentration does not contribute to the exposure and thus to the toxic effects of metals on ecological systems, and hence that only the anthropogenic added fraction should be regulated or controlled. Therefore, it must also assume that metals comprising the natural background are not physiologically available to organisms, and that background metals are irrelevant for either toxic or beneficial (in the case of essential elements) effects. However, organisms are not able to distinguish between the natural and the anthropogenic fractions of metals in the environment. Consequently, part of the natural background will be bioavailable and would therefore contribute to the total available metal concentration to which the organisms are exposed and could therefore cause toxicity to the organisms. The ARA may be employed as a pragmatic risk management solution to account for the impact of background concentrations if strict data conditions are fulfilled and when more robust approaches (e.g. the metallo-region approach) are not available. For example, the ARA has merits when high baseline concentrations compared to the EQS are observed.

In cases where the ARA is used, several important considerations need to be made and care should be taken in how the background correction is carried out. First, the ARA suggests correcting the toxicity data, which form the basis for the EQS derivation, for the background metal concentration of the culture medium to which the test organisms are adapted/acclimatised ($Cb_{culture\ medium}$). Indeed, for several metals (and especially essential metals) a relationship has been demonstrated between the sensitivity of the organism and the culture condition metal concentrations. Organisms cultured in media with a low metal concentration (which is often the case in standard media) are generally more sensitive than those cultured in higher metal concentrations. However, organisms are often cultured in media (e.g. treated tap water or natural water) with varying metal background concentrations or even with unknown metal concentrations, which therefore complicates the proper implementation of the ARA. It must be stressed that culture media often differ from the standard test media used in the toxicity tests, suggesting that test organisms may not be acclimatised to the metal background concentrations in toxicity tests. Furthermore, for proper application of the ARA and for compliance checking, the natural background of a particular location and/or a specific region (Cb_{site}) also needs to be established. Therefore, in the ARA the monitoring data should be corrected for the background metal concentration occurring at the investigated location/region. Current knowledge of the geographical distribution of metal background concentration in ecological systems is, however, often insufficient to correctly estimate the spatial/temporal variability in natural background concentrations and, hence, to implement the ARA in an appropriate way. It must be stressed that the establishment of a 'default background concentration' as currently carried out in generic risk assessment exercises will not contribute to the correct assessment of the risks posed by metals. Therefore, the FOREGS database represents the current best way forward to assess the variability of metal baseline concentrations at a regional scale.

7.3.3   HOW TO ACCOUNT FOR BIOAVAILABILITY?

The application of the '(bio)availability' concepts for the aquatic compartment consists of the translation of the conventional estimated effect thresholds (e.g. NOEC/EC_{10} values) towards conditions prevailing in a certain region or for a certain site-specific surface water, using either transformation to soluble fractions, speciation or bioavailability algorithms (e.g. the free metal ion, BLM). At TC NES III (September 2006) it was agreed that the level of refinement that can be obtained depends on the availability and applicability of the concepts and tools for the specific metal/metal compound under consideration, as outlined in Figure 7.3.2.

Level 1: Results (e.g. NOEC/EC_{10} values) from aquatic toxicity tests are usually expressed as total or dissolved metal concentrations. Most aquatic toxicity tests are conducted in artificial waters (without DOC) and therefore tend to maximise bioavailability and toxicity. In this first level (Level 1), conversion towards dissolved metal concentrations is aimed at. Hence, if dissolved concentrations are not given it is assumed

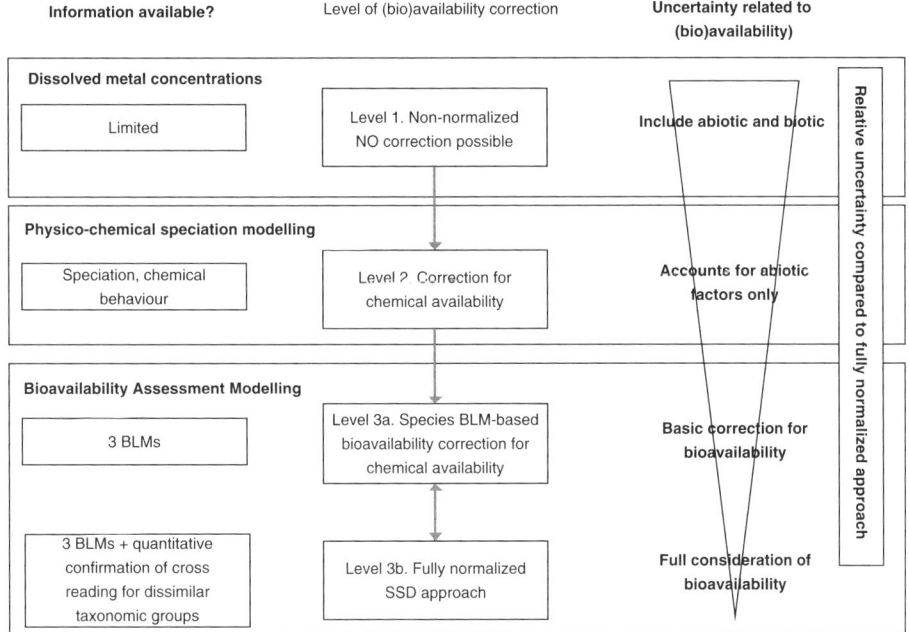

Figure 7.3.2 Overview (bio)availability correction strategy in the aquatic compartment for metals/metal compounds

that for these toxicity data, performed in artificial media, no additional conversion to a dissolved fraction has to be applied (i.e. the total concentration can be set equal to the dissolved concentration with reasonable accuracy). If natural waters are used, total concentrations can be recalculated to dissolved concentrations using partition coefficients, Kp. However, it must be stressed that in all cases the preference is for tests where dissolved metal concentrations are reported.

Level 2: In the second level (Level 2), physico-chemical speciation models are introduced in order to correct the toxicity data for chemical availability. Indeed, NOEC and/or EC_{10} values that are used in the effects assessment are generally generated in test media with varying physico-chemical characteristics (e.g. pH, hardness, DOC) known to alter metal availability and toxicity. In case metal concentrations are reported and appropriate speciation models (e.g. WHAM, MINTEQA2, etc.) and relevant input data (i.e. main physico-chemical parameters driving the availability of a metal such as pH, DOC, etc.) are available, NOEC and/or EC_{10} values should be expressed on the basis of the metal species of concern[1] in order to reduce uncertainty. For regulatory compliance purposes, the dissolved exposure concentrations should also be translated at the same level of availability (expressed in the same units) as the effects assessment,

---

[1] Most often this is the free metal ion but it should be noted that the free ion is not necessarily the best predictor for all metals, and other metal species such as neutral species (e.g. AgCl, HgS) and anionic species (e.g. SeO^{2-}, AsO_4^{2-}) may contribute to the observed toxicity (Campbell, 1995).

e.g. into free metal concentrations using the same speciation translator (e.g. WHAM, MINTEQA2). For that purpose, the physico-chemical parameters of the site-specific watershed driving the availability (e.g., pH, DOC) should be gathered or estimated.

Level 3: In this level the use of BLM is advocated. Recent software advances have facilitated the calculations and made the computations straightforward. This approach is based on the use of bioavailability assessment modelling, which considers both the effects of chemical speciation and biological factors on the toxicity of metals. With the use of validated BLMs, the NOEC and/or EC_{10} values should be expressed preferentially on a 'bioavailable' basis, assuming ambient dissolved metal concentrations are reported and appropriate bioavailability models and relevant input data (i.e. physico-chemical parameters) are available. BLMs have been developed in order to predict availability (speciation) and subsequent bioavailability of metals to certain aquatic organisms. These models can be used to reduce the intra-species variability in a large data set by accounting for bioavailability in the individual toxicity tests and normalising all the data to a specific set of physico-chemical conditions. This can ensure consistency of comparison across toxicity tests performed under a range of physico-chemical conditions and using a range of organisms. If such bioavailability models (e.g. BLM) are available, they exist mostly for a limited number of species, representing different trophic levels (algae, fish and invertebrates).

Level 3a: A baseline bioavailability correction (Level 3a) can be conducted if a BLM for algae, fish or invertebrates is available. Such baseline bioavailability correction, consisting of the application of a conservative bioavailability factor (Bio-F), is applicable if no direct evidence is available to support a cross-reading[2] or if mechanisms of toxicity differ across species within a specific trophic level. This method is based on the calculation of reference[3] NOEC/EC_{10} values (i.e. reference NOEC/EC_{10} dissolved) and normalised NOEC/EC_{10} values for the test organisms for which the bioavailability models were developed under typical site-specific conditions (i.e. site-specific NOEC/$EC_{10 \text{ bioavailable, dissolved}}$). Both normalised toxicity values are compared for every species for which the bioavailability models were developed, and the most conservative value (smallest correction for bioavailability,

$Bio - F = \dfrac{\text{site - specific NOEC/EC}_{10 \text{ bioavailable, dissolved}}}{\text{reference NOEC/EC}_{10 \text{ dissolved}}}$), is selected for the calculation

of the site-specific $EQS_{\text{bioavailable, dissolved}}$ $\left(Site - specific EQS_{\text{bioavailable, dissolved}} = \dfrac{reference\ EQS_{dissolved}}{BIO-F} \right)$. For certain circumstances (and certain metals) this may provide the most pragmatic and sensible option for the use of BLMs. This approach is not the ideal approach, as uncertainty regarding the bioavailability still exists. Indeed, applying the most conservative (lowest) Bio-F suggests that all organisms in the species sensitivity distribution (SSD) are exposed to similar and conservative levels of bioavailable metal, which can be severely questioned. Moreover, additional uncertainty is introduced through the setting of the reference EQS value, which should be representative

---

[2] The use of a bioavailability model developed/validated for a particular species towards other similar organisms for which no bioavailability model were developed/validated.

[3] A reference EQS should represent reasonable worst-case conditions and should therefore be protective for 90 % of the surface waters.

for reasonable worst-case conditions in the surface waters. However, this pragmatic approach makes the best possible use of the available knowledge on bioavailability in the situation, as described above.

Level 3b: If it can be justified that the bioavailability model of a species within the same trophic level can be applied to those species for which no specific bioavailability model has been developed (e.g. insects, amphibians, molluscs), the option of full read-across (Level 3b) should be preferred as it can be considered the most realistic assessment of bioavailability refinement. In case of a full read-across, the chronic *Daphnia magna* (water flea) BLM is used to predict metal toxicity to all other invertebrates, the chronic *Pseudokirchneriella subcapitata* BLM to predict metal toxicity to all other algae, and the chronic *Oncorhynchus mykiss* (rainbow trout) BLM to predict the toxicity to all other fish, for example.

If supportive evidence is present, it is clear that the full read-across approach is the preferred way forward. Application of the bioavailability models that have been developed for a number of metals has shown that a consistent and small number of water quality parameters need to be characterised: dissolved organic carbon (DOC), pH, alkalinity and the individual hardness (H) cations Ca^{2+} and Mg^{2+}. Similar information is needed on the media in which the tests were conducted.

7.3.4 PROPOSED APPROACH

Based on the above-mentioned information, the following approach is proposed for compliance checking. The general tiered framework proposed for compliance checking with metal (Me) EQS values for surface water is outlined in Figure 7.3.3. This framework particularly applies if information for a bioavailability correction at Level 3 is available. The general approach is based on the use of the total risk approach (TRA), as this was identified as being the most scientifically robust and defensible option. Tier 1 consists in comparing the dissolved monitoring results from a particular region or site (site-specific $C_{TRA, dissolved}$) with a reference $EQS_{TRA, dissolved}$ value. Compliance is reached if the exposure concentrations do not exceed the reference $EQS_{TRA, dissolved}$ value.

In case of non-compliance, further refinement of the assessment is needed. In Tier 2, further refinement of the TRA is achieved through the incorporation of the bioavailability concept (physico-chemical speciation models or toxicity-related bioavailability models). This suggests that the physico-chemical parameters of the investigated site/region (pH_{site}, H_{site}, DOC_{site}) affecting the metal bioavailability should be gathered and checked against the applicability domain of the bioavailability model. The use of the bioavailability models is only recommended if the pH_{site}, H_{site}, DOC_{site} values of the site/region are within the development/validation range of the model. In other cases, the use of such bioavailability models is only afforded on a case-by-case basis and only when strong scientific arguments can be formulated to support this application. The introduction of the bioavailability assessment modelling consists either in the application of a conservative bioavailability factor (Bio-F) or of the full read-across, depending on the availability of supporting evidence to extrapolate a bioavailability model from

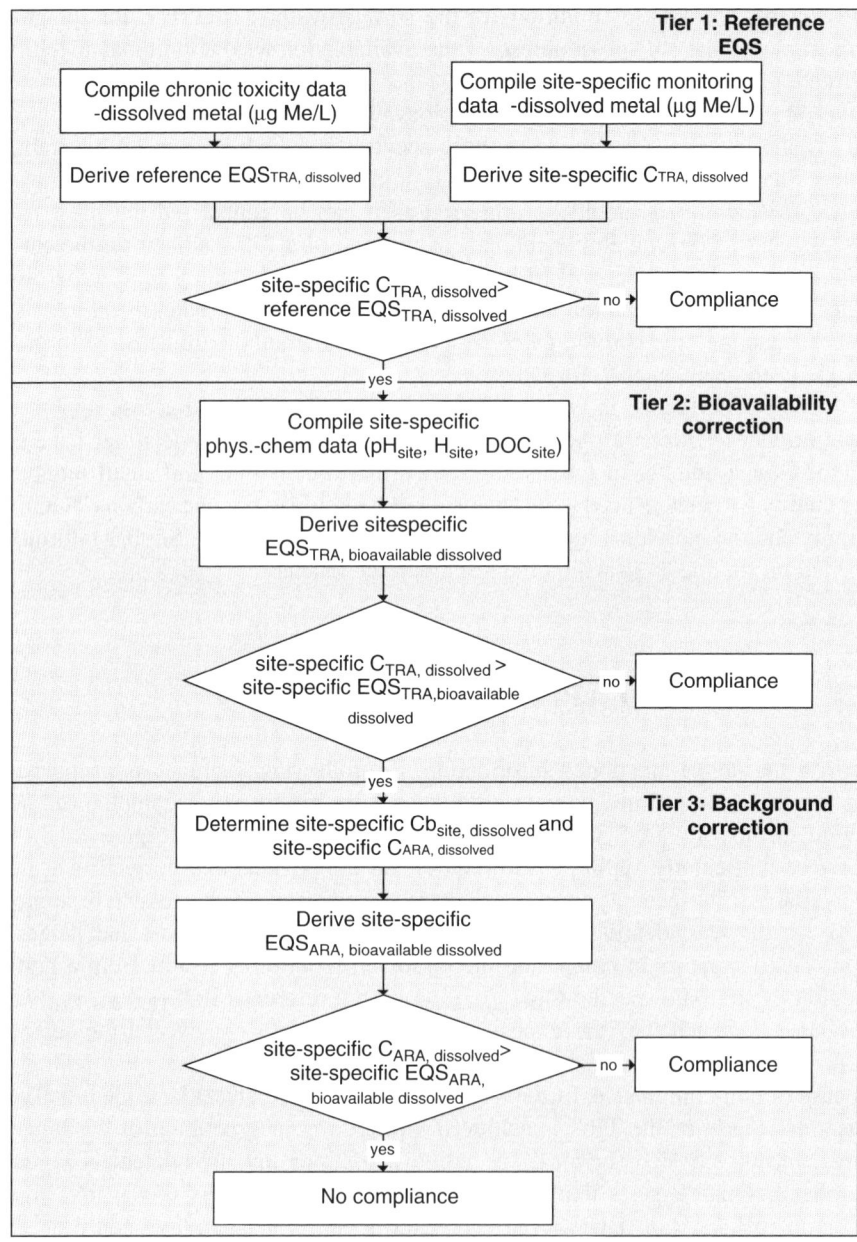

Figure 7.3.3 General tiered framework proposed for compliance checking with EQS values

a particular species to another one. In both cases, a normalised EQS value (site-specific EQS$_{TRA, bioavailable dissolved}$) towards the site-specific conditions occurring at the investigated site/region will be derived. Compliance is reached if the site-specific dissolved exposure concentrations (C$_{TRA, dissolved}$) do not exceed the site-specific EQS value.

In Tier 3, the added risk approach (ARA) is applied, suggesting that an additional background correction is applied on top of the bioavailability correction as carried out in Tier 2. In this tier the bioavailable added exposure concentrations at a particular site/region, $C_{ARA,\ bioavailable\ dissolved}$, are calculated from the $C_{TRA,\ bioavailable\ dissolved}$ − $Cb_{site,\ bioavailable\ dissolved}$. Compliance checking is subsequently performed by comparing the $C_{ARA,\ bioavailable\ dissolved}$ with the site-specific $EQS_{ARA,\ bioavailable\ dissolved}$. In this equation the site-specific $EQS_{TRA,\ bioavailable\ dissolved}$ is calculated from the site-specific $EQS_{TRA,bioavailable\ dissolved}$ − $Cb_{culture\ medium,\ bioavailable\ dissolved}$.

It must be emphasised that the above-mentioned approach could also be reversed. Indeed, if high $Cb_{site,\ dissolved}$ levels are observed in comparison with the calculated reference $EQS_{TRA,\ dissolved}$ value it could be more appropriate to apply the added risk approach (ARA) in the second tier. In that particular case, the added exposure concentrations at a particular site/region, $C_{ARA,\ dissolved}$, should be calculated from the $C_{TRA,\ dissolved}$ − $Cb_{site,\ dissolved}$. Compliance checking is then performed by comparison between $C_{ARA,\ dissolved}$ and the reference $EQS_{ARA,\ dissolved}$ (i.e. reference $EQS_{TRA,\ dissolved}$ − $Cb_{culture\ medium,\ dissolved}$). Tier 3 would then consist of the introduction of a bioavailability correction based on the added concentrations.

7.3.5   CASE STUDIES

Examples of the application of the full read-across normalisation using chronic BLMs (Level 3b) for copper and nickel towards specific European rivers and lakes in different EU regions (i.e. Sweden, Italy, The Netherlands, the United Kingdom and Spain) are provided hereunder. These rivers/lakes were selected to provide *examples* of typical conditions covering a wide range of physico-chemical conditions (pH between 6.7 and 8.2; hardness between 28 and 260 mg/L $CaCO_3$; DOC between 3 and 28 mg/L) occurring in EU surface waters. The different considered scenarios are summarised in Table 7.3.1. In this exercise, small (± 1000 m³/d), medium-sized (± 200 000 m³/d) and large (± 1 000 000 m³/d) alluvial (eutrophic) rivers were considered. For the lakes, the focus was on gathering physico-chemical data representative for sensitive systems, i.e. oligotrophic and acidic lakes.

An overview of the SSD and calculated EQS values[4] (defined as the median 5th % (median HC_5) of the SSD) for Ni and Cu for the different rivers and lakes is presented in Table 7.3.1. Figure 7.3.4 represents the SSD for Ni for the different rivers. Depending on the abiotic factors encountered in the rivers/lakes, the EQS values varied between 7.8 and 27.2 µg/L for Cu, and between 7.1 and 43.6 µg/L for Ni. The lowest EQS values for Cu and Ni are generally observed in aqueous systems with low DOC concentrations (e.g. River Otter, Lake Monate). For Ni, the effect of pH is more accentuated than for Cu. Indeed, for Ni, according to the BLM developments, all organisms tend to be more sensitive at high pHs, while for Cu a higher sensitivity at higher pH values was observed only for algae.

---

[4] The calculated EQS values do not include the influence of an assessment factor (AF).

Table 7.3.1 Summary of the physico-chemical characteristics and EQS values of Cu and Ni for the different selected scenarios. The EQS values (µg/L) are defined as the median 5th percentile (median HC_5) of the SSD for each metal

Type		Name	Country	pH	H (mg/L $CaCO_3$)	DOC (mg/L)	EQS Cu (µg/L)	EQS Ni (µg/L)
Rivers	Small (ditches with flow rate of ± 1000 m³/d)	/	The Netherlands	6.9	260	12.0	27.2	43.6
	Medium (rivers with flow rate of ± 200 000 m³/d)	River Otter	United Kingdom	8.1	165	3.2	7.8	8.1
		River Teme		7.6	159	8.0	21.9	19.0
	Large (rivers with flow rate of ± 1 000 000 m³/d)	River Rhine	The Netherlands	7.8	217	2.8	8.2	10.8
Lakes	Mediterranean rivers	River Ebro	Spain	8.2	273	3.7	10.6	8.7
	Oligotrophic systems	Lake Monate	Italy	7.7	48	2.5	10.6	7.1
	Acidic systems	/	Sweden	6.7	28	3.8	11.1	12.1

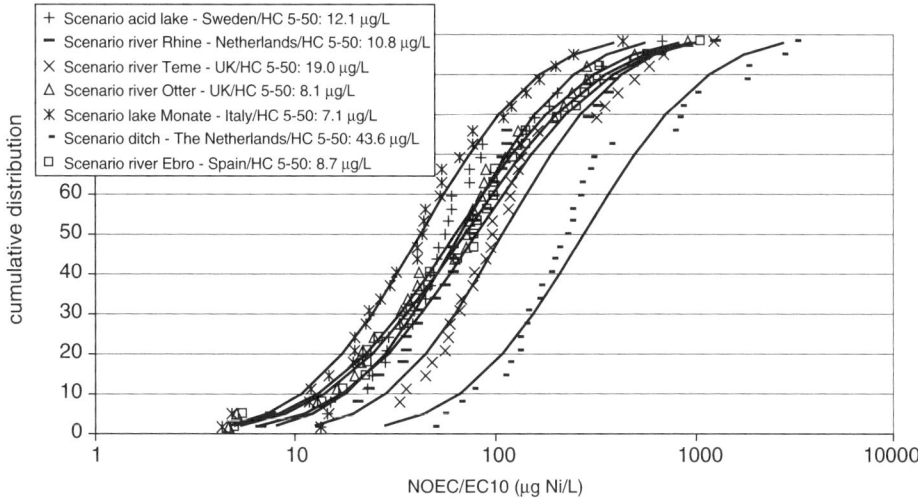

Figure 7.3.4 Species sensitivity distribution for Ni for the different selected EU scenarios

REFERENCES

Allen, H.E. and Hansen, D.J. (1996) The importance of trace metal speciation to water quality criteria, *Water Environ. Res.*, **68**, pp. 42–54.

Ankley, G.T., Di Toro, D.M., Hansen, D.J. and Berry, W.J. (1996) Technical basis and proposal for deriving sediment quality criteria for metals, *Environ. Toxicol. Chem.*. **15**, pp. 2056–2066.

Bergman, H.L. and Dorward-King, E.J. (1996) *Reassessment of Metals Criteria for Aquatic Life Protection*, SETAC Press, Pensacola, FL, USA.

Campbell, P.G.C. (1995) Interactions between trace metals and aquatic organisms: a critique of the Free-ion Activity Model, In: Tessier, A., Turner, D.R. (eds), *Metal Speciation in Aquatic Systems*, John Wiley & Sons, Ltd, New York, New York, pp. 45–102.

Crommentuijn, T. *et al.* (1997) Maximum permissible concentrations and negligible concentrations for metals, taking backgrounds concentrations into account, Netherlands, Institute of Public Health and the Environment, RIVM, Bilthoven, RIVM report No. 601501001.

Deleebeeck, N.M.E., De Schamphelaere, K.A.C. and Janssen, C.R. (2007) A bioavailability model predicting the toxicity of nickel to rainbow trout (*Oncorhynchus mykiss*) and fathead minnow (*Pimephales promelas*) in synthetic and natural waters, *Ecotoxicol. Environ. Safety*.

De Schamphelaere, K.A.C. and Janssen, C.R. (2002a) A biotic ligand model predicting acute copper toxicity for Daphnia magna: the effects of calcium, magnesium, sodium, potassium and pH, *Environ. Sci. Technol.*, **36**, pp. 48–54.

De Schamphelaere, K.A.C. and Janssen, C.R. (2004) Bioavailability and chronic toxicity of zinc to juvenile rainbow trout (*Oncorhynchus mykiss*): comparison with other fish species and development of a Biotic Ligand Model, *Environ. Sci. Technol.*, **38**, pp. 6201–6209.

De Schamphelaere, K.A.C., Vasconcelos, F.M., Heijerick, D.G., Tack, F.M.G., Delbeke, K., Allen, H.E. and Janssen, C.R. (2003) Development and field validation of a predictive copper toxicity model for the green alga *Pseudokirchneriella subcapitata*, *Environ. Toxicol. Chem.*, **22**, pp. 2454–2465.

Di Toro, D.M., Zarba, C.S., Hansen, D.J., Berry, W.J., Swartz, R.C., Cowan, C.E. *et al.* (1991) Technical basis for establishing sediment quality criteria for nonionic organic chemicals using equilibrium partitioning, *Environ. Toxicol. Chem.*, **10**, pp. 1541–1583.

European Community (2000) Directive 2000/60/EC of the European Parliament and of the Council of 23 October 2000 establishing a framework for Community action in the field of water policy, *Official Journal of the European Communities*, **L327**, pp. 1–72.

Heijerick, D.G., De Schamphelaere, K.A.C. and Janssen, C.R. (2002a) Biotic Ligand Model development predicting Zn toxicity to the alga *Pseudokirchneriella subcapitata*: possibilities and limitations, *Comparative Biochem. Physiol C*, **133**, pp. 207–218.

Heijerick, D.G., De Schamphelaere, K.A.C. and Janssen, C.R. (2002b) Predicting acute zinc toxicity for *Daphnia magna* as a function of key water chemistry characteristics: development and validation of a Biotic Ligand Model, *Environ. Toxicol. Chem.*, **21**, pp. 1309–1315.

Hoang, T.C., Tomasso, J.R. and Klaine, S.J. (2004) Influence of water quality and age on nickel toxicity to fathead minnows (*Pimephales promelas*), *Environ. Toxicol. Chem.*, **23**, pp. 86–92.

HydroQual (2003) Phase I development of a Biotic Ligand Model for cadmium, Batl1106, Technical Report, Mahwah, NJ, USA.

Keithly, J., Brooker, J.A., DeForest, D.K., Wu, B.K. and Brix, K.V. (2004) Acute and chronic toxicity of nickel to a cladoceran (*Ceriodaphnia dubia*) and an amphipod (*Hyalella azteca*), *Environ. Toxicol. Chem.*, **23**, pp. 691–696.

Landner, L. and Reuther, R. (2004) *A Critical Review of Current Knowledge on Metals in Society and the Environment*, Kluwer Academic.

Muyssen, B.T.A. and Janssen, C.R. (2002a) Zinc tolerance and acclimation of *Ceriodaphnia dubia*, *Environ. Pollut.*, **117**, pp. 301–306.

Muyssen, B.T.A. and Janssen, C.R. (2002b) Zinc accumulation and regulation in *Daphnia magna* Straus: links with homeostasis and toxicity, *Archives Environ. Contam. Toxicol.*, **43**, pp. 492–496.

Niyogi, S. and Wood, C.M. (2004) Biotic Ligand Model, a flexible tool for developing site-specific water quality guidelines for metals, *Environ. Sci. Technol.*, **38**, pp. 6177–6192.

Paquin, P.R., Di Toro, D.M., Santore, R.C., Trivedi, D. and Wu, K.B. (1999) A Biotic Ligand Model of the acute toxicity of metals: III application to fish and *Daphnia magna* exposure to silver, US Government Printing Office, EPA 822-E-99-001, Washington DC, USA.

Paquin, P.R., Gorsuch, J.W., Apte, S., Batley, G.E., Bowles, K.C., Campbell, P.G.C. *et al.* (2002) The Biotic Ligand Model: a historical overview, *Comparative Biochem. Physiol., Part C*, **133**, pp. 3–35.

Santore, R.C., Di Toro, D.M. and Paquin, P.R. (2001) A Biotic Ligand Model of the acute toxicity of metals: II application to acute copper toxicity in freshwater fish and *Daphnia magna*, *Environ. Toxicol. Chem.*, **20**, pp. 2397–2402.

TC NES III (2006) 11th Technical Committee on New and Existing Chemicals following Council Regulation (EEC) 793/93, Directive 67/548/EEC, Arona, September 2006.

US-EPA (2001) Update of ambient water quality criteria for cadmium, EPA-822-R-01-001.

7.4

Freshwater Ecosystem Responses to Climate Change: the Euro-limpacs Project

Richard W. Battarbee, Martin Kernan, David M. Livingstone, Uli Nickus, Piet Verdonschot, Daniel Hering, Brian Moss, Richard F. Wright, Chris D. Evans, Joan O. Grimalt, Richard K. Johnson, Edward Maltby, Conor Linstead and Richard A. Skeffington

The Water Framework Directive - Ecological and Chemical Status Monitoring   Edited by Philippe Quevauviller, Ulrich Borchers, Clive Thompson and Tristan Simonart © 2008 John Wiley & Sons, Ltd

7.4.1 INTRODUCTION

Although GCMs (General Circulation Models) vary in their projection of future climate change, all are in agreement that significant warming will occur within this century, principally as a result of a continued rise in the concentration of greenhouse gases, especially carbon dioxide (IPCC, 2007).

Against this concern it is important to examine the potential effects of future climate change on the functioning of the earth system. Here we consider the potential impact of climate change on freshwater ecosystems (streams, lakes and wetlands), using in particular recent results from the EU-funded project 'Euro-limpacs: global change impacts on European freshwater ecosystems' (http://www.eurolimpacs.ucl.ac.uk). The principal questions concern how climate change might affect the structure and function of freshwater ecosystems, how climate change might interact with concurrent trends in pollutant loading and land use, and what might be the implications of the projected responses for environmental policy and management.

Addressing such questions requires a large-scale collaborative and integrated research programme employing a range of complementary methodologies. Euro-limpacs combines:

(1) Experiments in the laboratory and field under controlled climate conditions using realistic predictions of future climate (over the next 50 to 100 years).

(2) Analysis of long-term observational physical, chemical and biological records to identify current trends, assess the role of climate change in explaining ecosystem change over recent decades and provide calibration and/or verification time series for model output.

(3) Palaeolimnological reconstructions to extend time series, assess natural variability and identify processes acting over longer (decadal) timescales.

(4) Space-for-time substitution to enable data from sites and regions with warmer climates today to be used to make predictions for cooler systems.

(5) Process-based modelling to explore mechanisms and run scenarios based on catchment-scale projections of climate, land use and pollution change.

In this chapter we summarise some of the results obtained to date from the Euro-limpacs project and draw upon the wider literature to illustrate changes that have taken place or might take place in the future as a result of global warming. First we consider some of the direct consequences of climate change on lakes, streams and wetlands. We then describe research designed to assess the interactions between climate warming and other stressors (hydromorphology, eutrophication, acidification and toxic substances). Finally we consider the relevance of the results for policy making and for the management of freshwaters, and describe modelling and other tools that need to be developed to aid decision making.

7.4.2 DIRECT IMPACTS OF CLIMATE CHANGE ON AQUATIC SYSTEMS

Projected changes in temperature, together with associated changes in precipitation and pressure systems, will have far-reaching direct impacts on the physical, chemical and biological characteristics of many, if not all, freshwater ecosystems in Europe and the wider world.

Long-term time series from freshwaters already show clear evidence for change associated with warming, especially increasing stream and lake surface water temperatures (Fang and Stefan, 1999; Livingstone, 2003), hypolimnetic warming in large lakes (Dokulil *et al.*, 2006), decreasing ice cover in northern and high-altitude lakes (Magnuson *et al.*, 2000), changes in seasonality of phytoplankton (Catalan *et al.*, 2002), an extension to growing seasons and increases in lake productivity (Blenckner *et al.*, 2007), changes in the geographical range of taxa, e.g. Odonota (Brooks *et al.*, 2007;

Hickling *et al.*, 2005), and threats to cold stenothermal taxa (e.g. Griffiths, 2007). In addition, recent results from the EU CLIME project have shown how the extremely hot summer of 2003 caused significant changes in the temperature, stratification and hypolimnetic oxygen concentration of deep lakes in Switzerland (Jankowski *et al.*, 2006), and hot summers are likely to be more frequent in future, particularly over central Europe (IPCC, 2007).

In Euro-limpacs there are many studies involving the direct impact of climate change on freshwater ecosystems. For lakes, these include examining the response to extreme events by modelling ecological thresholds at climatic extremes (Catalan *et al.*, in press) and the development and application of coupled hydrophysical modelling for large, deep lakes (Duwe *et al.*, 2007). River studies focus on the impacts on ecological functioning as a response to changing discharge regimes. In wetland ecosystems we examine changes in hydrology and biogeochemical processes (Bonnett *et al.*, 2006), plant communities (Clement and Aidoud, 2007), plant nutrient dynamics and productivity (Verhoeven *et al.*, 2005), riparian snow cover (Laudon and Bishop, 2006) and the relationships between marginal wetlands and adjacent surface waters (Vadineanu *et al.*, 2005). Here we present three examples with respect to surface water temperature, the melting of rock glaciers and the potential impact of increased wind stress on lakes.

7.4.2.1   Surface Water Temperature Trends

Some of the best long-term temperature data sets are for Swiss rivers and lakes. Hari *et al.* (2006) have collated and analysed data for river and stream temperatures measured at 25 stations in Switzerland. They show annual running mean temperatures based on daily means calculated from the original high-resolution measurements. The river temperatures show a high degree of regional coherence on inter-annual and inter-decadal timescales, implying a common, coherent response to regional climatic forcing on these timescales. The absolute differences in water temperature from river to river are primarily a result of the general decrease in water temperature that occurs with increasing altitude, but the degree of coherence also decreases somewhat as the mean altitude of the catchment area of the sampling station increases (and is disproportionately low under the influence of glaciers or hydroelectric power stations). On timescales exceeding the inter-decadal timescale, a coherent warming can be seen to have occurred at all altitudes. This coherent warming reflects a corresponding long-term increase in regional air temperature. Much of the long-term water temperature increase occurred as an abrupt increase between two approximately stationary periods from 1978 to 1987 and 1988 to 2002. This abrupt shift reflects a similar shift in air temperature and may be related in part to a shift in the North Atlantic Oscillation.

7.4.2.2   Hydrochemical Response to the Melting of Rock Glaciers

Chemical changes in rivers and lakes are more difficult to relate to climate warming than physical changes. For example, the rise in DOC observed from sites across Europe and parts of North America ascribed to climate change (Freeman *et al.*, 2001) may be associated with recovery of acidified sites from acidification (Monteith *et al.*, 2007). However, other long-term chemical trends are difficult to explain except as a result of warming. An example of this in the Euro-limpacs project is the work of Thies *et al.* (2007) in the Central European Alps.

Remote high-altitude lakes are sensitive indicators of environmental and climate change. In particular, a substantial rise in solute concentration has been observed at Rasass See (2682 m, Italy), a high alpine lake in a catchment of metamorphic rocks. During the past two decades, conductivity has increased by a factor of 18, while the most abundant ions, magnesium, sulphate and calcium, have reached 68-, 26- and 13-fold concentrations, respectively (Figure 7.4.1) (Thies *et al.*, 2007). The pronounced change in lake water composition is most likely caused by solute release from rock glacier outflows draining into the lake. This effect is expected to intensify with increasing air temperature and subsequent enhanced melt processes. In addition to major ions, unexpectedly high nickel concentrations ($243 \, \mu g \, l^{-1}$) have been found recently in Rasass See. The values exceed the limit for drinking water by one order of magnitude and cannot be related to catchment geology, leaving the source of nickel still unclear.

Similar but less intense processes have been observed at Schwarzsee (2796 m, Austria), where electrical conductivity rose by a factor of three during the past two decades (Figure 7.4.1) and nickel concentrations were just above the detection limit. This is attributed to the lower impact of glacial meltwater compared to Rasass See, where the area of active rock glaciers is larger and situated at a lower elevation.

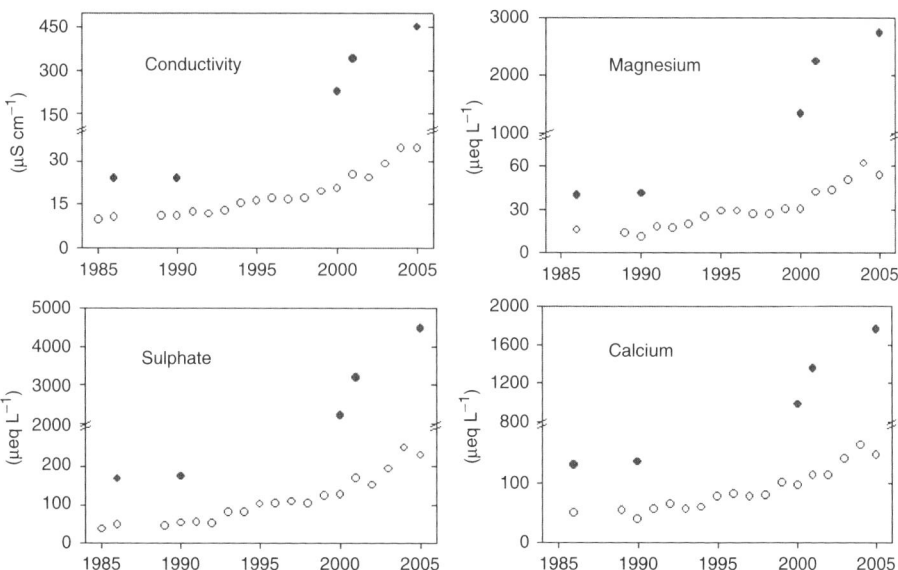

Figure 7.4.1 Conductivity, magnesium, sulphate and calcium concentrations in the water of two high-alpine lakes during the past 20 years (Rasass See: black dots; Schwarzsee: open circles). Note the break in the concentration axis due to the high change in solute concentrations. (Adapted with permission from Thies *et al.*, 2007. Copyright American Chemical Society, 2007)

There are still few data available on solute concentrations in outflows of active rock glaciers and their potential impact on freshwater systems, although comparable solute increases have been detected at other high mountain sites in the Austrian and Italian Alps, in the Rocky Mountains and in the Himalayas. Rock glaciers are widespread in high mountains around the world (Humlum, 1998), and as global climate models project a continuous rise in air temperature until the end of this century, high mountain freshwaters may become increasingly affected by solute release from active rock glaciers.

7.4.2.3   Increased Wind Stress on Lakes

In the climate change debate most attention is given to the probable impacts of future change in temperature and precipitation. However, lakes are also sensitive to changes in wind conditions. In response to the predicted increase in mean geostrophic winds in northern Europe as a result of the northward shift of cyclone activity, a whole-lake mixing experiment is being carried out in Euro-limpacs to assess lake response to increased input of mixing energy. In this experiment, which uses a submerged propeller to effect the mixing, the focus is on thermocline manipulation, biogeochemical cycling (including Hg) and physical modelling using the myLake model (Saloranta and Andersen, 2007). The results to date show that the impact of experimental mixing on Halsjärvi causes a depression of the thermocline to the extent that might be expected as a result of increased wind stress. The consequences of such a deepened thermocline include an increase in the volume of the mixed epilimnion, a decrease in the area of anoxic surface sediments, a longer ice-free season and an increase in the resuspension of littoral sediments.

7.4.3   INTERACTION BETWEEN CLIMATE CHANGE AND THE HYDROMORPHOLOGY OF STREAMS AND RIVERS

In many parts of Europe, hydromorphological alteration is the main stressor affecting rivers. Alterations include channel straightening, dam construction, disconnection of the river from its floodplain and destruction of riparian vegetation. Future climate change will introduce further stresses on channel hydromorphology, including the combined effects of changes in precipitation and climate-induced changes in land use patterns. These in turn may cause changes in catchment hydrology that will affect sediment transport and channel morphology, inundation frequency and extent, altering river ecosystems at both catchment and habitat scale (Verdonschot, 2000).

The changes that occur may not always be negative for biodiversity. As illustrated in Figure 7.4.2, we can hypothesise that climate change may cause biodiversity loss through intensification of land use or through a more variable discharge regime. Alternatively, improvements could occur if, as a result of increased flooding, human activity is withdrawn from floodplains, generating near-natural habitat structures.

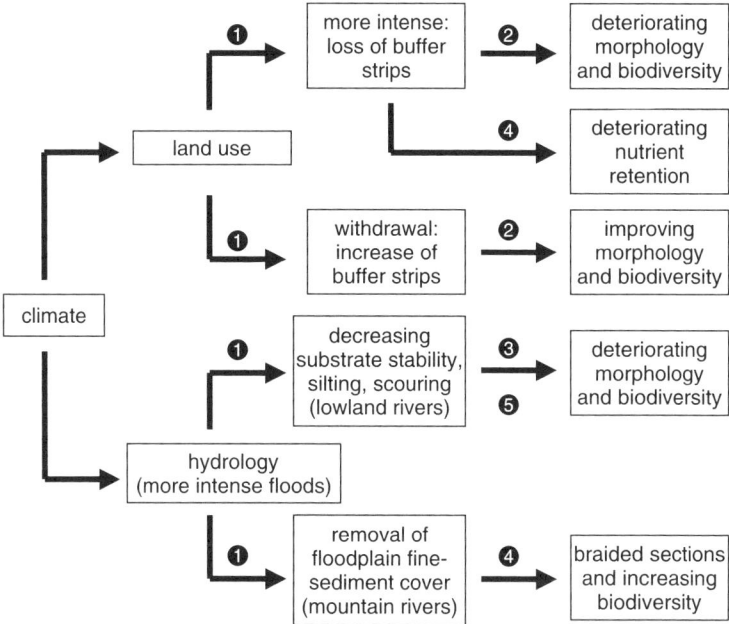

Figure 7.4.2 Cause–effect chain illustrating the potential impact of changes in climate on stream biodiversity

The Euro-limpacs project includes a number of case studies across Europe (Casalengo *et al.*, 2005) examining the relationships between climate change, flow regime, channel morphology and the distribution and biodiversity of taxa (Opatrilova *et al.*, 2005) under different hydromorphological conditions at the catchment, reach and habitat (Graf *et al.*, 2006a) scale at sites across Europe. In particular, it focuses on mountain streams comparing braided and non-braided reaches and meandering lowlands streams, comparing catchments spanning the north–south and west–east climate gradient in Europe, using both laboratory and field base experiments (Murphy and Nijboer, 2006). With regard to lakes, the project examines the effects of climate change on lake sediment accumulation rates (Rose *et al.*, 2007) and on the morphology of inflows to large lakes (Duwe *et al.*, 2008).

7.4.3.1 Climate-induced Changes in Flow Regimes

Climate change alters the dominant pattern of precipitation, which in its turn changes run-off, discharge regimes and hydraulic conditions in streams.

There has been considerable previous work during the 1970s and late 1990s assessing impacts of extreme flows on the biotic communities in the Lambourn chalk stream (e.g. Wright and Symes, 1999; Wright *et al.*, 1981, 2004). During both periods the river experienced sustained low-flow episodes (1976, 1997), and during the latter period there was also a phase of prolonged exceptionally high flows (2000/01).

About 90 % of the River Lambourn discharge is from the groundwater and only 10 % is from direct run-off. This means that the flow regime is relatively predictable and the catchment rarely, if ever, experiences spate flooding. The crucial phase in the annual hydrograph is the winter/spring recharge of the aquifer. A particularly dry winter, e.g. 2003/04, led to prolonged low-flow conditions through the following summer, with consequences for the in-stream flora and fauna. In drought years the usually dominant macrophyte *Ranunculus* tended to be restricted due to smothering by epiphytic algae and silt. *Callitriche* and other marginal emergent vegetation replaced the *Ranunculus* as the dominant vegetation under such conditions. Drought events appeared to have a more deleterious impact on macroinvertebrates than high-flow events, but in both cases the River Lambourn communities recovered within a year or less (Wright *et al.*, 2004).

Overall macroinvertebrate taxon richness tended to be greater in high-flow years. Many of these responses were influenced by changes in substrate composition and coverage of macrophyte beds in the river. *Ranunculus* thrives in sustained high-flow conditions and is capable of achieving a large biomass in the channel, which provides excellent habitats and refuges for invertebrates, and also traps and supports food resources. *Ranunculus* is generally a favoured habitat for Gammaridae, Baetidae, Ephemerellidae, Simuliidae, Rhyacophilidae and Hydropsychidae. Large macrophyte beds in the channel also increase the probability of exceeding bank-full discharge and flooding of adjacent land. This in turn dissipates the damaging energy of high flows, protecting the stream biota.

The key factor however is whether the magnitude, frequency and timing of extreme discharge events in the River Lambourn are likely to change under future climate change scenarios. Further research within Euro-limpacs will try to answer this question.

7.4.3.2   Climate-induced Changes in Mountain Streams

A potential consequence of increased flood intensity in mountain streams is the removal of sediment stored in flood plains and the development of multiple-channel streams. To assess whether an increase in multiple-channel reaches leads to an increase in habitat and in species diversity, single- and multiple-channel sections of several streams have been studied. In a paired-site study of mountainous streams in Germany, seven multiple- and seven single-channel sites (fourteen sections) were compared. The hypothesis that multiple-channel sections have larger and more diverse habitats and more diverse biota was tested. Various hydromorphological parameters (shore-line length, channel features, current velocity, water depth, substrate) were recorded, and data for floodplain vegetation, riparian ground beetles (Carabidae) and benthic macroinvertebrates were collected. Table 7.4.1 summarises the results. The hydro-morphological diversity of the multiple-channel sections is higher and is closer to the reference condition (Jähnig *et al.*, in press a). Habitat diversity is also higher and cross-sectional changes in stream sediments are more dynamic, but the effects of either substrate or the overall stream sections on the macroinvertebrate community are not detectable. At best they display a general tendency towards improvement (Jähnig *et al.*, in press b). However, taxonomic richness is clearly increased in floodplain vegetation

Table 7.4.1 Comparison of hydromorphological and biological indices in single- and multiple-channel sections. *U-test significant p < 0.05

Number/Share	Median		Minimum		Maximum	
	Single	Multiple	Single	Multiple	Single	Multiple
Meso habitats (channel features)*	3	9	2	8	7	12
Micro habitats (substrate)	10	10	8	10	11	11
Vegetation units*	4	6	2	5	6	7
Floodplain vegetation species*	80	125	60	73	101	173
Floodplain vegetation genus*	62	86	47	59	71	115
Floodplain vegetation families*	28	35	25	32	35	41
Carabidae species	5	12	2	9	17	15
Carabidae genus	4	7	2	6	10	8
Riparian Carabidae (%)*	29	75	7	67	51	95
Macroinvertebrate species	91	96	77	79	111	111
Macroinvertebrate genus	68	69	57	60	83	83
Macroinvertebrate families	44	46	40	42	53	53

and riparian ground beetles, with floodplain vegetation reacting most strongly (Hering *et al.*, in prep.). The stronger reaction of floodplain vegetation and ground beetles is mainly due to the generation of several additional habitats, such as gravel bars. Reasons for the close similarity of macroinvertebrate communities from single- or multiple-channel sections include the influence of large-scale catchment pressures, the relatively short length of restored sections and the lack of potential recolonisers. Other reasons explaining the different responses between organism groups include differences in dispersal abilities and source populations. Despite the mixed effects observed on organisms, the higher habitat diversity and the more dynamic hydromorphological environment of the multiple-channel sections are likely to have positive impacts on ecological quality.

7.4.3.3 Influence of Climate Change on Meandering Lowland Streams

A potential consequence of increased precipitation intensity in lowland streams is the increase in hydromorphological dynamics. An increase in flood frequency and intensity will cause an increase in flow regime dynamics and a higher stream bed instability. To assess whether a more dynamic discharge regime leads to a higher bed instability and to in-stream biodiversity loss, macroinvertebrate communities in natural, semi-natural and canalised stream sections in middle course of a Dutch lowland steam were compared. A total of 51 samples collected between 1981 and 2006, of which 35 were from the natural, 7 from the semi-natural and 9 from the canalised stream section, were analysed. In 2004 a hydrological restoration project took place in a branch of this stream, resulting in a hydrological stabilisation downstream of the junction. A total of 238 macro-invertebrate taxa occurred in the samples.

Detrended Canonical Correspondence Analysis (ter Braak and Smilauer, 2002) was used successively to compare the macroinvertebrate communities at the three

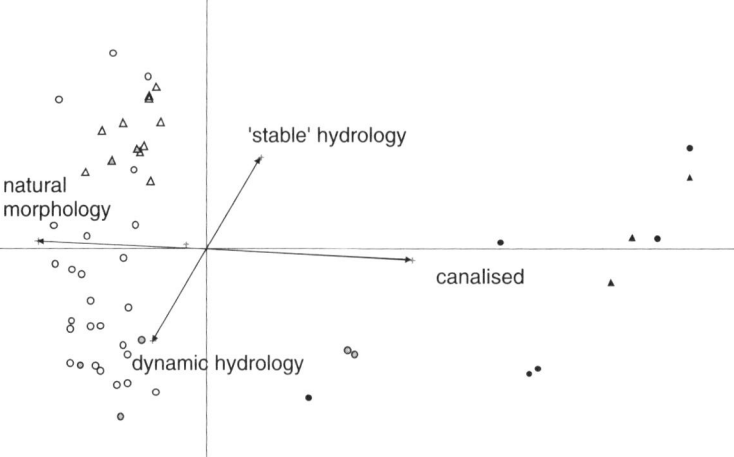

Figure 7.4.3   DCCA ordination diagram for the hydromorphological gradient analysis, with the morphological gradient indicated by black (canalised), grey (semi-natural) and open (natural) labels and with the hydrological periods indicated by circles (pre-hydrological restoration) and triangles (post-hydrological restoration)

hydromorphologically different stream sections (natural, semi-natural and canalised morphology) and the two hydrologically different periods in time (dynamic and 'stable' hydrological regime). The data (Figure 7.4.3) show that there was a strong effect of morphology and a lesser effect of hydrology upon the in-stream macroinvertebrate assemblage.

7.4.4   INTERACTIONS BETWEEN CLIMATE CHANGE AND EUTROPHICATION

There have been major successes in Europe in reducing eutrophication problems but eutrophication still remains a most serious environmental problem, especially for shallow lakes and sites where diffuse pollution sources dominate nutrient loading. Eutrophication, like climate change, affects the whole of an ecological system. Nutrient effects pervade all the trophic levels, just as temperature change affects the metabolic rates of all organisms and hence growth, reproduction, respiration and nutrient recycling. Likewise, climate change will affect ecosystem processes as well as potentially eliminating species whose niche limits are exceeded. In systems with hundreds of components and many interacting processes, predicting the consequences of simultaneous changes in climate, nutrient loading and biodiversity may well be impossible in any precise way.

The Euro-limpacs project focuses especially on eutrophication in shallow lakes and streams. It includes i) attempts to find analogues of how shallow lake and stream systems function across latitudes from the warm temperate to the arctic; ii) stable isotope analysis to reconstruct food web relationships; iii) experiments on parallel natural stream and wetland systems in Iceland that differ in temperature within a small area (Bakkers *et al.*, 2006); iv) mesocosm experiments to examine the effects of temperature increases on both shallow lake and littoral wetlands, with the former supplemented by molecular techniques to quantify micro-evolutionary responses to anthropogenic stress (Michels *et al.*, 2007); and v) analysis of long-term data sets and palaeoecological reconstruction to separate the effects of nutrient change and climate change. This latter task is particularly difficult, although there have been some successes with the simpler sub-systems such as the open-water plankton. Multivariate correlation analysis of data from Loch Leven, Scotland (May *et al.*, 2005) has demonstrated some changes in diatom community composition that may be more reliably attributed to a warming trend than to the reduction in phosphorus loading to the lake. Euro-limpacs has also been developing new palaeoecological methods for wetlands to examine long-term nutrient climate-change interactions from sediment records and long-term data sets in wetlands (Bennion 2006).

7.4.4.1 Geographical Comparisons

Lake systems differ in general ways with latitude, though exact analogues among future systems and current systems may not exist (Fox, 2007). The warmer systems of the tropics and Mediterranean-type warm-temperate zones have long growing seasons and have not been disrupted so completely by the last glacial period as systems in the glaciated temperate and polar regions. Warm-zone lakes have a more species-rich fish fauna that tends to combine omnivory with high fecundity and frequent reproduction (Meerhof *et al.*, 2007a, 2007b) compared with cold temperate lakes.

Processes involving fish operating through the food web are very important in determining how shallow lakes function (Scheffer *et al.*, 1993). In warmer, lower latitudes, fish predation on zooplankton is very intense, with the result that the zooplankton community is depauperate in those large species and individuals that are associated with efficient grazing on phytoplankton. Waters thus tend to be more turbid and more frequently dominated by algae, such as Cyanobacteria, with high temperature optima for growth. Increased nutrient load is thus associated strongly with increased algal crops and there is a greater risk of loss of littoral plant-dominated communities and their associated biodiversity (Gyllstrom *et al.*, 2005; Moss *et al.*, 2004).

Further north, there are shorter growing seasons, lower temperatures and a more limited fish fauna, still in the process of recolonising from southern refuges following the last glaciation. A mismatch in the earlier seasonal growth of zooplankton and the slightly later hatch of zooplanktivorous fish often leads to intense grazing on algae in the spring by a zooplankton community then untrammelled by fish predation. In turn, this creates clear water and allows early growth and development of plant communities. The established plant community can then survive the potentially increased growth of algae in summer through provision of refuges for zooplankton grazers against the increasing threat of predation by young-of-the-year fish. Northwards into the arctic regions (Jeppesen *et al.*, 2001) the influence of fish further decreases as growth seasons and fish reproduction become much reduced, and fish faunas are even poorer in species. Large zooplankters are able to keep waters clear, helped also by lower nutrient loading from the catchment in the often drier climates of the tundra and boreal forest regions. Euro-limpacs has been creating the evidence for these latitudinal scenarios using food web studies involving stable carbon and nitrogen isotope measurements and experiments.

7.4.4.2 Evidence from Experiments

Experimental studies are still very few (Liboriussen *et al.*, 2005; McKee *et al.*, 2002, 2003). They suffer from the inherent problem that an experiment set up with controlled treatments of water chemistry, sediment, temperature and nutrient loading may give divergent results in its replicates, or when carried out on different occasions, because of the consequences in a complex system of very small random differences in starting conditions and subsequent weather. Some general lessons are emerging, however. Growing seasons are extended by increased temperature; the timing of growth peaks of algae and zooplankton are altered. The reduced oxygen concentrations consequent on increased temperature, as well as increased respiration of biomass as nutrient levels are increased, may lead to fish deaths even of quite tolerant species. Figure 7.4.4 shows recent results from a UK Euro-limpacs mesocosm experiment. The experimental tanks contained a well-developed aquatic plant community and were rich in phosphorus. In a randomised block design, tanks were either heated by 4 °C above ambient and given one of three levels of nitrogen fertilisation. All tanks were initially stocked

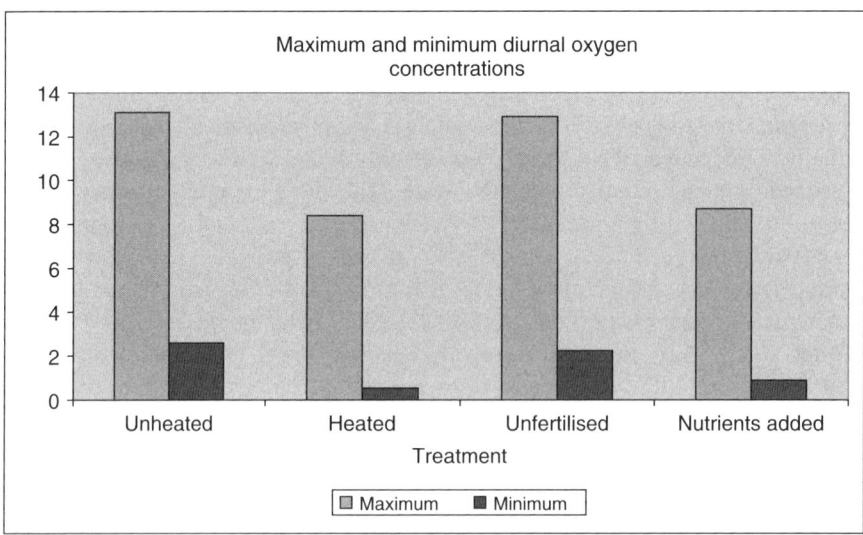

Figure 7.4.4 Average maximum and minimum oxygen concentrations (mg L−1) recorded during three 24 hour experiments carried out in a mesocosm system in late June and July 2007. For each treatment there were eight replicates. Both heating and addition of nutrients significantly reduced both maximal and minimal concentrations (p < 0.01). Unpublished data of Atkinson, D., Moss, B., Whitham, C., Moran, R. and Feuchtmayr, H

with a resilient fish, the three-spined stickleback, *Gasterosteus aculeatus*. With a 4 °C rise, there was a significant fall in both maximum and minimum oxygen concentrations recorded during three 24 hour sets of measurements in late June and July 2007. Nutrient addition also resulted in reduction in both maxima and minima. With either treatment, concentrations fell dangerously low during the night, and in the warmed treatments fish populations did not survive. Plant communities were also affected by warming. Exotic species, usually introduced from warmer regions, where there is a greater biodiversity, tend to proliferate over native species. Among native species, nutrients and warming together may lead to the predominance of floating plants, such as lemnids (duckweeds), and warming by 4 °C has been shown to increase the whole community respiration to total gross photosynthesis ratio of the systems by about 31 % (Atkinson, D., Moss, B., Whitham, C., Moran, R. and Feuchtmayr, H., unpublished data). This has major implications for positive feedback effects on carbon dioxide concentration in the atmosphere because of its relevance to the vast areas of wetlands in northern North America and Eurasia.

7.4.4.3 Future Projections

We can combine these generalisations into predicted scenarios, at present for lakes, and eventually for streams as Euro-limpacs data accumulate. We assume in Europe a future of increased temperature and reduced summer rainfall (IPCC, 2007), with nutrient additions to lakes most likely to be maintained as human populations and dietary aspirations increase, despite the best efforts at control under the Water Framework Directive. Experience has shown that many years are likely to elapse before the effects of nutrient control become significant.

Growing seasons will lengthen, algal biomass will be encouraged by warm, sunny weather, hypolimnetic oxygen conditions may deteriorate, and cyanobacterial blooms may become more extensive. For shallow lakes, food chain effects may exacerbate these problems, especially by a shift to more intense fish predation and reduced zooplankton populations in a warmer climate. Fish kills will, however, become more frequent, followed by a temporary abatement of the symptoms of eutrophication caused by build-up of zooplankton biomass. Exotic warm-water species of fish and water plants will proliferate with vigorous growth that will result in near monocultures, especially where fish kills allow greater zooplankton populations to persist. Where the exotic, thermally well-adapted common carp (*Cyprinus carpio*) has been introduced, we might expect it to reproduce effectively at northern latitudes, where presently it often fails. Carp cause increased damage to plant communities, associated with more turbid water, a consequence of both disturbance of sediment as they feed and of their mobilisation of sediment phosphorus into the water (Moss *et al.*, 2002).

On mainland Europe we might expect the steady movement northwards of warm-adapted systems, but almost certainly at rates far slower than the wave of temperature change. On islands like the UK, where immigration of new species is barred by the sea, this will not occur and fish kills will remove many native species from all but the northerly regions. However, there is a counter-risk that commercial pressures will result in introduction of warm-water species of plants and especially of fish, for the water garden and angling trades. Such species tend to be vigorous and tolerant of a wide range of conditions and may displace native species that have survived the warming.

One possible mitigating factor is that evolutionary change, in the smaller organisms at least, such as certain species of zooplankton, has been shown by work in Euro-limpacs (Van Doorslaer *et al.*, 2007) to be quite rapid, and thus existing systems might maintain some of their current structure and functioning as temperatures increase. This would, however, demand a rapid and synchronous set of evolutionary changes in a large number of different organisms, and this seems inherently unlikely on current ecological experience.

Effects on rivers and streams are even more difficult to predict. Flowing waters are subject to eutrophication no less than lakes, but their systems are driven much more by hydrological changes and the physical structures of their systems (see above), and only secondarily by nutrient supply. Fish predation is also less powerful in structuring the communities, though it certainly has an influence. Nonetheless, large beds of weed in lowland rivers and algal growths in slowly flowing stretches of floodplain rivers are greatly influenced by nutrient supply, and fish communities have similar effects to those in lakes. Only in the highly erosive, turbulent waters of the hills is it likely that combinations of temperature increase and current or changed nutrient load will not have many consequences for users of the water. Eutrophication still remains Europe's most serious freshwater problem and in almost all situations we expect climate change to make it a more difficult problem to control.

7.4.5   INTERACTIONS BETWEEN CLIMATE CHANGE AND ACIDIFICATION

Acid-sensitive freshwaters are found in upland and mountainous areas in which soils and overburden are poor in readily weathered minerals. Decades of deposition of acidifying sulphur (S) and nitrogen (N) compounds (acid deposition) have resulted in widespread and chronic acidification of sensitive freshwaters across Europe, with loss of fish populations and other biological effects. Since the mid-1980s, reduction in emissions of S and to a lesser extent N in Europe has resulted in recovery beginning to take place in lakes and streams (Stoddard *et al.*, 1999; Wright *et al.*, 2005). A critical question being addressed by the Euro-limpacs project is: to what extent is this recovery threatened by future climate change? Climate variability over many timescales (daily, weekly, seasonal, yearly, decadal) can affect surface water chemistry and biology and thus the acidification and recovery processes. Some key projected climate changes with the potential to impact water acidity are:

(1) higher temperatures

(2) increased incidence and severity of summer drought

(3) wetter conditions during winter

(4) increased frequency and magnitude of winter high flows

(5) reduced snowpack

(6) increased occurrence of sea salt deposition events.

Both indirect effects through impacts on the terrestrial catchments and direct effects on the water bodies themselves are important. Substantial inter-year variability in water chemistry is linked to climatic fluctuations. These include fluctuations in sea salt deposition, with higher loadings associated with a peak in the NAO Index causing short-term acidification in the early 1990s (Evans *et al.*, 2001). Nitrate and sulphate concentrations are also strongly influenced by climatic factors. The stability of invertebrate communities in acid-sensitive streams is sensitive to variations in the NAO Index (Bradley and Ormerod, 2002). Euro-limpacs examines the effects of episodic and seasonal variations in climatic factors on run-off water chemistry using a series of catchment scale experiments, including focusing on i) manipulations of snow cover, freeze–thaw cycles and soil wetness (Kaste *et al.*, 2007) and ii) simulation of hydrological and sea-salt extremes by experimental watering. State-of-the-art statistical techniques are being used to analyse episodic, seasonal and long-term effects of changing weather patterns on the chemistry of acid waters. Space-for-time substitution techniques are being used at the longer timescale to assess the potential impact of climate change on surface waters currently recovering from acidification. The results of the experimental work will be used in tandem with existing information to specify dynamic model input perturbations to simulate various scenarios combining future climate change with future acid deposition.

7.4.5.1 Acid Episodes and Climate

Short-term climate events can cause 'acid episodes' with biological damage and set back the recovery process. Acidic episodes are crucially important in terms of stream biota, as it is the severity of chemical extremes, rather than average conditions, which typically determines biological damage such as fish kills (Baker *et al.*, 1990; Hindar *et al.*, 1994) and loss of invertebrate species (Kowalik and Ormerod, 2006). Episodes can be caused by a variety of different drivers, including high rainfall events, snowmelt, sea salt deposition events, sulphate flushes after droughts and nitrate flushes after freezing events (Davies *et al.*, 1992; Evans *et al.*, in press; Wright, 2007). A consistent feature of all these drivers, however, is that they are associated with some form of climatic extreme. And scenarios of future climate generally project increased frequency and severity of extreme events.

A modelling study at the Afon Gwy monitoring catchment at Plynlimon, Wales indicated that the severity of high rainfall-driven acid episodes is declining in magnitude as S deposition is reduced (Evans *et al.*, in press). In areas subject to large annual snowpack accumulations, snowmelt events are a major cause of acid episodes (Laudon and Bishop, 2002; Laudon *et al.*, 2004). Projected decreases in snowfall will reduce the influence of snowmelt on run-off chemistry.

7.4.5.2   Sulphur and Climate

In wetland areas, anaerobic conditions in water-saturated soils lead to S storage, via reduction to organic S compounds and inorganic sulphides. Drought conditions lower the peat water table, allowing oxygen to enter the soil and the reoxidation of reduced S compounds to sulphate (Dillon *et al.*, 1997). This can generate extreme levels of acidity, which if flushed from the soil can cause major acid episodes in run-off, together with mobilisation of toxic metals (Tipping *et al.*, 2003). A long-term assessment of sulphate-driven acid episodes at Birkenes, Norway indicated that the severity of drought-induced acid episodes has decreased since the 1970s as the rate of S input has declined (Wright, 2007), but a modelling study for an Ontario wetland catchment by Aherne *et al.* (2006) suggested that repeated drought events, even at current levels of drought frequency, would be sufficient to severely retard recovery from acidification. Repeated droughts, together with reduced S deposition, will lead to the gradual depletion of peat S stores (Tipping *et al.*, 2003), but in more polluted regions this process may take many decades. Peat catchments containing large stores of anthropogenic S must therefore be considered highly sensitive to a projected increase in the frequency and severity of summer droughts, which could lead to the destabilisation of these stores, and consequently to an increased incidence of biologically damaging post-drought stream acidification events.

7.4.5.3   Nitrogen and Climate

N in terrestrial ecosystems is tightly cycled. Any climatic event which disrupts biological cycling is likely to result in nitrate leaching. In the UK this has been most clearly observed following soil freezing events, which typically occur when the winter NAO Index is negative (Davies *et al.*, 2005; Monteith *et al.*, 2000). Such events are likely to become less frequent under future climate change. Nitrate flushes also occur after droughts (Adamson *et al.*, 1998), and these events may increase in frequency. Extreme rain events may also transport nitrate directly to surface waters, where water bypasses biological sinks within the soil, e.g. as overland flow. Again, these events may be more common in future. Of greatest overall concern, however, is the long-term stability of the soil organic matter pool, as this contains most of the N accumulated over more than a century of elevated deposition. The Norwegian CLIMEX study, in which a small catchment was exposed to elevated temperature and CO_2, showed a marked increase in N mineralisation from the soil, which led to elevated nitrate leaching, effectively turning the catchment from an N sink to an N source (Wright 1998). Such a response to climate warming would clearly have grave consequences for the acidification and eutrophication of upland waters.

7.4.5.4   Dissolved Organic Carbon, Water Colour and Climate

DOC represents a large part of the carbon export of many upland catchments, and is the major source of water colour. It is a significant component of the upland carbon balance, contributes significant costs to water treatment, and impacts on aquatic ecosystems by altering light regime, nutrient transport, acidity, and metal transport and toxicity. Since the late 1980s, surface water DOC concentrations have approximately doubled across a large proportion of the UK (Evans *et al.*, 2005; Freeman *et al.*, 2001; Worrall *et al.*, 2004). The reasons for these increases have not been fully resolved, but there is growing evidence that a significant, and perhaps primary, driver has been the reduction in S deposition and subsequent recovery from acidification, which has increased the solubility of organic matter (Evans *et al.*, 2006a; Monteith *et al.*, 2007). As an integral part of the upland carbon cycle, however, there is little question that climate-related factors also impact on DOC export (Evans *et al.*, 2006a; Freeman *et al.*, 2001). Droughts also appear to have a strong effect on DOC release, generally decreasing DOC concentrations during the drought period itself, with increases observed thereafter (Clark *et al.*, 2005; Hughes *et al.*, 1997). Overall, it is possible that climatic changes have contributed to the DOC increases observed to date, and probable that they will contribute to further DOC changes in the future. However, it must be emphasised that other factors (S deposition, land management and possibly N deposition) are believed to have had as much, or more, influence on DOC trends during the last 30 years, and cannot be ignored in predicting future changes (Monteith *et al.*, 2007).

7.4.5.5 Modelling Effects of Future Climate Change

A number of studies have attempted to predict the impact of climate change on recovery from acidification. Wright *et al.* (2006) used the MAGIC (Model of Acid Groundwaters in Catchments) model to examine the sensitivity of future mean acidity to a range of projected climatic changes at 14 sites in Europe and North America. Sensitivity was highly variable both among different drivers and between sites, with climatic effects on organic acid leaching and N retention identified as the areas requiring the greatest focus (Figure 7.4.5). The results showed that several of the factors are of only minor importance (increase in partial pressure of CO_2 in soil air and run-off, for example), several are important at only a few sites (e.g. sea salts at near-coastal sites) and several are important at nearly all sites (increased concentrations of organic acids in soil solution and run-off, for example). In addition, changes in forest growth and decomposition of soil organic matter are important at forested sites and sites at risk of nitrogen saturation. The trials suggested that in future, modelling of recovery from acidification should take into account possible concurrent climate changes and focus especially on the climate-induced changes in organic acids and nitrogen retention.

Research on the effect of climate on recovery of freshwaters is of direct interest in formulating new policy goals with respect to emissions of S and N compounds. Here the major research challenge is still the link between the C and N cycles in terrestrial ecosystems and how N deposition and global change will affect these cycles in the future.

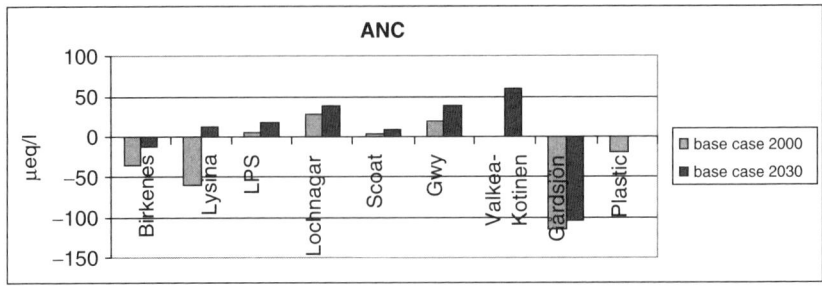

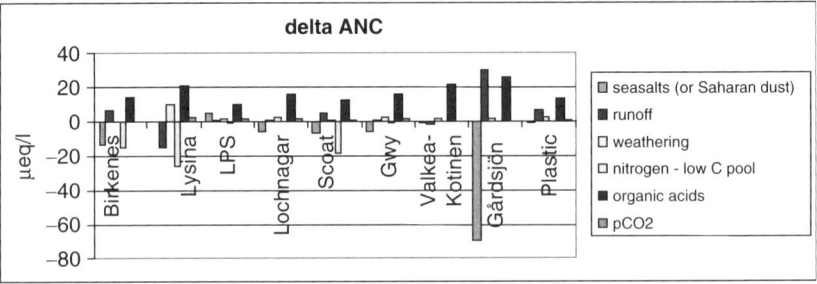

Figure 7.4.5   Top panel: ANC (volume-weighted annual mean) concentrations in run-off at nine sites (eight in Europe, one in Canada) for the calibration year (2000) and predicted (using the dynamic model MAGIC) for the year 2030 assuming no climate change (base scenario). Bottom panel: Change in ANC predicted for the year 2030 relative to the base scenario (Wright *et al.*, 2006)

7.4.6   INTERACTIONS BETWEEN CLIMATE CHANGE AND TOXIC SUBSTANCES

Although many of the most toxic substances introduced into the environment by human activity have been banned or restricted in use, many persist, especially in soils and sediments, and either remain in contact with food chains or can be remobilised and taken up by aquatic biota (Catalan *et al.*, 2004; Vives *et al.*, 2005a, 2005b). The high levels of metals (e.g. Hg, Pb) and persistent organic pollutants (PCBs, DDE, PBDEs) in the tissue of freshwater fish in arctic and alpine lakes (Grimalt *et al.*, 2001; Vives *et al.*, 2004a) attest to the mobility and transport of these substances in the atmosphere (Carrera *et al.*, 2002; Fernandez *et al.* 2003; van Drooge *et al.*, 2004) and their concentration in cold regions (Fernandez and Grimalt, 2003). For aquatic systems with long food chains, biomagnification can elevate concentrations in fish to lethal levels for human consumption. A major concern with respect to climate change for the Euro-limpacs project is the extent to which increased temperatures will cause changes in the accumulation of organic pollutants and metals in arctic and alpine freshwater systems (Fernandez *et al.*, 2005) and how these are transferred throughout the aquatic food web (Catalan *et al.*, 2004). With regard to changing precipitation patterns, a key issue is whether storm events and flooding might increase soil and sediment erosion and lead to the remobilisation of toxic substances from organic soils in upland catchments (Rose, 2005; Yang *et al.*, 2002), from glacierised alpine catchments and following reworking of river floodplain sediments. In the case of Hg, Euro-limpacs examines whether changing hydrology in boreal forest soils may lead to the enhanced production of MeHg (Munthe *et al.*, 2001).

7.4.6.1   Accumulation of Persistent Organic Compounds in High Mountain Lakes

In Europe, high mountain ecosystems are under the direct influence of the temperature-dependences of the accumulation of persistent organic pollutants (POPs) (Grimalt *et al.*, 2001). Lakes from these environments document the transfer mechanisms and impact of these compounds in the headwater regions of Europe's major river basins. Their accumulation patterns depend on many factors, including the time of their introduction into the environment (Gallego *et al.*, 2007). In a study of Pyrenean lakes, polybromodiphenyl ethers (PBDEs) in fish show higher concentrations at lower temperatures, as predicted in the global distillation model (Figure 7.4.6). Conversely, no temperature-dependent distribution was observed in vertical lake transects in the Tatra mountains (Central Europe, Figure 7.4.6), nor in fish from high mountain lakes

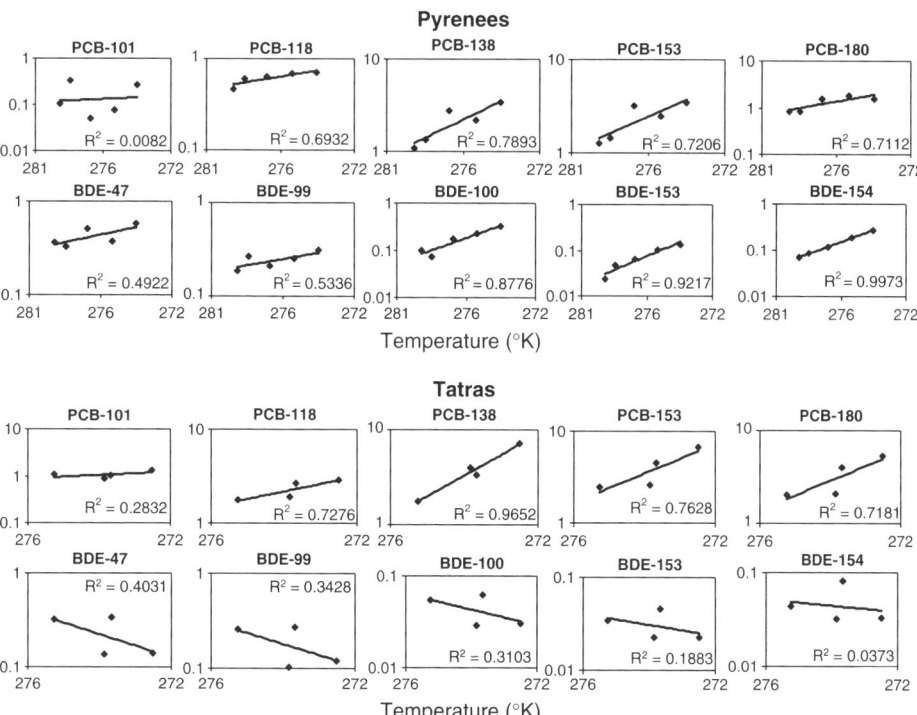

Figure 7.4.6   Lake-averaged muscle concentrations (n = 12–14) of selected PCB and PBDE congeners vs. reciprocal of annual mean air temperature

distributed throughout Europe (Vives *et al.*, 2004b). The fish concentrations of PCBs examined for comparison showed significant temperature correlations in all these studied lakes (Figure 7.4.6). Cold trapping of both PCBs and PBDEs concerned the less volatile congeners. These cases of distinct PBDE and PCB behaviour in high mountains probably reflect early stages in the environmental distribution of the former, since they have been under secondary redistribution processes over much shorter times than the latter (Gallego *et al.*, 2007).

The accumulation of these compounds, including PBDEs, is also subject to bioconcentration effects by organisms in high mountain aquatic systems. In a recent study of Tricoptera and Diptera it was observed that there is a non-selective enrichment of OCs and PBDEs from larvae to pupae (Bartrons *et al.*, 2007). These concentration increases may result from the weight loss of pupae during metamorphosis, as a consequence of mainly protein carbon respiration and lack of feeding. The concentration increases from larvae to pupae are very relevant for the pollutant ingestion of the higher predators. The intake of OCs and PBDEs by trout are between two and five-fold higher per calorie gained when predating on the latter. Since pollutant concentration, energy reward, predation susceptibility and duration of life stage are very different between these two insect stages, and none of them is irrelevant for the incorporation of OCs or PBDEs to higher levels, bioaccumulation food web models need to distinguish between the two sources (Bartrons *et al.*, 2007).

7.4.6.2   Climate Change and the Remobilisation of Metals from Polluted Soils

Although the emission of many toxic substances such as DDT, PCBs and trace metals has been controlled, soils and sediments throughout Europe remain contaminated as a result of decades of toxic substance accumulation. Remobilisation of such compounds occurs not only because of temperature-dependent distillation and condensation processes, but also as a result of soil processes through leaching and erosion. Any change in climate, such as an increase in storminess causing increased catchment soil erosion, or increased flooding causing disturbance of floodplain sediments, is likely to remobilise toxic substances, leading to changes in uptake and accumulation of these substances in freshwater food chains.

In Euro-limpacs we are examining these processes in several settings, including the remobilisation of metals from upland organic soils in Scotland and the impact of flooding on the remobilisation of metals from floodplain sediments of the River Elbe. The Scottish study builds on the work of Yang *et al.* (2002), who showed that Hg and Pb concentrations in the recent sediments of Lochnagar, a remote Scottish mountain lake, had not declined over recent decades despite a strong reduction in the emission of these toxic metals to the atmosphere. They hypothesised that the lake was continuing to receive toxic metals due to the erosion of catchment soils that were already contaminated by historic pollution deposition. In Euro-limpacs this hypothesis is being tested by comparing the sediment records of Scottish lakes with and without eroding catchment soils, in areas with relatively high and low pollutant deposition and with relatively high and low rainfall. The concern is that any increase in winter storminess in the future as a result of climate change might accelerate the transfer of pollutants from catchment soils to lakes and cause the already high levels of metal contamination in the fish tissue of remote lakes (Rose *et al.*, 2005b) to remain high or even to increase.

7.4.6.3   Climate Change and Mercury Mobilisation

In the boreal forest zone of Northern Europe there is a concern that climate change may lead to an increase in the concentrations of methyl mercury in fish. Already thousands of lakes in Scandinavia have mercury levels in fish that exceed health guidelines of 0.5 mg/kg making them unsuitable for human consumption. In a region of Europe where climate models predict an increase in winter precipitation an increase in groundwater levels will cause more water to flow through organic-rich soil horizons where a large fraction of the soil-bound mercury is present, potentially causing direct mobilisation of mercury and methyl-mercury. Changing redox conditions and release of DOC and nutrients may enhance this process and cause on increase in methyl mercury on aquatic ecosystems (Munthe *et al.*, 2001). In Euro-limpacs this process is being tested by manipulating precipitation and hydrology at Gårdsjön, an experimental lake site in south-west Sweden. Provisional results of the experiment show that increased soil wetness does indeed lead to anaerobic conditions, sulphate reduction and the generation of significant amounts of methyl-mercury and total mercury in runoff.

7.4.7   DEVELOPING INDICATORS OF CLIMATE CHANGE FOR FRESHWATER ECOSYSTEMS

Monitoring programmes of freshwater ecosystems in Europe have traditionally focused on water quality, either by direct chemical measurement or by using biological indicators (De Pauw and Hawkes, 1993; Knoben *et al.*, 1995). As water quality improved, additional indicator systems were developed to monitor habitat quality e.g. Raven *et al.* (1997). The Water Framework Directive (WFD) now requires an integrated approach to freshwater monitoring in Europe, using a range of different biological groups, including phytoplankton, benthic diatoms, aquatic macrophytes, macroinvertebrates and fish, either singly or together, to assess the ecological status of rivers, lakes, coastal and transitional waters. According to the WFD, the ecological status of a water body is defined by comparing the present-day biological community composition with near-natural reference conditions. These WFD guidelines on ecological water quality assessment have generated an urgent need to develop new or revised indicator systems.

Most new indicator systems are based on metrics, i.e. attributes of the biotic assemblages, which reflect community composition and abundance, taxon richness, the proportion of sensitive and tolerant taxa or functional attributes (Hering *et al.*, 2006). They are designed to assess overall ecological stress, e.g. deviation from reference conditions, or to assess single stressors, e.g. eutrophication (Kelly *et al.*, 1995), organic pollution (Zelinka and Marvan, 1961), hydromorphological degradation (Statzner *et al.*, 2001) or acidification (Townsend *et al.*, 1983).

However, the indicator systems defined for the WFD have shortcomings, limiting the applicability, reliability and interpretation of the assessment results. Problems include: i) important ecosystem types, such as wetlands which are not part of water bodies and small streams particularly at risk from climate change, are not included in the WFD; ii) reference conditions are poorly defined or have been derived with different methodologies; iii) scores used in the metric systems are rarely based on experimental evidence or on a thorough examination of the literature; iv) many assessment systems are not capable of discriminating between stressor types; and v) the system does not allow for new types of stressor. This is particularly the case for climate change. Although climate change interacts with many 'traditional' stressor types, such as eutrophication, it adds additional stress, e.g. through increases in water temperature or changes in stream discharge.

In addressing these problems, the Euro-limpacs project aims to assess how current chemical, biological and functional indicator systems need to be modified to be suitable for climate change monitoring. To do this it is necessary to:

- Generate hypotheses on how climate change will directly and indirectly affect freshwater ecosystems and their communities.

- Build databases that contain reliable ecological background information on species or higher taxa inhabiting different types of European freshwater ecosystem. Information on traits which demonstrate sensitivity to direct and indirect effects of climate change (e.g. temperature preferences, resilience to droughts) is especially important.

- Demonstrate that species or metrics changes with direct or indirect effects of climate change can be incorporated into standard assessment systems.

To derive hypotheses on the effects of climate change on freshwater ecosystems, around 1000 papers have been evaluated. From these, 'cause–effect chains' have been generated to categorise the effects for which indicators are needed. The literature evaluation has revealed a strong bias towards abiotic components of the different ecosystem types. For example, of the 317 papers dealing with effects on lakes, 179 are related to abiotic effects, such as water temperature or water column stratification, 90 papers deal with plankton communities, but less than 50 are concerned with other biota. Of the 713 papers dealing with the effects of climate change on rivers, 616 are related to abiotic effects, mainly on hydrology (397 papers). Only 97 are related to biotic communities and processes.

In Euro-limpacs, data on ecological characteristics and distribution patterns for six organism groups widely used in freshwater assessment, diatoms, fish and four groups of benthic invertebrates (Ephemeroptera, Plecoptera, Trichoptera and Chironomidae) have been compiled. For the invertebrates, we aimed to cover the complete literature, including 'grey' literature such as diploma and PhD theses. In total, more than 8000 literature references were consulted. The resulting data are stored in an online database (http://www.freshwaterecology.info) and include data on more than 22 000 taxa. Data on individual attributes differ in their degree of completeness. For example, for European caddisflies, completeness of the data for the individual attributes ranges from 0.2 % to 99.7 %. Besides the attribute 'distribution in ecoregions' (99.7 % of all taxa classified), high proportions of taxa were also classified for the attributes 'current preference' (84.2 %), 'stream zonation preference' (71.1 %) and 'substrate/microhabitat preference' (66.9 %) (Graf *et al.*, 2006b).

The sensitivity of European caddisfly species to the impacts of climate change differs greatly between ecoregions. One criterion, which leads to a high vulnerability, is 'restricted distribution'. Endemic taxa are often characterised by a restricted ecological niche and limited dispersal capacity. These taxa are more severely threatened by climate change than widely distributed species, as shown for vascular plants (Malcolm *et al.*, 2006) and as suggested for benthic invertebrates (Brown *et al.*, 2007). The number of endemic caddisfly species and subspecies is highest on the Iberian Peninsula (141 taxa), followed by Italy (118 taxa) and the Hellenic Western Balkan (76 taxa), while in 14 out of the 23 European ecoregions less than five endemic taxa occur (Hering *et al.*, submitted).

Other ecological characteristics leading to a high sensitivity of species to climate change include the preference for springs, preference for cold water temperatures, short flight periods and restricted ecological niches. On the community level, the following attributes are expected to be affected by changing climate: the proportion of spring-preferring taxa (or the longitudinal preferences of the taxa in general), the proportion of cold-stenothermic taxa and the proportion of taxa with a long life cycle. By including such metrics in assessment systems the impact of climate change can be better represented in overall appraisals of ecological quality (Figure 7.4.7).

7.4.8 IMPLICATIONS OF CLIMATE CHANGE FOR AQUATIC ECOSYSTEM RESTORATION

Attempts to restore aquatic ecosystems in Europe are now subject to the requirements of the European Water Framework Directive, which requires freshwaters to be maintained in or returned to good ecological status, a status that is defined as 'slightly' different from the pristine or reference state. Whilst the concept is simple, defining the reference state is or can be problematic, and as climate change increasingly influences all aspects of the structure and functioning of ecosystems, use of the reference state as a target for restoration may become increasingly difficult. In the Euro-limpacs project, therefore, we focus on how climate might affect or confound the use of reference conditions in ecological assessment. This is done first by recognising that a reference condition is not a static state, but changes over time, albeit within the confines of its ecological envelope, defined as: i) the physical conditions, such as geographic location, catchment geology and hydrology, that determine the physical constraints of the site; ii) the biological potential of the site, constrained by those populations whose distributions overlap at the site; and iii) the interactions between populations that result in complex networks and feedback loops (e.g. O'Neill, 2001). Thus the dynamic reference

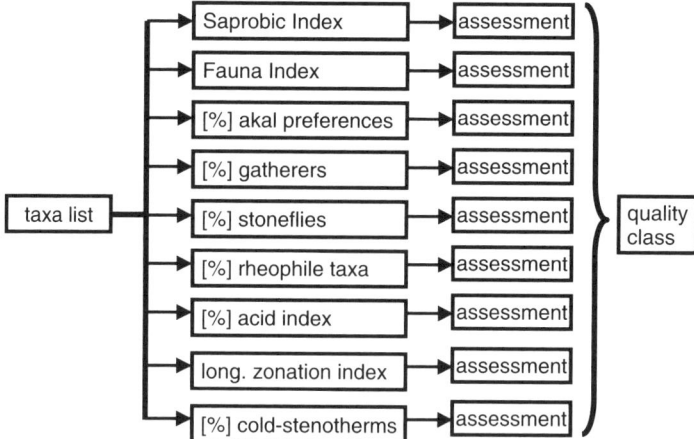

Figure 7.4.7   Scheme of integrating climate change effects into a multimetric assessment system for river invertebrates. Nine metrics, which are individually assessed by comparison to reference conditions, indicate four different stressor types: organic pollution (saprobic index), hydromorphological degradation (fauna index, [%] alkal preferences, [%] gatherers, [%] stoneflies, [%] rheophile taxa), acidification (acid index) and temperature (longitudinal zonation index, [%] cold stenotherms)

condition concept can be used to modify restoration targets in systems where changing climate may shift baseline conditions beyond those to which it can reasonably be expected to be restored. A range of techniques are available for establishing reference conditions, including temporally-based approaches using historical data or palaeoreconstruction, spatially-based approaches using survey data, modelling approaches such as hindcasting and expert judgement. Euro-limpacs uses these approaches to identify reference conditions and quantify the errors or levels of uncertainty associated with using them.

7.4.8.1   Detecting Recovery: Confounding Effects of Climate?

The reference state is dynamic, varying both chemically and biologically on seasonal, inter-annual and decadal timescales in response to natural fluctuations in climate, internal ecological processes and random events. Human modification of river and lake catchments through land use change and hydromorphological alteration may also alter the reference state and now there is concern that ecosystems thought to be in equilibrium with their surroundings (see O'Neill 2001), may already be undergoing unprecedented changes due to global warming (see Catalan *et al.*, 2002; Smol *et al.*, 2005). As warming continues to increase in future to temperatures not seen before in the Holocene, new boundary conditions for ecosystem processes will be created. Reference conditions are then not only dynamic but also undergoing directional change in which the baseline is continually changing (Battarbee *et al.*, 2005; Figure 7.4.8). It consequently prompts work into better understanding the drivers behind long-term

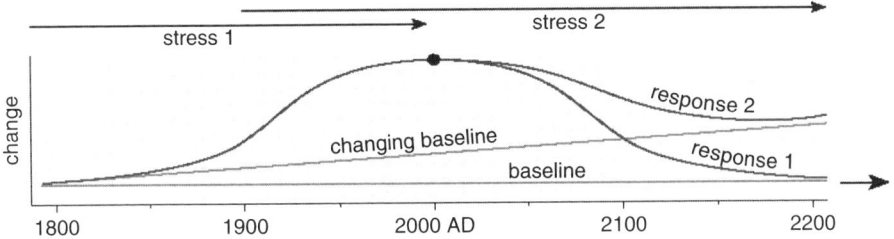

Figure 7.4.8 Conceptual diagram illustrating ecosystem response to increasing and decreasing stresses in relation to climate change, expressed as a changing baseline (from Battarbee *et al.*, 2005)

patterns and processes, with particular focus on the analysis of long time series from observational records, the use of sediment records from lakes (e.g. Bennion and Battarbee, 2007) and the understanding of processes and functions at extant analogue systems with good ecological status.

In addition to understanding the reference state and the longer-term changes in baselines that help to define restoration targets, studies designed to assess the success of restoration also need at the outset a consideration of factors, such as the choice of indicator and habitat, that maximise the chance of detecting responses to mitigation if and when it occurs. Moreover, because the success or failure of restoration is often judged as a deviation from an expected condition (target state), the ideal design should also include pre- and post-restoration monitoring (Downes *et al.*, 2002). In Euro-limpacs we address these issues, attempting to show how climate change might affect restoration strategies for different system types. Here we present three examples, on lake acidification, river hydromorphology and landscape connectivity.

7.4.8.2 Lake Acidification Restoration and Climate Change

As a result of a major reduction in the emission of acidifying S and N compounds since the 1970s in Europe (see Section 7.4.5), acid deposition on boreal lakes and their catchments has also strongly declined. Recently, Stendera and Johnson (2008), assessing the recovery of boreal lake ecosystems from acidification, used different indicators (water chemistry, biota), different trophic levels (primary producers and consumers) and different habitats within the lakes (pelagic, benthic) to compare the response of both acidified and reference lakes to the reduction in acid deposition.

Several of the indicators showed positive trends, whereas others showed surprisingly negative trends during the 16 year study (Figure 7.4.9). Although decreasing acidity and the changes in phytoplankton and littoral invertebrate assemblages indicated that there are signs of recovery from acidification, other changes unrelated to changes in acid deposition occurred. For example, unexpected changes in the reference lakes occurred in phytoplankton and littoral invertebrates, probably caused by climatically-related changes in water colour and temperature, and decreases in the richness of sublittoral and profundal invertebrate assemblages may have been the result of climate-related change acting on habitat quality, such as ambient oxygen

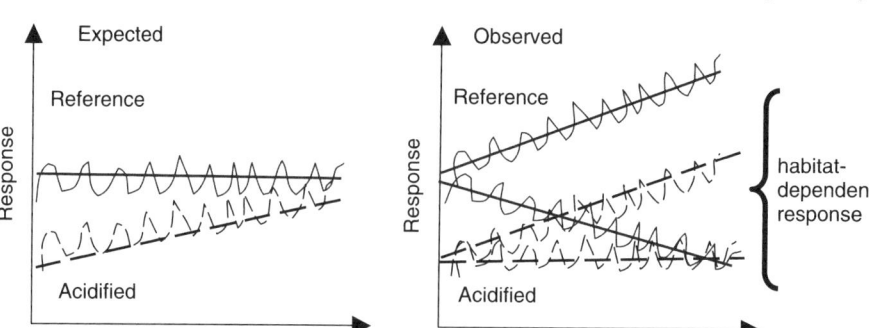

Figure 7.4.9 Conceptual diagram illustrating a comparison of expected and observed responses of macroinvertebrates (expressed as taxon richness) in acidified and reference lakes in southern Sweden to a reduction in sulphur deposition. (Reproduced by permission of Wiley-Blackwell)

concentrations and temperature, rather than changes in lake acidity. Taken together, the results of the study indicated that recovery is a complex process influenced by multiple spatial and temporal factors. It demonstrates the importance of including multiple organism groups and trophic levels in studies designed to detect and understand the processes and changes associated with recovery, especially in situations where there are complex interactions between pollutant behaviour and climate.

7.4.8.3   Stream Habitat Restoration and Climate Change

Concern for the loss of stream and floodplain habitats and biodiversity over recent decades has provided the impetus for extensive programmes of stream rehabilitation and restoration in many European countries. Many techniques are being used, including reforestation of the floodplain, remeandering, and removal of dams and bank supports. New, more innovative approaches include the adding of coarse woody debris (Gerhard and Reich, 2000; Gippel and Stewardson, 1996), the removal of sediment deposits in floodplains and various methods to combat gullying.

In order to make the proper choices in stream restoration, the complex spatial and temporal interactions between physical parameters, habitat diversity and biodiversity have to be understood, and consideration needs to be given to how the success of restoration schemes might be influenced by future climate-related changes in hydromorphology. When a stream has been restored, the success (increase in biodiversity) depends on the recolonisation of the original (indicator) species. Whether these species will be able to recolonise the restored stream depends on the distance to remaining populations, dispersal barriers in between the remaining populations and the restored stream, and the dispersal ability of the species. Establishment of an invasive or non-native species may also hinder recolonisation, and biodiversity may in general be threatened by invasive species replacing the native ones. Climate change affects all these processes and influences hydromorphology at all scales. Under changing climate

Table 7.4.2 Hydromorphology characteristics of the River Lahne in a single and multiple channel situation

	Single Channel	Multiple Channel	Factor
shore length (m)	432	1408	+3.3
CV current velocity	0.62	1.1	+1.8
CV depth	0.59	0.99	+1.7
wet area (m^2)	76	124	+1.6

conditions either the measures adopted to meet restoration targets in such schemes need to be adapted, or the targets themselves must be adjusted.

In Euro-limpacs the development of hydromorphology and macroinvertebrates was studied in a multiple channel restoration of the River Lahne. Table 7.4.2 shows that the shore length, current velocity variation, depth variation and wet area increased 1.6–3.3 times. However, the analysis of the macroinvertebrates of both the single and multiple channels showed no change at all in composition and abundances, illustrating the need for studies on longer timescales and the need for restoration on a catchment rather than reach scale.

7.4.8.4 Landscape Scale Restoration and Climate Change

Restoration endeavours in the past have often focused on individual sites (e.g. lakes) or stream reaches, largely ignoring the importance of connectivity such as channel movement for terrestrial–aquatic linkages (e.g. river–floodplain exchanges) and biotic dispersal, and ignoring the fact that connectivity may be increasingly threatened in future by the effects of climate change. Awareness that large-scale processes determine the structure and function of aquatic ecosystems and that these properties and principles need to be considered in restoration prompted Verhoeven *et al.* (submitted) within the Euro-limpacs project to propose the Operation Landscape Unit (OLU) concept. The OLU is defined as 'combinations of landscape patches with their hydrological and biotic connections' and is used to identify a parsimonious set of landscape elements and their configuration in order to understand better the constraints (e.g. fragmented landscapes) which may impede the effectiveness of restoration efforts. This implies that regional conservation strategies need to be spatially coherent, based on the relevant physico-chemical and biological processes, to ensure that restoration measures are successful and sustained.

Future climate-related impacts are expected to profoundly affect hydrological cycles. Practical management of lakes and streams needs to take into account catchment scale issues, and large-scale, landscape-level planning is needed if restoration projects are to be cost-effective and ecologically successful. Restoration targets need to be chosen carefully, taking into consideration the difficulty of defining a fixed reference state. High quality pre- and post-restoration monitoring is needed to understand how climate may confound recovery trajectories, so that lessons from past experiences can be used to improve future plans.

7.4.9   MODELLING THE IMPACTS OF CLIMATE CHANGE ON FRESHWATER ECOSYSTEMS AT THE CATCHMENT SCALE

The need for a catchment-scale approach to freshwater ecosystem management is recognised by the EU Water Framework Directive, where the basic unit of management is referred to as the 'river basin district' (European Commission, 2000). The complexity of the interactions between aquatic and terrestrial systems at the catchment scale necessitates a modelling approach also at the catchment scale. With respect to climate change, existing or new models need development to represent climate, soil, land use, lakes, rivers and coastal waters, so that the responses of whole catchment systems can be simulated and the models used to assess the impacts of alternative catchment management decisions.

The principal catchments being modelled in Euro-limpacs are shown in Table 7.4.3. The areas of these catchments vary over seven orders of magnitude, from the 5200 m^2 Gårdsjön experimental catchment to the Garonne-Adour catchments, which occupy the whole of south-western France. A wide variety of climates and ecoregions are also represented.

The most widely used models in the Euro-limpacs project are the INCA suite of models. The first INCA model (INCA-N) was aimed at understanding the response of rivers to changes in nitrogen inputs and catchment nitrogen metabolism (Wade *et al.*, 2002; Whitehead *et al.*, 1998). Since then the INCA framework has been developed to encompass phosphorus (INCA-P), particulates (INCA-SED), dissolved organic carbon (INCA-C) and mercury (INCA-Hg). Other dynamic models being used in Euro-limpacs include the Mike11-TRANS model. This model is a suite of ecological sub-models linked to a hydrodynamic model, and has been used in conjunction with the rainfall-run-off model NAM to model N fluxes in lowland Danish catchments (Anderson *et al.*, 2006). Another development of existing models is the use of a landscape-based mixing model (PEARLS), coupled with the acidification model MAGIC, to simulate the recovery from acidification of the Conwy in North Wales, a large heterogeneous river basin (Evans *et al.*, 2006b).

7.4.9.1   Example Model Application

Here (Figure 7.4.10) we provide an example of the use of INCA-N (Whitehead *et al.*, 2006). It is the first attempt to model adaptation strategies to mitigate the effects of climate change on nitrate concentrations in rivers. The example is from the River Kennet in southern England for the period 1961–2100, with precipitation scenarios downscaled from three general circulation models. These generate different effects in detail, but all show a general increase in nitrate concentration due to enhanced microbial activity. Figure 7.4.10 shows the results of attempting various mitigation strategies on this overall picture using one precipitation scenario (from the HadCM3 model). The baseline scenario (no mitigation attempted) shows a steady increase in nitrate. The peak towards the end of the period is due to a simulated drought. 'Deposition'

Table 7.4.3 Principal areas used in the Euro-limpacs project for catchment-scale modelling. Acid.: acidification; C: carbon; N sat.: nitrogen saturation; Sed: sediment. Updated from Wade (2006)

Country	Study Area	Area km^2	River	Lake	Wetland	Key Issues	Climate	Ecoregion
Austria	Piburger See	2	✓	✓		Eutrophication	Cool Continental	Mountainous
Denmark	Gjern	110	✓	✓		Eutrophication, Sed.	Maritime	Atlantic
	Odense	486	✓			Eutrophication, Sed.	Maritime	Atlantic
Greece	Cheimadidita	35		✓	✓	Eutrophication	Mediterranean/Cool Continental	Mediterranean
Finland	Savijoki	15.4	✓		✓	Eutrophication, Sed.	Sub-arctic	Boreal
	Simojoki	3160	✓	✓	✓	Acid., N sat.	Sub-arctic	Boreal
	Tueronjoki	439	✓	✓	✓	N sat, Eutrophication	Sub-arctic	Boreal
France	Garonne-Adour	56 500	✓	✓	✓	Eutrophication	Maritime	Atlantic
Norway	Bjerkreim	685	✓	✓	✓	Acid., N sat., C	Maritime	Boreal
	Tovdalselva	1855	✓	✓	✓	N sat, Eutrophication	Maritime	Boreal
Romania	Lower Danube Wetland	210			✓	Eutrophication	Cool Continental	Continental
Spain	La Tordera	124	✓		✓	Eutrophication	Mediterranean	Mediterranean
Sweden	Gårdsjön	0.005	✓	✓		Acid., N sat., C	Maritime	Boreal
	Svartberget	0.5	✓	✓		Acid., Hg	Sub-arctic	Boreal
UK	Conwy	590	✓		✓	Acid., Eutrophication, C	Maritime	Atlantic
	Kennet	1030	✓		✓	Eutrophication, Sed.	Maritime	Atlantic
	Lambourn	263	✓		✓	Eutrophication	Maritime	Atlantic
	Endrick/Falloch	781	✓	✓		Eutrophication	Maritime	Atlantic
	Tamar	917	✓		✓	Eutrophication	Maritime	Atlantic
	Wye	4140	✓	✓		Acid., Eutrophication, C	Maritime	Atlantic

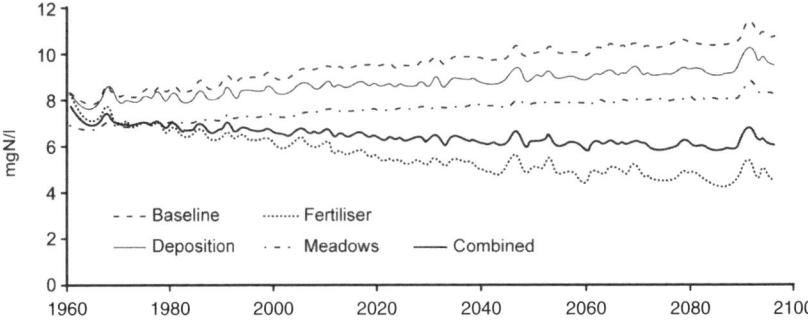

Figure 7.4.10   Modelled nitrate concentrations in the lower River Kennet, UK, 1960–2100, given various management treatments (see text) and downscaled climate scenarios from the HadCM3 GCM, medium-high emissions scenario. (After Whitehead *et al.* (2006), with permission from Elsevier)

represents a reduction of reactive nitrogen deposition by 50 %. 'Meadows' involves the construction of water meadows adjacent to the river, which are allowed to flood and remove nitrogen by denitrification. An area four times the river surface area is assumed, and this almost stabilises the nitrate concentration. 'Fertiliser' (reducing N fertiliser application by 50 %) is the most effective intervention, leading to a decrease in nitrate concentration, but this reduces agricultural intensity in the catchment to that of the 1950s, which seems unlikely. Finally a 'combined' strategy is modelled, in which each of the single strategies is applied at half intensity. This is not quite as effective as a 50 % reduction in fertiliser, but still leads to a decrease in concentration. Further work, using a version of INCA-N modified to account for the transport of nitrate through the unsaturated zone of the underlying chalk rock, predicts that reducing fertiliser inputs today will have a short-term impact on in-stream nitrate concentrations, but a clear long-term reduction will not occur until between 2060 and 2080. This is because of the large mass of nitrate that has accumulated in the chalk aquifer (Jackson *et al.*, 2007). Thus, some kind of in-stream intervention by construction of water meadows may be the best option to reduce in-stream nitrate concentrations within the timescale of the Water Framework Directive. This is an interesting result and suggests that in catchments with a long residence time, pollutants in the groundwater may continue to confound attempts to improve the chemical and ecological status between now and 2015.

7.4.9.2   Model Chains and Integration

Modelling catchment responses often requires the integration of separate models covering components of the catchment, such as soils, vegetation, groundwater, rivers, lakes, etc., and also the integration of models which predict the response of driving variables such as precipitation or temperature. This integration can occur in two ways: chains of models can be produced in which the output of one model is used as input to the next, following the pathways of water through the catchment; or the component models can be integrated so that data are passed from one model to another inside the larger model. The latter is more convenient for the modeller, especially when performing large numbers of runs, but much more difficult to achieve given that the component models have normally not been designed with this integration in mind. A number of chaining and integration projects are under way in Euro-limpacs as part of model toolkit development. One completed example is the chaining of various models to predict the response of the Bjerkreim river and fjord system in southern Norway to climate change (Kaste *et al.*, 2006). The models used were: two GCMs (ECHAM4 and HadAM3H), to predict the effects of climate change on meteorology on a large spatial scale; a regional climate model (HIRHAM), to downscale these predictions to the catchment scale and to a daily temporal scale; the hydrological model HBV, to translate the meteorological variables into water fluxes through the catchment and water status in catchment components; the water quality models MAGIC and INCA-N, which use this information to predict nitrogen fluxes and concentrations in catchment components; and the NIVA FJORD model, to use the predicted N discharge of the Bjerkreim River to predict annual variation of nitrate in the Egersund fjord, which receives the run-off. The results driven by the HadAM3H model indicated the possibility of increased productivity and eutrophication in the fjord by 2080, whereas those driven by the ECHAM4 model did not. Irrespective of whether either of these predictions is correct, the model chaining exercise allows the exploration of the possible consequences of climate change on the internal dynamics of the catchment system, such as the seasonal patterns of water flow, snowmelt changes, acidification, and the possible ecological consequences.

7.4.9.3 Model Uncertainty

All model predictions are uncertain to some degree, but quantifying and preferably reducing this uncertainty is one of the priorities for Euro-limpacs. The sources of this uncertainty are many (e.g. Morgan and Henrion, 1990). It might be thought that since catchment hydrochemical models can include more than 100 parameters, some of which are very imperfectly known, the overall uncertainty will be so large as to render them essentially useless, but this is not necessarily the case. The calculation of critical loads, deposition thresholds used in pollution control policy, involves models which are combinations of 10–20 uncertain parameters, but uncertainty propagation studies show that the uncertainty in the calculated critical loads is typically less than the uncertainty in *any* of the input parameters (e.g. Skeffington *et al.*, 2007). A number of techniques for estimating uncertainty in environmental models are currently being applied by Euro-limpacs participants. Partners who have predicted the future nitrogen status of their catchments using INCA-N are using a Monte Carlo-based tool developed within the project, which performs a general sensitivity analysis (e.g. Wade *et al.*, 2001) to identify key parameters and puts confidence limits on model predictions. At the same time, less formal approaches are being used, such as applying different models to the same problem and comparing the results. By the end of the project it is hoped we will be in a better position to assess the uncertainty in predictions of the effects of climate change on freshwaters, and to identify the steps necessary to reduce it.

7.4.10   IMPLICATIONS OF PREDICTED CLIMATE CHANGE IMPACTS ON FRESHWATER ECOSYSTEMS FOR POLICY AND MANAGEMENT

The general lessons that can be drawn from our improved understanding of how ecosystems may respond to climate change in the future have important consequences for their management and for the development of related environmental policy. The uncertainties inherent in this process need to be built into the policy and management process. These uncertainties do not just derive from climate and catchment models themselves (as above), but also relate to the uncertainty in predicting future patterns in the emissions of greenhouse gases, land use change and pollutant loading, and the response of society to such changes.

In Euro-limpacs we are considering these issues at the global, regional and local (catchment) scales, and engaging with a wide range of stakeholders to provide advice and to develop a decision support system (DSS) of practical value to the user community.

7.4.10.1   Socioeconomic Pressures and Global Change

In assessing the potential impact of climate change on freshwater ecosystems in the future, we need to take into consideration how current policies, protocols and socioeconomic pressures might influence other drivers of change on freshwater ecosystems, especially how scenarios for climate change itself might alter future patterns of pollutant loading and land use patterns. With the signing of the Kyoto protocol, climate change has become the most important driving force for the reduction of pollutant emissions to the atmosphere and will lead to significant reductions in emission and deposition patterns of both acids and nutrients over and above those necessary to reduce current problems of acidification. Likewise, under climate change the economic costs and returns of many agricultural practices will change. However, subsidies, taxation and tradable permits can also influence land management decisions within the agricultural sector. In Euro-limpacs we are developing an economic model that integrates the impact of global change on crop yield, growing costs (e.g. fertilisers) and the impacts of economic incentives. Information from this model will be used to assess the likely impact of changes in agriculture under different global change scenarios on surface waters and marginal wetlands. An integral component of this process is the contingent valuation of the impacts of climate change, land management and atmospheric deposition on lakes, rivers and marginal wetlands, especially the need to establish non-market values of habitat change brought about by the impacts of global change (Birol *et al.*, 2005).

7.4.10.2   Analysis of Policies Influencing Catchment Management

Policies for the protection of freshwater ecosystems are implemented at the catchment scale and catchment managers need to be aware not only of policies designed specifically for water bodies but also of those that influence other processes operating in catchments. This is often complicated by the fact that different responsibilities are vested in different administrative bodies which operate at different scales, including European, national and local scales, and that administrative boundaries are often not coincident with catchment boundaries at these different scales. In Euro-limpacs we are drawing up an inventory of policies and international agreements affecting catchment management, systematically assessing the agendas and resources of these authorities and examining their congruence with catchment management issues, highlighting especially those situations where policies in other sectors (notably agriculture but also transport) may be in conflict with catchment management. Differences between member states within the EU are of special concern and the DSS will need to be designed to apply in different national settings. To this end, workshops with national experts are being used for assessment purposes, taking specific catchments included in Euro-limpacs as a focus.

7.4.10.3   Stakeholder Engagement and the Decision Support System

A central element of Euro-limpacs is the involvement and integration of stakeholders to ensure that the scientific results of the project are of the most practical value to the user community. The user community includes national governments, the European Environment Agency, the European Commission, and especially government agencies charged with the responsibility of implementing policy at the catchment scale.

The DSS itself (Figure 7.4.11) aims to provide a framework for the integration and analysis of data from disparate sources related to a wide range of issues into a single holistic and GIS-based analysis of catchment management in the context of climate change (Maltby *et al.*, 2008). It also aims to bridge the gap between science and decision making, to achieve integrated catchment management and provide a mechanism by which the range of services provided by ecosystems can be evaluated in a holistic assessment, using multi-criteria analysis (MCA).

The DSS incorporates outputs from across the Euro-limpacs project to inform the management decisions made using the system. As the DSS is intended for application right across Europe, the expert knowledge and scientific results from the project that have a generic applicability are embedded within the DSS software, in the form of models, databases or documents. However, many of the outputs from the project are detailed scientific studies that are being carried out at a site or catchment scale and cannot often be generalised or extrapolated beyond the specific conditions of the study. This makes them unsuitable for being embedded within a DSS that is intended to be generic. However, these results are important outputs from the project, and the DSS framework therefore allows for the integration of site-specific data into applications of the DSS.

The resulting framework addresses specific and targeted management questions such as: i) will climate change affect some parts of a catchment more than others?; ii) what measures should be taken to mitigate the effects of climate change?; iii) which part of a catchment should resources be targeted towards?; and iv) which measures most effectively tackle the defined problem?

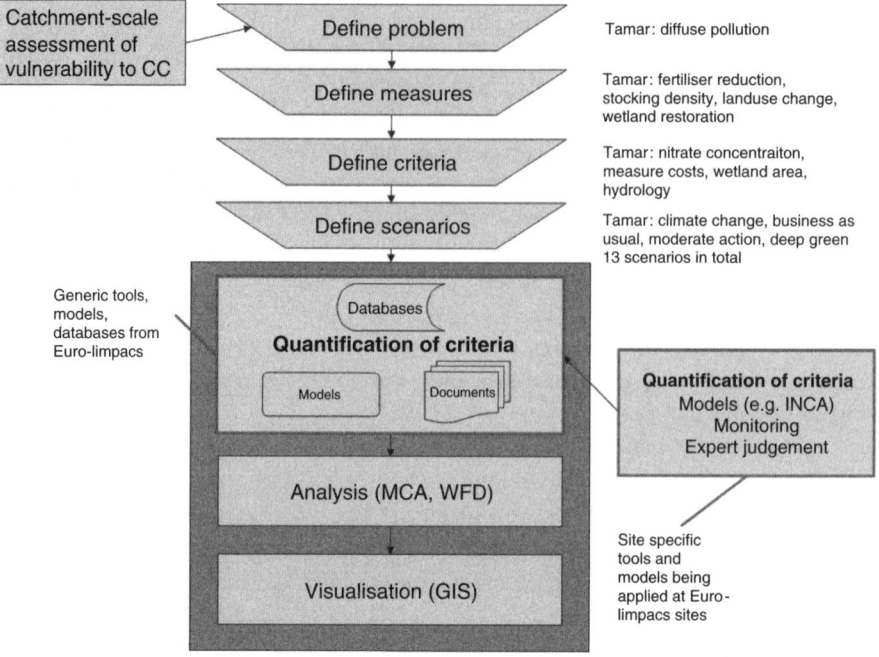

Figure 7.4.11   Conceptual structure of the Euro-limpacs decision support system

The DSS is being tested at seven catchments across Europe, which form a core, common set of catchments for the Euro-limpacs project (see also Table 7.4.3). These are:

- Tamar (UK)

- Danube sub-basin (Romania)

- Bjerkreim/Tovdal/Vannsjo-Hobol (Norway)

- Odense (Denmark)

- Inn (Austria)

- Tordera (Spain)

- Vecht (Netherlands)

- Cheimaditida (Greece).

These seven case studies will, firstly, act as demonstrations of how the DSS can be applied to different types of issue and management problem and, secondly, integrate the site-specific data being generated by the Euro-limpacs project in these catchments, within the application of the DSS. The case studies include the Tamar catchment (UK), which is being developed as the pathfinder case study. In the context of the application at the Euro-limpacs case study sites, much of the site-specific data comes from Euro-limpacs itself, but in the application of the DSS by external users, once it is completed and distributed it is anticipated that these additional site-specific data will come from the users themselves and will reflect the particular issues of concern in a user's catchment and the availability of data from studies and models specific to that catchment.

The DSS is implemented as an extension of the ArcGIS software, which facilitates the integration of the DSS framework with existing GIS information available to water managers. The geographical area under consideration is divided into spatial units for comparison. Spatial units can be individual water bodies (in the sense used in the Water Framework Directive), water body types (rivers, lakes wetlands) or other spatially delineated areas, such as sub-catchments, depending on the particular problem being addressed. These spatial units are represented in the DSS as shapes or linear features within a GIS layer. In the Tamar example total MCA scores are given, which reflect the status of the sub-catchments in 2050 for all climate and management scenarios. The results demonstrate how the implementation of different management scenarios affects the status of different sub-catchments under climate change, thereby enabling management measures to be targeted towards those parts of the catchment where intervention would be most cost-effective.

The prototype application of the DSS to the Tamar catchment has demonstrated the functionality of this approach and the contribution it can make to strategic planning in catchment management. Its flexibility and potential for integrating modelled and empirical data from across the environmental, social and economic dimensions of sustainability are key attributes of the DSS. By providing a mechanism for integrating data from disparate sources, it acts as a complementary tool to detailed, process-driven models that, because of their complexity and data requirements, can only focus on a single or small number of variables. In doing so it allows the outputs from models and empirical studies to be used in an holistic analysis of catchment management issues that takes into account the complex interactions between climate change and management actions and their effects on the environment, society and the economy.

7.4.11 CONCLUSIONS

Current mean temperatures in the northern hemisphere may be as high as if not higher than during the Medieval Period but have not yet exceeded those of the Holocene climate optimum of approximately 7500 years ago, when July temperatures were approximately $2\,°C$ higher than the twentieth century mean. Global climate models strongly suggest that temperatures in the next few decades will soon rise above this value, reaching levels that have not occurred naturally for over 100 000 years.

The impact on natural ecosystems is likely to be profound. Freshwater ecosystems are especially vulnerable to climate change, both through the direct effects of changing temperature and precipitation patterns, and indirectly through climate-driven changes in terrestrial ecosystems, land use and pollution loading. Some of the likely future impacts can be assessed from an examination of ecosystem response to past warm periods recorded in lake sediments and from long-term observational records. Additional insights can be gained from experiments designed to test hypotheses about future change under controlled conditions and from time–space comparisons, where lower latitude or lower altitude systems are used as analogues for the future. Information from paleoecological studies, long-term data records and experiments can be used to parameterise ecosystem models, which can then be driven with scenarios of future climate to project ecosystem response. The Euro-limpacs project is using all these approaches to better understand the processes controlling freshwater responses to climate change. Of special interest are the interactions with other stressors, such as artificial alteration of stream and river channels (hydromorphology) and pollution from acid deposition, nutrient enrichment and toxic substances. A major objective is to develop models both for specific ecosystem types and processes and also for scenario assessment at the catchment scale.

Provisional results from Euro-limpacs described here indicate that future climate change will affect the structure and functioning of European rivers, lakes and wetlands. Among phenomena likely to occur are:

(1) Loss of ice cover and strengthening of summer stratification in lakes.

(2) Reduction in river flow and lake water level, especially in southern Europe.

(3) Melting of rock glaciers, with associated solutes and pollutants in high mountain regions.

(4) Threats to biodiversity from increased river and lake temperatures.

(5) Changes to channel hydromorphology, affecting river and stream biodiversity.

(6) Increased algal growth, a decrease in piscivorous fish and a temporary increase in benthic and forage fish to the point where oxygen depletion results in overall decline and a decline in hypolimnetic oxygen concentrations in eutrophic lakes.

(7) Delay in recovery from acid deposition and threats to stream organisms from increased frequency and magnitude of high-discharge, low-pH events in upland regions of Northern Europe.

(8) Change in the transport, deposition and food chain uptake of volatile persistent organic pollutants.

(9) Remobilisation of toxic substances from increased flooding and storminess.

(10) Enhanced production of MeHg in boreal lake ecosystems.

The impact of climate change on freshwaters will necessitate thorough re-evaluation of national, EU and UNECE policies related to environmental protection. For freshwater ecosystems there are potentially major implications for the EU Habitats Directive, the Urban Wastewaters Directive and the Water Framework Directive, and for the UNECE Convention on Long-range Transboundary Air Pollution (CLTRAP).

The CLTRAP addresses emissions of pollutant gases to the atmosphere (mainly sulphur and nitrogen), and seeks to reduce emissions such that adverse effects on ecosystems can be prevented or remedied. Work within the CLTRAP is now attempting to incorporate future climate changes and their effects, and a possible revision of the Gothenburg protocol may take climate change into account.

EU polices relating to biodiversity and conservation will need to make allowance for geographical shifts in the range of species, and for changes in the nature of aquatic habitats, both chemically and morphologically. This might include the need for EU countries to work more closely together by assigning conservation value at the continental rather than national scale, integrating activities to provide migration corridors, improving habitat connectivity at all scales but especially at the catchment scale, and being alert to the impacts of climate change on the potentially disruptive effect of invasive alien taxa and pathogens.

For the WFD and other policies associated with environmental restoration, climate change has serious implications. In particular i) the reference state, although valuable as a concept, may be unstable for many freshwater systems over the longer term as reference sites themselves are subject to change; and ii) restoration targets for disturbed systems may not simply be achieved by removing stresses, as how those stresses interact or might interact with climate change in future will determine the directions in which ecosystems trend.

To understand better how future climate change will affect freshwater ecosystems in future, it will be necessary to:

- Continue to generate high-resolution climate models that project probable future climate change at the regional scale and can be down-scaled to project future climate for individual catchments.

- Generate realistic scenarios for future changes in pollution and land use that are influenced both by climate change and by societal responses to other drivers of change, especially economic ones.

- Continue to perform critical experiments at both the field and the mesocosm scale that aim to test specific hypotheses in controlled conditions.

- Continue to develop system-specific models to simulate probable hydrochemical and ecological responses to climate change.

- Continue to develop coupled or integrated models that are able to simulate catchment-scale responses to climate change.

- Continue to invest in high-quality monitoring programmes to provide early warning of future changes and to provide long-term data sets for model calibration and verification.

- Establish an array of appropriate chemical and biological indicators for detecting climate change effects.

- Continue to promote integration amongst the freshwater science community in Europe to maintain research capacity and enable coherent responses to emerging problems.

- Invest in central databases for hydrological, hydrochemical and hydrobiological data to safeguard data sets, especially long time series, and to enable model development and upscaling to the regional and continental levels.

- Improve the interaction between water managers and freshwater scientists to enable intelligent data analysis and decision making in restoring ecosystem quality.

As the years go by, the evidence that human activity is contributing significantly to global warming becomes more compelling, and as European freshwater ecosystems continue to recover from problems caused by nineteenth and twentieth century pollution, global warming will become the dominating influence on freshwaters. Some of the responses, such as the loss of winter ice cover on northern lakes or the reduction in summer flow in southern streams and rivers, can be predicted, but other responses, for example those involving interactions with future land use change and pollutant behaviour or the effects on ecological processes, are complex and uncertain. However, it is too late to prevent some of these responses. Indeed, it is probable that many of the changes are already occurring, as outlined in this chapter. In cases where we have adequate understanding of the consequences, policies and management practice will need to be adapted quickly to accommodate the threats. In other cases there is an urgent need to continue research programmes designed to address the many uncertainties. Throughout and underpinning everything is the importance of maintaining and developing long-term monitoring networks that are able to identify trends and alert both scientists and decision makers to future threats.

7.4.12   ACKNOWLEDGEMENTS

We would like to thank all Euro-limpacs participants for contributing directly or indirectly to this chapter and apologise to those whose specific work is not included here. Euro-limpacs is an EU FP6 Integrated Project (Project no. GOCE-CT-2003-505540) and we are grateful in particular to Christos Fragakis, our programme manager, for his support. We would also like to thank Miles Irving, Cath d'Alton, Katy Wilson and David Hunt for help in the preparation of the manuscript.

REFERENCES

Adamson, J.K. *et al.* (1998) *Environ. Pollut.*, **99**, p. 67.

Aherne, J.T. *et al.* (2006) *Sci. Total Environ.*, **365**, p. 186.

Anderson, H.E. *et al.* (2006) *Sci. Total Environ.*, **365**, p. 223.

Baker, J.P. *et al.* (1990) Biological effects of changes in surface water acid-base chemistry, State of Science and Technology Report 9, National Acid Precipitation Assessment Program, Washington, DC, USA.

Bakkers, I. *et al.* (2005) Report on the paired wetland sites (year 3), *Euro-limpacs Deliverable No. 235.*

Bartrons, M. *et al.* (2007) *Environ. Sci. Technol.*, **41**, p. 6137.

Battarbee, R.W. *et al.* (2005) *Freshw. Biol.* **50**, p. 772.

Bennion, H. *et al.* (2006) Further description of wetland sediment core analysis, *Euro-limpacs Deliverable No. 100.*

Bennion, H. and Battarbee, R.W. (2007) *J. Paleolimnol.*, **38**, p. 285.

Birol, E. *et al.* (2005) *Wat. Sci. Technol.: Wat. Supp. Journal*, **5**, p. 125.

Blenckner, T. *et al.* (2007) *Global Change Biol.*, **13**, p. 1314.

Bennion, H. *et al.* (2006) Further description of wetland sediment core analysis, *Euro-limpacs Deliverable No. 100.*

Bradley, D.C. and Ormerod, S.J. (2002) *Freshwater Biol.*, **47**, p. 161.

Brooks, S. *et al.* (2007) *British Wildlife*, **19**, p. 85 .

Brown, L.E. *et al.* (2007) *Global Change Biol.*, **13**, p. 958.

Carrera, G. *et al.* (2002) *Environ. Sci. Technol.*, **36**, p. 2581.

Caselegno, C. *et al.* (2005) Climate-hydromorphology interactions through changes in land-use and discharge: review of information relating selected study catchments across Europe, *Euro-limpacs Deliverable No. 12.*

Catalan, J.M. *et al.* (2002) *J. Paleolimnol.*, **28**, p. 129.

Catalan, J.M. *et al.* (2004) *Environ. Sci. Technol.*, **38**, p. 4269.

Catalan, J.M. *et al. Freshwater Biol.*, in press.

Clark, J.M. *et al. Global Change Biol.*, **11**, pp. 791–809.

Clement, B. and Aidoud, A. (2007) Report on plant functional groups and plant assemblages in response to change in environmental drivers (hydrology, nutrient, land use), *Euro-limpacs Deliverable No. 83.*

Davies, J.J.L. *et al.* (2005) *Environ. Pollut.*, **137**, p. 27.

Davies, T.D. *et al.* (1992) *J. Hydrol.*, **132**, p. 25.

De Pauw, N. and Hawkes, H.A. (1993) In: Walley, W.J. and Judd, S. (eds) *Proceedings of the 'Freshwater Europe Symposium on River Water Quality Monitoring and Control'*, Birmingham, UK, pp. 87–111.

Dillon, P.K. *et al.* (1997) *Environ. Monitoring and Assessment*, **46**, p. 105.

Dokulil, M.T. *et al.* (2006) *Limnol. and Oceanogr.*, **51**, p. 2787.

Downes, B.J. *et al.* (2002) *Monitoring Ecological Impacts – Concepts and practice in flowing waters*, Cambridge University Press, Cambridge, UK.

European Commission (2000) Directive 2000/60/EC of the European Parliament and of the Council of 23 October 2000 establishing a framework for Community action in the field of water policy, *Official Journal of the European Communities*, **L327**, 22.12.2000, p. 1.

Duwe, K. *et al.* (2007) Impact of Climate Change on Large Deep Lakes, *Euro-limpacs Deliverable No. 155.*

Evans, C.D. *et al.* (2001) *Sci. Total Environ.*, **265**, p. 115.

Evans, C.D. *et al.* (2005) *Environ. Pollut.*, **137**, p. 55.

Evans, C.D. *et al.* (2006a) *Global Change Biol.*, **12**, p. 2044.

Evans, C.D. *et al.* (2006b) *Sci. Total Environ.*, **365**, p. 167.

Evans, C.D. *et al.* (2008) *Hydrol. Earth Syst. Sci.*, **12**, p. 337.

Fang, X. and Stefan, H.G. (1999) *Clim. Change*, **42**, p. 377.

Fernandez, P.G. and Grimalt, J.O. (2003) *Chimia*, **57**, p. 514.

Fernandez, P.G. *et al.* (2003) *Environ. Sci. Technol.*, **37**, p. 3261.

Fernandez, P.G. *et al.* (2005) *Aq. Sci.*, **67**, p. 263.

Fox, D. (2007) *Science*, **316**, p. 823.

Freeman, C. *et al.* (2001) *Nature*, **412**, p. 785.

Gallego, E. *et al.* (2007) *Environ. Sci. Technol.*, **41**, p. 2196.

Gerhard, M. and Reich, M. (2000) *Internat. Review. Hydrobiol.*, **85**, p. 123.

Gippel, C.J. and Stewardson, M.J. (1996) In: Leclerc, M., Capra, H., Valentin, S., Boudreault, A. and Côté, Y. (eds), *Ecohydraulics 2000, Proceedings of the 2nd International Symposium on Habitat Hydraulics*, INRS-Eau, Québec, Canada.

Graf, W.F. *et al.* (2006a) Report on the design of a sampling framework for monitoring at the habitat scale, *Euro-limpacs Deliverable No. 86.*

Graf, W.F. *et al.* (2006b) Indicator value database for Trichoptera, *Euro-limpacs Deliverable No. 117.*

Griffiths, D. (2007) *Freshwater Biol.*, **52**, p. 1957.

Grimalt, J.O. *et al.* (2001) *Environ. Sci. Technol.*, **35**, p. 2690.

Gyllstrom, M. *et al.* (2005) *Limnol. and Oceanog.*, **50**, p. 2008.

Hari, R.E. *et al.* (2006) *Global Change Biol.*, **12**, p. 10.

Hering, D. *et al.* (2006) *Freshwater Biol.*, **51**, p. 1757.

Hering, D. *et al.* (in prep.).

Hering, D. *et al.* (submitted).

Hickling, R. *et al.* (2005) *Global Change Biol.*, **11**, p. 502.

Hindar, A.A. *et al.* (1994) *Nature*, **372**, p. 327.

Hughes, S.B. *et al.* (1997) *Hydrol. Earth Syst. Sci.*, **1**, p. 661.

Humlum, O. (1998) *Permafrost and Periglac, Process.*, **9**, p. 375.

IPCC (2007) *Climate Change 2007: The Physical Science Basis: Summary for Policymakers.*

Jackson, B.M. *et al.* (2007) *Ecological Modelling*, **209**, p. 41.

Jankowski, T. *et al.* (2006) *Limnol. and Oceanog.*, **51**, p. 815.

Jähnig, S.C. *et al.* (in press a) *Aquatic Conservation: Marine and Freshwater Ecosystems*

Jähnig, S.C. *et al.* (in press b) *Aquatic Conservation: Marine and Freshwater Ecosystems*

Jeppesen, E. *et al.* (2001) *Hydrobiologia*, **442**, p. 329.

Kaste, O. *et al.* (2006) *Sci. Total Environ.*, **365**, p. 200.

Kaste, O. *et al.* (2008) *AMBIO*, **37**, p. 39.

Kelly, M.G. *et al.* (1995) *Hydrobiologia*, **302**, p. 179.

Knoben, R.A. *et al.* (1995) Biological assessment methods for watercourses, UN/ECE Task Force on Monitoring and Assessment, Volume 3, RIZA report no. 95.066., Lelystad.

Kowalik, R.A. and Ormerod, S.J. (2006) *Freshwater Biol.*, **51**, p. 180.

Laudon, H. and Bishop, K.H. (2002) *Geophysical Research Letters*, **29**, p. 1594.

Laudon, H. and Bishop, K.H. (2006) Report on effects of altered soil temperature on the bioavailability of C and N in the riparian zone Euro-limpacs, *Deliverable No. 46.*

Laudon, H. *et al.* (2004) *Environ. Sci. Technol.*, **38**, p. 6009.

Liboriussen, L. *et al.* (2005) *Limnol. and Oceanog. – Methods*, **3**, p. 1.

Livingstone, D.M. (2003) *Clim. Change*, **57**, p. 205.

McKee, D.J. *et al.* (2002) *Aq. Biol.*, **74**, p. 71.

McKee, D.J. *et al.* (2003) *Limnol. and Oceanog.*, **48**, p. 707.

Magnuson, J.J. *et al.* (2000) *Science*, **289**, p. 1743.

Malcolm, J.R. *et al.* (2006) *Conserv. Biol.*, **20**, p. 538.

Maltby, E. *et al.* (2008) In: Kernan, M., Battarbee, R.W. and Binney, H.A. (eds) *Climate Change and Aquatic Ecosystems in the UK: science policy and management - Proceedings of a meeting held at the Environmental Change Research Centre, University College London May 2007*, pp. 29–34.

May, L. *et al.* (2005) Cross comparison of long-term contemporary data and palaeolimnological data from lakes - Interim Report, *Euro-limpacs Deliverable No. 19.*

Meerhoff, M. *et al.* (2007a) *Global Change Biol.*, in press.

Meerhoff, M. *et al.* (2007b) *Freshwater Biol.*, in press.

Michels, L. *et al.* (2007) Report on genetic changes in the resident Daphnia magna population in relation to habitat deterioration and restoration of Lake Ring, Denmark - a resurrection ecological and paleogenetical approach, *Euro-limpacs Deliverable No. 236.*

Monteith, D.T. *et al.* (2000) *Hydrological Processes*, **14**, p. 1745.

Monteith, D.T. *et al.* (2007) *Nature*, **450**, p. 537.

Morgan, M.G. and Henrion, M. (1990) *Uncertainty: A Guide to Dealing with Uncertainty in Quantitative Risk and Policy Analysis*, Cambridge University Press, Cambridge, UK.

Moss, B. *et al.* (2002) *Aquatic Conservation: Marine and Freshwater*, **12**, p. 229.

Moss, B. *et al.* (2004) *Freshwater Biol.*, **49**, pp. 1633–1649.

Munthe, J. *et al.* (2001) *Water Air and Soil Pollution Focus*, **1**, p. 385.

Murphy, J. and Nijboer, R. (2006) Hypotheses, experimental design, and analytical methods for artificial stream experiments, *Euro-limpacs Deliverable No. 91.*

O'Neill, R.V. (2001) *Ecology*, **82**, p. 3275.

Opatrilova, L. *et al.* (2005) Report – review of existing information on key taxa and functional groups relevant to the eight study catchments, *Euro-limpacs Deliverable No. 13.*

Raven, P.J. *et al.* (1997) In: Boon, P.J. and Howell, D.L. (eds) *Freshwater Quality: Defining the indefinable?*, The Stationary Office, Edinburgh, UK, p. 215.

Rose, N.L. (2005) The main transfer mechanisms for POPs and trace metals from soils to sediments: Sites, methodologies and preliminary data, *Euro-limpacs Deliverable No. 68.*

Rose, N.L. *et al.* (2005) In: Huber, U.M., Bugman, H.K.M. and Reasoner, M.A. (eds) *Global Change and Mountain Regions: A State of Knowledge Overview*, Springer, Dordrecht.

Rose N. L. *et al.* (2007) Statistical analyses on the sediment accumulation rate database, *Euro-limpacs Deliverable No.159.*

Saloranta, T.M. and Andersen, T. (2007) *Ecological Modelling*, **207**, p. 45.

Scheffer, M. *et al.* (1993) *Trends in Ecology and Evolution*, **8**, p. 275.

Skeffington, R.A. *et al.* (2007) *Sci. Total Environ.*, **382**, p. 199.

Smol, J.P. *et al.* (2005) *Proceed. Nat. Acad. Sci.*, **102**, p. 4397.

Statzner, B. *et al.* (2001) *Basic and Applied Ecol.*, **1**, p. 73–85.

Stendera, S. and Johnson, R.K. (2008) Tracking recovery trends of boreal lakes: use of multiple indicators and habitats. Journal of North American Benthological Society, **27**: 529–540

Stoddard, J.L. *et al.* (1999) *Nature*, **401**, p. 575.

ter Braak, C. and Šmilauer, P. (2002) *CANOCO Reference Manual and CanoDraw for Windows User's Guide: Software for Canonical Community Ordination (version 4.5).*

Thies, H. *et al.* (2007) *Environ. Sci. Technol.*, **41**, p. 7424.

Tipping, E. *et al.* (2003) *Environ. Pollut.*, **123**, p. 239.

Townsend, C.R. *et al.* (1983) *Freshwater Biol.*, **13**, p. 521.

Vadineanu A. *et al.* (2005) Report on existing hydrological, climatological and limnological data in a Danube sub-basin including identification of gaps in the data and new analyses to be undertaken, *Euro-limpacs Deliverable No.50.*

van Doorslaer, W. *et al.* (2007) *Global Change Biol.*, in press.

van Drooge, B.L. *et al.* (2004) *Environ. Sci. Technol.*, **38**, p. 3525.

Verdonschot, P.F.M. (2000) *Hydrobiologia*, **422**, p. 389.

Verhoeven J. *et al.* (2005) Specification of wetland productivity model, *Euro-limpacs Deliverable No. 53.*

Verhoeven, J. *et al.* (submitted) *J. Applied. Ecol.*

Vives, I. *et al.* (2004a) *Environ. Sci. Technol.*, **38**, p. 690.

Vives, I. *et al.* (2004b) *Sci. Total Environ.*, **324**, p. 67.

Vives, I. *et al.* (2005a) *Environ. Pollut.*, **133**, p. 343.

Vives, I. *et al.* (2005b) *Environ. Toxicol. Chem.*, **24**, p. 1344.

Wade, A.J. (2006) *Sci. Total Environ.*, **365**, p. 3.

Wade, A.J. *et al.* (2001) *Water Resources Research*, **37**, p. 2777.

Wade, A.J. *et al.* (2002) *Hydrol. Earth Syst. Sci.*, **6**, p. 559.

Whitehead, P.G. *et al.* (1998) *Sci. Total Environ.*, **210/211**, p. 547.

Whitehead, P.G. *et al.* (2006) *Sci. Total Environ.*, **365**, p. 260.

Worrall, F. *et al.* (2004) *Sci. Total Environ.*, **326**, p. 95.

Wright, J.F. *et al.* (1981) *Ecol. Entomology*, **6**, p. 321.

Wright, J.F. and Symes, K.L. (1999) *Hydrological Processes*, **13**, p. 371.

Wright, J.F. *et al.* (2004) *River Research and Applications*, **20**, p. 775.

Wright, R.F. (1998) *Ecosystems*, **1**, p. 216.

Wright, R.F. (2007) *Hydrol. Earth Syst. Sci. Discuss.*, **4**, p. 2945.

Wright, R.F. *et al.* (2005) *Environ. Sci. Technol.*, **39**, p. 64.
Wright, R.F. *et al.* (2006) *Sci. Total Environ.*, **365**, p. 154.
Yang, H. *et al.* (2002) *Environ. Sci. Technol.*, **36**, p. 1383.
Zelinka, M. and Marvan, P. (1961) *Archiv für Hydrobiologie*, **57**, p. 389.

Section 8
Ensuring Data Quality

8.1

NORMAN – Network of Reference Laboratories for Monitoring of Emerging Substances

Jaroslav Slobodnik and Valeria Dulio

The Water Framework Directive - Ecological and Chemical Status Monitoring   Edited by Philippe
Quevauviller, Ulrich Borchers, Clive Thompson and Tristan Simonart © 2008 John Wiley & Sons, Ltd

8.1.1 INTRODUCTION

Emerging *pollutants*, or – taking a wider perspective – emerging *substances*, are raising increasing concern among scientists, regulators and the public. These substances are not necessarily new chemicals and some of them have long been present in the environment, but their presence and significance are only now attracting close attention. Emerging substances often originate from consumer products and byproducts used every day in homes, on farms or by industry. Household cleaning products, nonprescription and prescription drugs, veterinary medicines, disinfectants, pesticides, flame retardants, plasticisers, fuel additives and nanoparticles are just some examples of frequently discussed emerging substances today. These substances are usually not included in routine monitoring programmes, despite their presence in the environment, and they are obvious candidates for future regulations – as long as research proves their toxic effects and/or widespread occurrence. In some cases, such as in the drinking water sector (European Community, 2004), the substance can be ranked as 'emerging' simply as a consequence of changing public perceptions of water quality (e.g. when a pharmaceutical detected in drinking water is not toxic but has passed through another person's body).

One key responsibility of environmental policies is to minimise existing, and prevent future, exposure risks. However, when implementing this in practice a series of questions arise, such as: *Do all substances we detect pose a risk? Should we screen the whole universe of chemical substances and their possible mixtures? If not, which substances should be monitored and how should we prioritise them? Are the substances we have found of local, national, European or even global importance? Can we trust the available data?* Correct and timely replies to these questions are of great importance for the competent authorities, as they need to have a solid scientific basis for making decisions that can incur high costs related to the monitoring of the emerging substances and their eventual removal from the environment.

8.1.1.1 Emerging Substances and the EU Water Framework Directive (WFD)

The WFD is undoubtedly one of the major driving forces in the research on emerging substances. Next to the obligatory monitoring of the 41 Priority Substances in each EU Member State, as requested under Annexes X and IX of the WFD (European Community, 2006a), the Directive strictly requires that '*other pollutants* also need to be monitored if they are discharged in significant quantities in the river basin or sub-basin'. It can be found further in the Directive that 'No definition of "significance" is given, but quantities that could compromise the achievement of one of the Directive's objectives are clearly significant and, as examples, one might assume that a discharge that . . . caused a biological or *ecotoxicological effect* in a water body would be expected to be significant' (European Community, 2000).

The legislative message is clear, but a typical situation often encountered in environmental laboratories is demonstrated in Figure 8.1.1. Of 200–300 compounds detected in a single river sediment sample, only some 40–50 % are provisionally identified and,

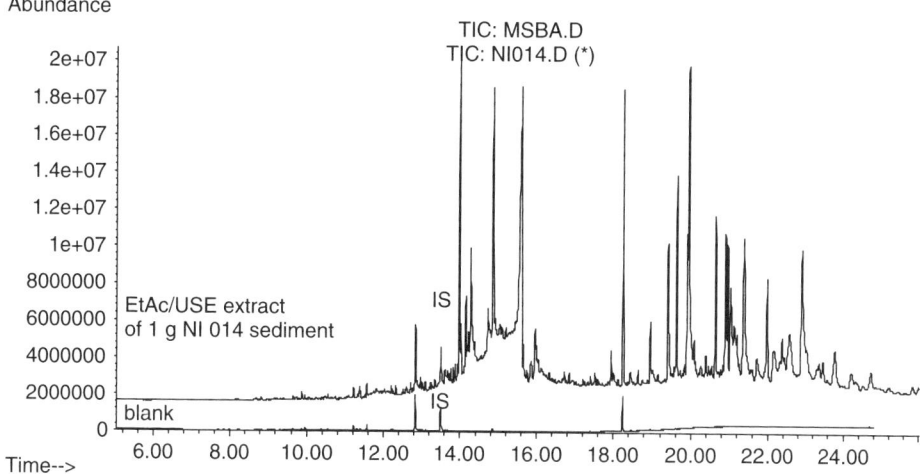

Abundance

Time-->

Figure 8.1.1 Gas chromatography–mass spectrometry (GC-MS) chromatogram of a typical river sediment sample (upper trace), where each peak represents a different compound. Some 200–300 organic substances are usually detected in each sample, of which more than 50 % remain unidentified and fewer than 10 % are regulated

of these, only a few are regulated according to national or EU standards. Submitting the sample to ecotoxicological analysis usually shows that compounds contained in the sample are causing an ecotoxicological effect, but it is far from clear which substance or group of substances is responsible for this effect and should therefore be 'blacklisted'. Also, little is known about the distribution of these substances among the sediment, water, biota and suspended particulate matter compartments, or about their temporal and spatial occurrence. A clear need therefore arises for a pan-European or even global platform dealing with these highly complex issues.

In order to address the issues raised above, the NORMAN project (http://www.norman-network.net) was funded by the European Commission (EC) with the aim of promoting the creation of a permanent network among reference laboratories and research centres dealing with emerging substances in collaboration with all involved stakeholders, including regulators, industry, standardisation bodies and NGOs. The current consortium brings together 17 partners from European research centres, coordinated by INERIS (for an overview see Figure 8.1.2). The project started in September 2005 and is supported by Commission funding until September 2008.

This paper sets out the achievements of the project so far, and its current activities and future plans in the wider context of the implementation of the WFD in Europe.

8.1.2 MAIN OBJECTIVES OF NORMAN

In operational terms, the first goal is to create a platform allowing for *the EU-wide exchange of information on emerging substances* among monitoring experts, environmental agencies and standardisation and regulatory bodies. This enables

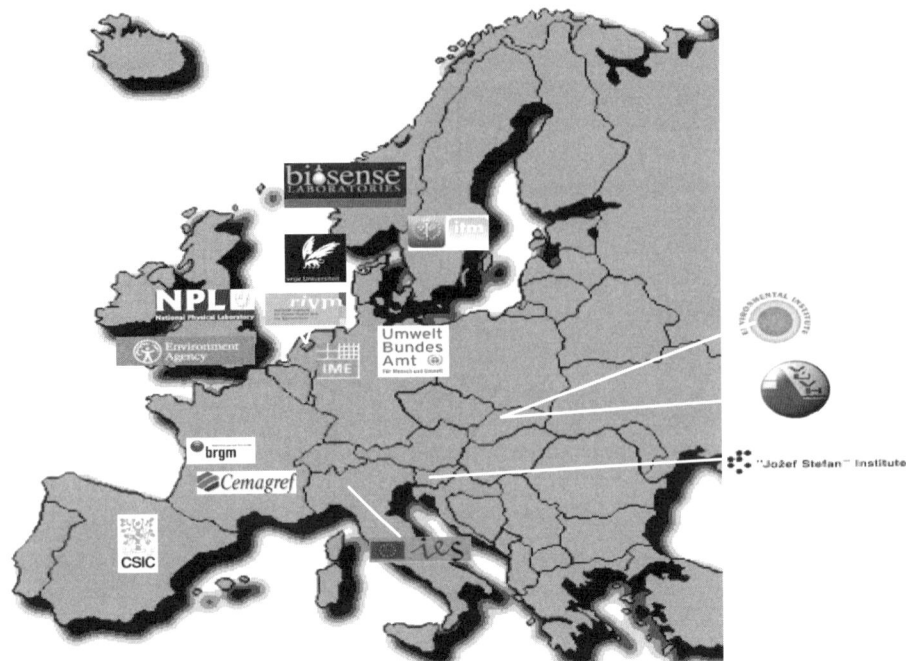

Partner	Abbreviation	Country
Institut National de l'Environnement Industriel et des Risques	INERIS	FR
Bureau de Recherches Géologiques et Minières	BRGM	FR
Centre national du machinisme agricole, du génie rural des eaux et des forêts	CEMAGREF	FR
Institute for Environmental Studies, Faculty of Earth and Life Sciences, Vrije Universiteit	VUA	NL
Umweltbundesamt - Federal Environmental Agency	UBA	DE
Water Research Centre	IWW	DE
Consejo Superior de Investigaciones Científicas	CSIC	ES
Jožef Stefan Institute	JSI	SL
Biosense Laboratories AS	BIOSENSE	NO
Department of Applied Environmental Science	ITM	SE
The Environment Agency, United Kingdom — National Laboratory Service (NLS) and Science Group	UKEA	UK
Water Research Institute	VUVH	SK
Environmental Institute	EI	SK
NPL Management Ltd	NPL	UK
Joint Research Centre — Institute for Environment and Sustainability	JRC-IES	IT
Fraunhofer Institute for Molecular Biology and Applied Ecology	Fh-IME	DE
The Netherlands National Institute for Public Health and the Environment	RIVM	NL

Figure 8.1.2 Current NORMAN partners – 17 organisations from 9 European countries, led by INERIS

regular assessment of the needs and requirements for analysis, and harmonisation of methodologies for monitoring and biomonitoring of emerging substances.

The second goal is to facilitate *access to and evaluation of existing data and information from EU and/or national research and monitoring programmes*, by developing web-based databases of i) leading experts, organisations and projects dealing with emerging substances; ii) monitoring data on target emerging substances; and iii) mass spectrometric information on provisionally identified and unknown substances.

The third goal is to develop *protocols for the validation and harmonisation of chemical and biological methods*, to meet EU demands for monitoring of emerging substances.

8.1.3   NORMAN NETWORK

In its first two years of operation the NORMAN network proved its capabilities in all key areas foreseen in the project-planning period. A brief overview of the project achievements so far is given below.

8.1.3.1   Scientific Watch and Information Exchange Activities

Contact Points, considered to be the important building blocks of the NORMAN network, have so far been appointed in 24 European countries, including Norway, Ukraine, Serbia, Croatia and Moldova. The goal is to have full coverage of all EU Member States and associated countries in 2008. The contact points facilitate the regular exchange and gathering of information on emerging substances from national reports and research initiatives, including information that appears in the so-called grey literature. A regularly upgraded overview of information reported by the contact points, with important links to the source documents, is available in the EMPOMAP database (see below), whereas the full reports produced at the national level are in the 'Library' section of the NORMAN project web site.

In parallel with the contact points, a network of NORMAN *Reference Laboratories* is being built (Figure 8.1.3). A set of criteria defining the profile of a reference laboratory in the field of emerging substances has been developed and key environmental laboratories in each European country have been invited to join the network. A careful selection procedure has been put in place to ensure that the backbone of the network will be formed only by laboratories with, on the one hand, a solid track record of activities in the field of emerging substances and well-developed analytical quality control systems and, on the other hand, close links to the national regulatory bodies, such as Ministries of Environment or Environment Agencies. The network benefits from close connections with the Joint Research Centre (JRC) of the European Commission in Ispra, Italy, with their excellent laboratory facilities, and the European Environment Agency (EEA) – an obvious final recipient of all data on environmental pollution in Europe in the longer term.

Industry representatives and regulators (Van Wijk, 2007) are often heard to say that 'It is information, not data, that we need'. Clearly, success in establishing a reliable

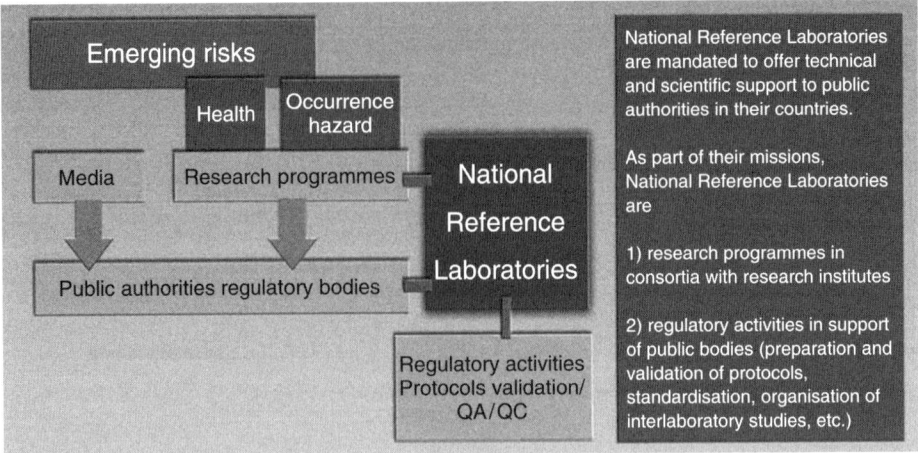

Figure 8.1.3 The role of national reference laboratories is crucial in the appraisal of emerging risks

process for converting data into reliable and meaningful information will represent a major advance in the early identification and resolution of potential problems. Addressing this, NORMAN has established a system in which appointed scientists carry out a regular literature search and compile critical reviews of recent scientific publications on targeted (groups of) emerging substances. The target audience is mainly decision makers within public administrations and institutional bodies. The aim is to foster wider sharing of the results of scientific work as an aid to early identification of priority areas of potential risk. Several such reviews are currently available in the NORMAN *Newsletter*, which is published regularly throughout the project (for downloadable copies see http://www.norman-network.net). The *Newsletter* also provides summaries of the latest relevant European research initiatives, environmental policy developments and EU-funded projects. The newsletter will be converted into a *Scientific Watch Bulletin* as part of the activities of the future NORMAN network.

NORMAN organises regular workshops for a wider audience, so as to share the results and visions of the project team with all stakeholders interested in emerging substances issues. Examples of thematic workshops are 'Emerging pollutants: key issues and challenges' (Stresa, Italy, June 2006), 'Chemical analysis of emerging pollutants' (Menorca, Spain, November 2006), 'New tools for biomonitoring of emerging pollutants' (Amsterdam, the Netherlands, October 2007) and 'Integrated chemical and biomonitoring strategies for risk assessment of emerging substances' (Lyon, France, March 2008). It is no surprise to find among the workshops' participants both 'problem owners and problem solvers', including strong representation of key European scientists, industry experts, national, EC and EEA officials. As with all NORMAN outputs, the results and conclusions of the workshops are widely shared across Europe via either the project web site or dedicated publications (Hanke *et al.*, 2007).

8.1.3.2 Collection and Evaluation of Data

NORMAN facilitates international cooperation and access to the existing data via three web-based databases: i) EMPOMAP: a database of leading experts, organisations and projects dealing with emerging substances; ii) EMPODAT: a database of geo-referenced monitoring data accompanied by the ecotoxicological information from bio-assays and biomarkers; and iii) EMPOMASS: a database of mass spectrometric information on provisionally identified and unknown substances. The databases have been developed and integrated so as to bring together the existing knowledge on emerging substances in air, water and soil compartments and set up a framework for systematic collection, elaboration and scientifically sound evaluation of future data. A list of the most relevant emerging substances has been compiled by the members of the NORMAN network and its latest upgrade can be viewed at http://www.norman-network.net ('About us/Emerging substances' section). An overview of major groups of emerging substances identified so far is given in Table 8.1.1.

The EMPOMAP database, accessible via the project web site, ('Databases' section) is designed to assist in the definition of the current state and future needs of research on emerging substances and to help to coordinate national research programmes in order to avoid duplication of research. The database allows for targeted identification of experts and stakeholders concerned with a specific emerging substance, and expedites the process of finding related monitoring data. Since 2007, registration for the database has been open to all experts or organisations interested in joining the network (see 'Join the NORMAN network' section on the project web site). The database is being updated regularly during the project and special care is being taken to establish links with other European projects and databases dealing with related topics, such as WEKNOW (drinking water), EUGRIS (water and soil), MODELKEY (ecotoxicology), the OECD database on research projects on nanoparticles, etc.

The EMPODAT database collects monitoring and occurrence data on emerging substances which are already known to be present in the environment but which are not yet included in routine monitoring programmes. The basic information consists of the concentration of the pollutant, sample matrix, geographical coordinates of the sampling site, sampling date, detailed information on the analytical methodology used and

Table 8.1.1 List of major groups of emerging substances. For the latest full list of individual substances see http://www.norman-network.net

Algal toxins	Fragrances
Anticorrosives	Industrial solvents
Antifoaming agents	Nanoparticles
Antifouling compounds	Perfluoroalkylated substances and their transformation products
Antioxidants	Personal care products
Biocides	Petrol additives
Bio-terrorism/sabotage agents	Pesticides
Complexing agents	Pharmaceuticals
Detergents	Trace metals and their compounds
Disinfection byproducts (drinking water)	Wood preservatives
Flame retardants	Other

related QA/QC measures. A separate module on bioassays and biomarkers provides additional information on the ecotoxicity of the compounds recorded in the database. The structure of the database is fully compatible with the requirements of WISE (http://water.europa.eu/) and aims to be linked to the IUCLID database (http:// ecbwbiu5.jrc.it/) collecting data within the REACH programme (European Community, 2006b).

In the construction of the database, special attention was paid to the harmonisation of formats of data and metadata reporting, such as measurement units, in order to improve inter-comparability of the data among all EU Member States. The data collection formats recently proposed for harmonisation at the EU level (INERIS, 2007) are already incorporated in the database structure. Recommendations made by other international bodies such as IUPAC (e.g. Egli *et al.*, 2003) were also taken into account.

Since 2007, monitoring data available through NORMAN project participants, NORMAN contact points, reference laboratories and other network members have started to be entered in the database. Whenever possible, the existing databases are uploaded directly into the project database via specific software interfaces. For collection of the other data, simple pre-programmed Excel sheets matching the structure of the database are used.

A protocol is being developed for the evaluation of the existing data, including their validity/fitness for purpose, comparability, time and spatial representativity, matrix coverage, etc., using the GIS technology to the maximum possible extent. The most relevant parameters allowing for assessment of the degree of validation of the specific analytical method used, according to the NORMAN validation protocols (see below), are included on the list of the database metadata.

The agreed formats of data and metadata reporting, the minimum requirements for establishment of new databases of emerging substances, and the set of monitored/measured parameters needed for risk assessment (including requirements with respect to data precision and spatial and temporal resolution) will be disseminated across Europe as a guidance document.

The EMPOMASS database is focused on the collection of geo-referenced data of 'unknown' and 'provisionally identified' substances in various environmental compartments. Mass spectra obtained from GC-MS screening in electron impact mode are widely accepted as unique fingerprints of individual organic compounds and therefore form its basis. However, with one eye on the latest scientific achievements in the field of mass spectrometry, the database format is also designed to accommodate information on LC-MS(MS) and accurate mass measurement data. The database consists of two modules: one containing numeric information extracted from the screening results, such as major ions in the mass spectrum of the compound, retention characteristics, match factor, proposed structure, CAS number, molecular mass, etc., and the other containing raw mass spectra organised and searchable in the specific emerging substances library.

A specific protocol has been developed for the rough estimation of the concentrations of 'unknown' substances, for which the standard chemical is not available. The database will be used as a tool to define new emerging threats in various environmental compartments and allows the user to trace occurrence and pollution trends for substances which are currently not included in major monitoring schemes because of the lack of knowledge of their identity. A set of criteria is being developed to judge whether additional targeted research is needed to identify the detected unknown substances, e.g. based on the frequency of occurrence, concentrations or evidence of biological impact in the vicinity of the sampling site(s). Extensive European research resources have already been invested in the identification of substances not having mass spectra in the commercially available libraries. It is expected that the EMPO-MASS database will bring all this knowledge together and also aid the interpretation of historical data by simple reprocessing of 'old' mass spectra of as yet unidentified compounds. In the initial setup of the database, large data sets of the Dutch GC-MS database (RIZA, the Netherlands), the Water Quality Database of the International Commission for the Protection of the Danube river (ICPDR; http://www.icpdr.org) and the DOPS (Database of Organic Pollutants in the River Systems of the Slovak Republic) were used.

The EMPODAT and EMPOMASS databases allow for comprehensive information to be obtained quickly on the occurrence of a specific emerging substance and its effect-based monitoring. Such information will pinpoint the need for further research, including development, improvement, validation and harmonisation of the used methodologies.

8.1.3.3   Validation and Harmonisation of Methods

Comparability and reliability of monitoring data are essential for any meaningful assessment and for the management of environmental risks. For emerging substances, there is concern regarding the comparability of data at the European level. Methods used for the monitoring of emerging substances have often not been properly validated either in-house (i.e. within a single laboratory) or at the international level. Such methods are not well established in the scientific community, and are therefore far from being harmonised or standardised. In addition, those methods developed by different institutions and organisations may only be applicable to specific conditions (matrix, organism), which may further complicate data comparability.

NORMAN is therefore developing *a common European framework for the validation of chemical and biological methods* for the respective monitoring and biomonitoring of occurrence and effects of emerging pollutants in a broad range of matrices. The protocol addresses three different validation approaches, in increasing order of complexity: i) method development and validation at the level of *research laboratories*; ii) method validation at the level of *expert/reference laboratories*; and iii) method validation at the level of *routine laboratories*. The concept of these three approaches is strictly hierarchical, i.e. a method must fulfil all the criteria of the lower level before it can enter the validation protocol of a higher level (see also Figure 8.1.4).

The endpoint of the first level is a method with complete internal validation for the intended purpose at the level of a single research laboratory. A method meeting the requirements of the second level can successfully be transferred to another laboratory possessing sufficient expertise and experience. Ranking the method at the third level means that the method possesses sufficient interlaboratory performance and is applicable for use at the level of routine laboratories. This also comprises the development and control of key aspects of method documentation and method usability. Having

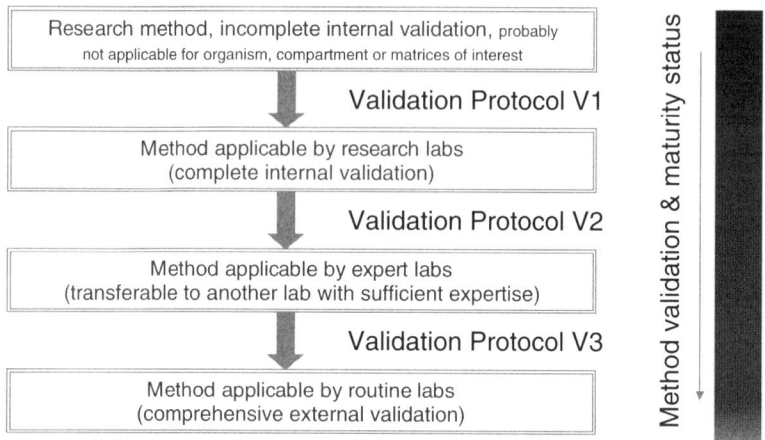

Figure 8.1.4   Common European protocol for method validation

successfully satisfied the third-level procedures, a method should be fit for standardi-sation at the European level. Detailed description of each of the levels is available for public review and comment in the document 'Protocol for methods validation' on the project web site.

As part of the project work plan, this protocol was tested in three case studies designed to match the three levels of validation, before preparation of the final ver-sion. The interlaboratory studies included analyses of i) selected natural and synthetic oestrogens and oestrogenic activity in wastewater, ii) pharmaceuticals in water and iii) brominated flame retardants in dust. The design of the interlaboratory studies, analyt-ical protocols and final reports from the interlaboratory studies are available on the project web site. The final aim is the implementation of the protocols in the fields of European standardisation and European legislation. To this purpose, negotiations will be initiated to launch New Work Item proposals at CEN level (see Figure 8.1.5).

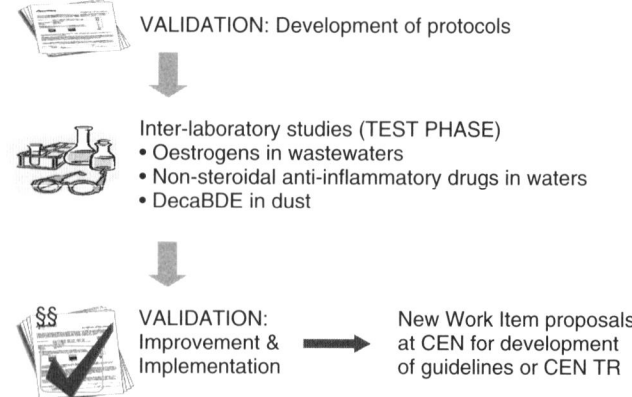

Figure 8.1.5   Procedure for development and implementation of NORMAN protocols for method validation

As NORMAN pays a lot of attention to the (eco)toxicological effects of the emerging substances, *a protocol for the integration of biological and chemical test methods* in a broader context, for which no standardised guidelines are available, has been developed. The protocol is downloadable from http://www.norman-network.net (QA/QC issues/validation framework). For the discovery of emerging pollutants through effect directed analysis (EDA) – indicating the integrated use of chemical and biological techniques – this protocol provides a set of recommendations for the various aspects of EDA studies.

8.1.4   NORMAN: A SELF-SUSTAINING PERMANENT NETWORK

The NORMAN project financed by the EC will end its activities in autumn 2008. We can, however, already state that this will not mean the end of the activities of the established network. A call for expressions of interest to become a member of the future permanent network, sent around Europe in November 2007, received excellent feedback from a large number of organisations.

8.1.4.1   Why a Permanent Network?

Since there is no way to stop progress in scientific knowledge, competent authorities have to anticipate regular discoveries of unregulated substances in the environment and, consequently, prevent future risks. Many of today's emerging substances will probably be part of tomorrow's regulated substances. The experience gathered throughout the project has shown that there is a need for an *independent and competent forum* in the field of emerging substances, with official recognition from institutional bodies at the EU level and on a wider international scale, in order to facilitate an exchange of information, debate and research collaboration at a global level.

8.1.4.2   Who Should Be Part of the Network?

The permanent network proposal puts no restrictions whatsoever on membership of the future network. It would ideally embrace all interested stakeholders dealing with emerging substances – whether in studying their occurrence and effects or risk assessment and risk management. Still, the key members are expected to be mainly i) competent authorities/reference laboratories, i.e. institutes/organisations designated by the competent authorities at the national level to offer technical and scientific support in specific fields related to environmental protection; ii) research centres and academia; iii) industry stakeholders; and iv) government institutions and standardisation bodies.

Since its launch, the NORMAN network has been recognised as an important player in the European environmental arena and discussions are taking place on the form of the future cooperation with various institutional bodies and international organisations, such as the European Commission's Directorate General (DG) Environment and DG Research, the OECD and EURAU.

8.1.4.3   What Will Be the Activities of the Network?

The network will continue to maintain and further upgrade its successful tools in the area of information exchange, mainly the three databases – EMPOMAP, EMPODAT and EMPOMASS – and the NORMAN *Scientific Watch Bulletin*, for wider sharing of recent results of scientific work and rapid identification of areas where potential health, safety or environmental issues arise. The network of NORMAN national contact points will guarantee access to the reports and research initiatives conducted in different countries in the field of emerging substances.

The network will create a platform for the expert group meetings, at which a limited number of experts will elaborate on high-priority topics selected by NORMAN members. The results and conclusions of the expert group meetings will lead to position papers summarising the position of the NORMAN network on these issues. Wider-scope workshops will be organised by the network on a yearly basis.

In the field of QA/QC the network will check regularly on the availability of methods for high-priority substances and their validation status, and initiate, whenever necessary, the organisation of interlaboratory studies for specific emerging substances. QA/QC information, such as available reference materials, proficiency testing schemes and training programmes for laboratories will be circulated.

An important goal of the network is to initiate and organise *EU-wide assessment of the occurrence of selected emerging substances*, inviting the network members to contribute to the analytical work. This assessment will include the definition of common sampling and analytical protocols and, when necessary, the organisation of inter-calibration exercises among the selected partners, in order to ensure harmonisation of the protocols before the start of the monitoring campaigns.

8.1.5   FINAL WORD

Considering the importance and legal implications related to the world of emerging substances, it comes almost as a surprise to discover that a network harmonising all such issues in a wider European context did not exist until recently. After the first two years of the NORMAN project it is possible to state that the foundations of the permanent network dealing with major aspects of emerging substances in the environment have been laid and its operational tools have been developed and thoroughly tested. The WFD, especially the parts relating to the 'other substances' specific to each European river basin and their impact on ecological status, has arguably been the best possible test field for the network to show its strength to date.

In the end it will be left up to the network members to make use of this platform. One can easily imagine that the greatest potential benefits, such as active involvement at an early stage in the debate on strategic topics, will be observed in the development of the list of tomorrow's regulated substances. Moreover, the network will provide opportunities for collaborative research between institutes sharing common interests, and will allow for more effective support to national regulators. NORMAN, as an independent and competent forum, is predesigned to be able to provide technical and scientific advice to the EC, thereby providing an overview of the latest topics of concern, research needs and priorities for future regulations. Using its close links with the scientific community and contact points/reference laboratories in the different countries, the network will carry out systematic collection (and conversion into a common format) of the information that is currently held in many different places and not included in national databases. The network will ensure that, as soon as an emerging substance is identified as a pollutant of concern – thereby requiring regular monitoring – there is sufficient capability across the EU for measuring it at the routine level. To this purpose, in the field of water policy, a link will be established between NORMAN and DG ENV in the framework of the Chemical Monitoring Activity (CMA) expert group beyond 2009.

8.1.6 ACKNOWLEDGEMENTS

The authors would like to acknowledge the European Commission's support of the NORMAN Coordination Action project, Contract No. 018486, financed under the 6th Framework Programme, thematic sub-priority 1.1.6.3 Global Change and Ecosystems, and all partners of the project whose inputs were used in this overview.

REFERENCES

Egli, H. *et al.* (2003) Minimum requirements for reporting analytical data for environmental samples, *Pure Appl. Chem.*, **75**(8), pp. 1097–1106.

European Community (2000) Directive of the European Parliament and of the Council 2000/60/EC establishing a framework for Community action in the field of water policy, as of 23 October 2000.

European Community (2004) Council Directive 98/83/EC on the quality of water intended for human consumption as of 3 November 1998 (Drinking Water Directive; DWD) and 'The Bonn Charter for Safe Drinking Water', Water Safety Plans, IWA.

European Community (2006a) Directive of the European Parliament and of the Council on environmental quality standards in the field of water policy and amending Directive 2000/60/EC, COM (2006) 397 final, as of 17 July 2006.

European Community (2006b) Regulation (EC) No. 1907/2006 of the European Parliament and of the Council of 18 December 2006 concerning the registration, evaluation, authorisation and restriction of chemicals, REACH, http://ec.europa.eu/environment/chemicals/reach/reach_intro.htm.

Hanke, G. *et al.* (2007) Proceedings of the international workshop on 'Emerging pollutants: key issues and challenges' in Stresa, September 2007.

INERIS (2007) Implementation of requirements for priority substances within the context of the Water Framework Directive, common template for data collection', Final version 20 March 2007.

Van Wijk, D. (2007) Editorial, *NORMAN Network Newsletter*, **2**, http://www.norman-network.net.

8.2
Data Quality Assurance of Sediment Monitoring

Ulrich Förstner, Susanne Heise, Wolfgang Ahlf and Bernard Westrich

8.2.1   INTRODUCTION

The European Water Framework Directive (WFD) so far does not consider sediment quality and quantity as a major issue (Förstner and Owens, 2007). Objections against *compliance monitoring for sediment*, based on Environmental Quality Standards (EQS) for sediment, were caused by analytical limitations and anticipated costs involved in

---

The Water Framework Directive - Ecological and Chemical Status Monitoring   Edited by Philippe Quevauviller, Ulrich Borchers, Clive Thompson and Tristan Simonart © 2008 John Wiley & Sons, Ltd

obtaining full spatial coverage (WFD/AMPS, 2004).[1] On the other hand, the strategies against chemical pollution of surface waters (WFD Article 16), i.e. establishment of a programme of measures until 2009, have to examine all generic sources that can result in releases of priority substances and priority hazardous substances – including the specific source/pathway 'historical pollution from sediment' (WFD/EAF, 2004).

Theoretical considerations and a pragmatic three-step approach for risk assessment of contaminated sediments in river basin have been presented by Heise and Förstner (2007), and monitoring sediment quality using toxicity tests as primary tool for any risk assessment is described in Chapter 6.2 of this book by Ahlf *et al.* The aim of the present contribution is to demonstrate that the procedures replacing a universal EQS approach for sediments, while being tailor-made for typical WFD objectives, will require individual considerations with respect to quality assurance and quality control.

8.2.2 PARTICULATE MATTER MONITORING IN RIVERS

The use of particulate matter as an assessment medium has several advantages, at least compared to the water phase (Bergmann and Maass, 2007):

- Through their formation over a period of time, sediments may serve as a 'long-term memory' of contaminant loads in a water body. Investigating sediments provides a retrospective history (trend) of contamination that cannot be obtained with water samples.

- A large number of dissolved hazardous substances (heavy metals, organic contaminants) have a strong tendency *to be adsorbed* onto the surface of solids. As a result, these substances may be difficult to detect in the water phase at a given time, possibly rendering an incomplete picture of the contamination situation.

- The relation of contaminants in sediments to those in *biota* living on or in the water bottom (benthos) is often different from that in the water phase. Bioaccumulation and other adverse effects on benthos can only be studied and assessed by including sediments in the investigations.

The reasons why monitoring sediments is more difficult than monitoring water may in part stem from the way sediments are formed, and from their various characteristics, which differ considerably from those of water (e.g. Bergmann and Maass, 2007):

- Disposal and resuspension (erosion) of sediment occur reversibly depending on hydraulic conditions, in particular current velocity. These hydraulic conditions may vary within a short time and distance, making sediment layers much more inhomogeneous than water samples.

---

[1] According to a proposal of 21 June 2007 for a Directive of the European Parliament and of the Council on environmental quality standards in the field of water policy and amending Directive 2000/60/EC, Member States may opt to apply EQS for sediment and/or biota instead of those laid down in Annex I, Part A, in certain categories of surface waters. (for the actual state of discussions, see Foerstner, 2008)

- The inhomogeneous character of sediment layers renders representative sampling as difficult as water sampling. However, sedimentation rate and grain size distribution are two of the important differentiating characteristics from water.

Despite these obvious drawbacks, in particular the inhomogeneous character of sediment layers and sediment/SPM grain size, particulate matter can be the preferred medium in a wide spectrum of assessment objectives and programmes (Thomas and Meybeck, 1992). The objectives are listed in increasing order of complexity, with each step requiring more sampling and measurement:

- *Type 1:* To assess the present concentrations of substances including pollutants found in the particulate matter and their variations in time and in space (*basic surveys*), particularly when pollution cannot be definitely shown from water analysis.

- *Type 2:* To estimate past pollution levels and events (e.g. for the last 100 years) from the analysis of deposited sediments (*environmental archive*).

- *Type 3:* To determine the bioavailability of substances or pollutants during the transport of particulate matter through rivers and reservoirs (*bioavailability assessment*).

- *Type 4:* To determine the fluxes of substances and pollutants to major water bodies (e.g. regional seas, oceans) (*flux monitoring*).

- *Type 5:* To establish the trends in concentrations and fluxes of substances and pollutants (*trend monitoring*).

Generally, the following type of information relating to a specific assessment objective can be obtained through the study of the particulate matter (Thomas and Meybeck, 1992):

- Suspended particulate matter: present concentrations of substances and pollutants (Type 1, 3) and pollutants and nutrient fluxes to seas and lakes (Type 4, 5).

- Bottom deposits: present concentrations of pollutants (Type 1, 3) and past concentrations of pollutants in some cases (Type 2, 3).

Especially when particle bound contaminants are supposed to be monitored for a river basin assessment, the transport of suspended matter and the load of contaminants transported with it deserve special concern. Its monitoring in times of floods or low water periods enables assessment of maximum risk potential of contaminant depots along rivers.

Table 8.2.1 shows a scheme for three levels of sophistication (A, B, C) of sediment analyses for bottom sediment and total suspended solids in rivers. With respect to basin-scale information, full cover of suspended particulate matter quality throughout flood stage and sediment cores at selected sites where continuous sediment may have occurred (both level C) is needed.

Table 8.2.1 Development of particulate matter quality assessment in rivers in relation to increasing levels of monitoring sophistication (modified after Thomas and Meybeck, 1992)

	Monitoring level		
	A	B	C
Suspended matter (SPM)	Survey of SPM quantity throughout flood stage (mostly when rising)	Survey of SPM quality at high flow (filtration or concentration)	Full cover of SPM quality throughout flood stage
Deposited sediment	Grab sample at station (end of low flow period)	Longitudinal profiles of grab samples (end of low flow period)	Cores at selected sites where continuous sedimentation may have occurred

Level A: simple monitoring, no requirement for special field and laboratory equipment
Level B: more advanced monitoring requiring special equipment and more manpower
Level C: specialised monitoring which can only be undertaken by fully trained and equipped teams of personnel

8.2.2.1   Need for Quality Control in Different Sediment-related Activities

With respect to data quality control in water, a European thematic network, METROPOLIS (2004), has identified a few typical problem areas, for example:

- Data are not representative for a certain purpose; for example, drinking water standards cannot be used as a yardstick for the quality of a river ecosystem.

- A too-high level of uncertainty of the selected data may endanger the whole decision-making process.

- Lack of traceability. The concept of traceability implies that measurement data are linked to stated references through an unbroken chain of comparison, all with stated uncertainties (Quevauviller, 2004).

Considering the complex systems of natural particulate matter and large river basins, a closer look is needed at quality requirements related to chemical, biological and hydraulic sediment data. In the view of the traceability concept, 10 typical actions, properties and functions can be distinguished (Förstner, 2004).

A basic sequence of measurements consists of three steps:

(1) *Sampling and sample preparation.* Project planning, sampling stations, sampling devices, handling and storage, and QC are not standardised, but are well documented in all aspects (Mudroch and Azcue, 1995).

(2) *Grain size* as a characteristic sediment feature. Sampling on fine-grained sediment and grain-size normalisation with 'conservative elements', such as Cs, Sc, Li and Al (reflecting clayey material content) is recommended as a standard approach (Förstner, 1989; Horowitz, 1991).

(3) *Analytical.* Reference sediment materials are commercially available. While direct species analysis is still limited, standardised extraction schemes for metals and

phosphorus in sediments as well as CRMs for comparisons were developed under the auspices of BCR/IRMM (Quevauviller, 2002; Rauret *et al.*, 2001).

Further steps in chemical sediment analysis are split up with regard to specific purposes: sediment quality assessment including biological effects (4, 5), coupling of sediment quality data with erosion risk evaluation (6, 7), chemical changes following resuspension of anoxic sediments (8, 9), and modelling of chemical sediment data (10).

Sediment quality assessment is part of the sustainable sediment management of the European Sediment Research Network Sediment (SedNet, 2003). In recent years, a paradigm shift has taken place to give priority to biological data (Den Besten *et al.*, 2003). Chemical sediment analysis for monitoring studies is usually combined with biological investigations and could be supported by pore water analysis:

(4) *Biological/chemical approach.* In a comprehensive sediment-assessment approach, five basic components should be considered (Krantzberg *et al.*, 2000): i) benthic community structure, ii) laboratory bioassays for evaluating the toxicity of in-place pollutants, iii) bioaccumulation information, iv) knowledge of site stability, and v) physico-chemical sediment properties.

(5) *Pore water.* Tests on pore water (interstitial water) were considered suitable for several types of regulatory framework but unsuitable for other, e.g. as stand-alone pass/fail methods or as a substitute for a solid phase test (Carr and Nipper, 2001).

Because of the particular dynamics of fluvial processes, *hydraulic parameters*, such as the critical shear stress of erosion processes, form the primary input factors for investigating and predicting the risks connected with the depositing of contaminated solids in downstream regions. Short-term issues include the fate of sediment-associated contaminants when sediment is deposited upland and a better understanding of the impact on ground water, water and soil ecosystems. Medium/long-term issues will focus on integration of quality and quantity aspects, and determination of the sediment transport processes at the river basin scale as a function of land and water use and hydrological (climate) change (Brils *et al.*, 2003).

(6) *Erosion effects.* Sediment physical parameters and techniques form the basis of any risk assessment in this field. Sampling of flood plain soils and sediments is affected by strong granulometric and compositional heterogeneities arising from the wide spectrum of flow velocities at which the sediments were eroded, transported and deposited. Sediment quality issues should include experimental designs for the study of chemical and biological effects during erosion and deposition (e.g. Haag *et al.*, 2001).

(7) *Aging effects.* 'Diagenetic' effects, which, apart from chemical processes (sorption, precipitation, occlusion, incorporation in reservoir minerals and other geosorbents such as char, soot and ashes), involve enhanced mechanical consolidation of soil and sediment components by compaction, loss of water and mineral precipitations in the pore space, may induce a quite essential reduction of the reactivity of solid matrices (e.g. Luthy *et al.*, 1997).

The dramatic effects of storm water events on particle transport can coincide with rapid and *far-reaching chemical changes*, in particular, the effects of sulphide oxidation on the mobilisation of toxic metals. The objectives of this research fall under the category 'process studies', usually involving a relatively high degree of complexity. Such field and laboratory studies, as well as the models using these data, are indispensable for long-term prognosis of erosion and chemical mobilisation risks arising from subaqueous deposition and capping, both favourable technologies for dredged material and in situ sediments (US EPA, 2005).

(8) *Anoxic/oxidised samples.* Changes of the forms of major, minor and trace constituents cannot be excluded when the sediment is transferred from its typical anoxic environment to chemical analysis via normal sample preparation. However, a comparison of extraction data from the original and oxidised samples could be used for worst-case considerations in respect of potential metal release during sediment resuspension, or subsequent to upland deposition of dredged material (Meyer *et al.*, 1994).

(9) *Capacity controlling properties.* Both pH and redox potential in sediment/water systems are significant parameters for mobilisation and transformation of metals or phosphorus. Criteria for prognosis of the middle- and long-term behavior of these and other substances should, therefore, include the abilities of sediment matrices to produce acidity and to neutralise such acid constituents (Kersten and Förstner, 1991; Salomons, 1995).

(10) *Modelling.* Transport and reaction models consider advective, dispersive and diffusive transport mechanisms as well as ad- and desorption processes (Fritsche *et al.*, 2007; Jacoub and Westrich, 2007; Karnahl and Westrich, 2007).

8.2.2.2 Examples of Typical Traceability Issues of Sediment Analyses

From a practical point of view, the traceability concept for quality control of chemical sediment analysis comprises three categories of investigation (Figure 8.2.1; Förstner and Heise, 2006):

- *Memory effect,* mainly in dated sediment cores from lakes, reservoirs and marine basins, as historical records reflecting variations of pollution intensities in a catchment area. As regards the traceability concept, the basic sequence of measurements consists of three steps, which can be considered as an unbroken chain of comparisons.

- *Basic characterisation,* i.e. sediment as ecological, social and economic value, as an essential part of the aquatic ecosystem through forming a variety of habitats and environments. A system approach is needed, comprising biotests and effect-integrating measurements due to the inefficiency of chemical analysis in the assessment of complex contamination.

- *Secondary source,* i.e. mobilisation of contaminated particles and release of contaminants after natural or artificial resuspension of sediments. On a river basin scale,

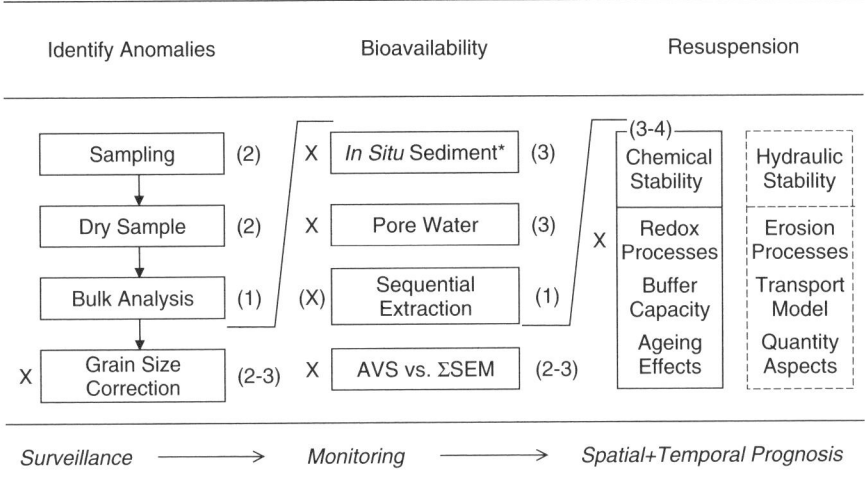

Figure 8.2.1 Schematic overview on traceability aspects of chemical sediment analysis (modified from Förstner & Heise, 2006). X = sediment-specific property. Numbers relate to levels of methodological standardisation (section 8.3.4): 1) certified reference material, 2) standardised method, 3) well-documented method, and 4) special method. AVS/Σ SEM = Acid Volatile Sulfide/Sum of Simultaneously Extractable Metals (DiToro *et al.* 1992); *Wet Sample: sub-sampling for tests under oxygen-free atmosphere (pore water, sequential extraction, etc)

i.e. when applied in a conceptual river basin model (CBM), chemical and ecological information need a strong basis of sediment quantity data. In a dynamic system, this assessment should include not just those materials that are currently sediments, but also materials such as soils, mine tailings, etc. that can reasonably be expected to become part of the sediment cycle during the lifetime of a management approach (Apitz and White, 2003).

Surveillance Investigations (1st Column in Figure 8.2.1)

Surveillance is a 'continuous specific observation and measurement relative to control and management' (UNESCO, 1978); the primary objective is to trace and observe sources and pathways of specified hazardous substances. If a simple aim of a study is to determine the presence or absence of a specific contaminant in bottom sediment at a given area, then the sediment can be sampled at one or a few sampling stations at fine-grained sediment deposition sites. However, after confirmation of the presence of the contaminant in the sediment, the study may be expanded to determine the extent of sediment contamination by the specific compound or element, with the area, the contaminant's sources, history of the loading of the contaminant, its transport, bioaccumulation, etc. (Mudroch and Azcue, 1995).

For the given purpose, namely to detect anomalies and to trace typical sources of pollution ('hot spots'), the standard scheme, from 'sampling', 'sample preparation' (with particular emphasis on grain size correction), 'chemical analysis' (use of dry bulk sediment reference material) to the 'interpretation of results', does not involve major uncertainties. Minor uncertainties, which will not affect the general applicability of the present approach, could arise from variations of typical matrix constituents and can be narrowed down by analysing parameters such as organic matter, carbonate and iron oxide contents.

Monitoring for Potential Biological Impacts (2nd Column in Figure 8.2.1)

Monitoring of environmental risks from particle bound substances is a complex challenge as their bioavailability and bioeffectivity change due to surrounding chemical conditions that affect the dissolved pore water concentration. The behaviour and physiology of the organisms themselves play an important role in the effect that contaminated sediment can have: behaviour determines length and pathways of exposure, while physiological processes govern the sensitivity of organisms to contaminants, such as uptake and detoxification processes.

In most cases, sediments contain a 'cocktail' of various compounds, which may even act in a synergistic way. Contaminants in different stages of aging and residual formation are bioavailable to different extents (Alexander, 2000; Reid *et al.*, 2000). The uncertainty of the risk for river biota increases because of chemicals of unknown toxicological effects. Hence, chemical monitoring alone cannot be sufficient for estimating risks. To base assumption on the impact of contaminants on observations of the biological community on the other side can be misleading, as a number of natural environmental factors can significantly alter a healthy community. Additional pillars of environmental risk assessment focus therefore on the impact of contaminants on organisms. One of those pillars comprises the analysis of sediments in order to predict the bioavailability of substances. Sequences of different extraction steps simulate the mobility of adsorbed substances, for example, while the co-occurrence of compounds that are known to render contaminants unavailable (e.g. sulphides in anoxic sediments) is used to estimate the bioavailable fraction (Figure 8.2.1).

Empirical methods that measure an integrated response of contaminant mixtures to organisms expose test organisms to sediments under standardised conditions. Combinations of different biotests are usually employed to cover a range of sensitivities and exposure pathways. Classification of ecotoxicological responses in order to facilitate monitoring is being developed. These approaches need to address different sensitivities and the variability of biological systems, which respond to the same stress in different ways and to a different extent, before a statement on sediment quality can be justified (Ahlf and Heise, 2005, 2007; see also Chapter 8.1).

Hydrological and Hydraulic Sediment Data (3rd Column Right in Figure 8.2.1)

Sediment hydrological and hydraulic properties are subject to great variety and uncertainty (Westrich *et al.*, 2007). In most cases there are only sporadic data available on the grain size and fall velocity distribution of suspended sediments, and almost no information about sediment erosion stability. Various instruments have been developed and applied for in situ or on-site particle size and fall velocity measurements. Advanced experimental techniques, including optical methods with digital image processing, are presented and discussed by Eisma *et al.* (1997); a comprehensive overview of in situ erosion measurement devices is reported by Cornelisse *et al.* (1997); the different performances of selected erosion devices is described by Gust and Müller (1997), Sanford (2006) and Jepsen (2006). Hence, a conclusive comparison of the various methods is not yet available.

The spatial and temporal variability of the physical, geochemical and biological sediment properties is the major source of uncertainties causing large variability of the critical erosion shear stress and erosion rates, which are key parameters of resuspension processes (Gerbersdorf *et al.*, 2005). For the risk assessment of contaminated sediment resuspension, further sources of uncertainties must be considered: i) the probability of occurrence of a certain discharge known as the hydrological risk and ii) the uncertainty associated with the determination of the spatial distribution of the contamination of the sediments, in particular the depth profile, which is required to quantify the mass flux of particulate and dissolved pollutants. Physically-based numerical transport models can account for uncertainties by using the statistics of the input data and the model parameters (Westrich, 2007).

8.2.3 TREATMENT OF UNCERTAINTIES IN SEDIMENT ANALYSES

8.2.3.1 Reducing Uncertainties by Following a Weight-of-evidence Approach

In a study on historical contaminated sediments in the Rhine catchment, different lines of evidence were discussed before drawing a conclusion on the risk of a site with regard to the situation in the Port of Rotterdam (Heise and Förstner, 2006; Heise *et al.*, 2004):

- The hazard class of the site.

- The capability of exceeding CTT values in the port upon resuspension of the sediment.

- The indication for resuspension.

CTT stands for 'chemistry-toxicity test'. These action levels are set by the Dutch authorities for the permit to relocate dredged material from Rotterdam harbour to the North Sea (Stronkhorst, 2003). The indication for resuspension was considered in dependence of the hydrological situation and was based on measurements of erosion potentials and observations of increased contaminated suspended matter downstream of the site.

A three-step approach was followed in the Rhine study (Heise *et al.*, 2004), in which the hazards of 'substances of concern' and of 'areas of concern' could be determined with higher certainty than the areas which posed environmental risks to downstream sites. Description of the latter required the combined information from critical erosion thresholds, indications that resuspension took place and the extent by which particle bound contaminant concentrations exceeded risk thresholds. Addressing the uncertainty that increases with the dynamic of a system, 'hexachlorobenzene' (HCB) in sediments in the barrages of the Higher and Upper Rhine could be identified as a 'high risk with high certainty' to the Rhine downstream, even at annual flood situations.

8.2.3.2   Information from Dated Sediment Cores

A significant increase in the weight of evidence for risks to downstream target areas could be expected from the precision of the term 'indications that resuspension occurred' (Westrich and Förstner, 2005). Under favourable conditions – e.g. in areas exhibiting continuous sedimentation – the study of dated sediment cores has proven particularly useful, as it provides a historical record of the various influences on the aquatic system by indicating both the natural background levels and the anthropogenic accumulation of substances over an extended period of time (Alderton, 1985).

Best locations for such historical records are within or close to the critical target areas (harbour basins, lakes, depressions, lowlands, flood plain soils and sediments, etc.). Additional information on the source areas of specific pollutants that are analysed in the target sediment cores can be gained from indicator substances or from typical isotopes (e.g. lead isotopes) and patterns of congeners (e.g. for dioxins/furans; Götz and Lauer, 2003).

8.2.3.3   Combination of Chemical and Hydrological Data

Sediment management on a river basin level needs to have a strong focus on contaminant loads that are transported, in addition to concentration. Contaminant depots that release little quantities of highly contaminated material may become comparatively less important in the course of the river than large volumes of suspended matter from a different but less stable site with a considerably lower level of pollution.

Both kinds of data on which such load calculations are based – amount of suspended matter transported with time and contaminant concentration per volume of SPM – have a high degree of uncertainty. Analysing contaminants in suspended matter or freshly deposited sediment is an analytically difficult task due to the relatively high content of organic carbon, the difficult sampling conditions and the analytical challenge that, especially, the determination of organic contaminants at low concentrations presents. Concentration of suspended matter is probably even more difficult because it varies strongly horizontally and vertically in rivers. Mass determinations strongly vary with the applied sampling method.

Any calculations of such combinations of data with high variabilities, such as particle-bound contaminant loads in rivers, should consider these variabilities, which express themselves as uncertainties of specific data.

A mathematical tool which has gained importance in environmental assessments is fuzzy logic (Adriaenssens *et al.*, 2004; Hollert *et al.*, 2002; Silvert, 1997; Tran *et al.*, 2002). Fuzzy sets which describe different categories, such as 'low contamination' and 'moderate contamination' (Figure 8.2.2), can overlap and hence enable the membership

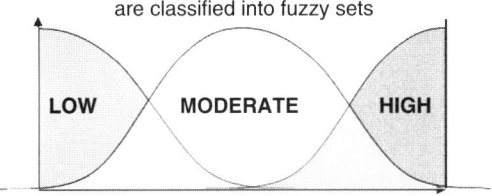

Hydrodynamic data
e.g. resuspension due to water discharges

Chemical data
e.g. contamination of SPM

are classified into fuzzy sets

LOW   MODERATE   HIGH

Outcomes are combined with 'IF....THEN ...rules'

IF river discharge is HIGH
AND contaminant load in SPM is HIGH
THEN the risk to the watershed is HIGH

Figure 8.2.2 Overlapping categories of fuzzy sets, which can be modified according to the different kinds of data and the uncertainties. The higher the uncertainties, the larger the overlap between sets. The degree of membership of a data point to one or more classes is the input into rule-based expert systems, which enable the combination of different kinds of data (Heise and Förstner, 2007). Reproduced by permission of the Royal Society of Chemistry

of certain data points to both categories. Where, for example, analytical precision does not allow one to identify the contamination of a sediment as being low or moderate, the use of fuzzy sets will allow a statement that it is both low and moderate to a certain extent. It therewith avoids the reduction or simplification of information in an early stage of assessment. The 'membership' of the data to potentially more than one class is the information that is used in further calculations or assessment of data. It can also be used when combining various kinds of data, such as hydrological data and chemical data (Figure 8.2.2).

Data of different units and uncertainties can be combined using 'IF... THEN...' rules, based on expert knowledge. Recently, the application of the traceability concept on ecotoxicological studies has been described (Ahlf and Heise, 2007). A suggestion for an ecotoxicological classification system for sediments based on fuzzy sets and fuzzy expert systems is under development (see Chapter 6.2).

8.2.4 CONCLUSIONS AND RECOMMENDATIONS

The different objectives of risk assessment and monitoring on solid material involve specific techniques favouring different media (SPM, sediments, biota). As a consequence, conditions for quality assurance will also differ. Below, practical recommendations are given. These are based on key references for reducing uncertainties in typical sediment issues under the WFD monitoring programme and programme of measures. In the three categories of sediment investigation (Figure 8.2.1), four levels of methodological standardisation can be distinguished: 1) certified reference material; 2) standardised method; 3) well-documented method; and 4) special method (references in Section 8.2.2):

I *Surveillance*, i.e. source screening and preliminary site characterisation. Media: sediments (biota). Methodologies: normalisation for grain size (2–3) and sorptive components (4); sample drying (2–3); and digestion (2–3). Additional tools: tracer substances, isotopic and congener patterns (4); (bio)accumulation on organic matrices (4); and chemometric evaluation (3).

II *Survey*, i.e. identification of anomalies and basic characterisation on the regional-to-river basin scale. Media: suspended particulate matter, sediments and biota. Methodologies: defined sampling patterns (3), hydrological and morphological conditions during sampling; sample preparation (see 'Surveillance'). Additional tools: sediment cores for measuring temporal changes (3), mechanical erosion (4) and chemical leaching experiments (1); comparison of dry and wet samples, e.g. in leaching tests (1); ecotoxicological test systems for screening and monitoring (2); metal-AVS-bioavailability (2–3).

III *Mass balance* on the regional-to-river basin scale. Media: suspended particulate matter, water. Methodologies: three levels of monitoring sophistication, as presented in Table 8.2.1 (3–4). Additional tools: sediment traps (4), transport models (3–4), mechanical erosion data (4) and sediment cores (3); morphological features, e.g. for temporal retention of contaminated solids in barrages and floodplains (4). Risk-based assessments on river basin levels using combined methodological approaches (Section 8.2.3), e.g. 'weight of evidence' approaches, fuzzy logic and expert systems; three-step pragmatic approach ('substances of concern', 'areas of concern' and 'areas of risk', sediment erosion thresholds and hydrological exceedance probability, etc.).

A fourth sediment monitoring issue under the WFD will be *assessing risks and functioning of measures*, in particular monitoring before and after remediation of contaminated sediments. In this field, initial recommendations have been presented in a guidance document of the US Environmental Protection Agency (2005) and for remediation dredging by the US National Research Council (2007): how to assess and monitor the five 'R's – the *risks* arising from *residuals, resuspension, release and recontamination*? Pre-remediation assessment differs for the major groups of technologies: i) remediation dredging and relocation; ii) in situ capping and monitored natural recovery (Förstner and Apitz, 2007); and iii) confined disposal facilities. Post-remediation monitoring will generally be based on the study of residual contamination, surface water quality and bioaccumulation effects.

8.2.5 LIST OF ABBREVIATIONS

AMPS   Expert Group on Analysis and Monitoring of Priority
        Substances
AVS   Acid Volatile Sulfide
BCR   Bureau Communautaire de Référence (BCR®)
CBM   Conceptual River Basin Model
CRM   Certified reference material

CTT   Chemistry Toxicity Test
EAF   Expert Advisory Forum on Priority Substances and
           Pollution Control
EQS   Environmental Quality Standards
IRMM   Institute for Reference Materials and Measurements,
           Joint Research Centre, Belgium
QC   Quality Control
SPM   Suspended Particulate Matter
WFD   Water Framework Directive

REFERENCES

Adriaenssens, V., Baets, B.D., Goethals, P.LM. and Pauw, N.D. (2004) Fuzzy rule-based models for decision support in ecosystem management, *Sci. Total Environ.*, **319**, pp. 1–12.

Ahlf, W. and Heise, S. (2005) Sediment toxicity assessment: rationale for effect classes, *J. Soils and Sediments*, **5**, pp. 16–20.

Ahlf, W. and Heise, S. (2007) Quality assurance of ecotoxicological sediment analysis, In: Westrich, B. and Förstner, U. (eds) *Sediment Dynamics and Pollutant Mobility in Rivers: An Interdisciplinary Approach*, Springer, Berlin, Germany, Chapter 10.1, pp. 380–391.

Alderton, D.H.M. (1985) Sediments, In: MARC Technical Report 31, *Historical Monitoring*, Monitoring and Assessment Research Centre, University of London, pp. 1–95.

Alexander, M. (2000) Aging, bioavailability, and overestimation of risk from environmental pollutants, *Environ. Sci. Technol.*, **34**, pp. 4259–4391.

Apitz, S. and White, S. (2003) A conceptual framework for river-basin-scale sediment management, *J. Soils and Sediments*, **3**, pp. 132–138.

Bergmann, H. and Maass, V. (2007) Sediment regulations and monitoring programmes in Europe, In: Heise, S. (ed.) *Sediment Risk Management and Communication*, Sustainable Management of Sediment Resources, Volume 3, Elsevier, Amsterdam, pp. 207–231.

Brils, J., Salomons, W. and van Veen, J. (eds) (2003) *SedNet Recommendations for Research Priorities Related to Sediment*, Den Helder (see SedNet, 2003).

Carr, R.S. and Nipper, M. (eds) (2001) Summary of a SETAC Technical Workshop on Porewater Toxicity Testing: Biological, Chemical, and Ecological Considerations with a Review of Methods and Applications, and Recommendations for Future Areas of Research, Society of Environmental Toxicology and Chemistry, Pensacola, FL, USA.

Cornelisse, J.M., Mulder, H.P.J., Houwing, E.J., Williamson, H.J. and Witte, G. (1997) On the development of instruments for in-situ erosion measurements, In: Burt, N., Parker, R. and Watts, J. (eds) *Cohesive Sediments*, John Wiley and Sons, Ltd, Chichester, UK, pp. 175–186.

Den Besten, P.J., de Deckere, E., Babut, M.P., Power, B., Angel DelValls, T., Zago, C. *et al.* (2003) Biological effects-based sediment quality in ecological risk assessment for European waters, *J. Soils and Sediments*, **3**, pp. 144–162.

DiToro, D.M., Mahony, J.D., Hansen, D.J., Scott, K.J., Carlson, A.R. and Ankley, G.T. (1992) Acid volatile sulfide predicts the acute toxicity of cadmium and nickel in sediments, *Environ. Sci. Technol.*, **26**, pp. 96–101.

Eisma, D., Dyer, K.R. and van Leussen, W. (1997) The in-situ determination of the settling velocities of suspended fine-grained sediment: a review, In: Burt, N., Parker, R. and Watts, J. (eds) (1997) *Cohesive Sediments*, John Wiley and Sons, Ltd, Chichester, UK, pp. 17–44.

Förstner, U. (1989) *Contaminated Sediments*, Lecture Notes in Earth Sciences, Springer, Berlin, Germany.

Förstner, U. (2004) Traceability of sediment analysis, *Trends Anal. Chem.*, **23**, pp. 217–236.

Förstner, U. (2008) Differences in policy response to similar scientific findings – examples from sediment contamination issues in River Basin Management Plans. New perspectives with sediment issues (monitoring and measures) under the European Water Framework Directive. *J. Soils and Sediments*, **8**, pp. 214–216.

Förstner, U. and Apitz, S. (2007) Sediment remediation: US focus on capping and monitored natural recovery, *J. Soils and Sediments*, **7**, pp. 351–358.

Förstner, U. and Heise, S. (2006) Assessing and managing contaminated sediments: requirements on data quality: from molecular to river basin scale, *Croatica Chem Acta*, **79**, pp. 5–14.

Förstner, U. and Owens, P.N. (2007) Sediment quantity and quality issues in river basins, In: Westrich, B. and Förstner, U. (eds) *Sediment Dynamics and Pollutant Mobility in Rivers: An Interdisciplinary Approach*, Springer, Berlin, Germany, Chapter 1.1, pp. 1–15.

Fritsche, A., Börnick, H. and Worch, E. (2007) Equilibrium and kinetics of sorption/desorption of hydrophobic pollutants on/from sediments, In: Westrich, B. and Förstner, U. (eds) *Sediment Dynamics and Pollutant Mobility in Rivers: An Interdisciplinary Approach*, Springer, Berlin, Germany, Chapter 6.3, pp. 241–249.

Gandrass, J. and Eberhardt, R. (2000) 'New' substances: substances to watch, In: Gandrass, J. and Salomons, W. (eds) *Dredged Material in the Port of Rotterdam: Interface between Rhine Catchment Area and North Sea*, GKSS Research Centre, Geesthacht, Germany, pp. 289–305.

Gerbersdorf, S., Jancke, T. and Westrich, B. (2005) Physico-chemical and biological sediment properties determining erosion resistance of contaminated riverine sediments: temporal and vertical pattern at the Lauffen reservoir, River Neckar, Germany, *Limnologica*, **35**, pp. 132–144.

Götz, R. and Lauer, R. (2003) Analysis of sources of dioxin contamination in sediments and soils using multivariate statistical methods and neural networks, *Environ. Sci. Technol.*, **37**, pp. 5559–5565.

Gust, G. and Müller, V. (1997) Interfacial hydrodynamics and entrainment functions of currently used erosion devices, In: Burt, N., Parker, R. and Watts, J. (eds) *Cohesive Sediments*, John Wiley and Sons, Ltd, Chichester, UK, pp. 149–174.

Haag, I., Kern, U. and Westrich, B. (2001) Erosion investigation and sediment quality measurement for a comprehensive risk assessment of contaminated aquatic sediment, *Sci. Total Environ.*, **266**, pp. 249–257.

Heise, S. and Förstner, U. (2006) Risks from historical contaminated sediment in the Rhine basin, *Water Air Soil Pollut., Focus*, **6**, pp. 625–636.

Heise, S. and Förstner, U. (2007) Risk assessment of contaminated sediments in river basins: theoretical considerations and pragmatic approach, *J. Environ. Monit.*, **9**, pp. 943–952.

Heise, S., Förstner, U., Westrich, B., Jancke, T., Karnahl, J., Salomons, W. and Schönberger, H. (2004) *Inventory of Historical Contaminated Sediment in Rhine Basin and its Tributaries, Report on Behalf of the Port of Rotterdam*, Hamburg/Stuttgart.

Hollert, H., Heise, S., Pudenz, S., Brüggemann, R., Ahlf, W. and Braunbeck, T. (2002) Application of a sediment quality triad and different statistical approaches (Hasse diagrams and fuzzy logic) for the comparative evaluation of small streams, *Ecotoxicology*, **11**, pp. 311–321.

Horowitz, A. (1991) *A Primer on Sediment-Trace Element Chemistry*, 2nd ed., Lewis Publ., Chelsea, MI, USA.

Jacoub, G.K. and Westrich, B. (2007) Two-dimensional numerical module for contaminant transport in rivers, In: Westrich, B. and Förstner, U. (eds) *Sediment Dynamics and Pollutant Mobility in Rivers: An Interdisciplinary Approach*, Springer, Berlin, Germany, Chapter 4.1, pp. 118–129.

Jepsen, R.P. (2006) Uncertainty in experimental techniques for measuring sediment erodibility, *Integr. Environ. Assess. Manag.*, **2**, pp. 39–43.

Karnahl, J. and Westrich, B. (2007) Two-dimensional numerical modeling of fine sediment transport behavior in regulated rivers, In: Westrich, B. and Förstner, U. (eds) *Sediment Dynamics and Pollutant Mobility in Rivers: An Interdisciplinary Approach*, Springer, Berlin, Germany, Chapter 4.2, pp. 130–142.

Kersten, M. and Förstner, U. (1991) Geochemical characterization of the potential trace metal mobility in cohesive sediment, *Geomarine Lett.*, **11**, pp. 184–187.

Krantzberg, G., Hartig, J.H. and Zarull, M.A. (2000) Sediment management: deciding when to intervene, *Environ. Sci. Technol.*, **34**, pp. 22A–27A.

Luthy, R.G., Aiken, G.R., Brusseau, M.L., Cunningham, S.D., Gschwend, P.M., Pignatello, J.J. *et al.* (1997) Sequestration of hydrophobic organic contaminants by geosorbents, *Environ. Sci. Technol.*, **31**, pp. 3341–3347.

METROPOLIS (2004) Evaluation of current gaps and recommendations for further actions in the field of environmental analysis and monitoring, Position Paper, March 2004, Metrology in Support of EU Policies, Verneuil-en-Halatte, France.

Meyer, J.S., Davison, W., Sundby, B., Oris, J.T., Laurén, D.J., Förstner, U. *et al.* (1994) The effects of variable redox potentials, pH, and light on bioavailability in dynamic water-sediment environments, In: Hamelink, J., Landrum, P.F., Bergman, H.L. and Benson, W.H. (eds) *Bioavailability: Physical, Chemical and Biological Interactions*, Lewis Publ., Boca Raton, FL, USA, Synopsis Chapter, pp. 155–170.

Mudroch, A. and Azcue, J.M. (1995) *Manual of Aquatic Sediment Sampling*, Lewis Publ., Boca Raton, FL, USA.

Quevauviller, P. (ed.) (2002) *Methodologies for Soil and Sediment Fractionation Studies*, The Royal Society of Chemistry, Cambridge, UK.

Quevauviller, P. (2004) Traceability of environmental chemical measurements, *Trends Anal. Chem.*, **23**, pp. 171–177.

Rauret, G., Lopéz-Sánchez, J.F., Lück, D., Yli-Halla, M., Muntau, H. and Quevauviller, P. (2001) Certification Report of BCR-684, EUR 19776 EN, European Commission, Brussels, Belgium.

Reid, B.J., Jones, K.C. and Semple, K.T. (2000) Bioavailability of persistent organic pollutants in soils and sediments: a perspective on mechanisms, consequences and assessment, *Environ. Pollut.*, **108**, pp. 103–112.

Salomons, W. (1995) Long-term strategies for handling contaminated sites and large-scale areas, In: Salomons, W. and Stigliani, W.M. (eds) *Biogeodynamics of Pollutants in Soils and Sediments: Risk Assessment of Delayed and Non-Linear Responses*, Springer, Berlin, Germany, pp. 1–30.

Sanford, L.P. (2006) Uncertainties in sediment erodibility estimates due to a lack of standards for experimental protocols and data interpretation, *Integr. Environ. Assess. Manag.*, **2**, pp. 29–34.

SedNet (2003) The SedNet strategy paper, Demand Driven European Sediment Research Network (SedNet), Proposal No. EVK-2001-00058 to EU Key Action 1 'Sustainable management and quality of water', 1.4.1 'Abatement of water pollution from contaminated land, landfills and sediments' (1/2002–12/2004), Den Helder.

Silvert, W. (1997) Ecological impact classification with fuzzy sets, *Ecol. Modelling*, **96**, pp. 1–10.

Stronkhorst, J. (2003) Ecotoxicological effects of Dutch harbour sediments, PhD Thesis, Free University of Amsterdam.

Thomas, R. and Meybeck, M. (1992) The use of particulate material, In: Chapman, D. (ed.) *Water Quality Assessment: A Guide to the Use of Biota, Sediments and Water in Environmental Monitoring*, Chapman and Hall, London, UK, Chapter 4, pp. 121–170.

Tran, L.T., Knight, C.G., O'Neill, R., Smith, E.R., Ritters, K.H. and Wickham, J. (2002) Environmental assessment: fuzzy decision analysis for integrated environmental vulnerability assessment of the Mid-Atlantic Region, *Environ. Manag.*, **29**, pp. 845–859.

UNESCO (1978) Water quality survey: a guide for the collection and interpretation of water quality data, Studies and Reports in Hydrology No. 23, UNESCO, Paris, France.

US Environmental Protection Agency (2005) Contaminated sediment remediation guidance for hazardous waste sites, Office of Solid Waste and Emergency Response (OSWER), 9355.0-85, EPA-540-R-05-012, Washington, DC, USA.

US National Research Council (2007) *Sediment Dredging at Superfund Megasites*, The National Academic Press, Washington, DC, USA.

Westrich, B. (2007) Sustainable sediment management, In: Westrich, B. and Förstner, U. (eds) *Sediment Dynamics and Pollutant Mobility in Rivers: An Interdisciplinary Approach*, Springer, Berlin, Germany, Chapter 2.1, pp. 35–49.

Westrich, B. and Förstner, U. (2005) Sediment dynamics and pollutant mobility in rivers (SEDYMO), assessing catchment-wide emission-immission relationships from sediment studies, *J. Soils and Sediments*, **5**, pp. 197–200.

Westrich, B., Li, C.-C., Hammer, D. and Förstner, U. (2007) Requirement on sediment data quality: hydrodynamics and pollutant mobility in rivers, In: Westrich, B. and Förstner, U. (eds) *Sediment Dynamics and Pollutant Mobility in Rivers: An Interdisciplinary Approach*, Springer, Berlin, Germany, Chapter 2.2, pp. 49–65.

WFD/AMPS (2004) Expert Group on Analysis and Monitoring of Priority Substances (AMPS), Discussion Document, 13 January 2004, Ispra, Italy. See also: WFD AMPS (2004) Sediment monitoring guidance discussion document, AMPS and SedNet, Draft Version 1 from 16 April 2004, Brussels, Belgium.

WFD/EAF (2004) Expert Advisory Forum (EAF) on Priority Substances and Pollution Control Concept Paper on Emission Control from 8 June 2004, Common Implementation Strategy (CIS) for the Water Framework Directive (2000/60/EC), 7th EAF Meeting at Brussels, 14–15 June 2004.

Section 9
Reporting Requirements

9.1

Reporting Requirements for Priority Substances

Valeria Dulio and Anne Morin

The Water Framework Directive - Ecological and Chemical Status Monitoring   Edited by Philippe Quevauviller, Ulrich Borchers, Clive Thompson and Tristan Simonart © 2008 John Wiley & Sons, Ltd

9.1.1   INTRODUCTION

Article 16 of the WFD, which sets out the European Union (EU) strategy against pollution of water by chemical substances, requires the Commission to identify priority substances of European relevance, for which priority action is needed at Community level. The first list of priority substances was adopted in November 2001 (Decision 2455/2001/EC). In order to assess the risk of failing the objective of 'good chemical status',[1] Member States are obliged to monitor these substances in all water bodies.

Article 16 also states that the list of priority substances needs to be reviewed by the Commission every four years. For the definition of the first list of priority substances, in accordance with the provisions of the WFD, a Combined Monitoring-based and Modelling-based Priority Setting scheme (COMMPS) was devised. The implementation of this scheme also involved a data collection exercise to assess the level of contamination by chemical substances in Europe. This led to the development of the COMMPS monitoring database. But this was a one-off exercise (the database was never updated) and it was therefore recognised as fundamental that a data collection process should be carried out on a regular basis, with data to be included in the Water Information System for Europe (WISE)[2] developed under the Common Implementation Strategy (CIS).

For this purpose the Commission developed a common template for regular reporting of chemical monitoring data in WISE. The template is proposed as a common template for the various data collection exercises on chemical substances to be carried out by EC DG ENV for the WFD, and by EEA for the preparation of the 'State of the Environment' (SOE) report.

The current format of the template is the result of year-long consultation with the representatives of the various Member States and major stakeholders involved in the implementation of the Water Framework Directive (WFD).[3]

This chapter attempts to provide an overview and an explanation of the main metadata (information about the data) that Member States should provide in order to allow a sufficient level of traceability and comparability of the monitoring data that will be collected in this database at European level.

9.1.2   THE IMPORTANCE OF METADATA

Monitoring data (measured data), together with calculated data (modelling estimation), are the basis for environmental exposure assessment through the calculation of a predicted environmental concentration (PEC). Compared to modelling data, measured

---

[1] According to the WFD, the assessment of the 'chemical status' concerns *exclusively* the substances listed in Annex IX and Annex X (substances of European relevance).

[2] WISE (http://water.europa.eu) is the system that organises the shared pool of common and timely data and information on the state of, and pressures on, Europe's water. The collected information is designed to meet the needs of all those organisations required to report and make assessments at a European level.

[3] The reporting template was developed by INERIS under its contract with DG ENV 'Implementation of Requirements for Priority Substances within the Context of the Water Framework Directive' (DGENV No. 07-010401/2006/432521/MAR/D2).

data have the advantage of providing information from direct observation of the status of the European environment. However, measured data need to be carefully checked for their reliability and comparability and their representativeness. This leads to considerations about the essential role of the metadata.

Ideally, an exhaustive set of metadata should accompany each data set as a means of helping evaluation of the acceptability and fitness for purpose of the data, their representativeness and their comparability with other data sets and, ultimately, as a means of deciding whether the collected data are adequate for use in the exposure assessment.

However, establishing minimum requirements for information to be reported by the Member States at the EU level is not easy; a balance must be struck between asking for the finest detail and minimum 'acceptable' requirements.

For the COMMPS process, the information that it was possible to collect for all data records was limited to the sampling site, year of sampling and limit of determination. The process had to cope with heterogeneous data of highly variable quality.

In general it is still difficult in the current monitoring programmes to access metadata such as the measurement methods employed, the detection limits of the methods and details of the quality assurance/quality control (QA/QC) procedures associated with the reported monitoring activities. What metadata are required for data reporting to the EU Commission under the WFD?

The data flow under the WFD, from the laboratory via the national competent monitoring authority to the EU Commission and the Water Information System for Europe is complex.

9.1.3 WHAT METADATA ARE REQUIRED FOR DATA REPORTING TO THE EU COMMISSION UNDER THE WFD?

Besides metadata such as the substance name and CAS number, the water body (name, type, etc.), the name of the sampling station and GPS coordinates, sampling date and type of matrix – which should indeed be regarded as mandatory information (i.e. Category 1 metadata) – it is also important to address in detail the characterisation of the sample matrix. As explained more fully in the following sections, without this information it will be difficult to compare results from, for example, different sites or samples collected at the same site but at different times.

Moreover, the QA/QC information about the laboratories which have carried out the sampling and the analytical measurements should be kept to the forefront in the gathering and interpretation of WFD data.

Although not all QA/QC information has to be available at each step of the information flow, suitable links should be kept between the data and the associated QA/QC information.

Whereas it is acknowledged that prime responsibility for assessment and demonstration of the quality and fitness for purpose of monitoring data should lie with the member country, a sub-set (summary) of the QA/QC data should be transmitted to WISE in order to allow an improved assessment of the between-country comparability of the data and, as a result, a more appropriate aggregation of the data sets at the EU level and ideas for possible use of the data in the future for different purposes. The common template for data collection was designed to meet these objectives.

9.1.4   THE COMMON TEMPLATE FOR MONITORING DATA COLLECTION

The metadata included in the EU data reporting template are reported in Table 9.1.1. Each field in the template is labelled according to the final use of the information collected:

- 'WFD': for information to be collected by DG ENV for WFD reporting purposes.

- 'SOE': for information to be collected by EEA for SOE reporting purposes.

- 'WFD, SOE': for information required by both of these data collection exercises.

The first and short-term application of the reporting template was made for the EU-wide collection of monitoring data launched in 2007 for the revision of the first list of priority substances.

A key issue here was to have information and data at the most comparable level possible, whilst taking into account that different Member States have for the moment different monitoring procedures and different rules for data collection and reporting. In order to encourage countries to respond, the 'WFD' and '**WFD, SOE**' fields (and the corresponding data/information) in the common reporting format were sub-divided into three different categories:

- *Category 1 data:* the corresponding fields identify the mandatory information needed for the prioritisation process.

- *Category 2 data:* the corresponding fields are seen as relevant in some specific cases, depending on the final use of the data.

- *Category 3 data:* the complete list of fields is needed to match high-quality requirements, but may not always be available.

For a more thorough assessment of comparability, Category 2 and Category 3 data will be requested in future data collections under WISE.

9.1.5   CHARACTERISATION OF THE SAMPLE MATRIX FOR DETERMINATIONS IN WATER (ALL TYPES OF CHEMICAL)

When completing the template for determinations in *water*, it is necessary that the template requires specification of whether the measurement was made on the dissolved fraction or on the whole water sample.

Table 9.1.1 Data collection template

Requested information	Format/Unit/Pick list	Label Obligatory fields (*)
Data source	• WFD mandatory data • National official data • Other monitoring data (national, regional, local, from public bodies) • Research and technical studies (from NGOs, industry, etc.)	**WFD, SOE** Cat. 1*
Country code	Abbreviation of EU or EEA Member or Collaborating Country (ISO 3166-alpha-2 code elements)	**WFD, SOE** Cat. 1*
RBD	Identifier of the River Basin District in which the Water Body is located Specify the code as reported in WISE	**WFD, SOE** Cat. 1*
Water Body (WB) ID	National Identification Code of the Water Body in which the station is located (if available) Name of the Water Body in which the station is located	**WFD, SOE** Cat. 2
Water Body Category Code	• SWB: Surface Water Body: (RW: river | LW: lake | CW: coastal water | TW: transitional water | MW: marine water) • GWB: Groundwater body (GW: groundwater)	**WFD, SOE** Cat. 1*
Sampling Station ID	National code of the sampling station Abbreviation of EU or EEA Member or Collaborating Country (ISO 3166-alpha-2 code elements)	**WFD, SOE** Cat. 1*
Name of the Sampling Station	National name of the station	**WFD, SOE** Cat. 1*
Latitude	International geographical coordinates in decimal degrees format. Provide latitude using the common geodetic datum ETRS89 (rivers, lakes, groundwater, water quantity) or WGS84 (TCM) Negative values should be used for coordinates west of the Greenwich Meridian (0 degrees) Please do not round coordinate values	**WFD, SOE** Cat. 1*
Longitude	International geographical coordinates in decimal degrees format. Provide longitude using the common geodetic datum ETRS89 (rivers, lakes, groundwater, water quantity) or WGS84 (TCM) Negative values should be used for coordinates west of the Greenwich Meridian (0 degrees) Please do not round coordinate values	**WFD, SOE** Cat. 1*
Precision of coordinates	In metres, 3 classes: precise (range 1–10 m), average (range 10–100 m), low (range 100–1000 m)	WFD Cat. 3
Type of station	For SWB (river, lakes, transitional coastal waters): • SWB5a: Operational monitoring • SWB5b: Surveillance monitoring • SWB5c: Drinking water • SWB5d: Investigative monitoring • SWB5e: Reference monitoring	**WFD, SOE** Cat. 1

(continued overleaf)

Table 9.1.1 (*continued*)

Requested information	Format/Unit/Pick list	Label Obligatory fields (*)
	For GWB:	
	• GW2b: Operational monitoring network chemical • GW2c: Surveillance monitoring network chemical • DW: Drinking water site • I: Industrial supply site • O: Other use site	

Additional information for groundwater stations	• Groundwater body area (km^2)
	• Main aquifer type
	• Detailed information on horizon and superpositioning
	• Maximum length (km)
	• Maximum width (km)
	• Long-term annual precipitation (mm)
	• Minimum, Mean and Maximum
	• Description of the stratigraphy
	• Description of the petrography
	• Thickness (m)
	
	   – Minimum, Mean and Maximum
	
	• Overlying strata (Free text). Information applicable to the groundwater body
	• Depth from the surface (m). Information applicable to the groundwater body
	
	   – Minimum, Mean and Maximum
	
	• Main recharge source
	• Hydraulic conductivity (kf-value). Information applicable to the groundwater body
	
	   – Minimum, Mean and Maximum
	
	• Annual GW level amplitude (m). Information applicable to the groundwater body
	
	   – Minimum, Mean and Maximum
	
	• Information about trans-boundary. Information applicable to the groundwater body
	• Capacity (m3)
	• Reference year above data
		SOE

Additional information for lake stations	• Surface of the area (km^2) • Average depth (m) • Average lake residence time in years. Calculated as volume/annual flow	SOE
Additional information for marine stations	• Distance from nearest mainland or coast • Purpose of monitoring station: – European/national/Marine convention • Environmental compartments • Average annual depth (m) • Mean tidal range (m)	SOE

Table 9.1.1 (*continued*)

Requested information	Format/Unit/Pick list	Label Obligatory fields (*)
	• Mean annual temperature ($^\circ$C) • Residence time (days) • Mixing characteristics • Salinity: (psu: practical salinity units) – Minimum, Mean and Maximum	
Additional information for river stations	• Altitude as regards sea level with the use of the ETRF93 altitude reference system (m) • Average width (m) • Average depth (m)	SOE
Proxy pressures	Proxy pressures associated with catchment of the station: name and description: • Aquaculture • Dam construction • Direct discharges from sewage treatment works and industry • Oil or gas extraction • Dredged spoil or waste disposal ground • Directly impacted by leachate from landfill disposal sites • Mariculture • Fishing • Marina • Port facilities • Other activities • Other discharges • Downstream river monitoring station • Water abstracted from groundwater body • Artificial recharge of the groundwater body • Main infrastructures affecting the dynamics of the groundwater body • Associated aquatic ecosystems	SOE
Date of sampling	Date in which sample was taken (yyyy/mm/dd)	**WFD, SOE** Cat. 1*
Sampling depth	For **spot sampling**: indicate the sampling depth in metres below water surface For **depth integrated sampling**: • Composite sample: indicate all sampling depths separated by '–' (e.g. 1-2-5-10, for samples taken at 1, 2, 5 and 10 m depths) • Continuous integrated sample: indicate the depth range (e.g. 1 to 10 for 1 to 10 m range) For depth integrated sampling, indicate the aggregation method used to calculate the depth-averaged concentration (text)	**WFD, SOE** Cat. 2
Sampling method	ISO/CEN code, or national code and additional description information	**WFD, SOE** Cat. 2

(continued overleaf)

Table 9.1.1 (*continued*)

Requested information	Format/Unit/Pick list	Label Obligatory fields (*)
Sample matrix	• W: Water • S: Sediment • B: Biota	**WFD, SOE** Cat. 1*
Fraction analysed (for water)	**For water:** • Dissolved fraction • Whole water with no separation of liquid and SPM phases • Whole water with determination on each separate phase (sum of all phases) • SPM	WFD Cat. 1*
Fraction analysed (for sediment)	**For sediment:** • (Whole) Fraction <2 mm • Fraction <63 µm • Fraction <50 µm • Fraction <20 µm • Other	WFD Cat. 1*
	For sediment: • Mud • Fine • Sand	SOE
Species group (for biota)	For biota: • M: Macrophyte • Mo: Molluscs • F: Fish • O: Other biota	WFD Cat. 1*
Species name (for biota)	Specify name of the species (Latin name)	**WFD, SOE** Cat. 1*
Additional information for determinations in water	SPM concentration: value in mg/l (This information is strongly recommended when the fraction analysed is 'whole water')	**WFD, SOE** Cat. 2
Additional information for bioavailability of metals in water	Ca (mg/l) Mg (mg/l) Na (mg/l) K (mg/l) SO_4 (mg/l) Cl (mg/l)	**WFD, SOE** Cat. 2
	S (mg/l)	WFD Cat. 2
	pH	**WFD, SOE** Cat. 2
	Temperature (°C)	**WFD, SOE** Cat. 2

Table 9.1.1 (*continued*)

Requested information	Format/Unit/Pick list	Label Obligatory fields (*)
	Dissolved organic carbon (mg/l)	**WFD, SOE** Cat. 2
	Hardness in mg/l of CaCO$_3$	**WFD, SOE** Cat. 2
Additional information for bioavailability of metals in sediments	Acidic volatile sulphides (AVS) in µmol/g of dry weight	**WFD** Cat. 2
Additional information for determinations in sediments	• TOC (% of total dry weight)	**WFD, SOE** Cat. 2
	• Concentration normalised for the particle size (yes | no | not known)	WFD Cat. 2
	• Grain size distribution:	WFD
	• % fraction <63 µm (or 50 µm)	Cat. 2
	• Upper Limit of particle size in the analysed fraction (µm)	SOE
Additional information for determinations in biota	• Tissue element of species monitored (Liver, Muscle, Soft Body, Whole Body, Not known)	**WFD, SOE** Cat. 2
	• Biota size:	
	– Mo : Molluscs (mm, cm)	
	– F : Fish (mm, cm)	
	– O: Other	
	• Number of organisms used	
	• Basis of measurement (Wet weight | Dry weight | Lipid (Fat) weight | Unknown)	
	• Dry Wet Ratio: Ratio of dry weight to wet (or fresh) weight (expressed as %)	
	• Fat content (% of total wet matter)	
Determinand/ measurand	English common name, upper case	**WFD, SOE** Cat. 1*
	When information on the **individual substance** exists, it should be provided. Only in the absence of such data should data relating to the **substance family** be given	
	When groups of substances are analysed together (e.g.: sum of 16 PAH), the name of the group of substances should be reported here (e.g.: sum 16 PAH)	
Substance unique code	Internationally agreed code: enter code and specify the code type (CAS | ELINCS | EINECS)	**WFD, SOE** Cat. 1*
Value	Concentration of determinand/measurand in the sample	**WFD, SOE** Cat. 1*
Unit (of determination)	Unit in which the measurement (determination) is expressed	**WFD, SOE** Cat. 1*
Uncertainty	% (value ± %)	**WFD** Cat. 2

(*continued overleaf*)

Table 9.1.1   (*continued*)

Requested information	Format/Unit/Pick list	Label Obligatory fields (*)
Date of analysis	Date on which the sample was analysed (yyyy/mm/dd)	WFD Cat. 2
Analytical method	ISO/CEN method or national or other method and description of the method used	**WFD, SOE** Cat. 2
Laboratory ID	Unique national code Abbreviation of EU or EEA Member or Collaborating Country (ISO 3166-alpha-2 code elements)	**WFD, SOE** Cat. 2
Laboratory name	Full name, upper case	**WFD, SOE** Cat. 1*
Limit of Detection (LoD)	Same unit as concentration value	**WFD, SOE** Cat. 1*
Limit of Quantification (LoQ)	Same unit as concentration value	**WFD, SOE** Cat. 1*
Uncertainty at LoQ	%	WFD Cat. 2
Is the laboratory accredited for the specified determinand?*	Yes/no/not known	**WFD, SOE** Cat. 2
Are the data controlled by a competent authority (apart from accreditation bodies)? *	Yes/no/not known	**WFD, SOE** Cat. 2
Is a field blank checked?*	Yes/no/not known	**WFD, SOE** Cat. 3
Extraction recovery*	%	**WFD, SOE** Cat. 2
Have the results been corrected for extraction recovery?*	Yes/no/not known/not applicable	**WFD, SOE** Cat. 2
Are control charts recorded for each determinand to test bias and reproducibility?*	yes/no/not known	**WFD, SOE** Cat. 3

Table 9.1.1 (*continued*)

Requested information	Format/Unit/Pick list	Label Obligatory fields (*)
Frequency of control*	X ≤ 10 samples	**WFD, SOE**
	X > 10 samples	Cat. 3
Does the laboratory participate in inter-laboratory studies for the given determinand?*	Yes/no/not known	**WFD, SOE** Cat. 3
Summary of performance of the laboratory in the inter-laboratory study for the given determinand*	z-score (according to ISO-13528) ≤ 3 z-score (according to ISO-13528) > 3 not known	**WFD, SOE** Cat. 3
Additional information on quality of laboratory	• EEA ETC/WTR data quality index value 1 to 12 • Data screening • Limit of Detection or Determination Flag (flag to indicate sample below analytical limit of detection or determination, in format <)	SOE

*This information addresses the application of QA/QC procedures employed by the laboratories that have carried out the sampling and the analytical measurements. The required information will need to be submitted by laboratory **and** by determinand rather than by single measurement.

Moreover, for measurements made on the whole water sample, in order to allow better data comparability, it should be specified whether the analysis was conducted with separation of the two phases and determination of the contaminant in the two separate phases (dissolved and suspended particulate matter (SPM)), or on the whole water sample without separation of the liquid and SPM phases. The concentration of SPM should also be indicated in order to assess the risk of underestimating the concentration of the contaminant when whole water is analysed without separation of the two phases.

This is crucial when sampling is conducted with high SPM concentrations, such as in large rivers and estuaries, and when highly hydrophobic substances are involved. It is clear, in fact, that solvent extraction of bulk water samples with high SPM concentrations will be much less efficient for substances with log $K_{ow} > 3$ (e.g. PCBs, PAHs, trichlorobenzenes, pentachlorophenol) than if performed on the SPM itself, using extraction methods designed for solid phases, such as sediment or soil (Coquery *et al.*, 2005).

The same considerations apply for metals. Heavy metals have a strong tendency to be adsorbed onto the surface of solids and for this reason the template requires the determination of the concentration of SPM in the sample for all measurements on whole water samples. However, it should be taken into account that the principal matrix for assessing compliance with respect to Environmental Quality Standards (EQS) (and therefore for reporting monitoring data) for priority substances is whole water, or for metals, the dissolved fraction obtained by filtration of the whole water sample (European Commission, 2007).

9.1.6 BIOAVAILABILITY OF METALS FOR EXPOSURE CONCENTRATIONS

Knowledge of whether dissolved or total metal was measured is the first crucial information. Moreover, in order to enable further investigation of the *bio-available fraction* from the measured concentration in the *dissolved fraction*, additional parameters have been included in the reporting template (see Table 9.1.2). They are classified as Category 2 metadata (i.e. non-mandatory information in the current prioritisation process). These parameters (metadata) are also mentioned in the EIONET-Water template, but actual data are not often reported and not readily available in most of the existing databases.

The determination of metals in the 'dissolved fraction' refers, by convention, to the dissolved concentration measured in the liquid fraction of a water sample obtained by filtration through a 0.45 µm filter. According to the scientific literature, toxicity of waterborne metals is primarily induced by free metallic ions, and possibly by some of the hydroxy complexes (formation of organic and inorganic metal complexes renders a significant fraction of the total metal non-bioavailable). But there exists no experimental method for accurate measurement of the quantity of free metal ions. This quantity may be estimated by speciation models, taking into account the environmental conditions such as pH, alkalinity, hardness and presence of complexing agents (Bonnomet and Alvarez, 2006).

Recent scientific research has led to the development of the Biotic Ligand Model (BLM). The BLM was designed to predict ecotoxicity from dissolved metal and given

Table 9.1.2 Information (metadata) required in the template for measurements of metals in water

Additional information for bioavailability of metals in water – waterborne exposure
Ca (mg/l)
Mg (mg/l)
Na (mg/l)
K (mg/l)
SO_4 (mg/l)
Cl (mg/l)
S (mg/l)
pH
Temperature ($^\circ$C)
Dissolved Organic Carbon (mg/l)
Hardness in mg/l of $CaCO_3$

water characteristics. The BLM is based on the assumption that toxicity is related to metal bound to a biochemical site (the biotic ligand).

For the application of the BLM it is necessary to provide a detailed ionic composition of the water in order to predict the concentration of free metallic ions by speciation models. The major ions required for the application of the BLM are: Ca, Mg, Na, K, SO_4, Cl, S.

Furthermore, water characteristics such as pH, alkalinity and temperature are required for the application of the speciation models.

Finally, the concentration of dissolved organic matter and amount of humic acids are required in order to take into account the interaction of free metallic ions with the dissolved organic matter, especially humic substances. This part of the model has not been fully validated for natural conditions (high diversity and complexity of natural organic matter). Furthermore, dissolved organic matter may interact directly with biological material.

The development of biotic ligand models is in progress. The possibility of calculating the bio-availability of metals in the aquatic compartment is still under discussion and the validity of this approach still needs to be assessed for more species and for chronic exposures (Bonnomet and Alvarez, 2006). However, the EU has used BLM in the framework of Regulation No. 793/93 for the risk assessment of zinc. Based on these considerations it was decided that these parameters should be included in the reporting template, although they are not mandatory at the present stage.

9.1.7 CHARACTERISATION OF THE SAMPLE MATRIX FOR DETERMINATIONS IN SEDIMENTS (ALL TYPES OF CHEMICAL)

It is widely recognised that most natural and anthropogenic substances (metals and organic contaminants) show a much higher affinity to fine particulate matter than to the coarse fraction. Constituents such as organic matter and clay minerals contribute to the affinity to contaminants in this fine material (OSPAR, 2002). As a result, a sediment containing a higher percentage of fine material will tend to display higher concentrations of pollutants than will a sediment with a higher percentage of coarse material. In a monitoring programme at any level, this uneven distribution of the contaminants makes it difficult to compare sediments from different sites or, indeed, samples taken at the same site but at different times.

This problem has been recognised since the earliest stages of the study of sediments and various normalisation methods have therefore been proposed for eliminating the heterogeneity attributable to particle size distribution or, more precisely, to the presence of relatively inert coarse material.

Normalisation is usually achieved by relating the contaminant concentration with components of the sediment that represent its affinity for contaminants (so-called 'normalisers'). Normalisation can be achieved by calculating the concentration of a contaminant with respect to a specific grain-size fraction such as <2 μm (clay), <20 μm or <63 μm (OSPAR, 2002). Aluminium and lithium can also be used as co-factors, since they are major elements of the clay fraction, whereas organic matter, usually represented by organic carbon, is the most common co-factor for organic contaminants, due to its strong affinity to this sediment component.

Another way to allow comparison of different sediment samples is to isolate the fine fraction by sieving (e.g. $<20\,\mu m$, $<63\,\mu m$). In this way the coarse particles, which do not usually bind anthropogenic contaminants and dilute their concentrations, are removed from the sample. Contaminant concentrations measured in these fine fractions can then be directly compared. This technique is applicable to both metals and organic contaminants (OSPAR, 2002). In the light of these considerations, indication of the size of the fraction analysed ($<2\,mm$ fraction; $<63\,\mu m$ fraction; $<20\,\mu m$) was set as an obligatory field in the reporting template.

Moreover, in reporting data for sediment monitoring, additional information about physico-chemical characteristics of the matrix is recommended:

- % of the fraction $<63\,\mu m$ in the sample or sediment type description (mud/fine sand/middle sand/coarse sand/gravel) if no analysis of grain size is available.

- Total organic carbon (expressed as % of total dry weight).

- Water content.

- Specify whether sediment concentration data have been normalised for the particle size.

9.1.8   CHARACTERISATION OF THE SAMPLE MATRIX FOR DETERMINATION OF METALS IN SEDIMENTS

In the case of determination of metals in sediments, it is recognised that metal in interstitial water may affect benthic fauna, but a certain amount of metal may be sequestered into anoxic sediments as insoluble metal sulphide, which is relatively non-available to organisms.

The application of the SEM/AVS approach has been included as part of the reporting requirements as a means to calculate the amount of metal sequestered in anoxic sediments and therefore the *bioavailable fraction of metals in SPM and sediments.*

Metal sorption to solid particles depends on various factors: e.g. the amount of Fe or Mn oxyhydroxides ($FeOOH$, $MnOOH$) mainly present in clay, the amount of organic carbon, pH and redox conditions. The equilibrium partitioning between the solid and water phase may be estimated (although with great uncertainties) by mathematical models.

The 'acid volatile sulphide' fraction (AVS), which represents the portion of the total sulphide concentration recovered and measured in a cold acid extraction of the bulk sediment, is compared to the 'simultaneously extracted metal' (SEM), which represents the fraction of metal measured in the same cold acid extraction. If the sum of SEM is lower than the amount of AVS, free metal is unlikely to occur in interstitial water.

The SEM/AVS concept is restricted to anoxic layers in sediments. Furthermore, its application is complicated by technical difficulties (specific sampling method) and uncertainties. It is recognised that further research is required before adopting this approach. Its use was not retained for the ongoing risk assessment for Zn. Nevertheless, the measurement of acidic volatile sulphides (AVS) in µmol/g of dry weight has been included as an optional part of the reporting requirements.

9.1.9 CHARACTERISATION OF THE SAMPLE MATRIX FOR DETERMINATIONS IN BIOTA (ALL TYPES OF CHEMICAL)

Measurement in biota is relevant for the most bioaccumulative substances. Concentrations in biota can vary depending among other things on biological factors such as the species, age, fat content, sex, etc. To reduce variability due to these factors and allow data comparison it is important to measure the most significant ones (factors) and normalise the concentrations against them before data can be assessed.

Most organic contaminants accumulate in the lipid tissue of the species studied. If the endpoint is human consumption, risk concentrations expressed as fresh weight should be considered. If the purpose is to gain insight into the partitioning among the different compartments, measurements on a fat weight basis are probably the best way forward.

Therefore, to allow the results to be used in a variety of ways, concentrations should be provided on a wet weight as well as lipid weight basis. In any case, the lipid content of the sample should be provided, together with the analytical results.

In conclusion, in reporting data of biota monitoring, the type of biota analysed and the species name are considered as mandatory reporting requirements in the template.

Moreover, additional information about physico-chemical characteristics of the matrix is recommended:

- tissue (e.g. muscle, liver)

- basis of measurement (dry, wet, fat)

- ratio of dry weight to fresh weight (expressed as %)

- lipid content (expressed as % of total dry matter).

The following information for each individual is useful for interpreting data: length; total weight; sex; age; reproductive status (GSI); total tissue weight of the dissected organ; the number of individuals and data specified for pooled samples; mean, minimum and maximum length or standard deviation for pooled samples. However, data should not be rejected if this information is missing.

9.1.10 APPLICATION OF QA/QC PROCEDURES BY THE LABORATORIES

The link between the monitoring data and the associated QA/QC information is essential if fitness for purpose (current and potential future purposes) is to be demonstrated.

However, the extent to which QA/QC information is linked with data appears, at present, to be relatively low. This information is difficult to find and is not readily available in existing databases. In most cases the laboratory in which analysis is performed remains the only link by which quality information may be retrieved. Once this link is lost (for example, when the data are put on a national database) it may not be possible to provide any supporting QA/QC data for the monitoring information (Gardner *et al.*, 2007).

The importance of QA/QC information for improving the level of reliability and comparability of the data is increasingly recognised at various levels.

This is also confirmed by the proposal for a 'Commission Directive laying down, pursuant to Directive 2000/60/EC of the European Parliament and of the Council, technical specifications for chemical analysis and monitoring of water status' (draft at the time of writing), which, based on principles concerning quality management systems set out in EN ISO/IEC-17025,[4] establishes minimum performance criteria for methods of analysis to be applied by Member States when monitoring water status, sediment and biota, as well as rules for demonstrating the quality of analytical results (European Commission, 2008).

In line with the above, a number of QA/QC metadata were selected in the common template for data reporting at EU level under the WFD, but not all of them are mandatory (i.e. they are not Category 1 data), as can be seen in Table 9.1.1.

Most of them have been labelled as either Category 2 or Category 3 in the reporting template. However, they are very important for a more thorough assessment of data comparability and they will probably become obligatory in the longer term for future collections of monitoring data.

The following sections consider some key parameters (metadata) included in the template for assessing the application of QA/QC measures by the laboratories.

9.1.10.1 Laboratory Name

The specification of the name of the laboratory is required in the template as mandatory information. As mentioned above, the name of the laboratory is important for data traceability and it is often the only way to trace information about how samples were collected and analysed. Unfortunately, at present this information is not always readily available in the databases.

9.1.10.2 Accreditation of the Laboratory for the Specific Determinand

Accreditation of laboratories is recognised today as an important tool of the quality management of a laboratory performing chemical measurements under the Water Framework Directive.

And accreditation for each determinand for which monitoring data are reported is crucial to demonstrate the competence of the laboratory in the analysis of those

---

[4] EN ISO/IEC 17025: 2005, General requirements for the competence of testing and calibration laboratories.

determinands. For this reason the specification of this information was included in the reporting template (see Table 9.1.1).

However, it is also recognised that accreditation alone cannot solve all the problems: the fact that a laboratory is accredited should be considered alongside the ability of the laboratory to achieve the required performance targets (e.g. achievement of a limit of quantification for the measurement of a given contaminant in line with the Environmental Quality Standard established for that contaminant). Common additional requirements should therefore be specified for accreditation and assessment of accredited 'WFD laboratories' across the EU (Morin *et al.*, 2007).

The current debate is also about the need for harmonisation of sampling procedures and demonstration of the competences of the samplers. For the moment, the reporting template requires information about the accreditation of the laboratory only for analytical work, because at present there is no system in place for accreditation of the sampling process.

9.1.10.3 Uncertainty, Limit of Detection and Limit of Quantification

Knowledge of the measurement uncertainty, together with the value of the limit of quantification (LOQ)[5] (and limit of detection (LOD)[6]) is the basis for checking compliance with Environmental Quality Standards (EQS) and assessment of fitness for purpose of the data against the predefined objectives of the monitoring programme.

The limit of detection and the uncertainty at the limit of quantification represent additional information to judge about the level of confidence of the data and for the interpretation of the so-called 'less-than values'.

The limit of detection and/or the limit of quantification are usually provided in the databases as metadata accompanying the concentration values. On the other hand, information about measurement uncertainty (and the method used for calculating the uncertainty) as well as the uncertainty at LOQ are not provided by default, unless this is requested by the customer.

A recent investigation conducted as part of the FP7 EAQC-WISE project (Gardner *et al.*, 2007) showed that in general, customers (i.e. competent authorities responsible for the monitoring programmes) at the moment do not require an explicit statement of uncertainty associated with the results.

In some cases, even if the uncertainty is provided by the laboratory to the competent authority, it appears that at present this information is actually not used as a basis for data assessment and decision making.

Change is afoot, however: the measurement uncertainty is going to become a mandatory requirement when reporting monitoring data.

---

[5] The limit of quantification expresses the lowest concentration of an analyte which can be quantitatively determined with a given uncertainty. It is recognised that LOQ is defined as 3 times the limit of detection.

[6] The limit of detection is defined as a concentration of substance for which there is an adequately high probability of detection when making a single analytical measurement. The LOD refers to the concentration equivalent to 3 times the standard deviation.

In the proposal of Commission Directive (European Commission, 2008) it is stated that Member States will need to ensure that the minimum performance criteria for all methods of analysis applied are based on a relative target uncertainty of 50 % (k = 2), estimated at the level of relevant Environmental Quality Standards and a limit of quantification equal to or below a value of 30 % of relevant Environmental Quality Standards.

For substances that are not yet defined as priority substances and for which EQS are not finalised, these performance targets cannot be applied. However, the knowledge of the measurement uncertainty and the limit of quantification in relation to predefined targets is important for judging the fitness for purpose of the data (for analysis of trends, comparison of data sets, classification of water bodies for the achievement of good ecological status, for example).

Finally, it should be stressed that at present only the measurement uncertainty associated with the analytical work is required in the common reporting template, but research regarding the influence of the sampling process is progressing and in the future the uncertainty associated with sampling will also have to be taken into account.

9.1.10.4   Extraction Recovery Percentage (and Correction of Results)

This is an important issue for the comparability of data, in particular for organic chemicals. Information about the percentage of extraction recovery, or at least, as a minimum requirement, information about whether or not correction for recovery has been applied to the results, should be provided to the customer.

The results of a recent investigation conducted as part of the FP7 EAQC-WISE project (Morin *et al.*, 2007) confirm that the approach used by different laboratories is very variable, even within the same country. Results are seldom corrected for organics. In any case, the information about correction of the results is rarely provided to the customer (unless requested), nor is it specified whether the extraction recovery is included in the measurement uncertainty.

The reporting template requires specification of the level of extraction recovery and whether or not results were corrected for extraction recovery.

9.1.10.5   Participation in Inter-Laboratory Studies

Participation in proficiency testing (PT) programmes is included as part of the requirements that the laboratories must fulfil in order to demonstrate their competences in analysing relevant physico-chemical or chemical measurands under the WFD.

Participation in PT schemes for analysis is generally requested by the competent authorities when contracting a laboratory and, in most cases, is part of the requirements for accreditation (and 'authorisation') of a laboratory. This is in fact nowadays recognised as a very powerful tool for QA/QC and the inclusion of 'obligatory participation' in inter-laboratory studies in the legal documents is regarded by experts in QA/QC and managers as an important requirement.

However, it has to be stressed that the aim is to use PT schemes as an educational tool, including the organisation of technical meetings by PT providers in order to discuss the results and allow laboratories to improve their performance.

PTs are, however, not available for all WFD determinands, matrices and concentration ranges (and there may be too few participants in a country).

Moreover, PT providers across the EU do not have a harmonised evaluation of the performance of the laboratories in these exercises (e.g. definition of the reference value). For this reason, an initiative is under way for harmonisation among WFD-PT providers, which could lead in the longer term to some kind of 'notification' from 'public authorities' in order to have a pan-European view on the minimum criteria for appropriate participation of laboratories in PTs (frequency of participation, scope of the PT scheme, equivalence of schemes), on what to do exactly with the outcomes of the participating laboratories (e.g. training, corrective actions) and, finally, on the criteria for suspending laboratories.

Here again a need is identified regarding the sampling process and the participation of laboratories in PTs for sampling. At this stage no requirement is included in the template regarding PTs for sampling because there are no regular, frequent and widely relevant PT programmes for sampling in the water field (in particular for fresh water, apart from some recent spot initiatives).

9.1.10.6 Summary of Performance of the Laboratory in the Inter-Laboratory Study for the Given Determinand

A summary of the performance of the laboratory in the inter-laboratory study for the given determinand (indicating whether the z-score is below or above '3' according to ISO-13528) is included in the reporting template as an optional requirement (Category 3).

However, it is well recognised that for the moment this information is very difficult to obtain without asking the laboratory directly. In general, success in PT schemes is assumed by the competent authorities to be covered by the accreditation system/authorisation granted by the Ministry of Environment (i.e. there are no selection criteria based on performance scores of the laboratories in inter-laboratory studies). Moreover, some competent authorities consider that verifying this type of information would take time and a technical background that they do not have available at present.

9.1.11   CONCLUSIONS

The information provided by monitoring data is one of the key elements in identifying the candidate priority substances of concern at community level and informing the substance prioritisation exercise, based on evidence of intrinsic hazard and widespread environmental contamination.

The collection of metadata is essential for the evaluation of the acceptability and fitness for purpose of the data, by judgement of their representativeness, their reliability and their comparability with other data sets.

Establishing criteria for the demonstration and assessment of data quality and fitness for purpose is not easy; a balance must be stuck between asking for the finest detail and not being too demanding.

The present data reporting template is a first attempt at harmonising these requirements at the European level for all the reporting exercises.

The template is subject to evolution since it reflects the current state of technical development in a field that is undergoing continuous changes through ongoing scientific research.

REFERENCES

Bonnomet, V. and Alvarez, C. (2006) Implementation of requirements for priority substances within the context of the Water Framework Directive: methodology for setting EQS: identifying gaps and further developments: background document, Ref. ENV.D.2/ATA/2004/01, INERIS, Office International de l'Eau, DG ENV Contract No. 07-010401/2005/4001371/MAR/D2.

Coquery, M., Morin, A., Bécue, A. and Lepot, B. (2005) Priority substances of the European Water Framework Directive: analytical challenges in monitoring water quality, *Trends Anal. Chem.*, **24**(2).

European Commission (2007) Proposal for a Directive of the European Parliament and of the Council on Environmental Quality Standards in the field of water policy and amending Directive 2000/60/EC.

European Commission (2008) Draft Commission Directive laying down, pursuant to Directive 2000/60/EC of the European Parliament and of the Council, technical specifications for chemical analysis and monitoring of water status.

Gardner, M., Morin, A. and Dulio, V. (2007) Current state of the art and existing gaps in the communication of QA/QC information for the Water Framework Directive, Report under EAQC-WISE project: European Analytical Quality Control in support of the Water Framework Directive via the Water Information System for Europe, Contract No. 022603 (SSPI).

Morin, A. *et al.* (2007) Towards an improvement of QA/QC in water monitoring within the Water Framework Directive, Workshop Report, Paris, 16–17 November 2006, Report under EAQC-WISE project: European Analytical Quality Control in support of the Water Framework Directive via the Water Information System for Europe, Contract No. 022603 (SSPI).

OSPAR (2002) JAMP guidelines for monitoring contaminants in sediments. OSPAR Commission, Reference No. 2002-16.

Section 10
Conclusions

10.1

Needs for an Operational Science–Policy Mechanism in Support of WFD Monitoring – National and Regional Examples

Philippe Quevauviller,[1] Bob Harris and Philippe Vervier

---

[1] The views expressed in this chapter are purely those of the author and may not in any circumstances be regarded as stating an official position of the European Commission

---

The Water Framework Directive - Ecological and Chemical Status Monitoring   Edited by Philippe
Quevauviller, Ulrich Borchers, Clive Thompson and Tristan Simonart © 2008 John Wiley & Sons, Ltd

10.1.1   INTRODUCTION

Policy development and implementation represents one of the increasing challenges for the private sector, NGOs, citizens' associations and professional organisations, with pressure for scientific evidence regarding policy orientations driving a more evidence-based approach, which requires a more sophisticated use of science (Scott *et al.*, 2005). A number of 'science-meets-policy' events have been held since 1998, reflecting an increased awareness of the need to improve the role that science plays in environmental policy making (see details in the last science-meets-policy report resulting from the London 2005 event; Scott *et al.*, 2005). These events have highlighted a number of remaining challenges to be addressed in order to ensure that science makes a full contribution to the development and implementation of robust policies; these can be summarised as follows (Scott *et al.*, 2005):

- Policy makers and scientists have different cultures, languages, motivations and constraints. A lack of understanding of these differences may lead to a breakdown in communication and mutual understanding.

- Better linkages between policy needs and research programmes are needed, requiring enhanced coordination regarding programme planning, project selection and management, and mechanisms for knowledge transfer, to ensure that outputs from research programmes really contribute to policy development, implementation and review.

- Research results are generally published in the scientific literature in a format which is hardly accessible to policy makers. There is hardly any systematic review and synthesis of the scientific evidence base that can put information into the hands of policy makers in a form, and on a timescale, that helpfully informs policy.

- The transparency of the science-into-policy process is seen as a necessary (although not sufficient) condition, which is related to 'good practices' on the use of expertise in policy making. This is very low at present, and the roles of 'translators' and 'boundary organisations' remain to be clarified and institutionalised as mechanisms for enhancing transparency and synthetising existing knowledge.

- A greater interactivity is considered necessary through the whole science-into-policy process, from question framing, through identification of research needs, to reviews of outcome and advice. Challenges include how to improve dialogue between policy makers, scientists and the public, and how to increase incentives for scientists to engage with the policy process.

- Finally, the nature of environmental problem solving requires contributions from many different disciplinary perspectives, which makes joint undertakings often difficult to manage. However, an inter-disciplinary approach is a key factor to enable science to make a greater contribution to policy.

The need to strengthen links among scientific outputs and policy-making activities and to better organise science–policy interactions is subject to ongoing discussions in the water sector (Environmental Science & Policy, 2005). One of the key conclusions of these discussions among scientists, policy makers and stakeholders underlined the possibility of developing a conceptual framework for a science–policy interface related to water, which would gather various initiatives and pieces of knowledge. In this context, possible scenarios regarding integrated environmental science and policy interface have been recently proposed (Quevauviller *et al.*, 2005, 2007). The issue is complex in that it involves many different disciplines, sectors and interests, as well as technicalities such as standardisation and its links with research and policies (Quevauviller *et al.*, 2007).

This chapter summarises ongoing discussions related to the development of an operational science–policy interface, with focus on monitoring features covered by the WFD.

10.1.1.1   The Key Importance of Monitoring in the Environmental Policy Chain

Key steps of the environmental policy chain related to protection against pollution can be summarised as follows:

- Describe what you want to protect.

- Measure or describe status.

- Define level of protection according to well-defined objectives.

- Identify pressures.

- Quantify relationship between pressure and environmental response.

- Quantify relationship between social and economic cost and pressure.

- Identify least cost pathway.

- Define policy instrument.

- Implement the policy instrument and assess response.

- Take appropriate measures (control, remediation).

- Review policy on the basis of scientific/technological progress.

Each link of this chain is based upon a scientific foundation and basic technical knowledge. In this context, the reliability of the overall chain will indeed depend upon the effectiveness of the integration of scientific and technological knowledge in a timely fashion at each step of the policy development, implementation and review. In particular, monitoring data are amongst the key elements of one or more decision-making steps, e.g. analysis and quantification of pressures, follow-up of measure's efficiency, evaluation of status, etc.

The increasing number of monitoring data (linked to EU policies and/or international programmes such as the European Environment Agency's State of the Environment programme) and the development of models provide a much better vision of the problems to be tackled and of the ways to approach them. In the medium term (5 years horizon), it will be possible to establish a much better holistic evaluation of 'environmental interfaces' (e.g. soil-sediment-water) and related pollution pathways. This will obviously have a direct effect on the implementation and review of policy steps. In the longer term (10–15 years), the increasing number (and quality) of environmental databases, models and other monitoring facilities (e.g. GMES) should enable us to look at the environment as a single entity instead of as a series of separate compartments.

As described in Chapter 1.1 of this book, monitoring data produced in 2007–2008 under the Water Framework Directive (European Commission, 2000) will form the basis for the design of programmes of measures to be included in the first river basin management plan (due to be published in 2009), and thereafter used for evaluating the efficiency of these measures. Monitoring data will hence obviously be used as a basis for classifying the water status, and they will also be used to identify possible pollution trends. This is an iterative process in that better monitoring will ensure a better design and follow-up of measures, a better status classification and a timely identification of trends (calling for reversal measures), which puts a clear accent on the need for constant improvements and regular reviews (foreseen under the WFD) and hence on the need to integrate scientific progress in an efficient way.

10.1.2 EU SCIENTIFIC FRAMEWORK IN SUPPORT OF WATER POLICIES

10.1.2.1 Water Policy Framework

The 6th Environment Action Programme (6th EAP) (European Commission, 2002) defines the environmental policy trends for the period 2002–2012. It strongly affirms that environment policies should be based on the best scientific evidence and also that its priorities need to figure prominently in the Community RTD (research and technological development) programmes. The complexity of environmental problems that we now face makes the science approach even more necessary – inter-linkages and trade-offs between problems are more apparent. Acting on one problem can harm or benefit the solution of other problems. Measures need to be assessed in an integrated fashion to avoid undesired side effects, and monitoring is one of the key elements in this respect. Significant research efforts have been devoted in recent EU research programmes to supporting the WFD, which requires 'good status' to be achieved for all waters by the end of 2015. This is based on clear milestones (e.g. characterisation, monitoring, river basin management plan, programme of measures), each representing technical challenges requiring scientific knowledge and exchange of expertise and best practice. In this respect, the Water Directors of EU Member States and Norway decided to launch a Common Implementation Strategy (CIS) in 2001, the aims of which are to develop a common understanding and approaches, elaborate informal technical guidance including best practice examples, share experiences and resources, avoid duplication of efforts, and limit the risk of bad implementation of the directive (European Commission, 2003a). This has resulted in the development and endorsement of 15 guidance documents, of which two deal with monitoring (European Commission, 2003b, 2006).

The WFD provides a well-established policy basis and a stable platform, which enables building up communication and best practice exchanges among different players (policy implementers, technology providers, scientific community, industrial stakeholders, NGOs, etc.). As described below, this is reflected in clear improvements within the last four years, with plans for developing joint initiatives (involving EC Research and Environment General Directorates, scientific actors and Member States through the WFD-CIS) in 2007–2009 for integrating scientific inputs and progress into the implementation process.

10.1.2.2   Water in the RTD Framework and LIFE Programmes

The EU RTD Framework Programme

The European Commission has been supporting research on water in the early days of the Framework Programmes (FP) for Research and Technological Development (RTD) (Schmitz *et al.*, 1994). The FP, as it has evolved, has aimed to foster scientific excellence, competitiveness and innovation through the promotion of better cooperation and coordination within the European Union. It also aims to produce advances in knowledge and understanding, and to support the implementation of related European policies (see details in Quevauviller *et al.*, 2005). Although it is not easy to establish clear frontiers between 'fundamental' (or 'basic') and 'applied' research, one may distinguish different types of R&D activity that support water policies:

- Research within the 'environment' theme contributes to medium-to-long-term policy objectives, e.g. developing scientific knowledge on hydrology and climate processes, ecological impact of global change, soil functioning and water quality, integrated management strategies and mitigation technologies, scenarios of water demand and availability, etc.

- Research of a more applied character is funded under the 6th Framework Programme (Scientific Support to Policy) to underpin the formulation and implementation of Community Policies. It hence enables publication of calls for proposals which accommodate specific research needs identified by the policy DGs. In this respect, a range of topics has been defined, in particular in support of the WFD implementation.

- Coordination of national research programmes is another type of instrument so-called ERA-NET scheme. Funding organisations like ministries and research councils may submit proposals for the networking of national or regional research programmes or innovation programmes in sectors of their choice. The Commission funds the coordination and the Member States finance the research activities. Typically, ERA-NET projects include exchanges of information on programmes and projects, exchanges of best practice, strategic analyses for future joint activities and programmes, joint calls for proposals, etc. This mechanism hence allows for the coordination of research programmes with relevance to environmental policies, including the Water Framework Directive, but also of bilateral or international (research) programmes. The ERA-NET scheme represents a very valuable mechanism for regrouping national funds at the level of programmes so that larger or more coordinated projects can be funded. Further, it allows increasing of the access to scientific expertise available at the regional or national level, as well as cross-border cooperation at the levels of programmes and of projects.

The 7th Framework Programme (2007–2013) covers priority areas reflecting EU research needs in sectors such as health, food and agriculture, information and communication technologies, nanosciences, energy, transport, socioeconomic sciences, space and security. Environment and climate change is one of these ten priorities. It focuses on knowledge of the interactions between the biosphere, ecosystems and human activities, and the development of new technologies, tools and services, with emphasis on:

- Improved understanding and prediction of climate, earth and ocean systems changes.

- Tools for monitoring, prevention and mitigation of environmental pressures and risks.

- Management and conservation of natural resources.

More specifically, the research areas will address pressures on environment and climate, impacts and feedback, environment and health, conservation and sustainable management of natural resources (including groundwater), evolution of marine environments, environmental technologies, understanding and prevention of natural hazards, forecasting methods and assessment tools, and earth observation. The overall environment (including climate change) theme has a budget of 1890 million euros for the period 2007–2013 (on a total budget of 50 521 million euros).

The 7th Framework Programme includes nine themes under the 'cooperation programme' (including environment), which cover 'research needed to underpin the formulation, implementation and assessment of EU policies', and will 'respond in a flexible way to new policy needs that arise during the course of the Framework Programme'. It indicates that 'In order to strengthen the diffusion and use of the output of EU research, the dissemination of knowledge and transfer of results, including to policy makers, will be supported in all thematic areas, including through the funding of network initiatives, seminars and events, assistance by external experts and information services, in particular CORDIS'.

The Joint Research Centre (JRC) Multi-annual Work Programme (MAWP)

EC-funded research has also been developed through the Joint Research Centre's Multi-annual Work Programme (MAWP), embedded into the EU RTD Framework Programme. Several key actions of the MAWP are directly relevant to water policies, e.g. actions related to chemicals in the aquatic environment, ecological water quality, marine ecosystems, the water quality information system, the European Measurement Infrastructure, and policies and human resources for research.

The LIFE Programme

Finally, besides FP funding, the LIFE programme – the Financial Instrument for the Environment – is an EC DG Environment's financial mechanism, specifically aimed at assisting the development of environmental policy through its co-finance of demonstration projects. LIFE has co-funded close to 1200 projects related to water policies since 1992, which were thus potentially in a position to help prepare the ground on location for a subsequent WFD implementation (see examples in LIFE, 2007). The rationale behind this instrument is simply that innovation, be it highly technical or more like a new approach to an old problem, needs to be demonstrated to persuade other potential users of their value, and to establish that any ideas proposed actually do work in the real world. This funding instrument thus represents in principle a natural continuation of research projects, aiming at demonstrating the applicability of innovative methods, solutions and techniques to real environmental cases.

10.1.2.3   Identification of Research Needs in the Water Monitoring Sector

As stressed above, it is not always possible to establish a clear cut between 'basic' (or 'fundamental') and 'applied' research. Also, the timing aspect (short-, medium- and long-term) is intimately linked to the way research instruments are operated. Discussions at science-meets-policy events (Scott *et al.*, 2005) have highlighted that policy makers often assume that the science community exists to support policy, while researchers are upset if their research is labelled as 'applied'. Furthermore, scientists often tend to focus on remaining uncertainties (and the next research grant) rather than on offering solutions or stating what they know with some certainty.

Regarding WFD monitoring, the identification of research needs is of course fed by advances in scientific knowledge, but it is also directly influenced by the Directive's agenda. In this respect, research needs in support of water monitoring may be defined according to timing considerations:

- *Short-term* ($\sim$1–2 years): Needs are basically concerned with the accessibility of research knowledge required on a short-term basis. In this case, the timing is often not adapted to develop new types of research activity (unless very specific needs are identified, which may be sorted out in a 6–12 months period). So the key issue is how to get efficient and user-friendly access to background scientific information and archives. The WFD monitoring programme, designed in 2006, now in the operational phase, typically falls under this category, i.e. implementation cannot wait any longer for research outputs, but monitoring will be all the more efficient as it will be based on an existing scientific foundation, hence the need for an efficient information transfer. In other words, specific research needs (e.g. the urgent need for method development) cannot be tackled through an FP call for proposals if the project – if retained – is to start the following year. The only possible way this could happen would be through either JRC action lines (see Section 2.2.3) or national research programmes (e.g. through the ERA-NET scheme).

- *Medium-term* ($\sim$2–5 years): Medium-term research is well adapted to policies with well-defined milestones, enabling them to anticipate research needs and respond in time. Typically for monitoring, the formulation of medium-term research needs will have to take into account the preparation phases (2007–2008) of the first river basin management plan, to be published in 2009, and future reviews of the monitoring programmes, including support for standardisation. It should be noted that policy makers often report difficulties in identifying their research needs over the medium term (i.e. 2–5 years) (Scott *et al.*, 2005) – precisely the timescale over which much academic research is conducted. This is often due to short-term pressures, turnover of staff and changing political priorities.

- *Long-term* ($\sim$5–10 years): Scientific progress in this respect supports either policy milestones which are clearly identified at the 10 years horizon, or the review process of the legislation. In the case of the WFD, long-term research needs are typically linked to the development of the programme of measures, which has to be operational in 2012. They may also concern the review process of the technical requirements detailed in the relevant annexes of the directive, including Annex V (monitoring programmes), which should be known at the time of the 2nd River Basin Management Plan in 2015.

10.1.3 INTEGRATION OF SCIENTIFIC OUTPUTS IN WATER MONITORING DEVELOPMENT

10.1.3.1 RTD and Networking Activities

Water monitoring can be looked at from different angles (e.g. method development, standardisation, quality assurance/control). This research area was prominent in previous framework programmes, as highlighted by Schmitz *et al.* (1994), and resulted in significant advances over the years 1990–2002 (Quevauviller, 2002). Within FP6, projects contributed to advances in monitoring sciences, but in a less focused way. Examples of general themes are given below, and specific examples are found in different chapters of this book:

- Series of research projects included monitoring features in pollution studies, e.g. development of assessment tools (models, measurement protocols, strategies), aiming to provide decision makers with specific tools that could be used for assessing risks linked to specific situations (e.g. contaminated sites, risks to biota from contaminant release, etc.). Large-scale projects also focused on integrated river basin management sciences, including integrated modelling of a river-sediment-soil-groundwater system. Examples are given in Chapters 4.1 and 4.2 (modelling) and 7.1 (risk-based management) of this book.

- More focused projects concerned the development of methods for screening (*in situ*, field measurements), as exemplified in Chapter 3.1. Other developments concerned emerging contaminants (Chapter 7.2) and analytical quality assurance (Chapter 8.1). Finally, links among ecological and chemical status, including monitoring elements, have also been studied in the framework of EU-funded research (see Chapter 1.3).

- Besides research, useful activities concern networking of scientific experts, policy implementers and stakeholders in the framework of coordinated actions (e.g. ERA-NET) or networks of excellence. These have resulted in recommendations that are reflected in several chapters of this book, e.g. on sediment monitoring.

10.1.3.2 Consultation Process (Involvement of Scientific Stakeholders)

With respect to WFD monitoring, the CIS (European Commission, 2003a) ensures that regular contacts and information exchanges take place among policy implementers, RTD project coordinators and the Commission through specific working groups, namely the ECOSTAT working group (see Chapter 1.3) and the Chemical Monitoring Activity (CMA, see Chapter 1.1).

10.1.3.3   Successes and Drawbacks

Short Analysis

The way research outputs have been integrated into WFD implementation has been discussed in a recent paper (Quevauviller *et al.*, 2005). While research needs were generally well considered in successive FPs, projects could often not anticipate the way monitoring programmes would be designed by Member States. This has sometimes led to results which, although they had potential, were not fit for policy development. In addition, the coordination among different actors has not been sufficient to allow a full integration of scientific inputs from RTD projects into the policy discussions, nor to allow all actors to know in time to meet policy needs. This has been due to a lack of clear 'science–policy interface'.

It should be noted that the developed science is not necessarily tailor-made to respond to policy needs. Tension may occur when attempting to achieve scientific 'independence' on the one hand, and policy 'relevance' on the other (Scott *et al.*, 2005). Scientists are often required to provide advice and/or conduct research in response to policy needs. Without a clear 'policy focus', however, the relationship between science and policy is felt to develop in a more messy, contingent and emergent way, while the upstream definition of policy-research needs is thought to lead to a tighter, more coherent relationship. Some worry has been expressed however that allowing policy to define research objectives may narrow the scope of the research, reducing benefits from 'free inquiry', which can produce unexpected findings or different options for policy that may ultimately be more useful than the results of policy-led research (Scott *et al.*, 2005).

It is also important to note that the relationship between science and policy is influenced by the nature of the scientific results or the stage that a particular scientific or policy field has reached. Many issues go through an 'attention cycle' where policy and scientific attention is initially low, then increases due to some discovery or political event, then receives resources in terms of scientific research and policy development, and then recedes if the problem is seen to be solved, or for other political reasons, perhaps to re-emerge later (Scott *et al.*, 2005). Strong science (i.e. a critical mass of clear evidence and consensual results from many scientists) is more readily taken up and utilised by policy makers, while emerging science (new problems identified, tentative and/or conflicting results) may need to be strengthened and more widely disseminated before being acted upon.

Recognition of Gaps in Knowledge Transfer

Interactions among policy and scientific communities vary according to the phases of the projects. To date, these interactions are not coordinated in a systematic way but rather function on an ad hoc basis (based on links among 'individuals', rather than 'institutional', structured links). On the other hand, the various policy departments have different formal and informal contact points in the Member States, coming from numerous research and 'policy' communities with discrete interests. These communities are often also competing internally (e.g. competition between the water quality/quantity research communities, or between responsible national authorities/local water managers), which creates an obstacle in the pursuit of effective collaboration.

10.1.4 INTERACTIONS BETWEEN THE POLICY AND SCIENTIFIC COMMUNITIES

10.1.4.1 Science in the Decision-making Process

Policy makers need to be able to persuade their constituencies that an issue is important. Therefore, they need to be presented with 'strong science' in a meaningful way, the costs of not addressing the problem, likely solutions to the problem, and the budget required for implementing these solutions. Not all of these can be supplied by researchers on their own. The need for strong science calls for processes that can bring evidence together (particularly where knowledge comes from different fields of inquiry using different methods), as well as for identifying areas of remaining uncertainty and ignorance (Scott *et al.*, 2005).

The social or policy relevance of research programmes and proposals needs to become a more central consideration in related decision making. Relevance criteria such as significance, urgency and dissemination effort need to be actively considered alongside scientific criteria.

Research that is aiming to be policy-relevant will inevitably have a focus on a real-world problem. This is a necessary distinction because academic disciplines often define research problems in a narrow way, leading to research that is abstract and possibly irrelevant to society's pressing challenges.

The need for greater inter-disciplinarity has consistently been identified as a priority for creating knowledge that is valuable for policy (Scott *et al.*, 2005). Yet there remain many institutional barriers to inter-disciplinarity within many research organisations and disciplines – and also policy organisations. Greater inter-disciplinarity can be facilitated through the use of inter-ministerial collaboration, international and national networks, and shared research budgets between funders with responsibility for different parts of the disciplinary pallet.

Researchers need to interact with those interested and affected parties that have a stake in the issue they are investigating. This is not merely to enhance communication, but to identify the right question to address in research. It is a tragedy when a completed research project is unable to answer the 'ordinary' questions that those in policy, the public or the media want to ask. Researchers need to ask relevant constituencies, 'if we are doing research in this area, what questions would you want answered?' By consulting all the relevant groups over this, researchers will build a more complete picture of the topic under investigation and the various interests involved (Scott *et al.*, 2005).

A problem with the interactive model of research is that researches can be accused of losing their independence (Scott *et al.*, 2005). First, all researchers are already affected in their research by external factors such as policy discourses and the priorities of funding sources, so it is not a question of *whether* they are affected but *how, by whom and to what extent*. It is probably healthier to face up to influences on research than to try to ignore them. Second, if researchers actively interact with the full range of interested groups, it enhances their ability to 'see the whole picture', identify salient questions and extreme views, and provide independent assessments of policy options. Thus the diversity of interactions is key to retaining and enhancing the independence so rightly valued by researchers.

It was noted that the sequence 'science-into-policy' does not always happen in the order suggested. In fact, 'policy-into-science' often better captures what happens in reality, e.g. an issue or question is presented by government departments to scientists for investigation (Scott *et al.*, 2005). This can be precipitated by a 'crisis', when a high official needs a 'position paper' or when a 'European-level policy is pending'. In other words, sometimes it is policy that drives the science, and at other times science alerts policy to the need for particular forms of legislation and regulations.

10.1.4.2   Matching Research and Policy Agendas

At the start of projects which have been identified as relevant to water policies in general and monitoring in particular, there is certainly a need to clarify specific issues by describing the aims, milestones and technical challenges to the RTD project coordinators so that they understand what the policy expectations are over the duration of their project. These exchanges of information/knowledge rarely occur, which may lead to divergent directions being taken by the projects in comparison to policy orientations.

It is common to find that multiple research agendas operate simultaneously in countries at the same time (Scott *et al.*, 2005). In addition to the research agendas set specifically for policy development, research not driven by policy concerns is conducted in organisations such as universities, academies and research councils. The definition of research agendas can be strongly influenced by the traditions, theories and methods of a single research discipline. This may often have a poor match with the shape of the 'real-world' social problems that policy is grappling with. Concerns were expressed, for example, that much of the environmental research in FP6 is driven by the interests of the science community rather than the policy community. This led to the question of how it is possible to ensure that the framing of problems for research takes into account the perspectives of the policy and research communities (Scott *et al.*, 2005).

The usefulness of a 'bottom-up' approach in addition to the more traditional 'top-down' approach was highlighted with regard to matching science and policy (Scott *et al.*, 2005). In the first case, the approach is based on a wide network consultation, recognising that policy processes are increasingly diffuse, involving a broad set of players (including industry and NGOs). This also implies that researchers are interactive in the way they carry out their research in order to ensure that outputs are relevant, timely and thorough. This contrasts strongly with the narrow, disciplinary and closed top-down style of research that is still strong in many parts of the research world. This is associated with fears that an interactive style of research can undermine the independence of the researchers involved, raising the question of how project selection, management and dissemination procedures can maximise policy relevance and uptake (Scott *et al.*, 2005).

Greater inter-disciplinarity is necessary. Many environmental policy challenges cut across organisational and disciplinary boundaries – sustainable development has social, economic and environmental dimensions (Scott *et al.*, 2005). Where research focuses on a real-world problem, and where it functions in 'silos' (vertical focus rather than horizontal), this often requires an inter-disciplinary style of enquiry, or at least cooperation between those with different disciplinary expertise. Assessments are becoming broader and more integrated, bringing in social aspects.

Great challenges remain in overcoming disciplinary barriers and encouraging researchers to focus on real-world problems (as compared to those thought interesting within the confines of their immediate discipline), leading to the question: how can researchers be encouraged to adopt a problem-focused, inter-disciplinary style of research?

10.1.4.3 Communication

Interactions need to take place throughout the research place. An important outcome of dialogue is the development of trust. But significant barriers still remain in building mutual familiarity among scientists and policy makers. All of the parties involved in the science–policy interface need to create opportunities for dialogue. This helps to improve not only communication but also mutual understanding between the policy and science communities (Scott *et al.*, 2005). Learning is an interactive, two-way process in which both decision makers and scientists stand to learn from each other. Dialogue helps with aspects of knowledge sharing that are widely underestimated in their importance: familiarity, the building of trust, and informal interaction.

It was widely expressed that more effort needs to be put into making research outcomes more accessible (Scott *et al.*, 2005). Different audiences need different outputs, and traditional forms of communication within the science community do not meet policy makers' needs. Many felt that two key stages are currently 'neglected': dissemination of scientific knowledge beyond the scientific world and translation/synthesis of findings for policy. In different science-meets-policy events, it was considered that the job is only half done on completion of a research project. Synthesis is needed across the research portfolio. Policy makers are also interested in work in progress; projects should be required to produce an annual summary of policy implications ('layman report').

It has been suggested that it should be obligatory for researchers to include dissemination plans and a set percentage of their budget for communication in project proposals (Scott *et al.*, 2005). Some agencies, such as the Department for International Development in the UK, have started to require researchers to spend 10 % of their budget on communication with non-academic partners. Examples made it clear that scientific knowledge rarely translates immediately into policy and that, while emerging problems may be identified by science, policy makers may not immediately see their relevance. The importance of providing options, scenarios and indications of risks was highlighted. At times, it is not until the costs and possible consequences of not addressing a problem become clear that the issue is taken up. The number of potential audiences, their different interests and needs, and the technical aspects of research communication mean that challenges still clearly remain, raising the question: how can researchers make their research findings available in accessible forms to key decision makers and the wider public? An example was given of an agency in Sweden which employs a 'scientific journalist' who has a background in both science and communications (this person is responsible for writing articles and liaising with the media).

Transparency is another key feature. Suggestions for enhancing transparency include 1) improving public participation in policy making, 2) clarifying the place of experts and stakeholders in the advisory chain and 3) ensuring that political targets are kept separate from scientific mechanisms (Scott *et al.*, 2005).

10.1.4.4   Synthesis Needs

At the end of the projects, the most critical issue is the way the scientific information is 'digested' or 'translated' so that it may efficiently be disseminated to policy end users and, possibly, applied. This integration phase is certainly the weakest link of the science–policy chain. Indeed, only a small percentage of RTD projects are known by policy implementers, which illustrates the need to improve awareness about RTD outputs, but also the need to encourage policy actors to reflect on research needs linked to their portfolios. This may be translated into needs to carry out synthesis works in the form of 'policy digests' (addressed to the scientific community from the policy implementer's side) and 'science digests' (prepared by the scientific community for the policy implementers).

Within science-meets-policy events, many thought that a new race of 'translators' was needed to help bridge science and policy (Scott *et al.*, 2005). Such translators would require different skills to those normally held by researchers. They would be more 'horizontal' than 'vertical' and will need to understand both the science and the needs of policy. The process of assimilating and synthetising relevant information and providing this information in user-friendly forms was widely thought to be a considerable challenge. Some believed that researchers often do not have the skills and inclination to undertake effective translation and dissemination (Scott *et al.*, 2005).

These roles need to be ongoing rather than only brought into play when an urgent problem emerges; it has to be someone's role to monitor the findings emerging from science, bring these together, and translate them into forms that are accessible to policy makers and practitioners. Translators need to be able to understand the languages and conventions of both science and policy. They require analytical synthesis and communication skills to assimilate large amounts of information, identify the essence and communicate this for non-specialists. The act of translation often means that the finer points of science are omitted; the risk of doing so needs to be carefully considered.

Many pointed out that scientists' core activity is doing science and publishing their work in traditional scientific journals. Scientists are not 'marketing people' and may not be dynamic communicators and presenters of their work. This means that it may be necessary to employ others to translate scientific findings for wider audiences, including policy makers.

About translation: it is important to involve scientists in the translation process, along with policy makers and other communication specialists, such as editors. Not only does material need to be copy-edited, it often needs to be rewritten in a meaningful way without the substance being changed or being over-simplified. It is important to note that, at times, scientific knowledge needs to be translated into other languages as well, and that this process presents yet another level of problems. This translation process should in principle be independent, i.e. the science should not be translated into 'what is acceptable to decision makers'. Also, scientists often do not know enough about the broader socioeconomic and political context in which the scientific findings might be used.

Different levels should be considered for different audiences: 1) extensive, detailed reporting for the expert community; 2) a 10 page summary plus detailed annex for decision makers; and 3) a 3 page folder plus video for the general public.

10.1.4.5 Exchange Platforms, Networking and Dissemination

As a follow-up to the RTD or LIFE projects, useful interactions may occur at the occasion of yearly meetings. Participation of policy officers in all project meetings may not be practicable due to a lack of resources but efforts are needed to organise regular joint meetings focusing on specific themes. This is already happening in the WFD sector (Quevauviller *et al.*, 2005) and should be systematised.

Dissemination is sometimes criticised for being based on an unrealistic and ineffective linear notion of knowledge creation, diffusion and uptake (Scott *et al.*, 2005). As policy processes and networks become more diffuse, open and consultative, there are more opportunities for researchers to make contributions – but this requires dissemination. The accessibility of scientific knowledge can be enhanced through various types of meeting and publication/web site, and the careful use of mass media. Translators of science – as well as scientists themselves – can play significant roles.

Problems arise where there is a feeling that the relationship between science and policy has been opaque and secretive. Transparency is where the workings of decision-making groups and discussions are made visible and accessible; openness is where these processes bring in a wider range of interested and affected groups than the traditional categories of 'experts' and 'policy makers'. Science-into-policy can be 'opened up' through the use of working groups and committees that include both scientists and end users. In addition, the use of the mass media and translators of science can foster a better understanding of why science is important for policy (Scott *et al.*, 2005).

The need for national 'receptors' was highlighted to ensure that countries are better prepared to act when new European legislation is coming up. A factor that influences this is the existence of dedicated structures responsible for liaising and communicating with the European level. We could go even further in recommending regional relays, an example of which is given in Section 10.1.8.

The extent to which scientists are able to network – both formally and informally – with their peers in other countries is another factor which promotes alignment between European-level and national policy. Collaboration through the Framework Programmes was reported to be extremely useful – especially for environmental science, where there are often 'no natural boundaries'. In certain niche areas, e.g. marine research, it was reported that the members of the scientific community are well connected and frequently work together.

10.1.5   TOWARDS A 'SCIENCE–POLICY' INTERFACE

As stressed above, at the present stage efforts are lacking for presenting results of research and demonstration projects in a form that policy makers may easily use, e.g. 'science-digested' policy briefs. On the reverse side, the consideration of research results by the policy-making community is not straightforward, mainly for political reasons and due to difficulties in integrating the latest research developments in legislation. The difficulty is enhanced by the fact that the policy-making community is probably not defining its role as 'client' sufficiently well. In other words, the dialogue and communication are far from being what one would hope to ensure an efficient flow of information. In this respect, improvements could be achieved through the development of a science–policy interface based on a coordination of relevant programmes/projects with direct relevance to the WFD implementation (Quevauviller *et al.*, 2005). In other words, strategies should identify needs for short-, medium- and long-term scientific developments and should establish an interface so that R&D results are synthesised in a way that can efficiently feed the implementation and further reviews of the policies. This interface might include:

- A screening phase evaluating which type of research is needed (background information or tailor-made research and demonstration) in accordance with the policy step of concern. In the case of water monitoring, this is already happening through regular contacts within Commission services and with the scientific community in the framework of expert groups (ECOSTAT or CMA) and/or research projects (e.g. EAQC-WISE).

- A mechanism to ensure that the most promising research projects in support of the policies are 'validated' through demonstration activities, disseminated efficiently and applied at the appropriate level (regional, national or EU). This is not yet or only rarely the case, but there are increasing examples of RTD projects which include a demonstration phase.

- A management scheme, involving both scientists and policy makers, to discuss the corresponding research and policy agendas from the very beginning in order to ensure a more structured communication at all appropriate levels of policy formulation, development, implementation and review. This is hardly operational to date.

More than dissemination and application, the interface should establish strong links among the different funding mechanisms existing at the EU level and the thematic policies. This should enable promotion of pilot projects combining the implementation of results of successfully completed EC-funded RTD or demonstration projects with the implementation of related policies. This would allow the formation of new and innovative partnerships by combining various EC (RTD, LIFE, COST, structural and cohesion funds (Interreg projects), agricultural funds, etc.) and regional/national funding mechanisms, and the establishment of a collaborative partnership involving scientists, policy makers, managers and other stakeholders, for the effective integration of science outputs into policy and management decisions. At present, however, examples

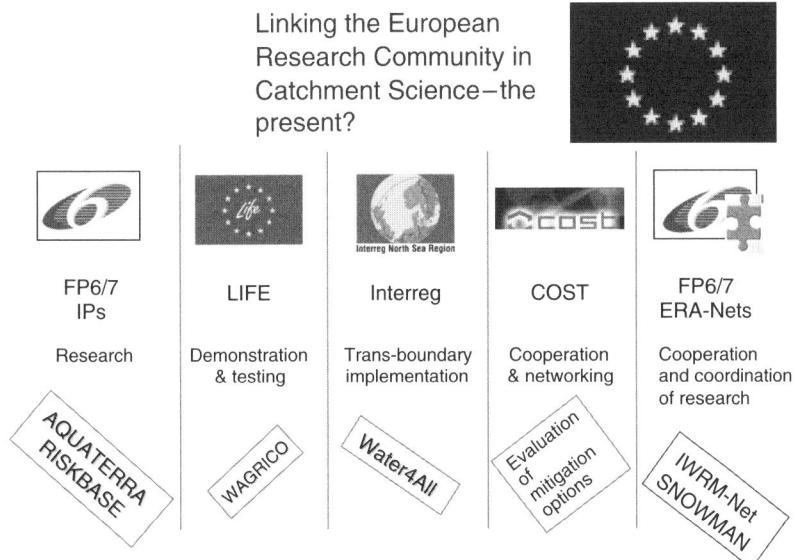

Figure 10.1.1 Catchment science projects from various funding mechanisms (courtesy of Bob Harris, UK Environment Agency)

show that such coordination is still not operational, e.g. Figure 10.1.1 shows different catchment-related projects funded by various mechanisms (FP6, LIFE, Interreg, COST, ERA-NET) for which no 'umbrella' coordination has been envisaged.

In response to the above considerations, a concept is under development, through collaboration between EC DG Environment, DG Research and LIFE, aiming to establish an efficient and sustainable science–policy platform linked to WISE (Water Information System for Europe) which has a direct relevance and link to monitoring development and data (d'Eugenio *et al.*, 2006). This concept is known as WISE-RTD (de Lange *et al.*, 2007) and is discussed in Section 10.1.6.

10.1.6 AN OPERATIONAL WEB INTERFACE: WISE-RTD

10.1.6.1 The Harmoni-CA Initiative

Harmoni-CA is a large scale *concerted action* supported by the DG-RTD under FP5 (HarmoniCA, 2001). One of the objectives of Harmoni-CA is to create a forum for communication, information exchange and harmonisation of information communication and technology (ICT) tools for integrated river basin management and the implementation of the WFD. This action has hence a clear dissemination component concerning activities relevant to monitoring, in particular modelling. The idea to link WFD requirements and RTD outputs through an 'interface' (initally through the building of a web portal) originated from this project. It primarily aimed to focus on operational managers (people responsible for practical aspects of policy implementation) and research and technology providers, and it developed into a wider information platform, as described in Section 10.1.4.2.

10.1.6.2   The WISE-RTD Web Portal

The WISE-RTD (http://www.wise-rtd.info/) has been conceived as a platform for accessing scientific information of potential use to water policy implementation (de Lange *et al.*, 2007). It will be progressively enlarged to cover specific scientific information for policy officers, RTD managers and scientific stakeholders, providing access to relevant scientific information. The web portal will be supported by manpower through a communication services centre (CSC) and it can be seen as a first step towards a sustainable communication process. The portal was made publicly available with the launching of WISE in March 2007 (http://water.europa.eu/), and will continuously develop within the forthcoming years.

In Figure 10.1.2 the water policy process (centre of the figure) is closely linked to a larger circle concerning *research* and technological developments (FP7 for EU research, ERA-NET for national research), *demonstration* (LIFE programme, link with pilot river basin of the CIS (Quevauviller *et al.*, 2005), possible INTERREG support), *interface* through a 'single web portal' (WISE-RTD and possible other linked web sites) and *research policy* development (reviews, integration, research needs).

This diagram however shows an ideal situation which is far from being operational. To make it workable, a strengthened coordination has to be developed among the responsible services at the horizon of the 7th Framework Programme (Figure 10.1.3).

The goal would be to operate a science–policy interface through WISE-RTD (providing access to RTD information), as displayed in Figure 10.1.3. In this configuration, a first phase would act as a 'steering filter' for RDT and demonstration projects (analysis of relevance to water) – Phase 1, RESEARCH OUTPUTS – followed by an EVALUATION (selection of the most relevant projects), a VALIDATION (demonstration of successful RTD) and COMMUNICATION in the form of technical guides translated into the EU languages (addressed to policy managers). It is most likely that only a limited number of projects would be selected until the last step. In other words,

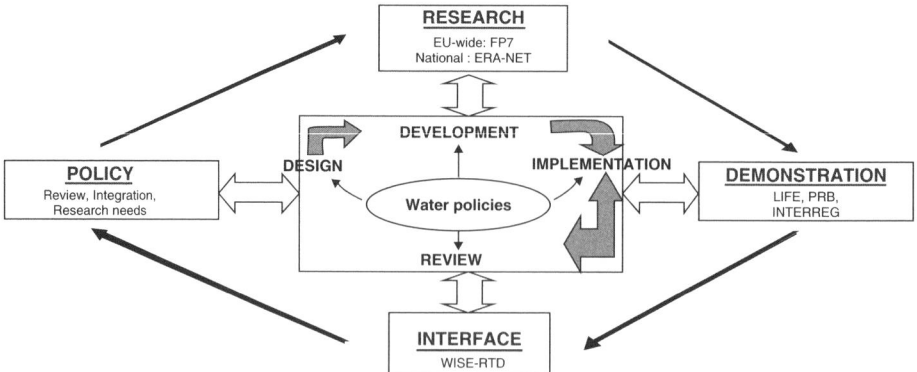

Figure 10.1.2   Links between water policies and research & demonstration

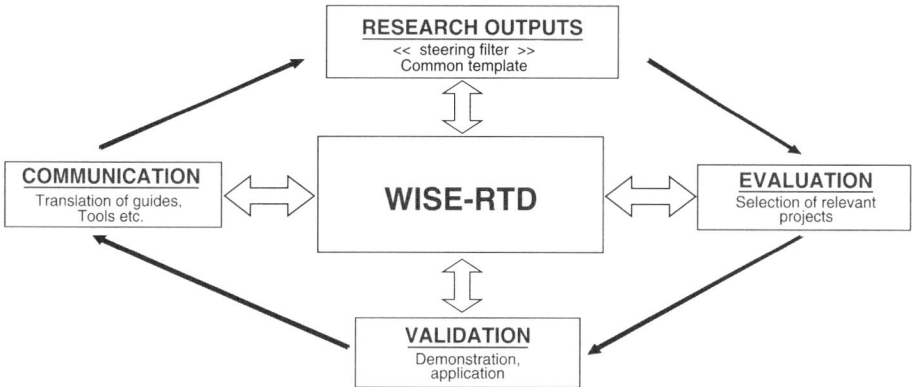

Figure 10.1.3 WISE-RTD and science–policy interfacing needs

there are many projects which may be of high scientific value without necessarily fulfilling concrete policy needs. The evaluation and demonstration steps would then be of critical importance in judging those projects which deserve to be transferred at wide scale to water managers in the EU and beyond.

The overall system depicted in Figure 10.1.3 could only be developed on the basis of operational national and/or regional relay platforms functioning in networks. The next sections provide examples of national and regional approaches.

10.1.7 NATIONAL CASE STUDY: THE UK SITUATION (WFD POLICY AND RESEARCH – WORLDS APART?)

10.1.7.1 Background

By using the concept of a healthy ecosystem as a surrogate measure for a healthy over-all environment, the WFD sets a challenge for society to develop ways of managing our land and water environments in a more integrated and holistic manner. We have to mit-igate the pressures affecting aquatic ecosystems, but can only do this by understanding how these pressures are promulgated through the landscape at the scale of river basins and catchments. This requires a more systemic understanding of how rivers and their ecosystems work and the interactions between the key environmental compartments. Until now these have largely been studied as separate entities and so there is a chal-lenge for researchers, research founders and end users alike to develop more integrated programmes and ways of working. How well is the UK rising to this challenge?

10.1.7.2 Fragmentation of Research Base

In the UK the research base is fragmented, with little overall coordination of the setting of the research agenda and the dissemination of results, i.e. identifying the

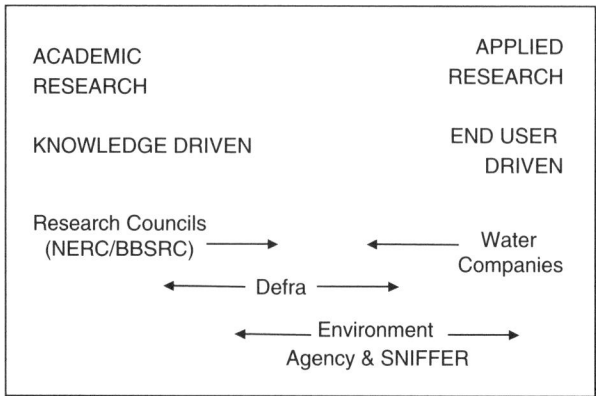

Figure 10.1.4   The range across the blue skies-applied spectrum of founders of water-related environmental R&D in the UK

needs of society and delivering scientific knowledge efficiently to the policy and end user communities. This is especially true for the emerging science area needed to underpin the WFD implementation that will enable us to move towards an integrated river catchment-scale approach to land and water management. Direct government research funding and commissioning is split between the government environment department[2] (Department of Environment, Food and Rural Affairs – Defra) and the research councils, mostly the Natural Environment Research Council (NERC), with to a lesser extent the Biotechnology and Biological Sciences Research Council (BBSRC), the Engineering and Physical Sciences Research Council (EPSRC) and the Economic and Social Science Research Council (ESRC). Indeed Defra and NERC between them fund nearly 70 % of all environmental research in the UK (Environmental Research Funders Forum, 2007). In government departments and research councils alike the issues surrounding integrated catchment science have been poorly defined and hence distinct cross-cutting programmes are difficult to identify.

A significantly smaller research budget is managed by government agencies such as the Environment Agency (EA) in England and Wales, and the Scottish Environmental Protection Agency (SEPA); the latter through the Scottish and Northern Ireland Forum for Environmental Research (SNIFFER). Figure 10.1.4 shows the range across a spectrum from blue skies (academic) to applied research covered by the major founders of water-related environmental R&D in the UK.

The EA and SEPA are government agencies charged with the direct implementation of the WFD and so can target their research at more applied issues. However, their budgets reflect a failure to acknowledge the importance of research and science. For example, the EA, with a total annual expenditure of €1500 million, only spends around €15 million on environmental research and associated training. The WFD-related Integrated Catchment Science Programme budget within this is only around £1 million/a

---

[2] The devolved governments of Scotland, Wales and Northern Ireland also each fund some research specific to their region.

for 2007/08 – i.e. 0.1 % of total budget investment in science to underpin what will become a dominant feature of the EA's future business activity in the next 5 years.

One initiative, set up in 2004, that is beginning to address the cohesion of environmental science research founders is the Environmental Research Founders Forum (ERFF). Its members are constituted from the main publicly accountable founders of such research[3] in the UK. However, it has taken 3 years to carry out a strategic analysis of UK environmental research activity (Environmental Research Funders Forum, 2007) and remains focused at a level strategically above that required for the WFD.

10.1.7.3   The Difficulties of Prioritising Cross-cutting Programmes

NERC consulted on its new 5 year strategy in March 2006.[4] The draft was divided into some 7 themes and WFD-related issues can be found within most of these. However, the management of river basins as a key environmental activity requiring scientific support is difficult to identify in the document, let alone as a major priority. In Defra, key WFD issues are also diffusely spread, in this case across an organisational structure that is largely aligned according to specific domestic and European legislation; for example in relation to drinking water supply, nitrates, flood risk, agricultural policy, etc. In both Defra and NERC the strategic science agenda is dominated by what are considered to be the current 'big issues' for the environment – climate change, mitigation of flood risks, GMOs, biodiversity loss, nanoparticulate matter, technological fixes, etc.

Some of these higher-profile issues such as climate change are driving a growing, but largely tacit, understanding that there is a need for better integration of science and scientists and more holistic thinking. However, in common with many matrix management approaches introduced into large organisations, the management of cross-cutting research programmes seems fraught with difficulty. In particular, the understanding needed to support the management of river basins is difficult to promote in the face of what seem to be bigger societal problems. WFD-related science is therefore not seen as sufficiently important to attract the funding profile needed. Despite the integration of scientific disciplines and systemic thinking being profoundly new and challenging, the external view is that river systems and land management are already well understood and do not warrant additional major funding. However, if we examine just two high-profile areas – climate change and flood risk management – it is easy to see that, in order to develop management solutions, both need a large amount of systemic understanding of land, water and ecosystem interactions; i.e. the very area encompassed by WFD-related or integrated catchment science.

---

[3] Environmental research is defined as research which is about any of the following: the inanimate natural world; the interactions of living things with the inanimate natural world; the interactions of living things with other species of living things; living things or manmade materials or products where the prime aim or the substantial result is to reduce adverse impacts on the environment, or promote positive environmental benefits.

[4] Due to be published in final form in late 2007.

10.1.7.4   Communication between Policy Makers and Research Providers

In many ways the breakdown in communication between researchers and founders results from the research community neither sufficiently expressing its vision of the future nor highlighting the lack of evidence to support the new methods and approaches that will be needed. It has also not made decision makers aware of the huge uncertainties in decision making and developing management solutions that will result from our current lack of understanding. Policy makers and river basin managers faced with the WFD implementation at both regional and national levels are firmly focused on delivering the first round of river basin plans and will remain so until 2009. As a consequence this first round is being undertaken on a 'best endeavours' basis, with little or no time for innovative thinking, new data, methods or indeed integrated approaches of any description. The research community in contrast is much further advanced in its thinking about the future needs of the WFD. With few exceptions there has been little dialogue between the research providers and the end users and there is a danger that science being undertaken now will not match with the management or policy directions being pursued by river basin programme managers.

10.1.7.5   Developing a Strategic Approach – the EA's Integrated Catchment Science Programme

One way of communicating better is to develop the research needs more strategically and try to link these more closely to the needs of the end user(s). An example is the EA's Integrated Catchment Science (ICS) Programme (Environment Agency, 2006). In common with many organisations, science to support river basin management was previously undertaken in functional or disciplinary chimneys in the EA and its predecessors. The historical research in this area has neither been coherent nor integrated and our understanding therefore remains patchy – we know a lot about some things but not enough about others and we have largely ignored the boundaries between environmental compartments. Data management, decision support and modelling tools have been produced and used by different groups in a highly ad hoc and uncoordinated manner in the EA. Although the different groups have common themes in the scales of operation, data sets, techniques, customers and engagement with the external research and modelling communities, there was previously little to no strategic direction to these activities.

A key aspect of the ICS programme strategy was therefore to bring this research together, supporting the science needs of policy managers in their development of integrated approaches for managing catchments, in particular focusing effort where policy is still developing. It set out a vision, objectives and a high-level structure for a five-year period (2006–2011), describing a logical approach to problem solving at the catchment scale, with a top-down framework for an integrated programme and a bottom-up approach to the issues that need to be addressed. It specifically addressed 'outcomes' set by the overarching policy questions, rather than being 'curiosity-driven', and was organised, for delivery, around multi-functional work packages.

The programme was intended to specifically provide the underpinning evidence base and the decision-support systems dictated by the implementation of the WFD. However, it aims to meet the needs of catchment managers by around 2015, rather than support the first round of river basin planning, as by 2015 alternative, knowledge-based solutions for the second iteration of river basin plans will be required. It focuses on providing tools for catchment managers to achieve their goals by applying knowledge-based approaches to problem solving, but anticipates the inevitable delay in the take-up of science and innovation by aiming their delivery at a 5–10 year horizon.

The programme vision implies an approach of 'better environmental management through knowledge'. Underlying it is the principle that catchments are social-ecological systems and that in all cases we need to understand economic and social aspects as well as environmental ones. The approach incorporates three types of work:

- *Improving the scientific understanding* of ecosystems, catchment processes and interactions through filling key fundamental research gaps.

- *Converting this understanding into knowledge* using an integrated catchment modelling framework and the creation of risk-based decision support tools.

- *Developing and testing management solutions* to reduce the impact of pressures and to ensure they are both effective and provide multiple benefits ('win-wins').

Figure 10.1.5 illustrates that these three areas are intimately linked. Where they come together, cross-cutting needs and activities are defined, such as monitoring and data, decision-support systems and management solutions. A network of demonstration catchments will be used to pull the programme together, test scientific understanding of how catchment processes (both social and environmental) interact and demonstrate management solutions. They will also be used to validate modelling and decision-support tools. Demonstration catchments also serve to focus scarce resources, maximise synergy, develop multi-disciplinary working and speed up the communication of results to policy makers and decision takers.

The EA's ICS programme is attempting to collaborate with the overall science community and other organisations to pool scarce research funding resources and align its activities better with both public and environmental priorities. For specific areas it is aiming to:

- Build capacity with key research communities, such as the Catchment Science Centre (CSC) at Sheffield (Surridge and Harris, 2007).

- Collaborate in joint partnerships with others involved in managing catchments, thus pooling funding and expertise and using local knowledge.

- Influence others, such as research councils, Defra and other government departments, and also regional government.

The types of collaboration need to vary according to whether they relate to fundamental science, knowledge transfer or catchment demonstration studies. Since most of the fundamental science required is best undertaken by the academic community,

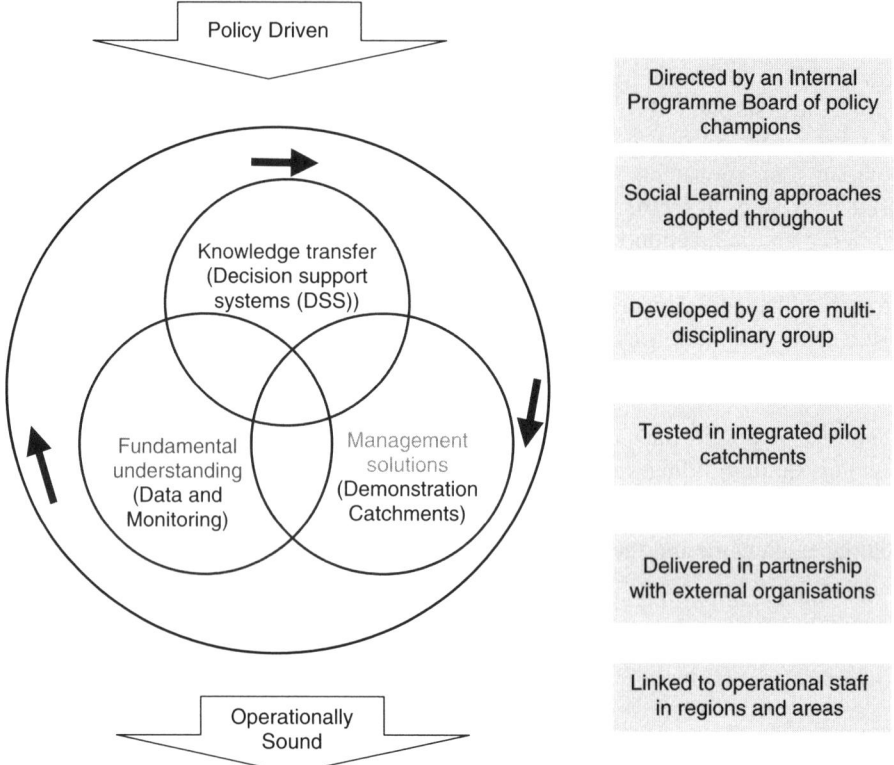

Figure 10.1.5 Better environmental management through sound knowledge – better integration through joined-up working practices

the EA, by using its ICS programme strategy as an influencing tool, is trying to influence research programme funders to allocate funds towards the priorities identified. To develop relationships with non-academic partners, it is necessary to align research agendas with other bodies engaged in policy development, land management or other stakeholders in land and water. However, a range of individual and institutional barriers to such partnerships exist (Surridge and Harris, 2007). Some of these bodies may have different policy agendas to the EA's and collaboration at the scientific level may be prejudicial to policy development. Staff involved in developing relationships therefore needs to consist of experienced, credible scientists who can see the bigger picture, be politically aware and be able to evolve the partnerships.

10.1.8 A REGIONAL PLATFORM: ECOBAG

In this section we give some feedback on an operational interface between science and policy in supporting integrated water management. This experience has been developed in the south-west of France, within the Adour-Garonne River Basin district. The interface started with the creation in 1994 of a scientific network and now involves implementation of a 'regional platform' called ECOBAG (Environment, Economy and Ecology of the Adour-Garonne River Basin District).

10.1.8.1 Background on the Need for Scientific Support for WFD Implementation

The previous sections have highlighted the need to integrate scientific and technical progress within the implementation of the WFD. However, if the WFD provides some indications of where and when, in the implementation process, it is necessary to integrate new scientific and technical knowledge (economic analysis, standards for monitoring of quality elements, review and update of river basin management plans, identification of priority hazardous substances, etc.), it makes no recommendations on how to do this.

The question of the integration of scientific knowledge into professional sectors and public policies has been considered differently over the 40 last years. During the sixties and seventies it was generally believed that useful results for industry and society would start flowing from basic research according to a so-called linear model of innovation (Erno-Kjolhede, 2001). The eighties and nineties saw a period of movement from the science push doctrine, corresponding to the self-governance of science, to society pull science, i.e. research was to be governed by the need to respond to the problems confronting societies and economies (European Commission, 1997). In the late nineties it became important to make industry, public policy and research cooperate sufficiently. This trend is referred to by Etzkowtiz and Leydesdorf (1997; Etzkowtiz *et al.*, 1998) as the 'triple helix' of science policy. This is a metaphor illustrating the fact that there are permanent interactions between these partners and that the development of each partner can be described as a spiral, since the global approach is based on networks and iterative process.

In the context of water policies, if a global approach is indispensable (at the EU level or at national level), local specificity (river basin district scale) has also to be taken into account. Therefore, scientific support for water policies implementation has to be considered at both levels. In this section, a trial of the triple helix method at the river basin district scale is illustrated.

10.1.8.2 Demonstration of the Triple Helix Method at the RBD Scale: the Adour-Garonne RBD

The 'triple helix' method has been trialled for 10 years at the Adour-Garonne RBD scale through ECOBAG. This approach was based on the growing awareness of the need to create links between knowledge producers, water managers and water 'users' in support of the implementation of 'innovative' water policies, such as the WFD. It became clear from the ECOBAG platform experience of making the triple helix approach operational that, in order to significantly improve the integration of research results with policies, significant investment of time and identification of the scientific support needs would be necessary.

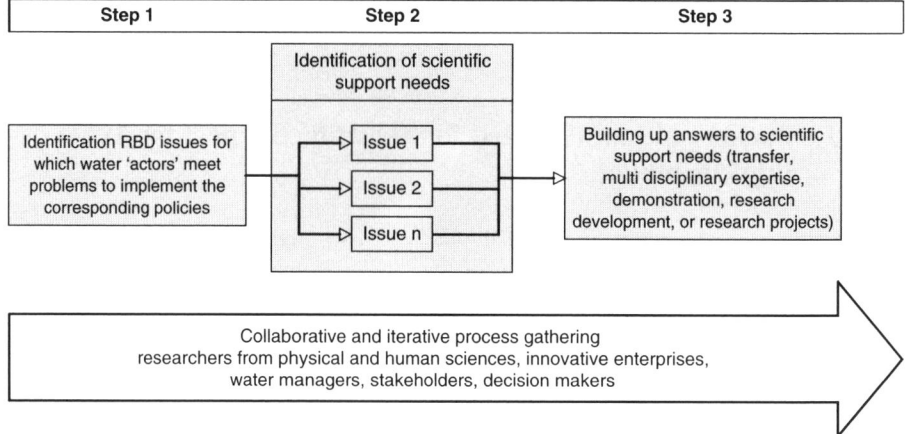

Figure 10.1.6 Description of the co-identification of the projects that aim to deliver scientific support to policies

Within the ECOBAG platform, the methodology used is based on several steps (Figure 10.1.6). The first step is to identify the major issues within the RBD for which water 'actors' face problems in implementing the corresponding policies. The second step is to organise an iterative and collaborative process of co-identification of problems that require scientific support to be solved. This process involves, from the beginning, representatives of decisions makers, water managers, water users, related water enterprises, innovative enterprises, technological platforms and researchers from physical and human sciences. This process allows 1) rewording of problems for water management or policy implementation into scientific questions, and 2) identification of the kind of scientific answer needed (transfer, multi-disciplinary expertise, demonstration, research and development or research projects). The third step is to build up the corresponding projects in the same collaborative approach.

One of the advantages of this time-consuming process is that it avoids mobilising researchers and engineers for 'wrong' questions. Of course, most of the time a research project will deliver results and the quality of these results will be linked to the quality of the scientific approach that is used. But a 'good' scientific result, one which can be published in well-known scientific journals, or which is an efficient tool for research, may be received as a 'poor' result if it is not usable by the end users, for examples water managers.

10.1.8.3 The Difficulties in Identifying and Promoting Policy Scientific Support Projects

To run the above-described methodology based on a collaborative and iterative process of co-construction of innovative projects to support water policies, four main difficulties have to be resolved:

(1) To build up and run projects that aim to deliver scientific support to policies, it is necessary to mobilise researchers for several types of work. The phase of co-identification is relatively long (from 1 month to 1 year) and mobilises researchers as experts. The phase of construction of the project may also be long since it is necessary to have a systemic approach and to develop collaborative engineering. Finally, running the project requires organisation and creation of synergy between academic research approaches, and know-how within research development. Even if one of the national centres of research is able to support researchers in this kind of global and collaborative approach, most researchers are actually evaluated on their academic research products.

(2) The development of scientific support for integrated water management most of the time requires setting up of multi-disciplinary approaches. Making physical and human sciences work together may be facilitated by collaborative identification of scientific support needs. In this process, researchers of different disciplines can very quickly find common objectives and it is not necessary for them to share their languages and their concepts. The way they reach the identified common objective will be based on the methodologies and tools of their science. In this kind of project, it can be useful to develop specific tools to integrate the results that will be produced by different approaches. This is the reason why it is important to involve innovative information technology enterprises in the process.

(3) The size and the budget of this kind of project represents a third difficulty. To develop scientific support for integrated water management, it is necessary most of the time to set up multi-disciplinary approaches. These kinds of project may require a much larger budget than mono-disciplinary researches.

(4) Finally, the fragmentation of the entities that are able to fund scientific support and research at the RBD scale is a fourth difficulty. Some of the scientific support needs are very local and some of them are of local, national and sometimes European priority.

Therefore, the lack of articulation between financial contributors may very often act as a limiting factor for project development. The organisation of common regulation for joint calls may be time-consuming and, sometimes, some research end users may prefer to develop smaller projects, with less multi-disciplinary approaches, to reach their objectives.

10.1.8.4 An Example of Scientific Support for WFD: Concert'eau Project

In carrying out the collaborative and iterative process of identification of research needs to support the WFD implementation, it came out that compliance with agriculture was one of the priorities of the Adour-Garonne RBD. Therefore, a pilot river basin (PRB), 'Gascogne Rivers', has been proposed to the Common Implementation Strategy (CIS, see Section 10.1.2). This PRB has a size of 6800 km^2 (6 % of the district), with a population of 263 000 inhabitants (4 % of the district's permanent population).

According to the local basins, 60–80 % of the total area is used for agriculture (maize, wheat, sunflower, soya beans, etc.) and breeding (ducks, cattle, etc.). In this PRB, nitrate and pesticides are widespread in surface waters.

In accordance with the WFD environmental objectives, river basin management plans, including summaries of programmes of measures, should be drawn up to reach 'good status' for of all waters (Article 4, Article 13 and Annexe VII (6) (7)). The programmes of measures may hence be considered as one of the major WFD implementation mechanisms. The WFD distinguishes between basic measures (minimum requirements) and supplementary measures. Among the basic measures, a number of parent legislations have to be effectively implemented, including the Nitrates Directive (91/676/EEC), with direct or indirect implication on water protection. If the basic measures are not sufficient, supplementary measures (see non-exclusive list of such measures in Annex VI Part B of the WFD) may be undertaken, including economic and fiscal instruments, negotiated environmental agreements, codes of good practice, voluntary agreements, demand management measures, efficiency and reuse measures, rehabilitation projects and research, and development and demonstration projects.

By referring to this point and to the works that have been conducted by the GRAMIP (Regional Group for Actions on Pesticides in Midi-Pyrénées Region) for five years, it came out that basic measures failed to be applied at large scale. For example, GRAMIP has supported four small experimental agricultural watersheds (from 100 to 1000 ha) to test the implementation of agro-environmental measures (AEM), which highlighted the difficulties in scaling up these measures to larger agricultural zones. Studies have demonstrated the importance of socioeconomic dimensions in the process of cooperation and willingness of farmers to participate in the implementation of water protection projects.

Concert'eau is a project funded by LIFE ENVIRONMENT and the Adour-Garonne Water Agency, proposing to demonstrate, from October 2006 to September 2009, that a collaborative technological platform (CTP), gathering scientists from a large range of disciplines, decision makers, water managers, and cooperatives and agriculture organisations, can support an integrated management of agriculture that matches WFD objectives in compliance with the Common Agriculture Policy (CAP) and national and local policies. This CTP will simulate and assess measures proposed by working groups that mix local actors (farmers, local water managers, users, stakeholders and scientists). This support decision system will be able to integrate environmental, social and economic dimensions of the scenarios of measures proposed by the working groups (Figure 10.1.7).

10.1.8.5   Facilitators of the ECOBAG Experience

The development of the ECOBAG science–policy interface has been supported by regional decision makers such as regional councils of the Aquitaine and Midi-Pyrénées Regions, the Water Agency of the Adour-Garonne River Basin District, and the Regional Directions of the Ministry of Environment and Ministry of Research. The universities of the main cities of the river basin (Bordeaux, Pau and Toulouse) and the

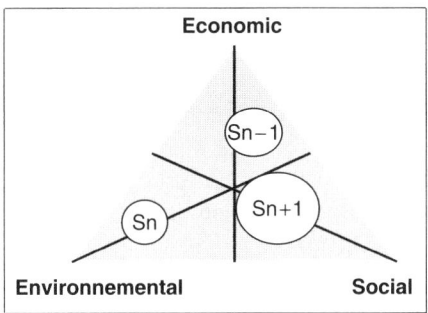

Figure 10.1.7 The Concert'eau CTP will deliver visual comparison of the scenarios of changes of farming practices (in this example, three scenarios called Sn, Sn − 1 and Sn + 1) by qualifying them according to their relevance for environmental, social and economic issues

regional institutes of research (CEMAGREF, CNRS and INRA) have been strongly involved in the construction of this interface.

10.1.9 RECOMMENDATIONS AND ACTIONS

A series of recommendations were expressed at the London 2005 workshop (Scott *et al.*, 2005), which are summarised below:

(1) Incentives should be offered to encourage both scientists and policy makers to engage with each other. Many researchers focus on blue skies research and only publish in the academic literature (which is rarely read by non-academics), while many policy makers do not specialise in a specific area and often move posts, making the generation of a stable working relationship difficult. It is therefore necessary to create specific and strong incentives, and to encourage those working on both sides of the relationship to engage. Specific recommendations include:

    – Resources should be provided to enhance science–policy engagement, e.g. through FP7 funding or national funding (possibly through the ERANET mechanism).

    – The engagement of researchers in policy-making processes should be better recognised in the context of assessment of scientists and research institutions, highlighting the value for society, an example of which is the UK 'Research Assessment Exercise' (RAE) of the Higher Education Funding Councils (Scott *et al.*, 2005).

    – Prizes should be awarded for inter-disciplinarity and dissemination (most prizes are now solely focused on scientific excellence and discovery of new knowledge).

(2) Dialogue throughout the research should be improved. The science–policy relationship is often imagined as a linear process in which the scientists deliver their

findings at the end of their research, and these findings then influence policy. This is an unrealistic representation of policy-making developments which are based on their own timescales and influenced by many other inputs, including research. This was also seen as an unrealistic model of learning, as it requires a building of familiarity and trust over time, and a mutual incremental exchange of knowledge, often informally and face-to-face rather than through documents. There is a need for constant interaction between science and policy from the start of research process, including at the stage of identifying research questions. Although this represents extra work, there is hardly any other alternative when research is attempting to be policy-relevant. Researchers can, by entering into dialogue with the full diversity of stakeholders interested in and affected by their research, come to understand the problem they are investigating in a rounded way, and this can actually help to enhance the relevance and quality of the research. There are several methods which have been identified:

– Within FP7, for projects considered to be policy relevant, the 'model' for project implementation would involve interaction between researchers and policy 'customers' throughout the project lifecycle.

– ERANET projects could explore this in several of the surveys, looking for examples of good practice and drawing out general lessons for future projects.

– Ad hoc workshops involving researchers and policy makers on different themes could be planned at regular intervals.

(3) Dialogue between policy makers and researchers is not only helpful in improving the research process; it is also critical to the policy process. Researchers can bring useful perspectives and evidence to bear on new policy debates. Systematic scientific evaluation of policy impacts is desirable, as well as review of long-standing policies in the light of new findings. In order to remain effective, policies also need to be reviewed regularly in the light of new evidence.

(4) Providing training and education is also a key aspect. Training for researchers on science communication and how to make effective contributions to policy debates should be built into scientists' education and career development. Equally, there is room for policy makers to draw more on scientific evidence. In some countries, this is becoming one of the skills that forms part of a civil servant's annual performance evaluation; an initiative to identify and share good practices and experience would be useful. There is a clearly identified need to develop specialists able to bridge the science–policy gap. One suggestion was the idea of a post-graduate course (Masters course) in 'science translation and synthesis', distinct from existing science journalism courses, which could be stimulated at EU level.

(5) Supporting inter-disciplinarity is becoming increasingly important for addressing current environmental issues due to their complex and inter-connected nature, and the need to investigate the 'natural' components together with the 'social' (including economic factors). This requires research founders to design multi-disciplinary research programmes and evaluate projects accordingly. Inter-disciplinary research

teams need to share a focus on a particular problem and develop a clear idea of what their goals are. 'Logical frameworks' are needed to link specific parts of the research to the functions and needs of policy organisations. This is a substantive challenge, and research funders and policy organisations do not only need to support inter-disciplinary research but also to review how effective it is.

(6) Many environmental challenges and related policies require long-term endeavours. On the research side, possible pragmatic responses include the development of trust funds to support long-term research and independent research organisations. Long-term monitoring is often neglected and considered unlikely to lead to significant discoveries (which is basically untrue as many environmental processes have been better understood on the basis of monitoring data, which in addition are useful for model validation). The European Commission, environment ministries and the EEA with its leading role in synthetising and presenting the results of environmental monitoring could play a key role in changing the perception about monitoring research needs.

(7) Planning and managing of research programmes in support of policies need to be improved. Processes for identifying research needs in support of policies should be developed, which could be helped by the drawing up of guidelines for best practices supported by case studies: where research is specifically developed to assist policy, criteria relating to policy relevance, timeliness and usefulness to evaluate project outcomes and their 'fit-for-purpose' character. This does not mean sacrificing scientific quality but rather fixing clear evaluation criteria that are tailormade to address policy needs. Finally, policy makers and others need to be closely engaged throughout research and policy processes. This includes joint framing and planning of the research, the presence of policy makers and stakeholders on research steering boards, policy mentors for research teams to enable linkages directly into policy networks, and regular exposure for scientists to the policy-making process. Engagement also helps researchers to build familiarity with the imperatives involved in policy making and the place of science in policy making; it should be required throughout decision-making processes, from problem framing, through option identification and evaluation, to implementation.

(8) The final set of recommendations relate to the accessibility and uptake of research into the policy process. Intermediaries and translators are needed to 'translate' research results into inputs that may be used by policy makers. In general, the findings of most research are published in academic papers, journals and books, few of which are read by policy makers. The language is often technical and the findings are not presented in ways that make clear the policy implications to non-specialists. It is thought that while scientists should be encouraged to interact more closely with policy makers and other stakeholders, there are many who lack the motivation and skills to do so. In any case, without a process of translation, much useful knowledge is likely to remain locked up in an inaccessible form, hence the need for intermediaries and translators to be found in either research or policy-making organisations or in the private sector. Such people may add great value to the science–policy interface and they need to be involved from

as early in the process as possible. One of their roles is to bring together the scientific findings being published on a particular topic (often scattered in different disciplinary journals) which are of policy relevance. In this respect, policy and research organisations should foresee appropriate funding to ensure such functions.

(9) New forms of communication are required, including policy or science briefs and work with media specialists. Finally, the need for better scientific databases is becoming more acute; they need to be easily searchable and written in plain language. There is a risk of a lack of memory of past research if no system enables people to trace research back. Electronic databases of scientific publications currently go back 10 or 20 years and mostly cover English-language publications.

10.1.10   CONCLUSIONS

This chapter highlights the need for a better integration mechanism of scientific findings into water policy implementation. Difficulties experienced to date stem from the fact that there is no sufficient streamlining of information from the scientific community to policy decision makers and vice-versa due to the lack of proper transfer and of appropriate 'relays' at international, national and regional levels. In this respect, the chapter describes efforts that are ongoing in the framework of various initiatives to examine how an efficient and operational science–policy interface could be developed in support of the implementation of the Water Framework Directive and other water policies. The aim in the long-term is to develop an operational science–policy interface in such a way that it could meet the demands of different levels of users (e.g. policy makers, industry, etc.) and stakeholders (e.g. the scientific community, academia, etc.), ensuring an efficient dissemination and use of research results. This interfacing goal is ambitious and involves many different actors, hence its complexity. It should also be seen in a wider framework, in liaison with other parallel activities. A possible framework, linking water policies in a broad sense to the R&D lifecycle, should include research development (links to FP7 and national research), demonstration (testing of R&D outputs at the most appropriate level, i.e. regional to international), communication (through the WISE-RTD web portal), and policy review (taking policy-related research needs into account when establishing research priorities). The challenge over the next few years will be to establish operational links among the different pieces of a puzzle composed of different types of instrument, actor and stakeholder.

   When considering possible 'relays', it should be kept in mind that the relationship between science and policy may be influenced by contextual factors such as the interest of the minister, political imperatives and the science culture in a country. Such country dimensions can be significant. The Defra study (Scott *et al.*, 2005) tried to detect whether there were systematic differences between big, small, Nordic and associated countries, considering that smaller countries can enjoy shorter lines of communication and Scandinavian countries are 'good at consultation with citizens and industry', but the size of a country, its location and culture may not be as important as participation in

EU Framework Programmes. They indeed provide knowledge of the ways in which the European level works and of the possibility of developing networks for collaboration with experts in both scientific and policy fields at the European level. Participation also strengthens the science base in countries, thus providing for 'strong science', which many consider important in policy development.

The issue of science–policy interfacing is actually part of ongoing efforts of the Commission for better governance. In this respect, the publication by the European Commission in 2001 of a White Paper on European Governance (European Commission, 2001) led to an Action Plan on Science and Society (European Commission, 2002b) and the publication of guidelines on the collection and use of expertise by the Commission (European Commission, 2002). The White Paper recognised the need to open up the EU policy making process to get more people and organisations involved in shaping and delivering policy, and to boost confidence in the way expert advice influences policy decisions. The Science and Society Action Plan had as one of several aims to 'put responsible science at the heart of policy making'. One specific action was to enhance mechanisms to provide scientific support to policy makers, leading to the SINAPSE initiative – a web-based communication platform enabling the exchange of information between the scientific community and policy makers (see http://europa.eu.int/comm/research/science-society/science-governance/synapse_en.html).

Very few countries use standard national guidelines or procedures for the science-into-policy process. The UK was one of the very few to report on a 'central government guide' for science–into-policy practices. However, these guidelines are rarely used on 'a day-to-day basis' and tend to be considered at a higher level than by staff in the system. Moreover, the guidelines address 'proprietary issues' rather than the full range of processes followed in using science to inform policy. Finally, they do not address how to work with other scientists in other organisations. This suggests that where organisations such as government departments and agencies have different organisational cultures, structures and processes in place, there are difficulties in communication, networking and collaboration.

There is hence a considerable room for improvements and, in the specific water monitoring field, a possible opening for a concrete science–policy interface development, which should be backed up by guidelines on the use of scientific outputs (at the most appropriate level, i.e. from regional to international) and demonstration activities.

10.1.11   ACKNOWLEDGEMENTS

Considerations expressed in this chapter are the fruit of many discussions with colleagues from the European Commission Research Directorate-General (among others Panagiotis Balabanis, Christos Fragakis, Elisabeth Lipiatou and Marta Moren-Abat) and the scientific community (among others Wim de Lange, Geo Arnold and Jos Brils). They are all gratefully acknowledged for sharing views and ideas. Many considerations are also issued from conclusions of science-meet-policy events, and in particular the London 2005 report (Scott *et al.*, 2005).

REFERENCES

De Lange, W. Arnold, G., Willems, P., Provost, F., Hatterman, F., Plyson, J. *et al.* (2007) WISE-RTD webportal: a gate to scientific information for WFD implementers and water managers, Poster, Int. Conf. Monitoring under the WFD, Lille, March 2007.

D'Eugenio, J., Haastrup, P., Jensen, S., Wirthmann, A. and Quevauviller, P. (2006) General Introduction into WISE, 7th Int. Conf. Hydroinformatics, Nice, September.

Environment Agency (2006) Integrated catchment science: a research strategy, Environment Agency, Bristol.

Environmental Research Funders Forum (2007) Strategic analysis of UK environmental research activity, ERFF Report 04.

Environmental Science & Policy (2005) Proceedings of the workshop on research and technology integration in support of the Water Framework Directive, *Envir. Sci. & Pol.*, **8**(3).

Erno-Kjolhede, E. (2001) *Managing Collaborative Research: Unveiling the Microdynamics of the European Triple Helix*, Copenhagen Business School Press.

European Commission (2000) Directive 2000/60/EC of the European Parliament and of the Council of 23 October 2000 establishing a framework for Community action in the field of water policy, *Official Journal of the European Communities*, **L327**, pp. 1–72.

European Commission (2001) Common implementation strategy for the Water Framework Directive, European Communities, Final CIS document available at http://europa.eu.int/comm/environment/water/water-framework/implementation.html.

European Commission (2002) 6th Environment Action Programme, European Commission, 2002–2012, http://ec.europa.eu/environment/newprg/index.htm.

European Commission (2003a) Guidance Document No. 7 on Monitoring under the Water Framework Directive, European Commission, Brussels.

European Commission (2003b) Common Implementation Strategy for the Water Framework Directive, European Communities.

European Commission (2006) Guidance Document No. 15 on Groundwater Monitoring, European Commission, Brussels.

Etzkowitz, H. and Leydesdorff, L. (eds) (1997) *Universities and the Global Knowledge Economy: A Triple Helix of University–Industry–Government Relations*, Pinter, London and Washington.

Etzkowitz, H., Webster, A. and Healey, P. (eds) (1998) *Capitalizing Knowledge: New Intersections of Industry and Academia*, SUNY.

Harmoni-CA (2001) Harmonised modelling tools for integrated river basin management, Harmoni-CA, EU-funded concerted action, Contract EVKI-2001-00192, http://www.Harmoni-CA.info.

LIFE (2007) LIFE and Europe's rivers, European Commission, Brussels, http://ec.europa.eu/environment/life/publications/lifepublications/lifefocus/nat.htm#.

Quevauviller, P. (2002) *Quality Assurance for Water Analysis*, Water Quality Measurement Series, John Wiley & Sons, Ltd, Chichester, UK.

Quevauviller, P. (2007) *Environ. Sci. Pollut. Res.*, **14**, p. 297.

Quevauviller, P., Balabanis, P., Fragakis, C., Weydert, M., Oliver, M., Kaschl, A. *et al.* (2005) *Environ. Sci. Pol.*, **8**, p. 203.

Quevauviller, P., Borchers, U. and Gawlik, B. (2007) *J. Environ. Monit.*, **9**, p. 915.

Schmitz, B., Reiniger, P., Pero, H., Quevauviller, P. and Warras, M. (1994) Europe and scientific and technological cooperation on water, European Commission, Report EUR 15645 EN.

Scott, A., Holmes, J., Steyn, G., Wickham, S. and Murlis, J. (2005) Report of the 'science-meets-policy' conference, London, November 2005, http://www.theknowledgebridge.com.

Surridge, B. and Harris, B. (2007) *Interdisc. Sci. Rev.*, **32**, p. 298.

10.2

Support for WFD Research Needs: Current Activities and Future Perspectives in the Context of RTD Framework Programmes

Andrea Tilche[1]

10.2.1 THREE DECADES OF EU WATER RESEARCH AND THE FP5 WATER KEY ACTION

Water research has been a major component of successive European Commission research Framework Programmes, primarily in the environment field, but also in other areas like agriculture, industrial technologies and international cooperation. EC-funded

---

[1] The views expressed in this chapter are purely those of the author and may not in any circumstances be regarded as stating an official position of the European Commission.

The Water Framework Directive - Ecological and Chemical Status Monitoring Edited by Philippe Quevauviller, Ulrich Borchers, Clive Thompson and Tristan Simonart © 2008 John Wiley & Sons, Ltd

research on water covers a very wide spectrum, spanning from hydrology to ecology, to technologies and to integrated aspects.

Earlier Framework Programmes were focused on the development of the scientific knowledge to support environmental quality standards and objectives, and on the development of technologies for end-of-pipe treatment.

In FP5 (1998–2002), a very large programme was established through a 'Key Action on Water', with emphasis on integrated approaches for sustainable water management, support for water policies (WFD, drinking water, urban wastewater, nitrate and IPPC directives, etc.) and management of global change.

The programme of the FP5 Key Action on Water was based on a broad consultation with the stakeholders to define action plans for EU collaborative research, and was intended to promote an increased participation of technology providers and end user communities.

About 180 project contracts, for an EC contribution of about 250 million euros, were awarded during the four years of the programme.

A major financial effort was put in support of the implementation of the Water Framework Directive. This Directive provides the overall policy objective of reaching a 'good' ecological status for all water bodies within a given timeframe, leaving the evaluation of the possible solutions to the river basin plans. It contains several scientific challenges, being unclear on what 'good ecological status' would even mean.

Research needs were identified at various levels:

- assessment methods

- pressures and impacts

- dynamic of recovery processes

- economic analysis

- monitoring tools

- modelling for assessment

- modelling for forecasting evolution

- modelling for checking cost-effectiveness of interventions

- decision support systems

- temporary water bodies

- highly modified water bodies.

A group of FP5 projects (AQEM, STAR, FAME) developed a first set of ecological status assessment methods applicable across a wide range of European eco-regions. Another group looked at the understanding of water quality issues in various surface water bodies, in order to give policy recommendations (ECOFRAME and EURO-LAKES), provide a framework for the analysis of pressures and impacts (DANUBS), develop decision support systems for areas where agriculture is competing for water uses (MULINO) and assess the sustainability of irrigation in the context of the combined application of the WFD and the new CAP (WADI).

Box 10.2.1 Taking informed decisions in complex river basins: the need for Operational Decision Support Systems

Addressing the complex multi-sectoral problem of water resources quantity and quality management at catchment and river basin scale – as required by the **Water Framework Directive** – can be very much facilitated by the development of flexible, adaptive and educational decision support tools, which provide a common foundation for interaction between researchers, water authorities and other interested stakeholders. This issue has been addressed by the EC-funded **MULINO** project. MULINO produced a software tool (**mDSS**) and an overall methodology within which the tool can be applied for an integrated approach to decision problems related to water management. The software incorporates integrated analysis modelling (**IAM**), multi-criteria analysis (**MCA**) and the European Environment Agency's **DPSIR** framework, adopting state-of-the-art data formats to guarantee interoperability. The system does not require additional software, which should improve the potential for its utilisation by water management administrations. Optional links with GIS packages, hydrological models and/or meta-models are provided, and in the most recent version of the software a full coupling procedure links mDSS3 with the CRASH hydrologic model. The software can also be coupled with any hydrological model which respects a standard input/output procedure. By combining a transparent definition of the decision making context, mDSS contributes to creating a common understanding and promoting communication on sustainable water resources management and provides a valuable tool for stakeholder engagement, social learning and **collaborative decision making**. The MULINO methodology has been applied at a variety of scales and issues, such as: evaluation of alternative agricultural production systems with a view to optimising environmental and socioeconomic parameters such as sediment loads, nitrate fluxes and farmers' income; assessment of strategic options for dam management in the context of water scarcity and competing uses of water from various economic sectors; and examination of various options for the sustainable allocation of water and the preservation of ecological flow requirements. MULINO was also tested at the European scale, evaluating alternative scenarios for agricultural policy and the implementation of the European Union's **Nitrate Directive**.

Among the most successful initiatives, the CATCHMOD (from Catchment Modelling) cluster has to be highlighted, grouping 10 projects with the aim of developing integrated modelling tools for the assessment, planning and management of river basins in support of the WFD. Its main achievements are in models benchmarking (BMW, EUROHARP), modelling systems (HarmoniQuA, HarmonIT) and uncertainty assessment (HarmoniRiB).

It is worth giving some more details on the HarmonIT project, responsible for the development of the OPEN Modelling Interface, a software environment that allows different models (OPEN-MI compliant) to dialogue between themselves, exchanging data and running iteratively. It was jointly developed by a wide consortium comprising the largest European hydrology software companies (Delft Hydraulics, HR-Wallingford and DHI-Water and Environment). The success of its output is witnessed by the great interest generated worldwide, and in particular in the US, and by the fact that today most commercial hydrological software packages are OPEN-MI compliant. OPEN-MI has already been used for the WFD in the river basin planning activity of the Scheldt River.

Box 10.2.2 Harmonisation, reliability and transparency for quantifying diffuse nutrient losses, in particular from agricultural land, to surface freshwater systems and coastal waters

Agriculture is an important driving force in the context of European economy and the quality of Europe's aquatic environments. In fact, diffuse nutrient losses from agriculture represent the major anthropogenic pressure on the water quality ecological status in a large number of surface waters in Europe. In the long run, this may threaten human health, industrial activities including food production, as well as various social activities including fishing, tourism and bathing. To deal with this problem, various political responses have already been initiated in many European countries, on the basis of various EU directives such as the Nitrates Directive and the Water Framework Directive. These important policy initiatives need to have a sound and transparent scientific basis in approaching the pressures from agricultural sources across Europe. The lack of appropriate methods and, particularly, the lack of transparency and comparability of methods actually used for identifying and quantifying diffuse agricultural nutrient losses in Europe has already been identified as a major problem that needs to be solved before progress can be made on this issue. The **EUROHARP** (EVK1-CT-2001-00096) project addressed this, providing end users with guidance on an appropriate choice of quantification tools (QTs) to satisfy existing European requirements on harmonisation, reliability and transparency in quantifying diffuse nutrient losses. The project included inter-comparison of different types of model, and assessment of the performance of individual models and the applicability of the same models in catchments with different data availability and environmental conditions throughout Europe. The basis for the performance and applicability studies was the compilation of a harmonised GIS/database for all catchment data and the analysis of these water quality data (e.g. trends, watercourse retention). The project developed an electronic decision support system (EUROHARP Toolbox) that synthesises the main outcomes of the work carried out on model testing and applications in the 17 European catchments, and facilitates a more efficient and harmonised approach in terms of quantifying and managing nutrient losses from diffuse sources in the context of integrated water resource management.

Box 10.2.3 Data and models uncertainty in water resources management

The Water Framework Directive (WFD) requires that integrated water resource management is carried out in international river basins and that the underlying Programmes of Measures and River Basin Management Plans are based on best possible confidence. Good data and credible integrated models are therefore crucial for the WFD implementation. Most of the WFD guidance documents emphasise that uncertainty analyses should be performed, but do not include recommendations on how to do so. Hence there is a need to develop guidelines and operational tools that can address uncertainties in data and models. The EU-funded **HarmoniRiB** (EVK1-CT-2002-00109) project addressed this need by developing methodologies for quantifying uncertainty and its propagation from the raw data to concise management information. More specifically, the project reviewed results on data uncertainty reported in the literature and compiled a guideline report for assessing uncertainty in various types of data (meteorological, soil physical and geochemical, hydrogeological, land cover, topography, discharge, surface water quality, ecological and socioeconomic data). A guidance document has also been prepared to support the framing of the selection of appropriate management measures to reach a certain goal. To illustrate the practical application of these results, a comprehensive report on the uncertainties in the economic analysis of the WFD has been prepared and applied in various case studies. Finally, the project developed a novel database which can store time-series data from many domains (i.e. rivers, lakes, groundwater, transitional and coastal waters) and spatial data (land use, pressures, socioeconomic data, system characteristics), as well as information about uncertainty on these data. The database is linked to the user-friendly Data Uncertainty Engine (DUE) software tool, which makes it easier to include uncertainty in the practical water management.

However, other EU water policies were also addressed by the programme, such as:

- The Nitrate Directive 96/676/EC, with a cluster of projects related to nutrient modelling (EUROHARP, INCA) and to agricultural best practices for minimising the release of nutrients from agriculture (AGRIBMPWATER).

- The Groundwater Directive 2006/118/EC (while it was being developed as a COM proposal), through a cluster of projects dealing with the baseline chemical composition of European aquifers (BASELINE), with groundwater risk assessment methodologies for contaminated sites (TRACE-FARCTURE, GRACOS, INCORE) and with the impact of pesticides use on groundwater quality (PEGASE).

- The Flood Protection Directive 2007/60/EC (while it was in preparation), through a cluster of projects (ACTIF) for the development and demonstration of new-generation flood forecasting methodologies which will advance the capabilities and accuracies of present forecasting systems (MUSIC, MANTISSA, FLOODRELIEF and others), and the elaboration of specific guidelines on how to use floodplains for flood risk reduction measures, with case studies illustrating the wide range of natural flood reduction schemes that can be carried out and the degrees of success that have been achieved (ECOFLOOD).

- The revision of the Drinking Water Directive 98/83/EC, through a cluster of projects coordinated by the concerted action WEKNOW, which developed position papers on chemistry, microbiology and sampling, monitoring and quality control, and a demonstration of the need to include the concept of Water Safety Plans; the other projects addressed a methodological framework for quantitative microbiological risk assessment (MICRORISK) and the problem of biofilms in distribution networks (SAFER).

- The strategy for water scarcity and droughts adopted as a Communication by the European Commission in July 2007, through the ARID cluster, composed of a group of projects on water management under scarcity (ARID, AQUADAPT, MEDIS, WaterStrategyMan), a centre of excellence (ASTHYDA) and a project that developed guidelines for wastewater reclamation and re-use (AQUAREC).

Box 10.2.4 Real-time flood forecasting

Real-time flood forecasting systems, which link weather forecasts, the state of the river catchment, river discharges and water levels, can be used to respond to floods as they occur and to reduce their costs in terms of lives, property and the breakdown of infrastructure. In comparison to construction of major flood protection works such as dams, dikes and polders, flood forecasting is cost-effective and the environmental impacts are minimal. More importantly, when used for flood warning these systems can save lives. Using new technologies to improve flood forecasting, we can forecast floods ahead of time, rather than clearing up afterwards. Current flood forecasting and warning systems have several limitations, such as insufficient lead-time to provide accurate flood warnings, inadequate spatial and temporal resolution of the real-time rainfall observations and forecasts for flood producing storm, and little integration of different sources of forecast information. Moreover, their ability in considering the uncertainties in estimating and forecasting precipitation and flood discharges is very limited, their application at regional level is also limited and the costs of improving forecasting may be prohibitive. European research projects, like **FLOODRELIEF (EVK1-CT-2002-00117)**, addressed these limitations by developing and demonstrating a new generation of flood forecasting methodologies which will advance present capabilities and accuracies. A new powerful and highly accessible Internet-based real-time decision support system has been developed. The system exploits and integrates different sources of forecast information, including improved hydrological and meteorological model systems and databases, radar, advanced data assimilation procedures and uncertainty estimation. These technologies have also been evaluated and tested in two highly flood-prone regional basins, one in Poland and one in the United Kingdom, demonstrating the benefits of flood forecasting innovations for saving lives and properties.

Box 10.2.5 Microbiological risk assessment: a scientific basis for managing drinking water safety from source to tap

Safe drinking water is a prerequisite for a high level of public health and economic development. Water suppliers that use the HACCP (Hazard Analysis & Critical Control Points)-based Water Safety Plan, developed in the food industry, are faced at several steps in this process with questions of a quantitative nature. The answers to these questions are usually based on semi-quantitative expert judgements and industry or legal standards. Quantitative microbiological risk assessment (QMRA) provides more objective, science-based and quantitative information to answer these questions, and hence a more precise basis for risk management. Quantitative assessment of the safety of EU drinking water requires quantitative data about the probability of exposure of drinking water consumers in the European Community to pathogenic microorganisms through drinking water from the tap. The overall objectives of the **MicroRisk (EVK1-CT-2002-00123)** EU-funded research project were to provide a harmonised protocol for quantitative microbiological risk assessment (QMRA), to collect scientific data from water supply systems in the EU to perform such a QMRA, and to determine the strengths and weaknesses of QMRA. The project developed a harmonised protocol for the site-specific assessment of the removal of pathogens by full-scale water treatment processes, a unique quantitative database on removal of microorganisms by water treatment processes, including analysis of the causes of major waterborne outbreaks across Europe, Australia and Canada since 1990, and finally an evaluation of the performance of analysis of QMRA as a scientific basis and tool for assessing the microbial safety of drinking water supplies. By concluding that the risk-based approach is the way forward for (EU) policy on safe drinking water, **MicroRisk** supports the incorporation of the risk-based approach (which is used in the EU policy in many consumer health (such as food safety) and environmental health areas) into the proposed revision of the current EU policy on drinking water (Drinking Water Directive).

Box 10.2.6 A manual for water reuse management

Alternative water sources are expected to play a significant role in areas suffering water shortages. In many places, waste water reuse is becoming a valuable component of sustainable water management practises. However, substantial and practical information is still needed to safely design, implement and operate waste water reuse schemes.

The project **AQUAREC**, 'Integrated concepts for reuse of upgraded wastewater', was funded by the Fifth Framework Programme of the European Commission. Its major aim was to investigate and develop concepts and methodologies supporting rational and knowledge-based waste water reuse strategies. Proper consideration of the state of the art and appropriate integration of implemented case studies and international expertise led to substantial achievements in tackling all water reuse matters. The project addressed particularly technologies for water reclamation and

reuse, successfully embracing monitoring and control issues, best management practises, economical aspects, water quality parameters, risk assessment methodologies,
institutional settings and governance, legislative frameworks and policy guidance.
One of the project's outcomes is a major publication* presenting practical information on waste water reuse concepts based on actual and proved management and
operational practises. The targeted users are mainly practitioners concerned with
implementation of waste water reuse schemes, but it is expected that this publication will also become 'the wastewater reuse manual' for decision makers, local
authorities, consultants and research bodies involved in the area.

* Bixio, Davide, Wintgens and Thomas (eds) *Water Reuse System Management
Manual – AQUAREC*, ISBN 92-79-01934-1.

10.2.2 THE 6TH FRAMEWORK PROGRAMME (2002–2006) AND ITS WIDER INTERNATIONAL DIMENSION

In the 6th Framework Programme, the focus of the environmental research activities on
water was slightly moved towards new concepts, strategies and tools for the mitigation
of global change impact (including climate change) on water resources in Europe
and worldwide. A broader attention to the international dimension came due to the
strong commitment of the European Union in Johannesburg – at the World Summit
on Sustainable Development – to support the UN Millennium Development Goals for
Water through a policy action – the EU Water Initiative – that contained a 'research
component', coordinated by DG RTD, which had the task of improving our scientific
cooperation with key areas of the world in the field of integrated water resources
management and sustainable water supply and sanitation.

This refocused attention on global issues, however, did not affect the research in support of water policies that was carried out through a dedicated sub-programme – named
'Specific Support to Policies' – specifically tailored to the research needs expressed
by various policy directorate generals of the European Commission.

The specific needs for water policies led to the selection of a group of projects
addressing the Water Framework Directive:

- For the further development and implementation of methodological approaches to
  assess the ecological status of rivers (REBECCA, EFI+).

- For the assessment and validation of existing screening methods for measuring
  WFD-relevant chemical and biological substances (SWIFT-WFD).

- For the development and testing of practical guidelines for the assessment of environmental and resource costs and benefits in the WFD (ACQUAMONEY).

- SPI-WATER, which provides synthesis and 'translation' of research results and contributes to the development of the WISE-RTD portal in a way that can efficiently
  feed the WFD implementation.

Other EU Water policies were also addressed through specific projects, such as:

- The project BRIDGE, for the development of a common methodology for establishing groundwater threshold values (maximum concentrations of pollutants) at appropriate national, regional or local levels, in relation to criteria linked to the good groundwater chemical status definition given in the Groundwater Directive 2006/118/EC.

Box 10.2.7 Relationships between ecological and chemical status of surface waters

The strategic objective of **REBECCA** has been to provide relevant scientific support for the implementation of the Water Framework Directive (WFD). The two specific aims of the project have been, first, to establish links between the ecological status of surface waters and physico-chemical quality elements and pressures from different sources, and, second, to develop and validate tools that Member States can use in the process of classification, in the design of their monitoring programmes, and in the design of measures in accordance with the requirements of the WFD.

The **REBECCA** project has collected existing data owned by **REBECCA** partners and from data providers outside the consortium. Thus, data sets with improved geographical (and also temporal) coverage have been produced. They include chemical and biological data from thousands of sampling sites in European lakes, rivers and coastal waters. A big challenge was to assure the data quality, due to the lack of harmonised sampling and analysing methods, as well as the lack of harmonised taxonomic resolution. These data sets have been used:

- To investigate the sensitivity of single species, other taxonomic groups, multimetric indices against different pressures and chemical indicators.

- To test the response of existing biological indicators (described in earlier literature) to pressures, and to adjust these relationships if necessary.

- To develop new biological indicators with a better response, and indicators for pressures not referred to in earlier literature.

- To assess reference conditions (in lakes).

- To evaluate whether ecological thresholds (points of no return) can be found across the pressure gradients.

- To develop new tools to be used in ecological classification.

The results of these analyses are all available at the project Toolbox (http://www.rbm-toolbox.net/rebecca).

Much attention has been paid to organising continuous information flow between the project and the end users of its results. At the Community level, the most important link has been between **REBECCA** and the **ECOSTAT** working group. **REBECCA** scientists have also been working together with **ECOSTAT Geographical Intercalibration Groups** to provide them with information in order to set the ecological class boundaries for surface waters.

However, the mainstream projects of the FP6 Global Change and Ecosystems programme, also carried out through new, very large funding schemes (frequently of more than 10 million euros of EC contribution) called Integrated Projects (IP), constituted the current backbone of water research. Those very large IPs, or clusters of more focused and smaller projects addressing the main water problems, are of very high policy relevance, in particular:

- A group of large IPs dealing with large-scale water issues, the assessment of the global water cycle (WATCH) and the definition of scenarios of water resources in Europe and the surrounding countries (SCENES).

- Two IPs that look in-depth at the functioning of water systems, either through ecological perspective (EUROLIMPACS) or through the interactions between surface water bodies, soil and groundwater.

- A group of IPs dealing with integrated water management and technological options at river basin level, with an accent on adaptive management (NEWATER), looking at water-stressed areas (AQUASTRESS), addressing the problems of megacities in the world (SWITCH), and addressing the problems of drinking water production and distribution (TECHNEAU).

- Clusters of technology-oriented projects on membrane bioreactor processes in the municipal wastewater sector (MBR-Network), on technologies for the artificial recharge of groundwater (GABARDINE, RECLAIM WATER), on source control options for Priority Hazardous Pollutants (SCOREPP, SOCOPSE), on innovative desalination technologies (MEDINA, AQUASOL), on innovative wastewater treatment technologies (INNOWATECH, NEPTUNE), on technologies for trenchless rehabilitation of buried water infrastructures (ORFEUS, WATERPIPE) and finally on irrigation systems (PLEIADES, FLOW-AID).

Most of the above-mentioned projects are still running, and they are delivering a considerable mass of results that will need some time to be validated and used for policy purposes.

At the end of FP6, about 60 contracts were awarded on water (and soil) research issues, corresponding to more than 180 million euros of European Commission contribution.

Box 10.2.8 Time to de-stress water stress

Water stress occurs when the water demand exceeds its availability during a certain period, or when poor quality restricts its use. It is a global problem detrimentally impacting on economical, societal and environmental well-being. It could be exacerbated in the future by population grow, economic development and trends of climate change. The 4 year research programme of the **AquaStress** integrated project* began in 2005 with an overall budget of 16 million euros. About 2000 person months are committed to delivering knowledge-based interdisciplinary methodologies and tools enabling involved actors to mitigate water stress problems.

The **AquaStress** project adopted a quite unique and definitely challenging case study stakeholder-driven approach proactively involving the acting community in the different stages of the decision-making process. This powerful and responsive mechanism is already providing first inputs for a real integrative water resources management.

The project is organised in three phases: i) characterisation of selected reference sites and relative water stress problems, ii) collaborative identification of preferred solution options and iii) testing of solutions according to stakeholder interests and expectations. So far the first phase is almost complete, a set of water stress indicators is available and suitable management or technological options have been identified and suggested for validation in the different case studies. To facilitate the use of and access to this huge amount of data, and to enable the combination of experts' tools with improved participatory approaches, dedicated pieces of software are being developed.

* *Additional information: http://www.aquastress.net.*

Box 10.2.9 Priority pollutants: end-of-pipe or source control?

SCOREPP and **SOCOPSE** are two FP6 projects that began towards the end of 2006. Both projects deal with source control of priority pollutants and aim at developing guidelines for stakeholders to help them achieve the requirements in this regard of the WFD. The two projects will interact and exchange results during their implementation.

SOCOPSE is looking mostly at hazardous priority pollutants to provide guidelines and decision support tools for their management at river basin scale. Its activities include material flow analysis, a series of detailed substances reports, identification of measures and management options, and application to four case studies. A strong interaction with industry research networks, authorities and NGOs is envisaged.

SCOREPP is looking at a wider selection of pollutants, including all the priority pollutants and an additional list of relevant pollutants. This project will develop appropriate strategies to reduce emissions from urban areas into the receiving waters. The activities include the identification of sources in urban areas, the identification of measures, assessing their cost-effectiveness, and the validation of the strategies on a number of selected case studies. Interaction with relevant stakeholders is also an integral part of the project.

10.2.3   THE 7TH FRAMEWORK PROGRAMME (FP7)

The debate for the preparation of the 7th Framework Programme (2007–2013) was mainly characterised by the recognition of the fundamental role of research in promoting innovation, and consequently the competitiveness of European industry. Along with the ambitious target of reaching an average of research investments of 3 % of GDP by 2010, and with the objective of promoting private R&D investments, a group of industry-led technology platforms was promoted, among which was the Water Supply and Sanitation Technology Platform (WSSTP).

Its launch in April 2004 was also connected with the Environmental Technologies Action Plan (ETAP). The final goal of WSSTP is to strengthen the potential for technological innovation and competitiveness in the European water industry, water professionals and research institutions through the development of a common strategic and visionary science and technological research agenda and a suitable implementation plan.

Its ambition is to promote step changes in the technological capacity of the European Water industry, consolidating and strengthening its position in the world market, while contributing to the global challenge of reaching the Millennium Development Goals of ensuring safe and secure water supply for different uses and sanitation services through the development of sustainable technologies and appropriate institutional frameworks.

The WSSTP Strategic Research Agenda identified:

- Four main challenges (increasing water stress and water costs; urbanisation; extreme events; and rural and under-developed areas).

- Five research areas (balancing demand and supply; ensuring appropriate quality and security; reducing negative environmental impacts; novel approaches to the design, construction and operation of water infrastructure assets; and establishment of an enabling framework).

- The integration of these through the concept of pilot programmes, articulated in six specific 'pilots', each one associated with a number of test sites in Europe and beyond, where they carry out targeted and prioritised research defined by and tested in real-life applications, covering generic research, enabling technologies development and full-scale demonstration in the various implementation cases, and mobilising different sources of funds – beyond solely FP7 resources – through a solid public/private partnership. The six pilots are:

- Mitigation of water stress in coastal zones.

- Sustainable water management inside and around large urban areas.

- Sustainable water management for industry.

- Reclamation of degraded water zones (surface and groundwater).

- Proactive and corrective management of extreme hydro-climatic events.

The European Commission has taken well into account the research strategy proposed by the WSSTP in the formulation of the FP7 programmes, and the activities are now in the phase of implementation.

European water policies – and in particular the Water Framework Directive – are deeply embedded in all the above-mentioned research subjects and pilot projects, which are and will become subject to FP7 funding throughout the seven years of the programme.

FP7 water research will, however, span beyond the subjects of more 'industrial' interest, in particular with the objective of better understanding the potential impact of global and climate changes on water resources and, consequently, on EU water policies.

10.2.4 CONCLUSIONS

FP5 and FP6 research projects have provided important knowledge of and policy insights into the role water plays in our environmental and socioeconomic welfare.

The application of results from those projects will certainly help the implementation of short-term water policy requirements, as well as long-term adaptation to global change and progress towards sustainable development.

However, it became clear that in order to better implement for policy purposes the impressive amount of data and results originated from European projects in FP5 and FP6, a targeted action at the science–policy interface would be necessary. The effort first put in place by the HARMONI-CA project, and later by SPI-WATER, led to the establishment of the research portal of the Water Information System for Europe, WISE-RTD, but additional effort is needed to make this interface work at its best and exploit all the existing potential of EC-funded research.

In view of the increasing complexity of water problems, the use of models in policy formulation and the need to understand and assess uncertainties associated with these models will become a fundamental issue in the future.

Research will continue in FP7, with more focus on the development of solutions to water problems (in connection with the Water Supply and Sanitation Technology Platform), and looking at the impact of climate change on water systems.

Water utilities have the possibility to access and make use of, through European research results, the most advanced available methodologies, technologies and tools in order to implement the Water Framework and other directives, minimising the costs of implementation. However, a higher participation of water companies in the research consortia would be extremely beneficial, allowing for a closer participation of 'problem owners' in the definition of research projects, and associating them with the evaluation of research results through case studies and pilot implementation tests. More and more, the most efficient companies are those which are more able to introduce innovative technologies and management methods.

10.2.5 ACKNOWLEDGEMENTS

I wish to thank my colleagues Panagiotis Balabanis, Elena Dominguez, Avelino Gonzalez-Gonzalez and Javier Peinado Lebrero, who constitute part of the team that managed the research projects in support of the Water Framework Directive and other EU water policies in DG RTD, for their contribution in revising the text and preparing the information boxes on selected projects.

USEFUL LINKS

(1) site where FP5 projects information, FP5 projects results and FP6 projects information on Water and Soil can be found: http://circa.europa.eu/Public/irc/rtd/eesdwatkeact/library?l=/projects_information&vm=detailed&sb=Title.

(2) The site of FP7 Environmental research on Europa: http://ec.europa.eu/research/environment/index_en.htm.

(3) The site of Environmental Technology newsletter: http://circa.europa.eu/Public/irc/rtd/eesdwatkeact/info/data/ETR_newsletter_issue_3.htm.

(4) The site of the Water Supply and Sanitation Technology Platform: http://www.wsstp.org/default.aspx.

(5) The CORDIS site for Environment (including climate change) research: http://cordis.europa.eu/fp7/cooperation/environment_en.html.

Index

The Water Framework Directive - Ecological and Chemical Status Monitoring   Edited by Philippe
Quevauviller, Ulrich Borchers, Clive Thompson and Tristan Simonart © 2008 John Wiley & Sons, Ltd